Classification, Elements and Properties of Soil

Classification, Elements and Properties of Soil

Edited by Donald Cronin

SYRAWOOD
PUBLISHING HOUSE
New York

Published by Syrawood Publishing House,
750 Third Avenue, 9th Floor,
New York, NY 10017, USA
www.syrawoodpublishinghouse.com

Classification, Elements and Properties of Soil
Edited by Donald Cronin

International Standard Book Number: 978-1-68286-579-8 (Hardback)

Cataloging-in-Publication Data

Classification, elements and properties of soil / edited by Donald Cronin.
p. cm.
Includes bibliographical references and index.
ISBN 978-1-68286-579-8
1. Soils. 2. Soils--Classification. 3. Soil mechanics. I. Cronin, Donald.
S591 .C53 2018
631.4--dc23

Printed in China.

TABLE OF CONTENTS

PREFACE

Over the recent decade, advancements and applications have progressed exponentially. This has led to the increased interest in this field and projects are being conducted to enhance knowledge. The main objective of this book is to present some of the critical challenges and provide insights into possible solutions. This book will answer the varied questions that arise in the field and also provide an increased scope for furthering studies.

The branch of science which focuses on the formation and classification of soil is known as soil science. Some of the branches of soil science are pedology, soil zoology, soil physics, soil ecology, soil chemistry, etc. With new techniques being introduced in the field, new discoveries and researches continue to expand our understanding of soil. Such selected concepts that redefine the subject matter have been presented in this book. It will prove to be immensely beneficial to agronomists, students and researchers associated with this field.

I hope that this book, with its visionary approach, will be a valuable addition and will promote interest among readers. Each of the authors has provided their extraordinary competence in their specific fields by providing different perspectives as they come from diverse nations and regions. I thank them for their contributions.

Editor

1

Pyrosequencing-Based Assessment of Bacterial Community Structure Along Different Management Types in German Forest and Grassland Soils

Heiko Nacke[1], Andrea Thürmer[2], Antje Wollherr[2], Christiane Will[1], Ladislav Hodac[3], Nadine Herold[4], Ingo Schöning[5], Marion Schrumpf[4], Rolf Daniel[1,2]*

1 Department of Genomic and Applied Microbiology, Institute of Microbiology and Genetics, Georg-August University Göttingen, Göttingen, Germany, **2** Göttingen Genomics Laboratory, Institute of Microbiology and Genetics, Georg-August University Göttingen, Göttingen, Germany, **3** Experimental Phycology and Culture Collection of Algae at the University of Göttingen, Göttingen, Germany, **4** Max Planck Institute for Biogeochemistry, Jena, Germany, **5** Institute of Ecology, Friedrich-Schiller-University, Jena, Germany

Abstract

Background: Soil bacteria are important drivers for nearly all biogeochemical cycles in terrestrial ecosystems and participate in most nutrient transformations in soil. In contrast to the importance of soil bacteria for ecosystem functioning, we understand little how different management types affect the soil bacterial community composition.

Methodology/Principal Findings: We used pyrosequencing-based analysis of the V2-V3 16S rRNA gene region to identify changes in bacterial diversity and community structure in nine forest and nine grassland soils from the Schwäbische Alb that covered six different management types. The dataset comprised 598,962 sequences that were affiliated to the domain Bacteria. The number of classified sequences per sample ranged from 23,515 to 39,259. Bacterial diversity was more phylum rich in grassland soils than in forest soils. The dominant taxonomic groups across all samples (>1% of all sequences) were *Acidobacteria*, *Alphaproteobacteria*, *Actinobacteria*, *Betaproteobacteria*, *Deltaproteobacteria*, *Gammaproteobacteria*, and *Firmicutes*. Significant variations in relative abundances of bacterial phyla and proteobacterial classes, including *Actinobacteria*, *Firmicutes*, *Verrucomicrobia*, *Cyanobacteria*, *Gemmatimonadetes* and *Alphaproteobacteria,* between the land use types forest and grassland were observed. At the genus level, significant differences were also recorded for the dominant genera *Phenylobacter*, *Bacillus*, *Kribbella*, *Streptomyces*, *Agromyces*, and *Defluviicoccus*. In addition, soil bacterial community structure showed significant differences between beech and spruce forest soils. The relative abundances of bacterial groups at different taxonomic levels correlated with soil pH, but little or no relationships to management type and other soil properties were found.

Conclusions/Significance: Soil bacterial community composition and diversity of the six analyzed management types showed significant differences between the land use types grassland and forest. Furthermore, bacterial community structure was largely driven by tree species and soil pH.

Editor: Jack Gilbert, Argonne National Laboratory, United States of America

Funding: The work has been funded by the DFG Priority Program 1374 "Infrastructure-Biodiversity-Exploratories" (DA 374/4-1). The funders had no role in study design, data collection and analysis, decision to publish, or preparation of the manuscript.

Competing Interests: The authors have declared that no competing interests exist.

* E-mail: rdaniel@gwdg.de

Introduction

Soils are considered to be the most diverse microbial habitat on Earth with respect to species diversity and community size. Bacteria are the most abundant group of microorganisms in soil [1]. The calculated number of distinct bacterial genomes ranges from 2,000 to 18,000 per gram of soil [2]. Although the importance of bacteria for ecosystem functions and maintaining soil quality in agriculturally managed systems has long been recognized, the influence of land use type and management type on soil bacterial communities is poorly understood. In a recent pyrosequencing survey, bacterial diversity of forest soil was more phylum rich compared to agricultural soils, which were more species rich [3]. Furthermore, it has been described that *Bacteroidetes* were more predominant in Pullman soil in agricultural systems than in the same soil under non-disturbed conditions, whereas the opposite trend was found for *Actinobacteria* [4]. It has been reported that land use indirectly affects the bacterial community structure by modification of soil properties [5]. Other studies also indicated that soil properties are important drivers of soil bacterial community structure [6], but soil pH appears to be a major factor influencing community composition [7]. This influence of soil pH has been recognized at coarse levels of taxonomic resolution [8], but also within individual phyla [9]. In addition, it has been shown that the type of plant species [10], soil type [11], soil texture [12], and nitrogen availability [13] can affect bacterial community structure. Tree species influences on soil bacterial communities are indicated by previous studies [14], but detailed information on the affected bacterial groups and degree of these influences is still lacking.

In most previous studies the effects of land use and soil properties on soil bacterial communities have been assessed by employing traditional molecular methods such as Sanger sequencing-based analysis of 16S rRNA gene libraries or fingerprinting methods [15]. These approaches are often limited to the analysis of a relatively small number of clones and a few different soil samples. Taking into account the large bacterial community size and the heterogeneity of soils, only a tiny fraction of the bacterial diversity was unraveled by these studies. Recently, high-throughput pyrosequencing of 16S rRNA gene fragments has been applied for in-depth analysis of soil bacterial communities [3,4]. However, most of the available pyrosequencing studies do not allow a statistical assessment of land use and management effects on soil bacterial communities, as analyses of replicates were often not performed.

In this report, we applied pyrosequencing of the V2-V3 16S rRNA gene region to analyze bacterial community structure in A horizons of forest and grassland sites, which varied in management type. A horizons are mineral soil horizons formed at the surface or below an O horizon, which is dominated by organic material consisting of undecomposed or partially decomposed litter. A horizons are often characterized by accumulation of humidified organic matter [16]. It has been shown that analysis of the V2-V3 region provides a taxonomic resolution ranging from the phylum level to the genus level [17]. Thus, it is possible to detect variations in bacterial communities at different taxonomic levels. We analyzed 18 different soil samples derived from the Schwäbische Alb, which is one of the three German Biodiversity Exploratories [18]. Schwäbische Alb is a mosaic of forest and grasslands with a higher proportion of grassland. This is due to traditional sheep herding. We determined soil bacterial community structure in A horizons of 9 forest and 9 grassland sites. The selected grassland and forest sites covered a range of 6 different management types. Triplicates of the different management types were analyzed, which is an important feature of this study, as it allows statistical analysis of management effects on soil bacterial communities. For each sample, the relative abundance and the distribution of bacterial groups were determined. Subsequently, we correlated variations in the relative abundances with land use type, management type, and soil properties.

Results and Discussion

General characteristics of the soil samples

In this study, we assessed and compared the composition of soil bacterial communities present in the A horizons of 18 soil samples derived from forest and grassland sites of the Schwäbische Alb (Germany) by large-scale pyrosequencing-based analysis of 16S rRNA gene sequences. The soil samples represented triplicates of 6 different management types, which encompassed spruce age class forest (SAF1-3), beech age class forest (BAF1-3), unmanaged beech forest (BF1-3), fertilized intensely managed grassland (FUG1-3), fertilized mown pasture grazed by horse and cattle (FMG1-3), and unfertilized pasture grazed by sheep (UPG1-3) (Tables 1 and S1). The soil groups of the forest soils and the grassland soils were Cambisols and Leptosols, respectively (Table 1). In addition, soil properties such as total nitrogen (N) content, organic carbon (OC) content, pH, and soil texture were determined. The soils had overall low sand (71 ± 64 g kg^{-1}) and highly variable clay contents with values ranging from 188 to 670 g kg^{-1} (average 412 g kg^{-1}). Similarly, OC contents showed a huge variability (68 ± 16 g kg^{-1}). Total N contents were on average lower in forest sites than in grassland sites and C/N ratios were accordingly higher (14 ± 1 forest and 11 ± 1 grassland)

Table 1. Physical and geochemical characteristics of the analyzed grassland and forest soil samples.

Management type	Sample	Soil group	pH	OC (g kg^{-1})	Total N (g kg^{-1})	C:N ratio	Gravimetric water content (%)	Particle size (g kg^{-1})		
								Sand	Silt	Clay
Spruce age class forest	SAF1	Cambisol	3.30	64.57	3.97	16.26	62.8	26	668	306
Spruce age class forest	SAF2	Cambisol	4.55	65.19	4.35	14.99	65.2	43	446	511
Spruce age class forest	SAF3	Cambisol	5.04	74.68	5.14	14.53	76.5	60	445	495
Beech age class forest	BAF1	Cambisol	6.38	78.50	6.01	13.06	75.1	70	534	396
Beech age class forest	BAF2	Cambisol	4.52	57.53	4.45	12.93	70.4	47	587	368
Beech age class forest	BAF3	Cambisol	5.36	39.05	3.15	12.40	50.8	107	575	318
Unmanaged beech forest	BF1	Cambisol	4.87	77.62	5.54	14.01	75.7	109	371	520
Unmanaged beech forest	BF2	Cambisol	5.10	105.00	6.77	15.51	96.6	34	296	670
Unmanaged beech forest	BF3	Cambisol	6.37	60.03	4.49	13.37	54.9	56	495	449
Fertilized intensely managed grassland	FUG1	Leptosol	6.71	77.09	7.58	10.17	66.2	38	543	419
Fertilized intensely managed grassland	FUG2	Leptosol	6.92	72.25	7.18	10.06	59.6	139	646	215
Fertilized intensely managed grassland	FUG3	Leptosol	6.32	53.74	5.18	10.37	57.2	25	449	526
Fertilized mown pasture, horse and cattle	FMG1	Leptosol	5.11	51.61	5.35	9.65	57.5	80	475	445
Fertilized mown pasture, horse and cattle	FMG2	Leptosol	6.36	85.16	7.87	10.82	76.4	56	694	250
Fertilized mown pasture, horse and cattle	FMG3	Leptosol	6.14	68.17	6.67	10.22	64.0	32	492	476
Unfertilized pasture, sheep	UPG1	Leptosol	7.24	40.85	3.65	11.19	46.7	282	530	188
Unfertilized pasture, sheep	UPG2	Leptosol	6.45	81.15	7.41	10.95	74.3	18	384	598
Unfertilized pasture, sheep	UPG3	Leptosol	6.65	68.89	5.82	11.84	67.6	44	684	272

(Table 1). The forest samples showed lower pH values than the grassland soils, which were all, except FMG1, near neutral. The analysis of differences of soil properties and management types by employing one-way analysis of variance and Tukey pair-wise comparisons showed that the analyzed management types did not vary significantly in OC, total N, and soil texture (Table S2). The only significant difference between management types was observed for the pH values, which were higher in unfertilized pastures grazed by sheep (6.9±0.4) than in spruce age class forests (4.7±0.9).

General analyses of the pyrosequencing-derived dataset

Profiling of pylogenetic diversity and community composition by large-scale pyrosequencing of 16S rRNA gene sequences provides more sequence information compared to traditional Sanger sequencing of 16S rRNA gene clone libraries [19]. Although the per-base error rate of pyrosequencing of 16S rRNA genes is not higher than that of Sanger sequencing, the intrinsic error rate of pyrosequencing might lead to overestimation of the number of rare phylotypes. Since each pyrosequencing read is treated as an unique identifier of a community member and correction by assembly and sequencing depth applied during genome projects is not feasible, errors can result in overestimation of diversity [20,21]. To minimize the overestimation of rare phylotypes, we used quality filtering of the pyrosequencing-derived dataset, and clustering and diversity estimates were performed at genetic divergences of ≥3% [21]. Alpha diversity analysis was performed at the same level of surveying effort (22,000 sequences per sample). In addition, denoising of each sequence subset was performed to avoid overestimation of operational taxonomic units (OTUs) and diversity [22,23]. The pyrosequencing-based analysis of the V2-V3 region of the 16S rRNA genes resulted in recovery of 599,284 high quality sequences with a read length of ≥200 bp across all 18 samples. The average read length was 255 bp. The number of sequences per sample ranged from 23,519 to 39,273 with an average of 33,275 (Table S1). We were able to assign 598,962 sequences to the domain Bacteria and to classify 474,868 (79.3%) of these sequences below the domain level. Taking into account the number of sequences per sample and the number of analyzed sequences, the size of this study exceeded other published studies on pyrosequencing-based determination of soil bacterial community composition [3,4,7].

Bacterial diversity and richness

To determine rarefaction curves, richness, and diversity, OTUs were identified at genetic distances of 3, 5, and 20% by using 22,000 randomly selected and denoised sequences per sample. At 20% sequence divergence most rarefaction curves reached saturation, indicating that the surveying effort covered almost the full extent of taxonomic diversity at this genetic distance (Figure S1). Comparison of the rarefaction analyses with the number of OTUs determined by Chao1 and ACE richness estimators revealed that 50.0 to 100% (20% genetic distance) of the estimated taxonomic richness was covered by the surveying effort (Table S3). At 3 and 5% genetic distance, the rarefaction curves were not saturated and the richness estimators indicated that 35.5 to 89.3% and 38.9 to 84.8% of the estimated richness, respectively, were recovered by the sequencing effort (Figures 1, 2 and S1, and Table S3). Thus, we did not survey the full extent of taxonomic diversity at these genetic distances, but a substantial fraction of the bacterial diversity within individual soil samples was assessed at species and genus level by the surveying effort (Figure 1 and Table S3). The comparison of mean Chao1 richness estimates of all forest soils with all grassland soils showed similar values at genetic distances of 3% (3,219 OTUs and 2,611 OTUs, respectively) and 5% (2,331 OTUs and 2,095 OTUs, respectively) but at a genetic distance of 20% (75 OTUs and 153 OTUs, respectively) the richness was higher in grassland ($P<0.05$). The analysis of differences of richness estimates at genetic distances of 3% and 20% and the six management types by employing one-way analysis of variance showed that the analyzed management types did not vary significantly in the predicted number of OTUs ($P>0.05$ in both cases). Comparing this result to previous studies is difficult, as the number of analyzed sequences per sample has an effect on the predicted number of OTUs. In addition, denoising of amplicon sequences was not performed in other studies employing soil-derived pyrosequencing datasets [3,24]. In our study, richness estimates at 3% sequence divergence were approximately 2-fold higher in non-denoised datasets than in the corresponding denoised datasets (data not shown). In addition, in most other studies far fewer 16S rRNA fragments derived from a few soil samples have been analyzed.

The Shannon index of diversity (H') was determined for all samples (Table S3). At a genetic distance of 3%, the Shannon index ranged from 4.96 to 5.92 in the grassland samples and from 4.74 to 5.99 in the forest samples. Comparison of the mean H' of the different management types revealed that the highest bacterial diversity at a genetic distance of 3% was found in unmanaged beech forest, followed by fertilized intensely managed grassland, fertilized mown pastures grazed by horse and cattle, beech age class forest, spruce age class forest, and unfertilized pastures grazed by sheep (Table S3). In forest soils, the sample with the lowest pH (SAF1; pH 3.3) showed the lowest predicted diversity of all forest samples at all analyzed genetic distances (Figures 1, 2 and S1, and Table S3). Similar results were obtained by Fierer and Jackson [25] but a peak of diversity in soils with near-neutral pH values (BAF1 and BF3) that has been found in other studies [7] was not recorded. The spruce forest samples SAF2 and SAF3 showed higher diversity and richness estimates at phylum level but lower richness estimates at species level than the beech forest samples (Figure 2 and Table S3). Thus, an influence of the tree species on bacterial diversity is indicated. In addition, the rarefaction curves and the H' values derived from beech age class forest soils and unmanaged beech forest soils were not separated at all analyzed genetic distances (Figures 1 and S1, and Table S3), indicating that harvesting type (age class forest or unmanaged forest) has a minor or no impact on overall bacterial diversity and richness.

In grassland soils, similar values for estimated bacterial richness were obtained for the three samples derived from fertilized mown pastures grazed by horse and cattle whereas the replicated samples from the other two management types showed strong variations in estimated bacterial richness (Figure 2 and Table S3). At a genetic distance of 3%, the highest average bacterial richness according to Chao1 richness estimator was predicted for fertilized intensely managed grassland (2,887 OTUs), followed by fertilized mown pastures grazed by horse and cattle (2,720 OTUs), and unfertilized pastures grazed by sheep (2,226 OTUs). Nevertheless, the soil sample UPG3 derived from an unfertilized pasture grazed by sheep showed the second highest OTU estimate of all grassland soils (3,413 OTUs). Thus, bacterial diversity showed strong variations within management types in grassland soils.

Distribution of taxa and phylotypes across all samples

The 474,868 sequences classified below domain level were affiliated to 17 bacterial phyla and 4 proteobacterial classes (Tables S4 and S5). The dominant phyla and proteobacterial classes across all samples were *Acidobacteria*, *Alphaproteobacteria*, *Actinobacteria*, *Betaproteobacteria*, *Deltaproteobacteria*, *Gammaproteobacteria*, and *Firmi-*

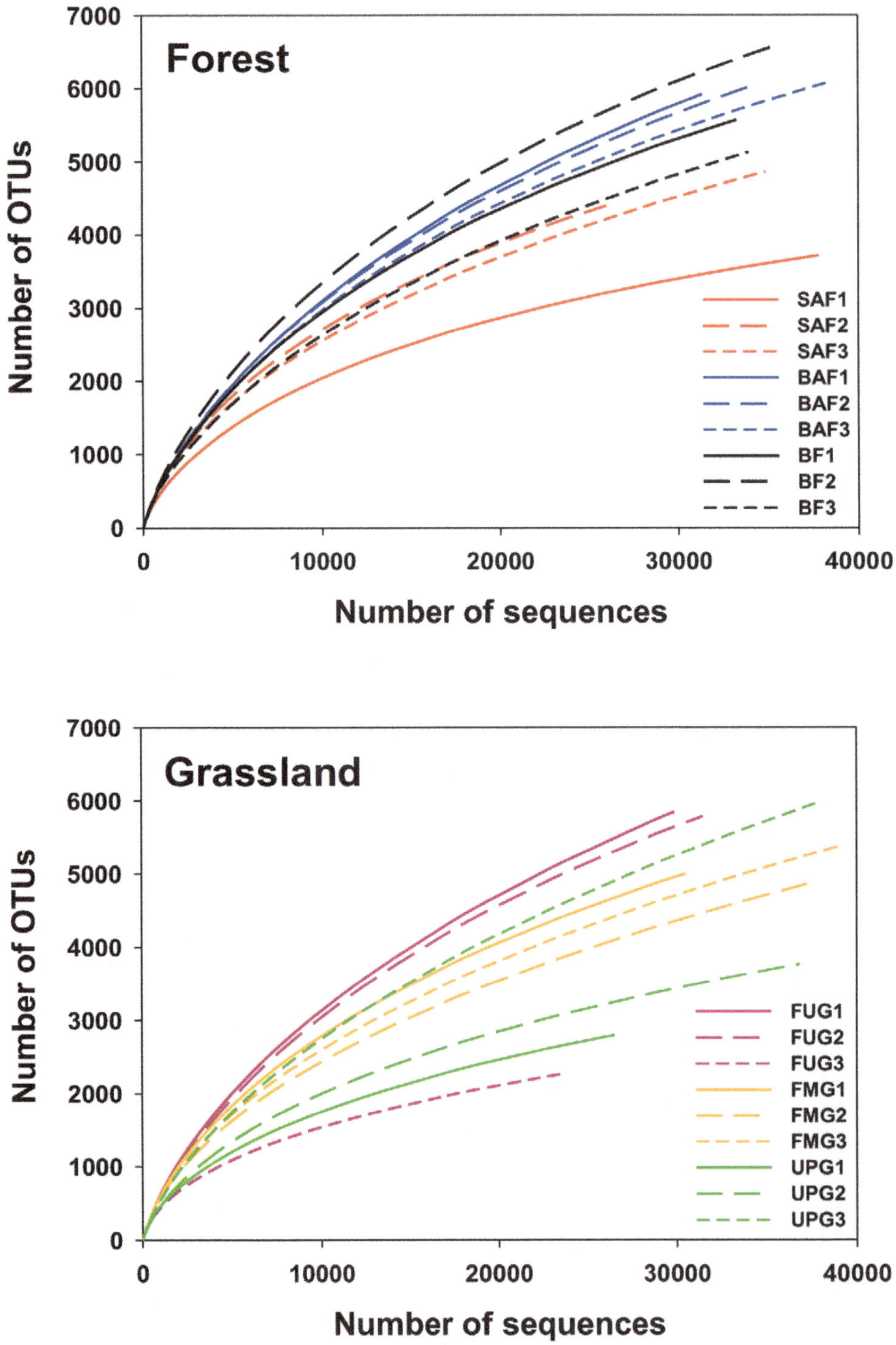

Figure 1. Rarefaction curves indicating the observed number of operational taxonomic units (OTUs) at a genetic distance of 3% in different forest and grassland soils. The spruce age class forest (SAF1-3), beech age class forest (BAF1-3), and unmanaged beech forest (BF1-3) sampling sites are marked by the red, blue, and black color, respectively. The fertilized intensely managed grassland (FUG1-3), fertilized mown pasture grazed by horse and cattle (FMG1-3), and unfertilized pasture grazed by sheep (UPG1-3) sampling sites are shown in purple, orange, and green, respectively.

cutes, representing 19.6, 18.3, 16.1, 5.9, 3.4, 2.9, and 1.2%, respectively, of all sequences that were assigned to the domain Bacteria. The dominant taxa were present in all samples and corresponded roughly with those reported in other studies on soil bacterial community composition [26]. The members of rare phyla (<1% of all classified sequences) included WS3, *Bacteroidetes*, TM7, *Chloroflexi*, *Verrucomicrobia*, *Cyanobacteria*, *Fibrobacteres*, *Spirochaetes*, *Gemmatimonadetes*, *Planctomyces*, OP11, *Deinococcus-Thermus*, and *Fusobacteria* (Figures 3 and 4, and Tables S4 and S5). The most abundant phylotype at a genetic distance of 3% across all samples was an unclassified member of the *Alphaproteobacteria*, representing 2.9% of all sequences. The most abundant phylotype at a genetic

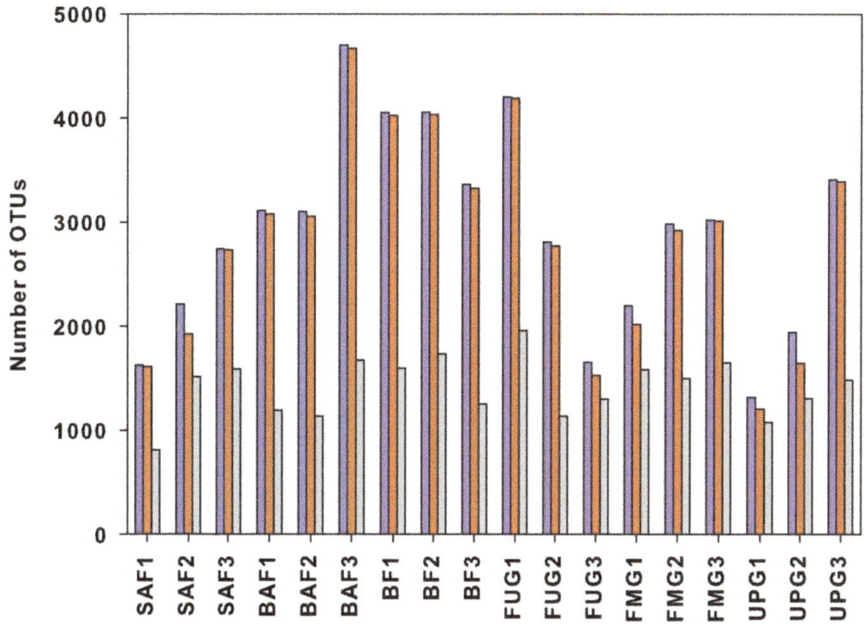

Figure 2. Bacterial richness estimates of German grassland and forest soils representing different management types at a genetic distance of 3%. Richness is expressed as number of observed unique OTUs. In addition, richness has been estimated by the abundance-based coverage estimator (ACE), which is a nonparametric richness estimator based on distribution of abundant (>10) and rare (≤10) OTUs, and the richness estimator Chao1, which is a nonparametric richness estimator based on distribution of singletons and doubletons. Richness prediction from Chao1 is colored in blue, richness prediction from ACE is colored in red, and richness observed is colored in grey. Sample numbers indicating the different management types are given below the graph. A description of the samples is shown in Table 1.

distance of 3% within one individual forest soil sample (SAF1) was a member of the family *Caulobacteraceae*, representing 7.9% of the sequences from that soil. In grassland, an unclassified member of the *Proteobacteria* was the predominant phylotype (22.5% of all sequences) within an individual soil sample (UPG2).

Differences in community structure between forest and grassland soils

The relative abundances of dominant taxa varied between grassland and forest soils. The dominant taxa in forest soils were *Alphaproteobacteria* (25.1±8.9%), *Acidobacteria* (20.4±3.0%), *Actinobacteria* (12.7±2.1%), and *Betaproteobacteria* (6.0±2.1%), whereas in grassland soils the predominant phylogenetic group was *Actinobacteria* (19.6±6.5%) followed by *Acidobacteria* (18.7±4.4%), *Alphaproteobacteria* (11.4±4.4%), and *Betaproteobacteria* (5.9%±1.2) (Figure 3, and Tables S4 and S5). The bacterial phyla and proteobacterial classes observed in our forest and grassland soils were also present in similar relative abundances in a meta-analysis of 32 bacterial 16S rRNA gene libraries derived from a variety of different soils, including samples from pristine forest, grassland and

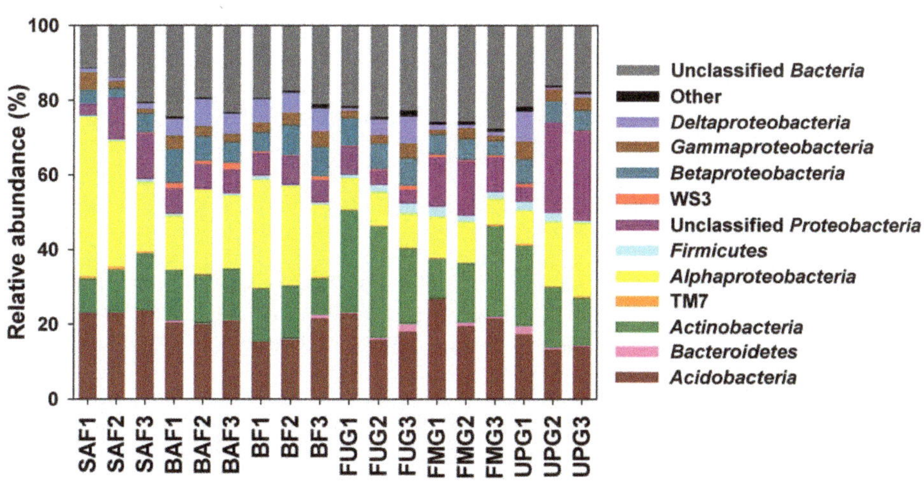

Figure 3. Relative abundances of phylogenetic groups in soils derived from the different grassland and forest sampling sites. Sample numbers indicating the different management types are given below the graph. A description of the samples is shown in Table 1. Phylogenetic groups accounting for ≤0.4% of all classified sequences are summarized in the artificial group 'others'.

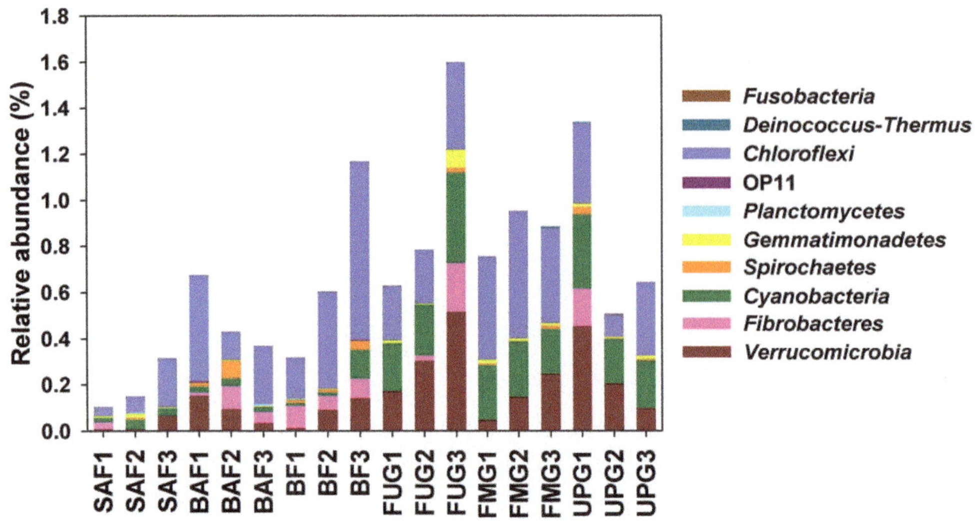

Figure 4. Relative abundances of rare phylogenetic groups of all sequences that were assigned to the domain Bacteria in soils derived from the different grassland and forest sampling sites. A description of the samples is shown in Table 1.

agricultural soils [26]. Principal components analysis (PCA) based on the relative abundances of the different bacterial phyla and proteobacterial classes confirmed that the bacterial communities in grassland soils, except the one in sample UPG3, differed from communities in forest soils (Figure 5). We observed significant higher relative abundances of *Actinobacteria*, *Firmicutes*, *Verrucomicrobia*, *Cyanobacteria*, and *Gemmatimonadetes* in grassland soils than in forest soils whereas *Alphaproteobacteria* showed the opposite pattern ($P<0.05$ in all cases) (Figures 3 and 4). Thus, the shifts in soil bacterial community composition correlated with a change from forest to grassland. A similar trend was also found by comparison of Typic Placandept soils derived from a forest site and a pasture grazed by cattle [27]. In addition, sequences affiliated to *Alphaproteobacteria* dominated in 16S rRNA clone libraries of a spruce-fir-beech forest soil in Austria as well as in a Canadian boreal forest soil [14,28].

Differences of bacterial community structure between grassland and forest soils were also found in the phylogenetic structure within individual lineages. Members of the phylum *Acidobacteria* were predominant across all samples and the second most abundant group in forest and grassland soils, representing approximately 20% of all classified sequences. Correspondingly, members of this phylum have been reported to constitute an average of 20% in bacterial communities derived from various soils [29]. Based on their abundance and the presence in various soil types, *Acidobacteria* appear to play an important role in ecosystem functions of soils, but little is known about physiology and metabolic functions of acidobacterial species. The phylum *Acidobacteria* is divided into 26 subgroups [30] with subgroups 1, 2, 3, 4, and 6 being most abundant within a variety of diverse soils [26,31]. Here, we detected 18 and 22 of these subgroups in grassland soils and forest soils, respectively. Most abundant in the grasslands soils were subgroups 16, 6, 4, 3, and 7, which represented 6.8, 4.4, 2.8, 1.8, and 1.4%, respectively, of all sequences that were classified in grassland. In forest soils, the dominant subgroups were 3, 16, 6, 1, and 4, representing 7.0, 3.0, 2.9, 2.9, and 2.1%, respectively, of all sequences that were classified (Tables S6 and S7).

Most of the sequences belonging to the second most abundant phylum *Alphaproteobacteria* across all samples were affiliated on the order level to the *Rhodospirillales* in forest soils and to *Rhizobiales* in grassland soils. *Actinobacteridae* and *Rubrobacteridae* were the most abundant subclasses within the *Actinobacteria* in both land use types, but the actinobacterial subclass *Coriobacteridae* was only detected in grassland (Tables S8 and S9). Taking into account that members of this subclass are frequently found in gut or rumen samples [32,33] it is possible that they were introduced in the grassland sites by cattle or sheep.

At the genus level, comparison of the relative abundances revealed significant differences between grassland and forest soil bacterial communities. *Mycobacterium* was the most abundant genus across all soil samples, representing 3.7% of all classified sequences in forest soils and 5.7% in grassland soils. *Mycobacteria* are free-living saprophytes and well adapted to a variety of different environments including soils [34]. The distribution of the other dominant genera *Phenylobacter*, *Bacillus*, *Kribbella*, *Agromyces*, and *Defluviicoccus* varied significantly between forest and grassland soils ($P<0.05$). *Phenylobacter* showed a higher relative abundance in forest soils than in grassland soils whereas *Bacillus*, *Kribbella*, *Agromyces*, and *Defluviicoccus* showed the opposite pattern (Figure 6). *Rubrobacter* and *Streptomyces* were present in higher proportions in grassland soils compared to forest soils ($P<0.05$) (Figure 6). Consistently, Acosta-Martínez et al. [4] found *Rubrobacter* and *Streptomyces* among the top 20 predominant bacteria in two non-disturbed grass systems derived from Texas High Plains.

In summary, significant differences of the community structure between the two analyzed land use types forest and grassland were visible. Here, the different analyzed management types in grassland and forest were not reflected by significant changes in bacterial community structure. Thus, soils derived from an identical management type, i.e., UPG1 to UPG3 do not necessarily harbor similar bacterial communities. An exception was the significant impact of tree species (beech or spruce) on community structure in our forest soils. The comparison of relative abundances of bacterial phyla and proteobacterial classes with respect to tree species revealed significant differences between soils derived from spruce and beech forests (Figure 5). Based upon two sample t-test analyses, *Deltaproteobacteria* were less abundant in spruce forest than in beech forests ($P<0.05$) (Figure 3). At the genus level, *Methylocapsa* and *Burkholderia* were more abundant in

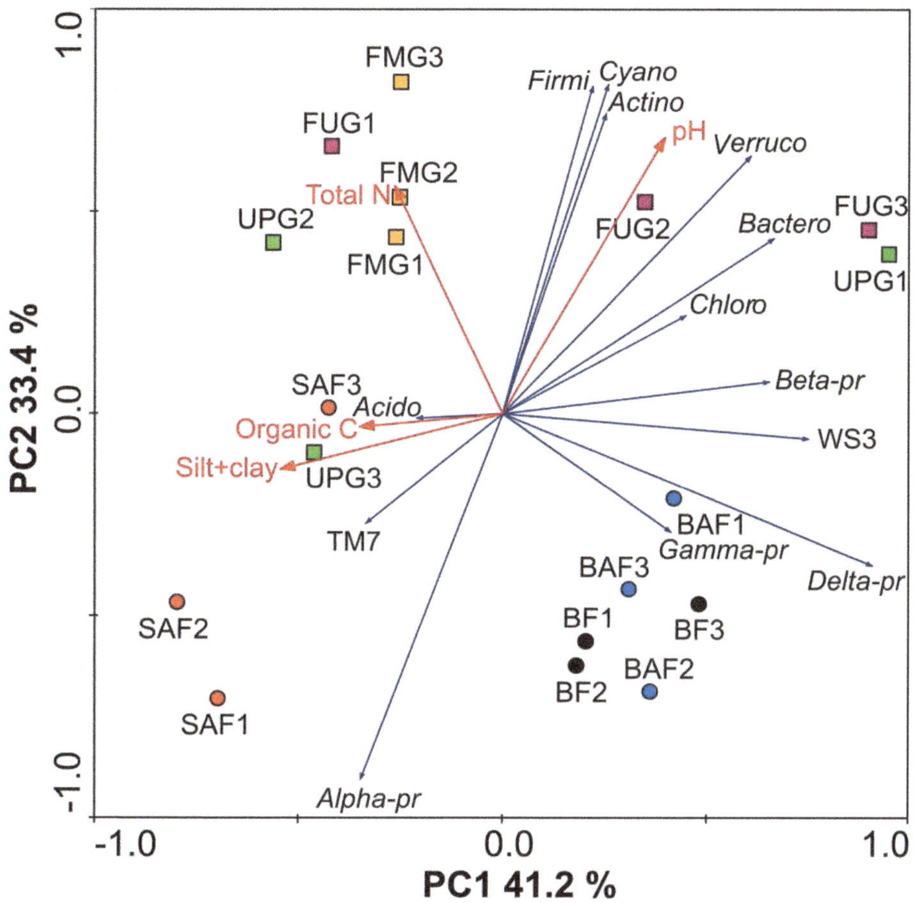

Figure 5. Principal components analysis of bacterial communities as affected by land use, based on the relative abundance of bacterial phyla and proteobacterial classes. Every vector points to the direction of increase for a given variable so that soil samples with similar bacterial communities are localized in similar positions in the diagram. The spruce age class forest (SAF1-3), beech age class forest (BAF1-3), and unmanaged beech forest (BF1-3) sampling sites are marked by the red, green, and black circles, respectively. The fertilized intensely managed grassland (FUG1-3), fertilized mown pasture grazed by horse and cattle (FMG1-3), and unfertilized pasture grazed by sheep (UPG1-3) sampling sites are depicted by red, green, and black squares, respectively. Abbreviations in figure: *Firmi, Firmicutes; Cyano, Cyanobacteria; Actino, Actinobacteria; Verruco, Verrucomicrobia; Bactero, Bacteroidetes; Chloro, Chloroflexi; Beta-pr, Betaproteobacteria; Delta-pr, Deltaproteobacteria; Gamma-pr, Gammaproteobacteria; Alpha-pr, Alphaproteobacteria; Acido, Acidobacteria.*

spruce forest soil than in beech forest soil, whereas *Nocardioides*, *Leptothrix*, and *Amaricoccus* showed the opposite pattern (Figure 6). Thus, tree species appear to be an important driver of soil bacterial community structure, but the type of harvesting (age class forest or unmanaged forest) does not significantly affect bacterial community composition (Figure 5).

Impact of soil properties on the relative abundances of bacterial taxa

Previous studies indicated that soil properties such as pH value or soil texture are important drivers of bacterial community structure [12,35]. We used correlation analysis to identify relationships between the relative abundances of bacterial groups and soil properties. The relative abundances of bacterial groups at different taxonomic levels responded strongly to soil pH. This is in accordance to other surveys on soil bacterial communities derived from different management types in which pH-dependent changes in abundance and distribution of bacterial phyla were observed [36,37]. At the phylum level, relative abundances of *Bacteroidetes* and *Actinobacteria* in the analyzed soils significantly increased with higher pH values ($P<0.05$ in both cases) (Table 2).

As described for a freshwater lake [38] and diverse soils [9], we also found strong correlations of pH and relative abundances of bacterial groups below the phylum level. The relative abundances of the proteobacterial classes *Alphaproteobacteria* and *Betaproteobacteria* were significantly correlated to pH ($P<0.05$). The abundances of *Alphaproteobactia* were negatively correlated with soil pH, whereas the abundances of *Betaproteobacteria* increased with pH (Table 2). Within the *Alphaproteobacteria*, the relative abundances of the order *Caulobacterales* and the family *Acetobacteraceae* showed similar correlations to soil pH as the *Alphaproteobacteria* in general ($P<0.05$ in both cases) (Figure 7). This result corresponded to a cultivation-dependent study of Jimenez-Salgado et al. [39], in which more members of the *Acetobacteraceae* were isolated from low pH soils than from high pH soils. Although relative abundances of *Gammaproteobacteria* showed no significant correlation to soil pH at the class level, the relative abundances of the gammaproteobacterial genus *Dyella* significantly increased with lower pH values ($P<0.05$) (Figure 7). The genus *Dyella* has been recently described by Xie and Yokota [40]. So far, it includes seven species isolated from soil, but no growth of these isolates below pH 4.0 was described [41,42]. In contrast, the highest relative abundances for sequences affiliated to the genus *Dyella* (0.6% of all classified

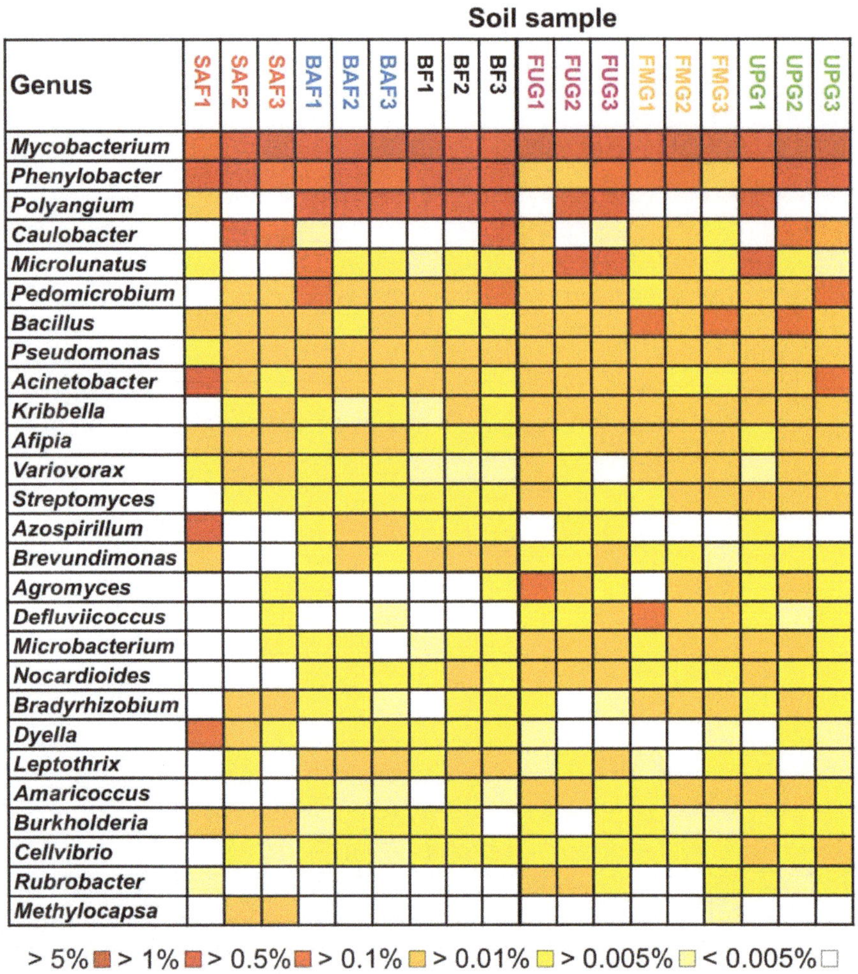

Figure 6. Relative abundances of the most abundant genera as affected by land use. Percentages below the map indicate the abundance of each genus relative to all bacterial sequences that were classified in each of the 18 soils. A description of the samples is shown in Table 1. Grassland and forest samples are separated by a bold line. Samples of different management types are colored in red (SAF1-3), blue (BAF1-3), black (BF1-3), purple (FUG1-3), orange (FMG1-3), and green (UPG1-3).

sequences) were found in sample SAF1, which exhibited the lowest pH value of all samples (pH 3.3). Furthermore, we obtained the highest relative abundances for genera *Azospirillum* and *Acinetobacter* (each representing more than 0.5% of all classified *Bacteria*) in soil sample SAF1 (Figure 6). Thus, our results might help to identify conditions that are best suited for a targeted cultivation of members belonging to these genera.

The occurrence of several subgroups of the *Acidobacteria*, which were predominant across all samples, was also dependent on soil pH. The relative abundances of acidobacterial subgroups 1, 3, 6, 13, 17, and 18 showed strong significant correlations to soil pH ($P<0.001$ in all cases). The relative abundances of subgroups 1, 3, and 13 decreased with pH whereas those of subgroups 6, 17, and 18 were positively correlated with pH (Figure 7 and Table S10). Similar correlations of soil pH and the abundances of acidobacterial subgroups 1, 3, 6, 13, 17, and 18 have been reported by Jones et al. [9]. In addition, the inverse relationship of soil pH on the abundance of members affiliated to subgroup 1 has been reported for soils derived from rotationally grazed perennial ryegrass and white clover pasture [43].

In general, more groups at different taxonomic levels showed significant correlations to soil pH in forest soils than in grassland soils (data not shown). This might be due to the different pH range covered by the analyzed forest and grassland soils. The pH in our forest samples ranged from pH 3.30 to 6.37 (Table 1) whereas the pH values of the grassland samples were all, except sample FMG1, near neutral. Thus, a relatively small pH range was covered by our grassland samples (Table 1), so there is simply less pH range from which to determine correlations. Significant correlations of relative abundances with other soil properties were found for *Deltaproteobacteria* and *Actinobacteria*. The *Deltaproteobacteria* showed a significant correlation to OC ($P<0.05$) with higher abundances in soils with low OC content, whereas *Actinobacteria* showed a significant correlation to total N ($P<0.05$) with higher abundances in soils with high total N content (Table 2), but a connection to the observed correlations was not evident.

Conclusion

The analysis of one of the largest bacterial 16S rRNA-based datasets from soils revealed statistically significant differences in soil bacterial diversity and community structure between the two land use types forest and grassland. Additionally, the occurrence of different tree species had statistically significant effects on soil bacterial diversity, richness, and community composition in forest. The analysis of influences of soil properties on bacterial community structure revealed that pH had the strongest effect

Table 2. Spearman's rank correlations between the relative abundances of the six most abundant bacterial phyla and proteobacterial classes and the soil properties in grassland and forest soils.

Taxonomic group	Correlation			
	pH	OC	Total N	Sand/Silt/Clay
Actinobacteria	**0.58**	0.26	**0.52**	0.02/−0.08/−0.02
Bacteroidetes	**0.71**	0.14	0.33	−0.08/0.17/−0.19
Alphaproteobacteria	**−0.68**	0.05	−0.44	−0.12/−0.13/0.22
Betaproteobacteria	**0.56**	0.22	0.35	0.04/0.04/0.00
Deltaproteobacteria	−0.10	**−0.48**	−0.55	0.43/−0.15/−0.04
Gammaproteobacteria	0.27	−0.04	−0.17	−0.13/0.19/−0.19

Correlations for *Acidobacteria* are shown at higher taxonomic resolution Table S10.
Bold numbers: $P<0.05$; Bold and underlined numbers $P<0.001$.

on bacterial community structure of the analyzed soil properties. Management type and other soil properties appear to have a minor impact on soil bacterial community structure and diversity.

In this survey, the correlations between land use type and community composition were obvious. The relative abundances of a number of taxonomic groups changed significantly between forest and grassland soils (e.g., *Actinobacteria*), but the abundances of other taxa (e.g., *Gammaproteobacteria*) were almost unaffected by land use type, indicating that the abundances of the latter groups are influenced by other factors. Specific bacterial groups such as *Amaricoccus* or *Methylocapsa* showed significantly higher abundances in beech or spruce forest soils. Finally, we cannot determine whether pH has a direct or indirect effect on community composition, as a number of soil properties (e.g., OC) are directly or indirectly related to pH [44]. Thus, the effect of a number of different factors is reflected by soil pH and these factors may also drive community composition.

Availability

The 18 pyrosequencing-derived 16S rRNA gene sequence datasets have been deposited in the GenBank short-read archive under accession number SRA022075.

Materials and Methods

Site description, sampling, DNA extraction, and soil characterization

In the frame of the German Biodiversity Exploratories, initiative soil samples were collected from 9 forest and 9 grassland plots of the German Biodiversity Exploratory Schwäbische Alb. The Schwäbische Alb covers more than 450 km×450 km in the state of Baden-Württemberg (southwestern Germany). Soil samples were collected in April 2008. The forest sampling sites included 3 spruce age class forests (SAF1-3), 3 beech age class forests (BAF1-3), and 3 unmanaged beech forests (BF1-3). Grassland sampling sites comprised 3 fertilized intensely managed grasslands (FUG1-3), 3 fertilized mown pastures grazed by horse and cattle (FMG1-3), and 3 unfertilized pastures grazed by sheep (UPG1-3) (Table S1). The dominant grasses included *Poa trivialis*, *Trisetum flavescens*, and *Arrhenaterum elatius* in sites FUG1-3, *Poa trivialis*, *Alopecurus pratensis*, *Trisetum flavescens*, *Dactylis glomerata*, *Festuca pratensis*, *Lolium perenne*, and *Arrhenaterum elatius* in sites FMG1-3, and *Brachypodium pinnatum*, *Bromus erectus*, and *Festuca guestfalica* in sites UPG1-3. A detailed description of the dominant grasses of the individual plots is provided in Table S11.

Soil samples were collected and classified at each of the grassland and forest sites as described by Will et al. [45]. Briefly, five soil cores (8.3 cm in diameter) were sampled with a motor driven soil column cylinder at each corner and in the center of each plot within a given area of 20 m×20 m. Composite samples of the five collected A horizons per plot were used for DNA extraction, after the soils were homogenized and coarse roots and stones (>5 mm) were removed. Total microbial community DNA was extracted from approximately 8 g soil derived from the A horizons of each plot by employing the MoBio PowerMax Soil DNA isolation kit (MoBio Laboratories, Carlsbad, CA, USA) as recommended by the manufacturer. DNA concentrations were quantified by using a NanoDrop ND-1000 UV-Vis Spectrophotometer (NanoDrop Technologies, USA) according to the manufacturer's protocol.

OC content, total N content, soil texture, and soil pH were measured as described by Will et al. [45]. To determine the gravimetric water content, 10 g of moist soil were dried to constant weight at 105°C for 24 h. The mass of water was calculated per mass of dry soil.

Amplification of 16S rRNA genes and pyrosequencing

The V2-V3 region of the 16S rRNA gene was amplified by PCR. The PCR reaction mixture (33 µl) contained 3.3 µl 10-fold reaction buffer (Fusion GC buffer, FINNZYMES, Espoo, Finland), 800 µM of each of the four deoxynucleoside triphosphates, 3% DMSO, 1.2 µM of each of the primers, 0.5 U of Phusion hot start high-fidelity DNA Polymerase (FINNZYMES), and 20 ng of isolated DNA as template. The V2-V3 region was amplified with the following set of primers containing the Roche 454 pyrosequencing adaptors (underlined): V2for 5′-GCCTCCCTCGCGCCATCAGAGTGGCGGACGGGTGAGTAA-3′ and V3rev 5′-GCCTTGCCAGCCCGCTCAGCGTATTACCGCGGCTGCTG-3′ (modified from Schmalenberger et al. [46]). The following thermal cycling scheme was used: initial denaturation at 98°C for 5 min, 25 cycles of denaturation at 98°C for 45 s, annealing at 68°C for 45 s, and extension at 72°C for 25 s followed by a final extension period at 72°C for 5 min. All samples were amplified in triplicate, pooled in equal amounts, and purified using the peqGold gel extraction kit as recommended by the manufacturer (Peqlab Biotechnologie GmbH, Erlangen, Germany). Quantification of the PCR products was performed using the Quant-iT dsDNA BR assay kit and a Qubit fluorometer (Invitrogen GmbH, Karlsruhe, Germany) as recommended by the manufacturer. The Göttingen Genomics Laboratory determined the sequences of the partial 16S rRNA genes by using a Roche GS-FLX 454 pyrosequencer (Roche, Mannheim, Germany) and the instructions of the manufacturer for amplicon sequencing.

Analysis of pyrosequencing data

Sequences that were shorter than 200 bp in length and reads containing any unresolved nucleotides were removed from the 18 pyrosequencing-derived datasets. For taxonomy-based analysis, the RDP Classifier of the Ribosomal Database Project (RDP) was used [47] at a confidence threshold of 80%. Pyrosequencing noise was removed for alpha diversity analyses by using the denoiser program [23]. For the determination of OTUs, we defined species, genus, and phylum level at 3, 5, and 20%, respectively, sequence divergence according to Schloss and Handelsman [48]. OTUs were determined for each denoised sequence dataset by using the uclust OTU picker version 1.2.21q of the QIIME software pipeline [49]. We calculated rarefaction curves as well as the

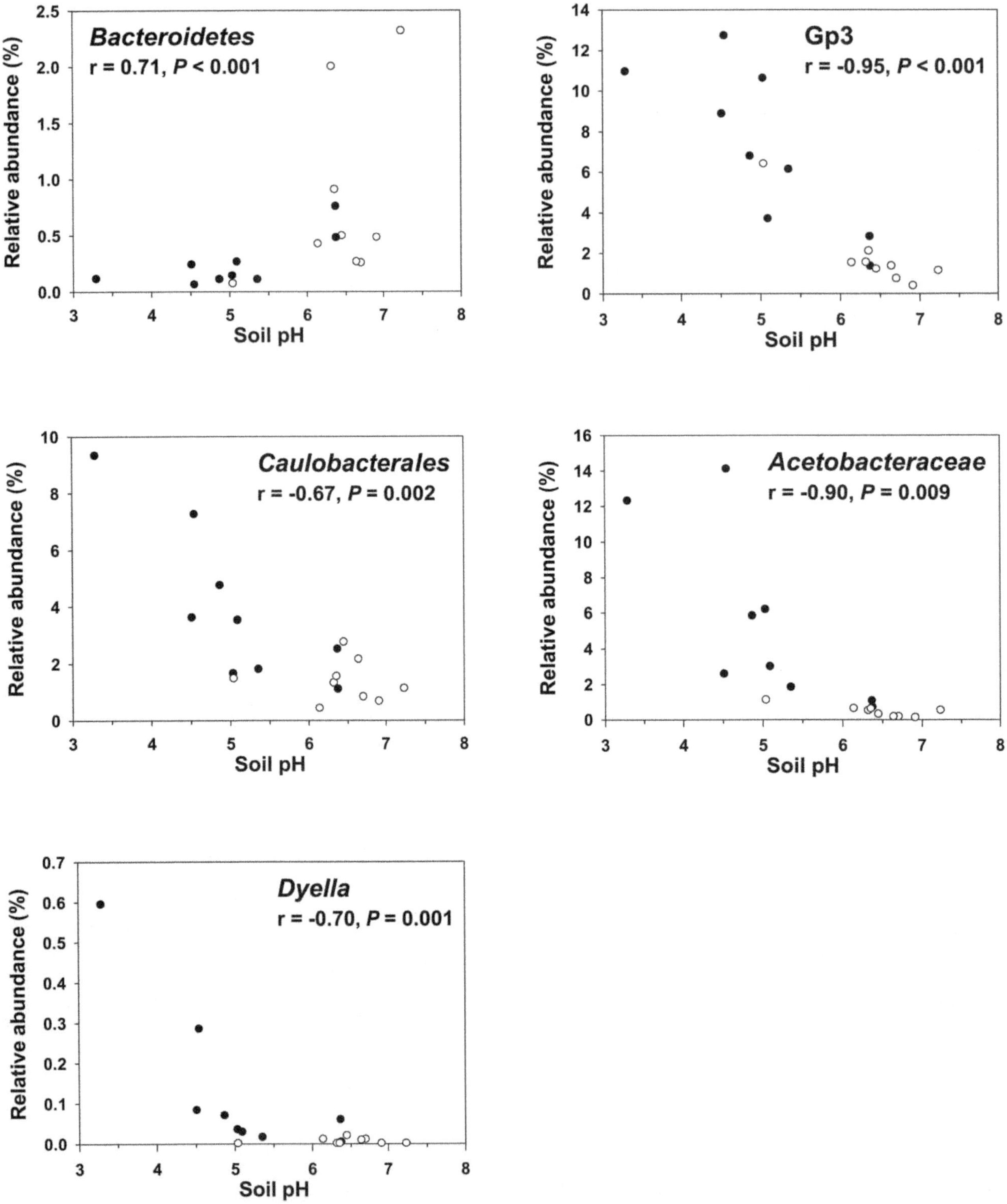

Figure 7. Correlations between relative abundances of different taxonomic groups and soil pH. Black circles represent forest sites and white circles represent grassland sites. Spearman's rank correlation coefficients (r) with the associated *P* values are shown for each taxonomic group. Abbreviation: Gp3, acidobacterial subgroup 3.

Shannon [50] index based on OTU picker data, by employing the RDP pyrosequencing pipeline [51]. ACE and Chao1 [52] indices were calculated using the EstimateS program version 8.2.0 (http://purl.oclc.org/estimates).

Statistical analyses

Normality tests (Shapiro-Wilk) were performed with data that were used for principal component analysis (PCA), and one-way analysis of variance. Data that did not pass normality test were log transformed and normality test was repeated. Only data that passed normality test were used for further analyses. For each soil attribute and each richness estimate at 3 and 20% genetic distance, one-way analysis of variance and Tukey pair-wise comparisons were used to determine the minimum significant difference ($P<0.05$) between management types by employing STATISTICA 8.0 (StatSoft, Inc., Tulsa, USA). To compare bacterial community structures across all samples based on the relative abundance of bacterial phyla and proteobacterial classes, PCA was performed by using CANOCO for Windows [53]. To correlate bacterial taxonomic groups with soil properties, Spearman's rank correlations were determined by using the SigmaPlot program version 11.0 (Systat Software, Inc., San Jose, CA). We used two sample t-test analyses and M-W-U-Test for non-parametric data to compare relative abundances of bacterial groups and richness estimates between grassland and forest, and on a second level between different management types using the software package PAST [54].

Supporting Information

Figure S1 Rarefaction curves indicating the observed number of OTUs at genetic distances of 5 and 20% in the different forest and grassland soils. The spruce age class forest (SAF1-3), beech age class forest (BAF1-3), and unmanaged beech forest (BF1-3) sampling sites are marked by the red, blue, and black color, respectively. The fertilized intensely managed grassland (FUG1-3), fertilized mown pasture grazed by horse and cattle (FMG1-3), and unfertilized pasture grazed by sheep (UPG1-3) sampling sites are shown in purple, orange, and green, respectively.

Table S1 Localization of the sampling sites and number of 16S rRNA gene sequences derived from the analyzed grassland and forest soil samples.

Table S2 Mean values of soil properties and standard deviation for each management type and ANOVA *P* values. Differences of soil properties between management types were analyzed by employing one-way analysis of variance and Tukey pair-wise comparisons. Significant ANOVA *P* values are shown in bold ($P<0.05$). Figures followed by different letters indicate differences among management types ($P<0.05$). Abbreviations: SAF, spruce age class forest; BAF, beech age class forest; BF, unmanaged beech forest; FUG, fertilized intensely managed grassland; FMG, fertilized mown pasture grazed by horse and cattle; UPG, unfertilized pasture grazed by sheep. Complete soil and site information for all 18 sampling sites is provided in Table 1.

Table S3 Bacterial diversity as assessed by Shannon index (H') and species richness estimation in all forest and grassland soils. The results from the rarefaction analyses are also depicted in Figure 1 and Figure S1.

Table S4 Relative abundances of bacterial phyla and proteobacterial classes in the analyzed forest soils. Values represent percentages of all sequences assigned to the domain Bacteria for all forest soils or individual forest soils. Groups labeled with asterisks could not be assigned to a specific phylum or a proteobacterial class.

Table S5 Relative abundances of bacterial phyla and proteobacterial classes in the analyzed grassland soils. Values represent percentages of all sequences assigned to the domain Bacteria for all grassland soils or individual grassland soils. Groups labeled with asterisks could not be assigned to a specific phylum or a proteobacterial class.

Table S6 Relative abundances of acidobacterial subgroups in the analyzed forest soils. Values represent percentages of all sequences assigned to the domain Bacteria for all forest soils or individual forest soils. Groups labeled with asterisks could be assigned to the phylum level only.

Table S7 Relative abundances of acidobacterial subgroups in the analyzed grassland soils. Values represent percentages of all sequences assigned to the domain Bacteria for all grassland soils or individual grassland soils. Groups labeled with asterisks could be assigned to the phylum level only.

Table S8 Relative abundances of taxonomic groups within the phylum *Actinobacteria* and within proteobacterial classes in the analyzed forest soils. Values represent percentages of all sequences assigned to the domain Bacteria for all forest soils or individual forest soils. Groups labeled with asterisks could be assigned to the phylum level only.

Table S9 Relative abundances of taxonomic groups within the phylum *Actinobacteria* and within proteobacterial classes in the analyzed grassland soils. Values represent percentages of all sequences assigned to the domain Bacteria for all grassland soils or individual grassland soils. Groups labeled with asterisks could be assigned to the phylum level only.

Table S10 Spearman's rank correlations between relative abundances of *Acidobacteria* subgroups and soil properties. Only relative abundances of acidobacterial subgroups that represented ≥0.029% of all analyzed sequences were considered.

Table S11 Dominant grasses of the analyzed grassland sites.

Acknowledgments

Field work permits were given by the responsible state environmental office of Baden-Württemberg. We thank the local implementation team of the Schwäbische Alb exploratory, the BEO (biodiversity exploratories office), and Fabian Alt for supporting the coordinated soil sampling campaign. In addition, we thank Steffen Boch for providing vegetation data of the analyzed study sites.

Author Contributions

Conceived and designed the experiments: RD. Performed the experiments: HN AT CW. Analyzed the data: HN AT AW LH. Wrote the paper: HN RD. Contributed data on soil properties and analysis of these data: IS NH MS.

References

1. Gans J, Wolinsky M, Dunbar J (2005) Computational improvements reveal great bacterial diversity and high metal toxicity in soil. Science 309: 1387–1390.
2. Daniel R (2005) The metagenomics of soil. Nat Rev Microbiol 3: 470–478.
3. Roesch LFW, Fulthorpe RR, Riva A, Casella G, Hadwin AKM, et al. (2007) Pyrosequencing enumerates and contrasts soil microbial diversity. ISME J 1: 283–290.

4. Acosta-Martínez V, Dowd S, Sun Y, Allen V (2008) Tag-encoded pyrosequencing analysis of bacterial diversity in a single soil type as affected by management and land use. Soil Biol Biochem 40: 2762–2770.
5. Jesus ED, Marsh TL, Tiedje JM, Moreira FMD (2009) Changes in land use alter the structure of bacterial communities in Western Amazon soils. ISME J 3: 1004–1011.
6. Wakelin SA, Macdonald LM, Rogers SL, Gregg AL, Bolger TP, et al. (2008) Habitat selective factors influencing the structural composition and functional capacity of microbial communities in agricultural soils. Soil Biol Biochem 40: 803–813.
7. Lauber CL, Hamady M, Knight R, Fierer N (2009) Pyrosequencing-based assessment of soil pH as a predictor of soil bacterial community structure at the continental scale. Appl Environ Microbiol 75: 5111–5120.
8. Bååth E, Anderson TH (2003) Comparison of soil fungal/bacterial ratios in a pH gradient using physiological and PLFA-based techniques. Soil Biol Biochem 35: 955–963.
9. Jones RT, Robeson MS, Lauber CL, Hamady M, Knight R, et al. (2009) A comprehensive survey of soil acidobacterial diversity using pyrosequencing and clone library analyses. ISME J 3: 442–453.
10. Wieland G, Neumann R, Backhaus H (2001) Variation of microbial communities in soil, rhizosphere, and rhizoplane in response to crop species, soil type, and crop development. Appl Environ Microbiol 67: 5849–5854.
11. Girvan MS, Bullimore J, Pretty JN, Osborn AM, Ball AS (2003) Soil type is the primary determinant of the composition of the total and active bacterial communities in arable soils. Appl Environ Microbiol 69: 1800–1809.
12. Sessitsch A, Weilharter A, Gerzabek MH, Kirchmann H, Kandeler E (2001) Microbial population structures in soil particle size fractions of a long-term fertilizer field experiment. Appl Environ Microbiol 67: 4215–4224.
13. Frey SD, Knorr M, Parrent JL, Simpson RT (2004) Chronic nitrogen enrichment affects the structure and function of the soil microbial community in temperate hardwood and pine forests. Forest Ecol Manage 196: 159–171.
14. Hackl E, Zechmeister-Boltenstern S, Bodrossy L, Sessitsch A (2004) Comparison of diversities and compositions of bacterial populations inhabiting natural forest soils. Appl Environ Microbiol 70: 5057–5065.
15. McCaig AE, Glover LA, Prosser JI (2001) Numerical analysis of grassland bacterial community structure under different land management regimens by using 16S ribosomal DNA sequence data and denaturing gradient gel electrophoresis banding patterns. Appl Environ Microbiol 67: 4554–4559.
16. FAO (2006) Guidelines for soil description, 4th edn. Rome, Italy: FAO.
17. Chakravorty S, Helb D, Burday M, Connell N, Alland D (2007) A detailed analysis of 16S ribosomal RNA gene segments for the diagnosis of pathogenic bacteria. J Microbiol Meth 69: 330–339.
18. Fischer M, Bossdorf O, Gockel S, Hänsel F, Hemp A, et al. (2010) Implementing large-scale and long-term functional biodiversity research: The Biodiversity Exploratories. Basic Appl Ecol 11: 473–485.
19. Sogin ML, Morrison HG, Huber JA, Welch DM, Huse SM, et al. (2006) Microbial diversity in the deep sea and the underexplored "rare biosphere". Proc Natl Acad Sci USA 103: 12115–12120.
20. Huse SM, Huber JA, Morrison HG, Sogin ML, Welch DM (2007) Accuracy and quality of massively parallel DNA pyrosequencing. Gen Biol 8: R143.
21. Kunin V, Engelbrektson A, Ochman H, Hugenholtz P (2010) Wrinkles in the rare biosphere: pyrosequencing errors can lead to artificial inflation of diversity estimates. Environ Microbiol 12: 118–123.
22. Quince C, Lanzen A, Curtis TP, Davenport RJ, Hall N, et al. (2009) Noise and the accurate determination of microbial diversity from 454 pyrosequencing data. Nat Meth 6: 639–641.
23. Reeder J, Knight R (2010) Rapidly denoising pyrosequencing amplicon reads by exploiting rank-abundance distributions. Nat Meth 7: 668–669.
24. Morales S, Cosart T, Johnson J, Holben W (2009) Extensive phylogenetic analysis of a soil bacterial community illustrates extreme taxon evenness and the effects of amplicon length, degree of coverage, and DNA fractionation on classification and ecological parameters. Appl Environ Microbiol 75: 668–675.
25. Fierer N, Jackson RB (2006) The diversity and biogeography of soil bacterial communities. Proc Natl Acad Sci USA 103: 626–631.
26. Janssen PH (2006) Identifying the dominant soil bacterial taxa in libraries of 16S rRNA and 16S rRNA genes. Appl Environ Microbiol 72: 1719–1728.
27. Nüsslein K, Tiedje JM (1999) Soil bacterial community shift correlated with change from forest to pasture vegetation in a tropical soil. Appl Environ Microbiol 65: 3622–3626.
28. Dimitriu PA, Grayston SJ (2010) Relationship between soil properties and patterns of bacterial β-diversity across reclaimed and natural boreal forest soils. Microb Ecol 59: 563–573.
29. Dunbar J, Barns SM, Ticknor LO, Kuske CR (2002) Empirical and theoretical bacterial diversity in four Arizona soils. Appl Environ Microbiol 68: 3035–3045.
30. Barns SM, Cain EC, Sommerville L, Kuske CR (2007) *Acidobacteria* phylum sequences in uranium-contaminated subsurface sediments greatly expand the known diversity within the phylum. Appl Environ Microbiol 73: 3113–3116.
31. Barns SM, Takala SL, Kuske CR (1999) Wide distribution and diversity of members of the bacterial kingdom *Acidobacterium* in the environment. Appl Environ Microbiol 65: 1731–1737.
32. Anderson RC, Rasmussen MA, Jensen NS, Allison MJ (2000) *Denitrobacterium detoxificans* gen. nov., sp. nov., a ruminal bacterium that respires on nitrocompounds Int J Syst Evol Microbiol 50: 633–638.
33. Collado MC, Sanz Y (2007) Quantification of mucosa-adhered microbiota of lambs and calves by the use of culture methods and fluorescent in situ hybridization coupled with flow cytometry techniques. Vet Microbiol 121: 299–306.
34. Ventura M, Canchaya C, Tauch A, Chandra G, Fitzgerald GF, et al. (2007) Genomics of *Actinobacteria*: tracing the evolutionary history of an ancient phylum. Microbiol Mol Biol Rev 71: 495–548.
35. Rousk J, Bååth E, Brookes P, Lauber C, Lozupone C, et al. (2010) Soil bacterial and fungal communities across a pH gradient in an arable soil. ISME J 4: 1340–1351.
36. Lauber CL, Strickland MS, Bradford MA, Fierer N (2008) The influence of soil properties on the structure of bacterial and fungal communities across land-use types. Soil Biol Biochem 40: 2407–2415.
37. Hartman WH, Richardson CJ, Vilgalys R, Bruland GL (2008) Environmental and anthropogenic controls over bacterial communities in wetland soils. Proc Natl Acad Sci USA 105: 17842–17847.
38. Newton RJ, Jones SE, Helmus MR, McMahon KD (2007) Phylogenetic ecology of the freshwater *Actinobacteria* acI lineage. Appl Environ Microbiol 73: 7169–7176.
39. Jimenez-Salgado T, Fuentes-Ramirez LE, Tapia-Hernandez A, Mascarua-Esparza MA, Martinez-Romero E, et al. (1997) *Coffea arabica* L., a new host plant for *Acetobacter diazotrophicus*, and isolation of other nitrogen-fixing acetobacteria. Appl Environ Microbiol 63: 3676–3683.
40. Xie CH, Yokota A (2005) *Dyella japonica* gen. nov., sp. nov., a gamma-proteobacterium isolated from soil. Int J Syst Evol Microbiol 55: 753–756.
41. Weon HY, Anandham R, Kim BY, Hong SB, Jeon YA, et al. (2009) *Dyella soli* sp. nov. and *Dyella terrae* sp. nov., isolated from soil. Int J Syst Evol Microbiol 59: 1685–1690.
42. Anandham R, Kwon SW, Indira Gandhi P, Kim SJ, Weon HY, et al. (2010) *Dyella thiooxydans* sp. nov., a facultative chemolithotrophic thiosulfate oxidizing bacterium isolated from rhizosphere soil of sunflower (*Helianthus annuus* L.). Int J Syst Evol Microbiol: doi:10.1099/ijs.0.022012-0.
43. Sait M, Davis KER, Janssen PH (2006) Effect of pH on isolation and distribution of members of subdivision 1 of the phylum *Acidobacteria* occurring in soil. Appl Environ Microbiol 72: 1852–1857.
44. Brady NC, Weil RR (2007) The nature and properties of soil, 13th edn. Upper Saddle River, New Jersey, USA: Prentice Hall.
45. Will C, Thürmer A, Wollherr A, Nacke H, Herold N, et al. (2010) Horizon-specific bacterial community composition of German grassland soils, as revealed by pyrosequencing-based analysis of 16S rRNA genes. Appl Environ Microbiol 76: 6751–6759.
46. Schmalenberger A, Schwieger F, Tebbe C (2001) Effect of primers hybridizing to different evolutionarily conserved regions of the small-subunit rRNA gene in PCR-based microbial community analyses and genetic profiling. Appl Environ Microbiol 67: 3557–3563.
47. Wang Q, Garrity G, Tiedje J, Cole J (2007) Naïve Bayesian classifier for rapid assignment of rRNA sequences into the new bacterial taxonomy. Appl Environ Microbiol 73: 5261–5267.
48. Schloss PD, Handelsman J (2005) Introducing DOTUR, a computer program for defining operational taxonomic units and estimating species richness. Appl Environ Microbiol 71: 1501–1506.
49. Caporaso JG, Kuczynski J, Stombaugh J, Bittinger K, Bushman FD, et al. (2010) QIIME allows analysis of high-throughput community sequencing data. Nat Meth 7: 335–336.
50. Shannon CE, Weaver W (1949) The mathematical theory of communication. MD Comput 14: 306–317.
51. Cole JR, Wang Q, Cardenas E, Fish J, Chai B, et al. (2009) The ribosomal database project: improved alignments and new tools for rRNA analysis. Nucl Acids Res 37: D141–145.
52. Chao A, Bunge J (2002) Estimating the number of species in a stochastic abundance model. Biometrics 58: 531–539.
53. ter Braak CJF, Smilauer P (2002) CANOCO reference manual and CanoDraw for windows user's guide: software for canonical community Ordination (version 4.5). Ithaca, NY, USA: Microcomputer Power.
54. Hammer Ø, Harper DAT, Ryan PD (2001) PAST: Paleontological Statistics Software Package for Education and Data Analysis. Palaeontologia Electronica 4: 1–9.

2

Historical Human Footprint on Modern Tree Species Composition in the Purus-Madeira Interfluve, Central Amazonia

Carolina Levis[1]*, Priscila Figueira de Souza[2], Juliana Schietti[1], Thaise Emilio[1], José Luiz Purri da Veiga Pinto[3], Charles R. Clement[4], Flavia R. C. Costa[3]

1 Programa de Pós-Graduação em Ecologia, Instituto Nacional de Pesquisas da Amazônia (INPA), Manaus, Amazonas, Brasil, **2** Programa de Pós-Graduação em Botânica, Instituto Nacional de Pesquisas da Amazônia (INPA), Manaus, Amazonas, Brasil, **3** Coordenação de Biodiversidade, Instituto Nacional de Pesquisas da Amazônia (INPA), Manaus, Amazonas, Brasil, **4** Coordenação de Tecnologia e Inovação, Instituto Nacional de Pesquisas da Amazônia (INPA), Manaus, Amazonas, Brasil

Abstract

Background: Native Amazonian populations managed forest resources in numerous ways, often creating oligarchic forests dominated by useful trees. The scale and spatial distribution of forest modification beyond pre-Columbian settlements is still unknown, although recent studies propose that human impact away from rivers was minimal. We tested the hypothesis that past human management of the useful tree community decreases with distance from rivers.

Methodology/Principal Findings: In six sites, we inventoried trees and palms with DBH≥10 cm and collected soil for charcoal analysis; we also mapped archaeological evidence around the sites. To quantify forest manipulation, we measured the relative abundance, richness and basal area of useful trees and palms. We found a strong negative exponential relationship between forest manipulation and distance to large rivers. Plots located from 10 to 20 km from a main river had 20–40% useful arboreal species, plots between 20 and 40 km had 12–23%, plots more than 40 km had less than 15%. Soil charcoal abundance was high in the two sites closest to secondary rivers, suggesting past agricultural practices. The shortest distance between archaeological evidence and plots was found in sites near rivers.

Conclusions/Significance: These results strongly suggest that past forest manipulation was not limited to the pre-Columbian settlements along major rivers, but extended over interfluvial areas considered to be primary forest today. The sustainable use of Amazonian forests will be most effective if it considers the degree of past landscape domestication, as human-modified landscapes concentrate useful plants for human sustainable use and management today.

Editor: John P. Hart, New York State Museum, United States of America

Funding: This work was supported by FAPEAM/FAPESP (project no. 09/53369-6, led by FRCC), CNPq/FAPEAM/PRONEX (project no. 673/2010 led by William E. Magnusson, INPA) and CENBAM/MCTI/CNPq/INPA. The funders had no role in study design, data collection and analysis, decision to publish, or preparation of the manuscript.

Competing Interests: The authors have declared that no competing interests exist.

* E-mail: carollevis@gmail.com

Introduction

The common belief that natural environments of the Americas were relatively untouched by humans before the European conquest is no longer accepted [1,2]. Starting at least 3000 years before present, pre-Columbian populations increased in size, density and duration of their occupations [3]. Landscapes and many plants were domesticated in different degrees to sustain these societies [4]. The extent and degree of Amazonian landscape domestication, however, are still controversial. While some archaeologists suggest extensive modification of the landscape [5,6], ecologists often argue the opposite, that most of the Amazon basin shows few signs of disturbance [7,8]. Bush & Silman [9] and McMichael *et al.* [10] suggest an intermediate hypothesis, in which the intensity of human impacts decreases exponentially with increasing distance from the major Amazonian rivers, especially in non-seasonal forests. Human impacts that are easily recognized include changes in soils and relief, such as anthropogenic soils and geoglyphs, as well as changes in forest composition, often recognized as anthropogenic forests [11].

Anthropogenic soils, called *Terra Preta de Índio* or Amazonian Dark Earth (ADE), are usually found on bluffs along the major rivers [12,13]. This led to the hypothesis that pre-Columbian settlement in Amazonia was mostly located on the bluffs of white water rivers [12]. The preference for these sites is explained by the concentration of food resources and more fertile soils in the adjacent floodplains [12]. Nevertheless, hundreds of geoglyphs have been found in the upper Purus-Madeira interfluve distributed over an area 250 km from south to north, encompassing both floodplains and interfluvial uplands [14]. In the Llanos de Mojos in Bolivian Amazonia most of the earthworks are on interfluves in a forest-savanna mosaic with many useful species [15]. The occurrence of these earthworks suggests the existence of complex societies and dense populations in interfluvial areas, environments previously described as unable to support large numbers of people

[12,14]. Aside from the geoglyphs, archaeological studies are rare on the interfluves, often thought to constitute the largest proportion of the Amazonian landscape [8]. This notion of hard-to-occupy interfluves requires caution, however, as the Amazonian landscape contains numerous drainage basins of rivers and streams that extend into the interfluves and are easily accessed. These seasonal and permanent wetlands represent more than 30% of the basin [16], so aquatic resources are often easily available.

Upon identifying useful species in the forest community, Native Amazonians frequently increased their abundance, creating oligarchic forests often associated with ADE [11,17]. These forests, known as anthropogenic forests, are dominated by one or more useful species due to human activity [11], but are commonly considered primary forests because of their diversity, stature, and closed canopies. In most cases, the only way to identify signs of past manipulation is by assessing the distribution and abundance of useful species [18]. Forest patches dominated by Brazil nut trees (*Bertholletia excelsa*), known as *castanhais*, and by the palm caiaué (*Elaeis oleifera*), known as *caiauezais*, are well-known anthropogenic forests [11,19]. Brazil nut trees with diameters greater than 220 cm are probably older than the colonization of Amazonia, counting from the establishment of Belém, Pará, in 1616 [20]. As caiaué is no longer intensely used by traditional communities, *caiauezais* can be considered pre-colonization also [21].

The evidence left by Native Amazonian populations reflects different degrees of landscape domestication practiced in pre-Columbian times [4]. In settled and cultivated landscapes, signatures can be found in the soils, such as ADE, charcoal and crop phytoliths, which may extend for several kilometers from the river edge. Around cultivated landscapes are managed anthropogenic forests, which extend further away. Managed landscapes have different human footprints, with little or no charcoal and crop phytoliths, but with many useful plants [18]. Further away the forests are used primarily for gathering and hunting. These forests are often promoted in the sense that hunter-gatherers often discard seeds along trails or may even plant seeds or cuttings at preferred sites, as reported for the Kayapó in southeastern Amazonia [11,22]. Since hunters and gatherers visited essentially all of Amazonia outside the settled, cultivated and managed landscapes [8], these promoted landscapes probably occupy most of the basin, but their identification requires botanical and ecological techniques, such as those used here, rather than the archaeological methods advocated by Bush & Silman [9] and McMichael et al. [10].

In this study, we examine past human modifications of the forest from the Solimões, Purus or Madeira Rivers into their interfluve, considering also the role of secondary rivers. The environmental conditions of these major rivers are favorable for human settlement and numerous ADE sites have been found on their riverside bluffs, especially along the Solimões and Madeira [13]. In many areas, the interfluvial forests are exposed to flooding during the rainy season [16], creating environments unsuitable for year-round human occupation and intensive agriculture. Even under these conditions, we show that interfluvial forests have signs of manipulation at different distances from rivers.

We assessed human intervention in the forest by the abundance, richness and basal area of useful arboreal species, mostly fruit trees and palms, and studied the hydrological conditions of the sites as a possible ecological factor influencing the distribution and abundance of useful palms. We considered the mass of charcoal in the soil of each site as another indication of past human cultivated landscapes [9,10]. Fire was the most powerful tool for landscape transformation [6] and the presence of abundant charcoal in the soil is an important evidence of human disturbance in tropical forests [9,23]. Due to the lack of archaeological data on the interfluve, we mapped archaeological evidence (ADE), and two types of anthropogenic forests, *castanhais* and *caiauezais*, around the study sites. These data were used to test the hypothesis that human modifications in forest landscapes decrease with distance from rivers, as suggested by Bush & Silman [9] and McMichael *et al.* [10].

Results

Archaeological evidences

Archaeological sites and anthropogenic forests were found inside and around the six study sites, far from major rivers (Fig. 1). All sites with ADE were on the banks of secondary rivers (>50 m wide) and had not previously been identified, e.g., WinklerPrins and Aldrich [13]. *Castanhais* and *caiauezais* are mostly near secondary river banks, but also occurred in the interior; for more information about archaeological evidence and anthropogenic forests see Text S2. In the sites closest to main rivers, there were shorter distances between the archaeological evidences and the sampling plots (Table 1), and more evidences in general, as expected from the hypothesis. These results highlight the need for more field archaeological investigation in this and other interfluves to continue testing the hypothesis.

Charcoal in the soil

Macroscopic charcoal particles were recorded in all areas and in all soil layers down to 50 cm (Fig. 2). Only in the 30–40 cm layer at site 6 was charcoal absent. Higher values of charcoal were detected in the sites closest to major and secondary rivers. At M1 the charcoal mass was high in the top 20 cm of soil. Charcoal particles were abundant in all layers at M2, 36 km from the main river, and the mean was much higher than the median value observed by Piperno and Becker [26] in upland forest soils 90 km north of Manaus in Central Amazonia, which is thought to be the value expected in soils without past human activities. All other sites had charcoal close to the median value.

Relationship between the useful tree community and the distance from rivers

The relative abundance, basal area and richness of useful trees and palms decreased with distance from major rivers (Fig. 3). The relationship between these variables was a highly significant negative exponential curve. Plots located from 10 to 20 km from a major river had 20–40% useful tree and palm species, plots with distances between 20 and 40 km had 12–23%, plots more than 40 km had less than 15%. In the first 20 km from major rivers there was a rapid decrease in useful tree species and individuals. Beyond 40 km, the proportions of useful individuals and species decreased slowly. The sites with higher concentrations of useful plants (M1, M2 and M6) were on *paleo-várzeas*, pre-Holocene floodplains [24,27] (Table 1). These higher abundances are not of single species, but suites of useful species that vary among sites (Table S1).

In order to focus on the influence of the secondary rivers, we analyzed the relationship between useful tree parameters and the distance to secondary rivers, excluding plots from the two sites closest to the main rivers, M1 and M6. We observed a negative relationship between abundance and basal area of useful trees and the index of distance from secondary rivers crossing the interfluve (Fig. 4). This relationship was linear and explained ≥50% of the variance. This analysis showed that the abundance and the basal

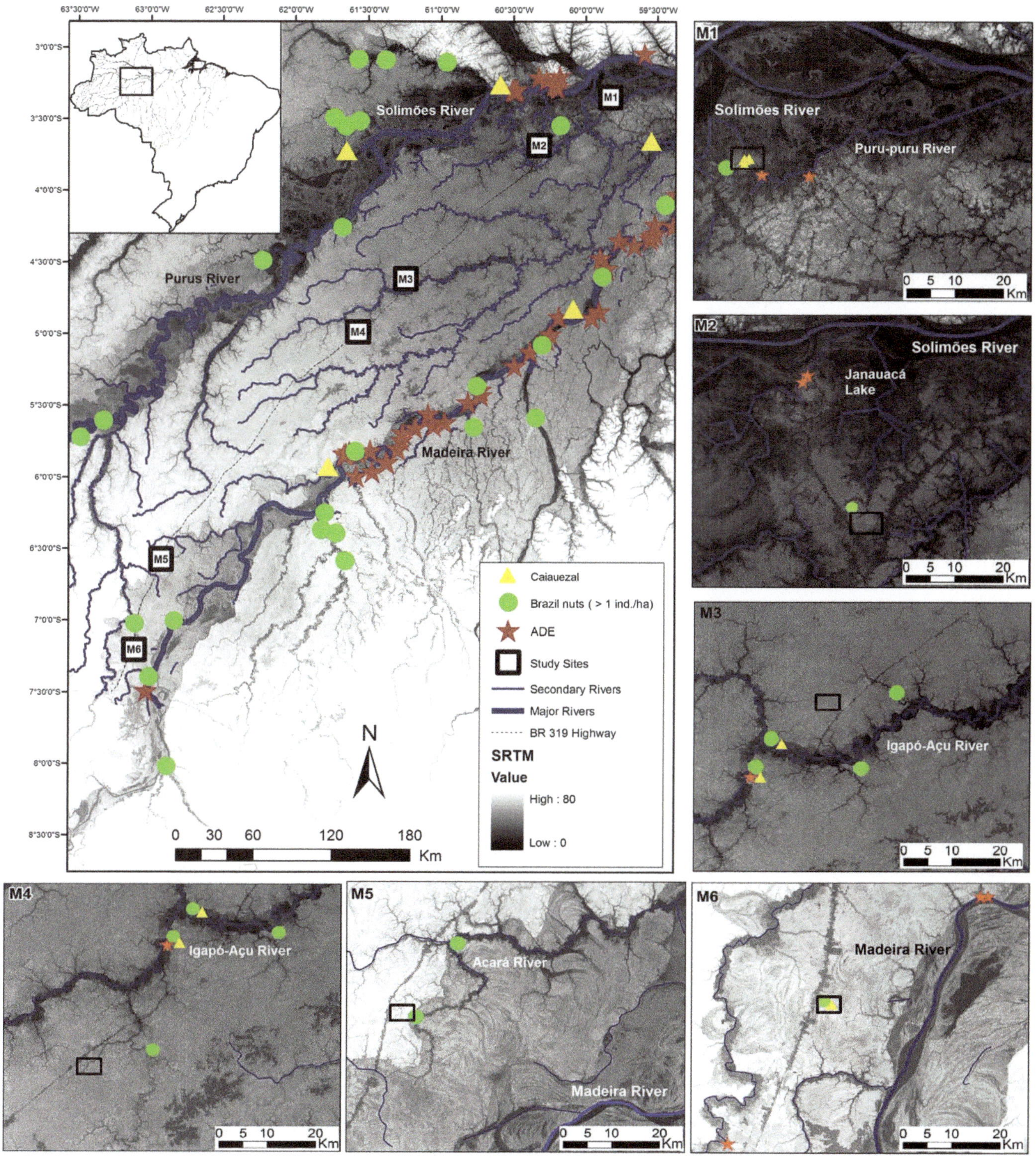

Figure 1. Map of the study area. The large map shows the central-northern Purus-Madeira interfluve. The empty squares are the six sampling sites at different distances from major rivers. In each site, two plots were sampled for floristic analysis and five for charcoal analysis. Yellow triangles are *caiauezais*, green circles are *castanhais* (Brazil nuts >1 individual/hectare) and red stars are ADE (Amazonian Dark Earth sites). ADE data from WinklerPrins and Aldrich [13], *castanhais* from RADAMBRASIL [24] and *caiauezais* from Moretzsohn *et al.* [25]. The other six maps show the newly identified Amazonian Dark Earth sites (ADE) and anthropogenic forests in the vicinity of each site, identified by interviews with local residents or by observation.

Table 1. Means and standard deviations of all botanical data and distances measured in the six study sites along the Purus-Madeira interfluve, Amazonas, Brazil.

	M1	M2	M3	M4	M5	M6
Abundance of all useful species[1]	195,5±36,5	131±13	75±19	42	69,5±10,5	132,5±5,5
Total abundance	520±39	614±39	661,5±1 28,5	588	551,5±41,5	531,5±76,5
Abundance of useful palms	133,5±27,5	84±26	19± 3	8	28±4	85,5±2,5
Abundance of useful dicotyledonous trees	62±9	47±13	56±22	34	41,5±6,5	47±3
Basal area of all useful species[2] (m^2/ha)	7,47±1,64	3,6±0,14	2,40±0,39	1,67	3,67±0,27	7,16±2,74
Total basal area (m^2/ha)[3]	19,99±2,03	22,27±0,31	23,37±2,01	26,92	24,22±2,67	25,35±3,13
Richness of all useful species	9±1	12±1	14	10	11,5±1,5	12±1
Total richness	44,5±4,5	142±9	164,5±4,5	164	76±1	90±2
Distance from main rivers (km)	11±0,5	36	91,5±0,5	80	39,5±0,5	18,5±0,5
Index of rivers distances	[-]0,06±0,04	0,58±0,03	0,87±0,01	0,93	0,79±0,02	0,77
Distance from ADE (km)	5±0,7	33,0	23	31	-	37
Distance from anthropogenic forests (km)	0,5	3,0	13±1,4	15	3±0,7	0,5±0,7

[1]The values of species abundance are the number of all useful trees and palms with DBH≥10 cm in 1 ha.
[2]The value of species basal area are the basal area of all useful trees and palms with DBH≥10 cm in 1 ha.
[3]The values of species richness are the sum of all trees and palm species with DBH≥10 and <30 cm in 0.5 ha and the largest in 1 ha plot.

area of useful species in areas away from major rivers is closely related to their distance to secondary rivers, suggesting that the secondary rivers were also occupied by pre-modern populations.

Palms

Palms were the most abundant useful family in all plots (Table 1, see also Table S1). The relationship between the relative abundance of palms and the distance to rivers was as strong as the relationship for all useful species together (Fig. 3B and 4B). Excluding the useful palms, the relationship between the abundance of dicotyledonous trees and the distance to major rivers was less significant ($p = 0.05$). A multiple regression analysis of the relationship between the relative abundance of useful palms, the distance from rivers and the hydrologic gradient indicated a strong negative effect of the distance from rivers (Fig. 5) and a significant negative effect of the hydrologic gradient on the abundance of useful palms (Fig. 5). The hydrologic gradient was not the major determinant of useful palm abundance in the interfluve, and a large fraction of variance is attributable to distance per se, which is our proxy for anthropogenic effects.

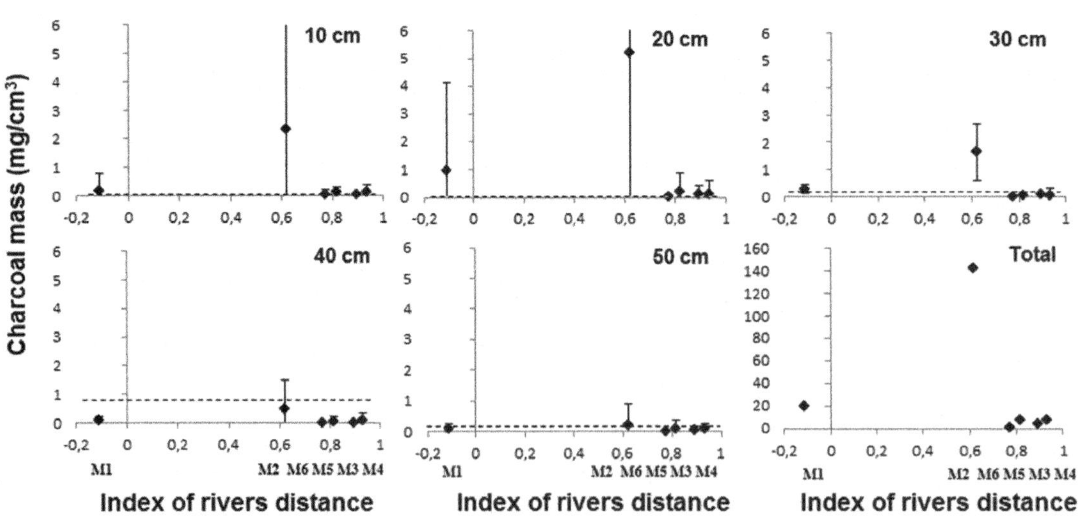

Figure 2. Mass of charcoal in the soil (mg/cm^3) for each soil layer from 10 cm to 50 cm in depth. The plots are in order of increasing distance from the sites to the rivers, expressed by the index of rivers distances. Each point is the average mass of charcoal in the soil from 14–15 samples in each area. The vertical lines represent the standard deviation. The dotted line represents the median value of the mass of charcoal found at each depth in the soil of forests north of Manaus in Central Amazonia, which is the value expected in soils without past human activities [26].

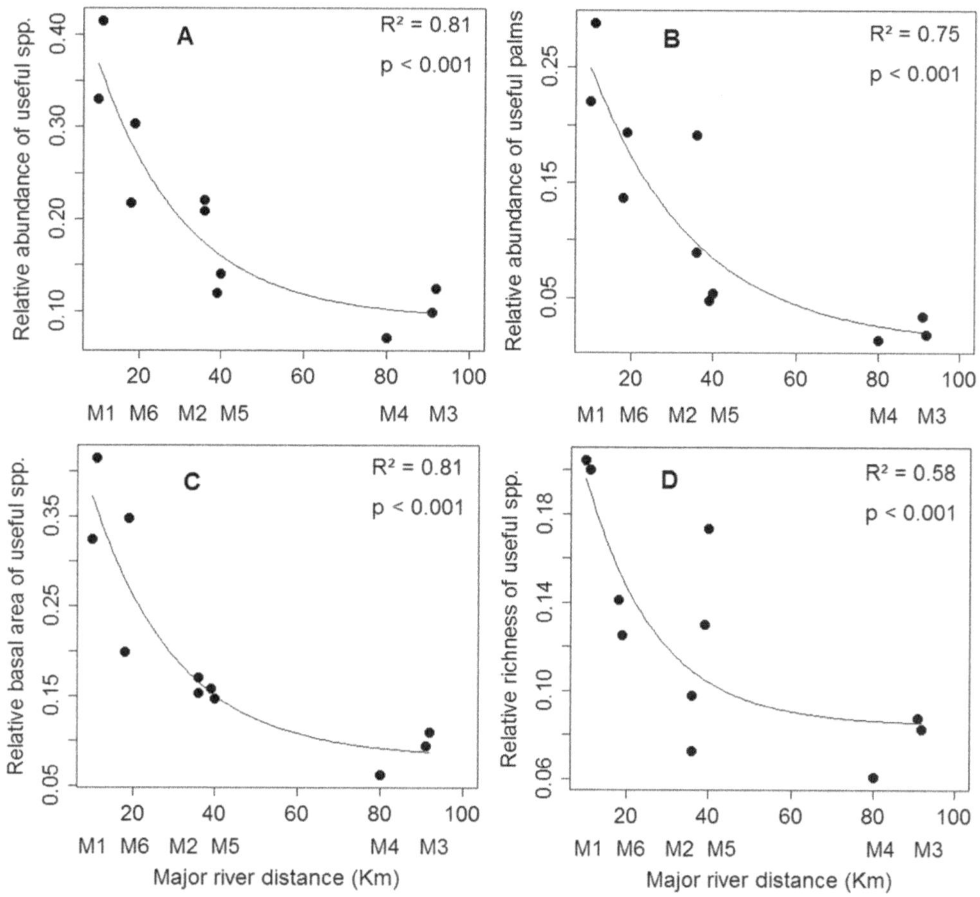

Figure 3. Relationships between useful tree parameters and the distance to major rivers. Relationships between the useful trees parameters and the distance to the relative abundance of useful species per plot (A), the relative abundance of useful palms per plot (B), the relative basal area of useful species per plot (C) and the relative richness of useful species per plot (D), with the distances from major rivers. Points are the plots of all sites, totaling 11 plots. The shortest straight line distance in km from the plot to the Solimões, Purus or Madeira River was evaluated.

Discussion

Human intervention in the landscape decreases with distance from rivers

Our data support McMichael *et al.*'s [10] hypothesis and the expectations of Bush & Silman [9] that human intervention in the landscape decreases with distance from major and secondary rivers. However, the extent of past human impact in the forest observed in our study is much higher than expected by these authors and the assumptions of Peres *et al.* [7] and Barlow *et al.* [8]. Using only simple regressions with distance to major and secondary rivers, which reflect the distance from possible pre-Columbian settlements, we explained 50–90% of the variation in the useful tree parameters. We found high abundances of useful tree species up to and beyond 20 km into the interfluve, and also the presence of anthropogenic forests and ADE far from the main rivers, but close to secondary rivers. Studies that only assessed past human disturbance in terms of charcoal, pollen and phytoliths of cultivated plants [9,10] failed to detect signals from less intensive interventions in the landscape, such as enrichment of forest through extractive activities and hunting. Less intensive activities also caused changes in the concentration of useful plants in the past [18,28] and even today contribute to increases in the concentration of certain plants, such as Brazil nut, along trails [22,29].

Archaeological evidence and useful species composition found at M1, the site closest to the major rivers, indicate forest management practices by different groups in different historical moments. The landscape surrounding site M1 includes numerous rivers and lakes (Fig. 1) that were and still are waterways used for movement and fishing. The existence of ADE, *castanhais* and *caiauezais* near plots are evidence of landscape domestication by indigenous groups before European conquest. On the other hand, all inventoried individuals of the rubber tree (*Hevea brasiliensis*) had marks of extraction. Rubber is usually rare in the forest [30]; however, we found 30 individuals in a one ha inventory (Table S1), the density of a very common species. Hence, some of these forests were reoccupied, exploited and transformed by rubber tappers in the early twentieth century, resulting in the increased abundance of rubber trees (and possibly other useful species) in the forest.

Distinguishing between pre-Columbian and post-conquest management events requires more historical and ethnographic studies in each locality. Except at M1, we didn't find signs of rubber tapper impacts. However, in 1970 the BR-319 Highway was constructed, allowing movement of modern migrants into the interfluve. Most local residents in the vicinity of the study sites have been there since this period. Current management practices in mature forest performed by these recent arrivals probably

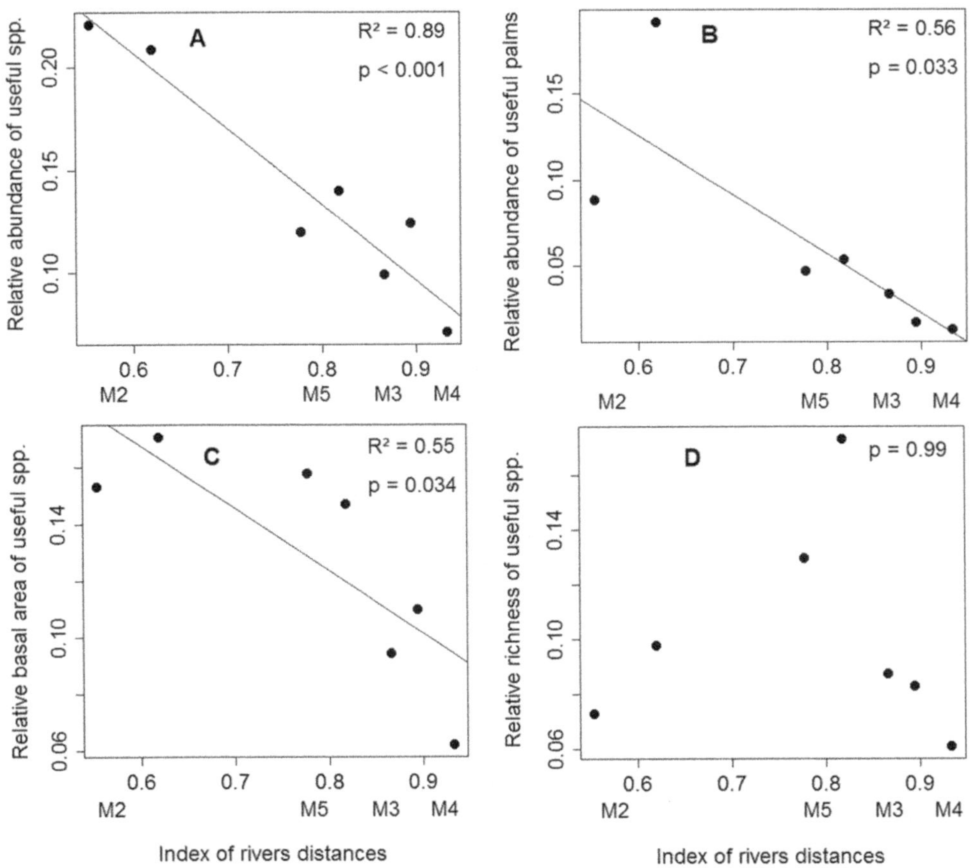

Figure 4. Relationships between useful tree parameters and the distance to secondary rivers. Relationships between the relative abundance of useful species per plot (A), the relative abundance of useful palms per plot (B), the relative basal area of useful species per plot (C) and the relative richness of useful species per plot (D), with the distances from secondary rivers. Points are the plots of sites M2, M3, M4, M5, totaling seven plots. The index of rivers distances is the sum of the inverse distances from each plot to all perennial rivers greater than 50 m wide in a 25 km diameter zone around the sites.

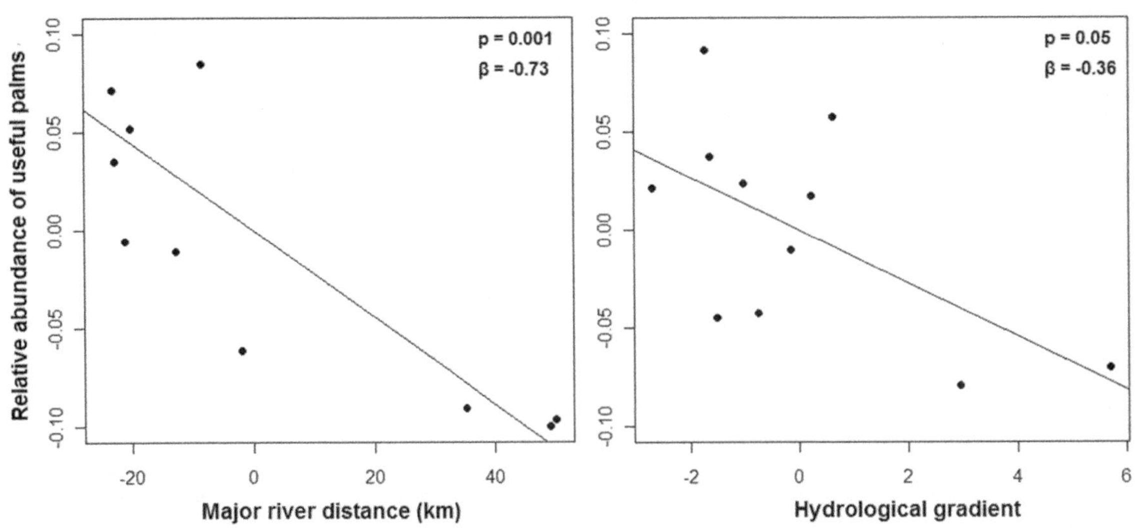

Figure 5. Multiple regression analysis of useful palms and the distance from rivers and the hydrologic gradient. Partial regressions between the relative abundance of useful palms and the distance of each plot to the major rivers (left) and the hydrologic gradient (right). The full multiple regression model has an $R^2 = 0.73$. Points are the plots of all sites, totaling 11 plots.

wouldn't significantly affect the tree community with DBH≥10 cm.

We expect that the footprints of past human activities in the forest will be found in other parts of Amazonia given results in other areas of the world. Around ancient Roman ruins in France, the composition and diversity of plants reflects the impacts of agriculture 1500 years after abandonment [31]. In the Maya forests in Central America, past human management of useful species was identified 1000 years after this civilization's decline [32]. In Central Africa, current tree species composition and diversity still reflects human disturbances after nearly four centuries [33]. Our results showing a gradient of human manipulation in the forest from the major rivers into the interfluve agree with these other studies, however, our study is preliminary and more field investigations that assess forest composition are needed to fully understand variation in landscape domestication across Amazonia.

It is too early to make basin-wide projections, such as Barlow *et al.*'s ([8], p. 4) suggestion of "a largely imperceptible footprint from subsistence hunting and resource extraction across vast tracts of Amazonian forests". This caution is especially true for the interfluves, all of which are insufficiently sampled [10,34]. Only the Tapajós-Xingu interfluve has a large number of ADE records [13,35], but the others have not been adequately surveyed. In the Purus-Madeira interfluve, we detected ADEs near black water secondary rivers and even in places susceptible to flooding in anomalous years. Geoglyphs were found in the upper Purus-Madeira interfluve [14], indicating that this region is unique and may not be representative of other interfluves. Moreover the abundant social, cultural, historical and ecological variation observed across the Amazon basin makes difficult to extrapolate our results to broad regional patterns. On the other hand, all of these observations suggest that if we look, we will find more and more evidences of past human activities on the interfluves.

Ecological factors and past human management influence useful tree and palm abundance and distribution

Ecological conditions can explain tree mono-dominance without invoking the need for human dispersal, especially of some palm species, e.g., *Mauritia flexuosa* in swamp forests [36]. A gradient of hydrological conditions runs from the floodplain to the interfluve, and in wetlands the flood-level is a determinant of plant distribution along this gradient [37]. Even in upland forests, some palm species respond to the hydrological condition of soils [38]. As expected, we found a significant effect of the hydrologic gradient on the abundance of useful palms in our multiple regression model. However, the effect of distance to rivers, after partitioning out the hydrologic gradient effect, was stronger, probably due to past human management.

If ecological conditions were the sole determinants of plant distributions in the interfluve, we should find roughly the same useful tree and palm communities on *paleo-várzeas*, which are pre-Holocene floodplains with similar geomorphological and hydrological conditions. Instead, we found unusual species-environment associations: *Bertholletia excelsa*, *Attalea speciosa*, which are naturally associated with upland forests, were found as dominants in M6, a site on the *paleo-várzeas* of the Madeira River (see also Table S1). In addition, in Central Amazonia, *Euterpe precatoria* and *Oenocarpus bataua* occur mainly in low areas with poorly drained soils [38]. However, we found these two species in the same plots with *Attalea maripa* and *Theobroma* spp. (Table S1), which are disturbance indicators and associated with archaeological sites [11,39,40] normally found in non-flooded areas [41]. The co-occurrence of these species, therefore, is likely to be due to their usefulness to humans. Considering that *Oenocarpus bataua* and *Attalea maripa* are usually more abundant in forests with more open canopies [42,43], the high density of these species at M2 and the presence of abundant charcoal particles in this area are probably associated with the historical presence of humans and fire in the region, which may have increased light penetration in the forest.

We observed forests dominated by a number of useful species with different environmental preferences, including species occurring outside of their natural environments. Thus, ecological conditions alone can't explain useful tree and palm dominance and distribution in the interfluve; pre-conquest and historical management must be considered also. Our results suggest that pre-Columbian populations of the interfluve affected the distribution and abundance of palms even in places that were not ideal for their establishment.

Charcoal and landscape modification

Paleoecological analyses in Amazonia found large amounts of charcoal associated with pollen of cultivated plants, indicating that fire was associated with past agricultural practices [44]. Charcoal, pollen and phytoliths of cultivated plants are more direct evidences of past agricultural practices than the modern useful arboreal community. We found charcoal in the soils of all areas at almost all depths, indicating that interfluvial forests were burnt at different moments in the past. If charcoal particles in the top 20 cm soil depth are often of modern origin [45], most of the charcoal at depths greater than 20 cm is probably pre-Columbian. Charcoal is common in soils of the interfluve; however phytoliths from agriculture and early successional growth taxa related to human disturbance were found to be scarce in the interior of the interfluve [45].

Despite the widespread occurrence of charcoal in the landscape, its abundance was high in only two sites (M1 and M2). The forest in M1 was used by rubber tappers, so the charcoal in the top 20 cm soil depth may be related to their activities. At M2 we observed even larger amounts of charcoal in all soil layers. This site is located 36 km from the Solimões River and 5 km from the Janauacá River, a wide river stretching into the center of this portion of the interfluve (Fig. 1). We also found high densities of useful species in these plots (Table 1). All this evidence of landscape domestication is indicative of intensive agricultural and other management activities in these two areas, even though we did not study pollen or phytoliths.

Since not all charcoal particles found in Amazonian forests can be attributed to past human intervention [26], the charcoal particles we found in other sites of the interfluve may be from natural fires or low impact human activities. Low intensity fires without another indication of intensive clearing are probably not a signal of extensive forest disturbance. In two of 13 sites studied on the Purus-Madeira interfluve, charcoal was found associated with phytoliths, confirming human agricultural activities at these sites, but little clearing of the interfluve forest in other sites far from rivers [45]. Our results also suggest that not all sites on the interfluve were affected by human fire management; however, further archaeological studies are needed for a more complete understanding of the locations of the pre-Columbian cultivated landscapes within this interfluve.

Charcoal analysis is not useful to detect human activities not related to fire, such as planting useful species on trails inside the forest [22] and discarding seeds while walking to extract fruit or while hunting [18,28,46]. Hunter-gatherers can travel 25–26 km from their settlement on a hunting trip [47,48], essentially putting all of Amazonia within the category of hunted landscapes [8] where landscape management is likely to happen. Stahl [49] reviewed Amazonian studies that demonstrate how the cumulative historical impact of small bands of hunter-gatherers is probably greater than that of traditional farmers, although the latter also manipulate the forests near their settlements. Contemporary hunter-gathers are sophisticated and dynamic managers of forest landscapes, subtly thinning some areas to make space for their preferred resources and creating "wild orchards" [50]. They exploit temporary gaps, disperse fruit trees and palms, and move their camps regularly, concentrating resources at preferred locations over wide areas of forest landscapes. Over thousands of years the results are dramatic, but the forests look mature, untouched, pristine to the untrained eye. Even though we did not find a considerable mass of charcoal in M6, useful species composition suggests a history of intense human management at this site, like that reviewed by Stahl [49]. *Attalea speciosa*, *Astrocaryum aculeatum* and *Bertholletia excelsa*, indicators of anthropogenic forests [11,19,29,39], were found in the same plots as *Astrocaryum murumuru* and *Elaeis oleifera*, both related to ADE [51].

Understanding the past to conserve and manage for the future

Our study suggests that past human impacts in the forest extend over large areas considered primary forest today, since we found a gradient of human manipulation in the forest from the main rivers into the Purus-Madeira interfluve. Because this interfluve is full of secondary and smaller rivers crossing its interior, the effect of human manipulation should also be directly related to the distance from these rivers. ADE in the vicinity of a secondary river more than 5 km away from the Madeira River contains a unique species composition related to past human activity when compared with non-anthropogenic soils [51]. As most ecological studies focus on the vicinity of the principal navigable rivers [31], these studies need to incorporate the effect of human history to better understand the patterns and mechanisms that explain biodiversity. This issue has been raised previously [52,53], but is still remarkably ignored by many biologists and ecologists [54]. Future research must associate floristic inventories with paleoecological and archaeological data to build a more complete view of the impact of pre-Colombian populations in Amazonia, especially in other interfluvial areas.

Our results have important implications for the conservation and sustainable use of forest resources today. Although Amazonia is mostly sparsely populated and filled with apparently empty areas today, such as the interfluves, people live in these forests in remote locations. These people, both indigenous and peasant, depend on the forest's resources for their well-being. We argue that the modifications left by ancient Native Amazonians in the landscape and in the useful tree community are extremely important to plan sustainable practices for the use of these forests today [55].

The strategies for Amazonian conservation, suggested by the Brazilian National System of Conservation Units [56], recognize the existence of people living within the forest and extracting non-timber forest products (NTFP). Accordingly, forests that were managed and enriched in the past have an important role for biodiversity conservation, as they concentrate NTFPs that should be sustainably managed by human populations [57]. From this point of view, the role of traditional populations, with their management practices, becomes crucial to ensure the conservation of the forest and the culture of the people who live there.

Materials and Methods

Study Area

The study was conducted in the interfluve between the Purus, Madeira and Solimões Rivers, in the state of Amazonas, Brazil (Fig. 1). The study was carried out in mature lowland forests along the BR-319 Highway in six previously installed sites of the Research Program in Biodiversity (PPBio) [58]; for more details about study area and the sites see: Text S1 and http://ppbio.inpa.gov.br/repositorio/dados. The PPBio obtained prior informed consent to install and study all sites. In the PPBio website the identification of each study site from north to south of the interfluve is: M01, M02, M05, M06, M10, M11. We changed the original identification of the sites to produce a more understandable sequence (see Table S2). The sites are located at different distances from the major rivers and in different environments.

Mapping Archaeological Evidence

Archaeological evidence and anthropogenic forests were identified and mapped around each research site. The main evidence of pre-modern human activities documented were Amazonian Dark Earths (ADE) and two types of anthropogenic forests, the *castanhais* and *caiauezais*. The identification of ADE and anthropogenic forests was obtained by questioning local residents and, when possible, GPS coordinates were recorded on site. After self-presentation and explaining that the project was studying the trees and palms of the local PPBio module, the single question was: where are the ADE, *castanhais* and *caiauezais* near here? Since no information related to use and management of ADE and anthropogenic forest sites was sought from the local residents, the approval of the Ethics Committee for Research with Humans at INPA was considered unnecessary.

Charcoal Data

We analyzed charcoal in 4–5 plots in each site, for a total of 29 plots surveyed. Two of them are the same plots used for botanical analysis at each site. At the beginning, middle and end of each of the five plots, small pits were excavated to 50 cm in depth. Using a Kopecky cylinder (100 cm^3), a horizontal collection of undisturbed soil was made at 10 cm intervals. The soil was dried and then visible charcoal was removed for weighing.

Useful Species

To create the list of useful species considered in this study, we used the most important papers in ethnobotany and previous inventories in anthropogenic forests [4,11,39,41,46,51,59,60,61] in Amazonian forests and archeological sites. The useful species mentioned in at least two studies were included in the list. We also considered their degree of domestication [4], their use as food resources in the daily diet of human populations during long periods in the forest for game hunting or other activities, and also their capacity to attract game. Species with commercial value in the post-colonial period, such as *Hevea brasiliensis* and *Carapa guianensis*, were also included in the list.

Botanical Data

Ten plots were installed in each site by PPBio, one km from each other. In five of them, all trees were marked and measured. After considering the topographic variation between plots, using SRTM (Shuttle Radar Topography Mission) images, we chose two of the five plots to sample tree composition. All trees and palms with diameter at breast height (DBH)$\geq$10 cm were sampled in plots of 0.5 ha (250×20 m). Trees with DBH$\geq$30 cm were sampled in 1 ha plots (250×40 m).

At M4, only two plots had been installed at collection time, and one of them is located less than 500 meters from the highway and on the edge of a shifting cultivation plot. We intended to work only in mature forest, so we excluded this plot from analysis.

We inventoried trees with DBH≥10 cm because they may be descendants of pre-conquest management, since old trees that may have been planted or promoted by ancient people will reproduce and their recruits will persist in old anthropogenic forests. Ross [32] found small individuals of useful tree species (>2.5 cm DBH) in ancient Maya forest gardens after a millennium of abandonment, confirming this expectation.

The botanical material was pre-identified in the field by parataxonomists and also collected for comparison with herbarium collections. After a preliminary identification in the field with the aid of parataxonomists, the botanical identification was confirmed by Priscila Souza, graduate student in Botany at INPA, specialists, identification guides and by comparing the vouchers collected to specimens at the INPA Herbarium (Manaus, Brazil) and virtual herbariums (http://fm1.fieldmuseum.org/vrrc/index.php, http://sciweb.nybg.org/science2/vii2.asp). Fertile specimens were deposited at INPA and sterile material will be deposited at the EAFM Herbarium (Herbarium of the Federal Institution of Amazonas State). Floristic and charcoal data will be available on the PPBio web site and may be requested from the first author.

Distance Measurements

The distances from botanical plots to rivers were calculated using Landsat Thematic Mapper (TM) images. We considered a straight-line distance from the plot to the closest major river (Solimões, Purus or Madeira). To measure the distance to smaller rivers, a buffer zone with a 25 km radius around each plot was traced. We considered 25 km the maximum distance that could be covered on foot leaving the center of occupation for long hunting activities [47,48]. Within each buffer, the shortest distance from the plot to perennial rivers (greater than 50 m in width) was calculated. Only 50 m wide rivers were chosen, as this width represents the minimum width of navigable rivers in the region detected using TM images. An index of rivers distances was calculated by the sum of all inverse distance values from each plot to perennial rivers inside the 25 km buffer zone (index of rivers distances = 1−(1/distance river 1+1/distance river 2+1/distance river n)).

Hydrological Measurements

The gradient of hydrological conditions was measured in two ways. In plots that flooded during the rainy season, we used the height of water marks on tree trunks left by the highest water level in the previous year. In plots that did not flood, we installed a piezometer from the soil surface to seven meters below the ground level. The distance from the soil surface to the highest groundwater level was measured in March 2011, as this is the period with the highest groundwater level during the rainy season.

Data Analysis

To evaluate the relationship between useful "tree" parameters and the distance to rivers we used simple linear regressions and nonlinear regressions. The useful tree parameters were: 1) the relative abundance of useful species – number of individual useful trees and palms as a percentage of the total number of individuals per plot; 2) the relative richness of useful species – number of useful species as a percentage of the total number of species per plot; and 3) relative basal area occupied by useful species – basal area of useful species as a percentage of the total basal area per plot. We used relative values of abundance, richness and basal area due to the high variation in total numbers from one plot to another (Table 1). The shortest distance from major rivers and the index of rivers distances, which reflect the ability of human movement within the interfluve, were the independent variables used in the regressions. For abundance and basal area parameters, the values of all trees and palms with DBH<30 cm in 0.5 ha were extrapolated to 1 ha.

Palm Analysis

To determine if the patterns of useful palm abundance observed are associated with environmental parameters, rather than the distance from possible occupation sites, we used multiple regression models. These models included the hydrological gradient as environmental predictor and the distance to rivers as a predictor of human activity, and relative abundance of palms as response variable. The predictor variables had a low Pearson correlation ($r = 0.25$, $p = 0.46$).

Supporting Information

Text S1 Details of the study area.

Text S2 Archaeological evidence and anthropogenic forests.

Table S1 List of useful species found in 11 plots and their respective abundances along the Purus-Madeira interfluve, Amazonas, Brazil.

Table S2 Correspondence between the identification of each study site in the PPBio website and the identification adopted in this study.

Acknowledgments

We thank the residents of the BR-319 Highway for their hospitality and patience, FAPEAM/FAPESP (project no. 09/53369-6, led by FRCC), CNPq/FAPEAM/PRONEX (project no. 673/2010 led by William E. Magnusson, INPA) and CENBAM/MCTI/CNPq/INPA for financial support, ICMBIO for logistical support, Alexandro Elias dos Santos and José Adailton Correa for field assistance, Francislaide da Silva Costa for help with charcoal analysis, Bernardo M. Flores, INPA, Mark M. Bush, Florida Inst. Technology, André B. Junqueira and Marielos Pena-Claros, Wageningen Univ., David Campbell, Grinnell College, William Denevan, Univ. of Wisconsin, and two anonymous reviewers for suggestions to improve earlier versions of this manuscript, CAPES for a Master's scholarship to CL, and CNPq for research fellowships to CRC and FRCC and for awarded stipends to TE.

Author Contributions

Conceived and designed the experiments: CL CRC FRCC. Performed the experiments: CL PFS JS TE JLPVP. Analyzed the data: CL PFS JS TE JLPVP CRC FRCC. Wrote the paper: CL CRC FRCC.

References

1. Willis KJ, Gillson L, Brncic TM (2004) How "virgin" is virgin rainforest? Science 304: 402–403.
2. Denevan WM (2011) The "Pristine Myth" revisited. Geogr Rev 101: 576–591.
3. Piperno DR, Pearsall DM (1998) The Origins of Agriculture in the Lowland Neotropics. San Diego: Academic Press.

4. Clement CR (1999) 1492 and the loss of Amazonian crop genetic resources. I: The relation between domestication and human population decline. Econ Bot 53: 188–202.
5. Heckenberger MJ, Kuiruro A, Kuikuro UT, Russel JC, Schmidt M, et al. (2003) Amazonia 1492: Pristine forest or cultural parkland? Science 301: 1110–1114.
6. Erickson CL (2008) Amazonia: The Historical Ecology of a Domesticated Landscape. In: Silverman H, Isbell WH, editors. Handbook of South American Archaeology. Berlin: Springer. pp. 157–183.
7. Peres CA, Gardner TA, Barlow J, Zuanon J, Michalski F, et al. (2010) Biodiversity conservation in human-modified Amazonian forest landscapes. Biol Conserv 143(10): 2314–2627.
8. Barlow J, Gardner TA, Lees AC, Parry L, Peres CA (2012) How pristine are tropical forests? An ecological perspective on the pre-Columbian human footprint in Amazonia and implications for contemporary conservation. Biol Conserv 151(1): 45–49.
9. Bush MB, Silman MR (2007) Amazonian exploitation revisited: Ecological asymmetry and the policy pendulum. Front Ecol Environ 5: 457–465.
10. McMichael CH, Bush MB, Piperno DR, Silman MR, Zimmerman AR, et al. (2012) Spatial and temporal scales of pre-Columbian disturbance associated with western Amazonian lakes. The Holocene 22(2): 131–141.
11. Balée W (1989) The culture of Amazonian forests. In: Posey DA, Balée W, editors. Resource Management in Amazonia: Indigenous and Folk Strategies. New York: The New York Botanical Garden. pp. 1–21.
12. Denevan WM (1996) A bluff model of riverine settlement in prehistoric Amazonia. Ann Assoc Am Geogr 86(4): 654–681.
13. WinklerPrins AMGA, Aldrich SP (2010) Locating Amazon Dark Earths: Creating an interactive GIS of known locations. J Latin Am Geogr 9(3): 33–50.
14. Pärssinen M, Schaan D, Ranzi A (2009) Pre-Columbian geometric earthworks in the upper Purús: a complex society in western Amazonia. Antiquity 83: 1084–1095.
15. Erickson CL, Balee W (2006) The Historical Ecology of a Complex Landscape in Bolivia. In: Balée W, Erickson CL, editors. Time and Complexity in Historical Ecology: Studies in the Neotropical Lowlands. New York: Columbia University Press. pp. 187–233.
16. Junk WF, Piedade MPF, Schöngart J, Cohn-Haft M, Adeney JM, et al. (2011) A classification of major naturally-occurring Amazonian lowland wetlands. Wetlands 31(4): 623–640.
17. Clement CR, McCann JM, Smith NJH (2003) Agrobiodiversity in Amazônia and its relationships with Dark Earths. In: Lehmann J, Kern D, Glaser B, Woods W, editors. Amazonian Dark Earths – Origin, Properties, and Management. Dordrecht: Kluwer Academic Publishers. pp. 159–118.
18. Peters C (2000) Pre-Columbian Silviculture and Indigenous Management of Neotropical Forests. In: Lentz D, editor. Imperfect Balance: Landscape Transformations in the Pre-Columbian Americas. New York: Columbia University Press. pp. 203–223.
19. Shepard GH, Ramirez H (2011) "Made in Brazil": human dispersal of the Brazil nut (Bertholletia excelsa, Lecythidaceae) in ancient Amazonia. Econ Bot 65: 44–65.
20. Chambers JQ, Higuchi N, Schimel JP (1998) Ancient trees in Amazonia. Nature 391: 135–136.
21. Cunha RNV, Lopes R, Rocha RNC, Lima WAA, Teixeira PC, et al. (2009) Domesticação e melhoramento de caiaué. In: Borém A, Lopes MTG, Clement CR, editors. Domesticação e Melhoramento: Espécies Amazônicas, Viçosa: Universidade Federal de Viçosa. pp. 275–296.
22. Posey DA (1985) Indigenous management of tropical forest ecosystems: the case of Kayapó Indians of the Brazilian Amazon. Agrofor Syst 3: 139–158.
23. Sanford RL Jr, Horn SP (2000) Holocene rain-forest wilderness: A Neotropical perspective on humans as an exotic, invasive species. USDA Forest Serv Proc RMRS 3: 1–15.
24. Brazil (1978) Projeto RADAMBRASIL Folha SB.20 Purus; geologia, geomorfologia, pedologia, vegetação e uso potencial da terra. Rio de Janeiro: Ministério de Minas e Energia, Departamento Nacional de Produção Mineral.
25. Moretzsohn MC, Ferreira MA, Amaral ZPS, Coelho PJA, Grattapaglia D, et al. (2002) Genetic diversity of Brazilian oil palm (Elaeis oleifera H.B.K.) germplasm collected in the Amazon Forest. Euphytica 124: 35–45.
26. Piperno DR, Becker P (1996) Vegetation history of a site in the central Amazon Basin derived from phytolith and charcoal records from natural soils. Quat Res 45: 202–209.
27. Irion G, Mello JASN, Morais J, Piedade MTF, Junk WJ, et al. (2010) Development of the Amazon Valley During the Middle to Late Quaternary: Sedimentological and Climatological Observations. In: Junk WJ, Piedade MTF, Wittmann F, Schöngart J, Parolin P, editors. Central Amazonian Floodplain Forests: Ecophysiology, Biodiversity and Sustainable Management. Berlin: Springer, Berlin. pp. 27–42.
28. Guix JC (2005) Evidence of old anthropic effects in forests at the confluence of the Caurés and Negro Rivers – NW Amazonia: The role of Indians and Caboclos. Grupo Estud Ecol Sér Doc 8(1): 1–27.
29. Scoles R, Gribel R (2011) Population structure of Brazil nut (Bertholletia excelsa, Lecythidaceae) stands in two areas with different occupation histories in the Brazilian Amazon. Hum Ecol 39: 455–464.
30. Souza AD, Oliveira RS, Furtado EL, Kageyama PY, Freitas RGS, et al. (2011) Rubber tree, seringueira (Hevea brasiliensis). In: Fruit Trees and Useful Plants in Amazonian Life. Shanley P, Cymerys M, Serra M, Medina G, editors. Rome: FAO-CIFOR-PPI. pp. 121–128.
31. Dambrine EJL, Dupouey J-L, Laüt L, Humbert L, Thinon M, et al. (2007) Present forest biodiversity patterns in France related to former Roman agriculture. Ecology 88: 1430–1439.
32. Ross NJ (2011) Modern tree species composition reflects Ancient Maya 'forest gardens' in NW Belize. Ecol Appl 21(1): 75–84.
33. van Gemerden BS, Olff H, Parren MPE, Bongers F (2003) The pristine rain forest? Remnants of historical human impacts on current tree species composition and diversity. J Biogeogr 30: 1381–90.
34. Pitman NCA, Widmer J, Jenkins CN, Stocks G, Seales L, et al. (2011) Volume and geographical distribution of ecological research in the Andes and the Amazon, 1995–2008. Trop Conserv Sci 4: 64–81.
35. Smith NJH (1980) Anthrosols and human carrying capacity in Amazonia. Ann Assoc Am Geogr 70(4): 553–566.
36. Kahn F, Mejia K (1990) Palm communities in wetland forest ecosystems of Peruvian Amazonia. For Ecol Manage 33/34: 169–119.
37. Wittman F, Junk WJ, Piedade MTF (2004) The varzea forests in Amazônia: flooding and the highly dynamic geomorphology interact with natural forest succession. Forest Ecol Manag 196(2–3): 199–212.
38. Kahn F, Castro A (1985) The palm community in a forest of central Amazonia, Brazil. Biotropica 11: 210–216.
39. Balée W, Campbell DG (1990) Evidence for the successional status of liana forest (Xingu River basin, Amazonian Brazil). Biotropica 22: 36–47.
40. Morcote-Rios G, Bernal R (2001) Remains of palms (Palmae) at archaeological sites in the new world: a review. Bot Rev 67: 309–350.
41. Cavalcante PB (2010) Frutas Comestíveis na Amazônia. Belém: Museu Paraense Emílio Goeldi.
42. Salm R (2005) The importance of forest disturbance for the recruitment of the large arborescent palm Attalea maripa in a seasonally-dry Amazonian forest. Biota Neotrop 5(1):1–7.
43. Svenning JC (1999) Recruitment of tall arborescent palms in the Yasuni National Park, Amazonian Ecuador: are large treefall gaps important? J Trop Ecol 15: 355–366.
44. Bush MB, Silman MR, Toledo MB, Listopad C, Gosling WD, et al. (2007) Holocene fire and occupation in Amazonia: records from two lake districts. Philos Trans R Soc Biol Sci 362: 209–218.
45. McMichael CH, Piperno DR, Bush MB, Silman MR, Zimmerman AR, et al. (2012) Sparse pre-Columbian human habitation in Western Amazonia. Science 336:1 429–1431.
46. López-Zent E, Zent S (2004) Amazonian Indians as Ecological Disturbances Agents: The Hotï of the Sierra Maigualida, Venezuelan Guayana. In: Carlson TS, Maffi L, editors. Ethnobotany and Conservation of Biocultural Diversity. New York: The New York Botanical Garden. pp. 79–112.
47. Shepard G (2002) Primates in Matsigenka subsistence and world view. In: Fuentes A, Wolfe LD, editors. Primates face to face: conservation implications of human and nonhuman primates interconnections. Cambridge: Cambridge University Press. pp. 101–136
48. Peres CA, Nascimento HS (2006) Impact of game hunting by the Kayapo of south-eastern Amazonia: implications for wildlife conservation in tropical forest indigenous reserves. Biodiv Conserv 15: 2627–2653.
49. Politis GG (1996) Moving to produce: Nukak mobility and settlement patterns in Amazonia. World Archaeology 27:492–511.
50. Stahl PW (2008) The contributions of zooarchaeology to historical ecology in the Neotropics. Quaternary International 180:5–16.
51. Junqueira AB, Shepard GH Jr, Clement CR (2010) Secondary forests on anthropogenic soils in Brazilian Amazonia conserve agrobiodiversity. Biodiv Conserv 19: 1933–1961.
52. Clark DB (1996) Abolishing virginity. J Trop Ecol 12(5): 735–739.
53. Wills KJ, Araújo MB, Bennett KD, Figueroa-Rangel B, Froyd CA, et al. (2007) How can a knowledge of the past help to conserve the future? Biodiversity conservation and the relevance of long-term ecological studies. Phil Trans R Soc B 362: 115–186.
54. Clement CR, Junqueira AB (2010) Between a pristine myth and an impoverished future. Biotropica 42: 534–536.
55. Heckenberger MJ, Russell JC, Toney JR, Schmidt MJ (2007) The legacy of cultural landscapes in the Brazilian Amazon: Implications for biodiversity. Phil Trans R Soc B 362: 197–208.
56. Brasil (2000) Sistema Nacional de Unidades de Conservação da Natureza – SNUC. Brasília: Diario Oficial da União.
57. Scoles R, Gribel R (2012) The regeneration of Brazil nut tree in relation to nut harvest intensity in the Trombetas River valley of Northern Amazonia, Brazil. For Ecol Manage 265: 71–81.
58. Magnusson B, Costa F, Lima A, Baccaro F, Braga-Neto R, et al. (2008) A program for monitoring biological diversity in the Amazon: an alternative perspective to threat-based monitoring. Biotropica 40: 409–411.
59. Balée W (2010) Contingent diversity in anthropic landscapes. Diversity 2: 163–181.
60. Junqueira AB, Shepard GH, Clement CR (2011) Secondary forests on anthropogenic soils of the Middle Madeira River: Valuation, local knowledge, and landscape domestication in Brazilian Amazonia. Econ Bot 65: 85–99.
61. Prance GT, Balée W, Boom BM, Carneiro RL (1987) Quantitative ethnobotany and the case for conservation in Amazonia. Conserv Biol 1(2): 296–310.

3

Characterizing Spatiotemporal Dynamics of Methane Emissions from Rice Paddies in Northeast China from 1990 to 2010

Yuan Zhang[1,2]*, Shiliang Su[3], Feng Zhang[4], Runhe Shi[1], Wei Gao[1]

1 Key Laboratory of Geographical Information Science, Ministry of Education, East China Normal University, Shanghai, China, **2** Research Center of Remote Sensing & Geoscience, Northeast Institute of Geography & Agroecology, Chinese Academy of Sciences, Changchun, China, **3** College of Environmental & Resource Sciences, Zhejiang University, Hangzhou, China, **4** Key Laboratory of Arid and Grassland Agro-Ecology, Lanzhou University, Lanzhou, China

Abstract

Background: Rice paddies have been identified as major methane (CH_4) source induced by human activities. As a major rice production region in Northern China, the rice paddies in the Three-Rivers Plain (TRP) have experienced large changes in spatial distribution over the recent 20 years (from 1990 to 2010). Consequently, accurate estimation and characterization of spatiotemporal patterns of CH_4 emissions from rice paddies has become an pressing issue for assessing the environmental impacts of agroecosystems, and further making GHG mitigation strategies at regional or global levels.

Methodology/Principal Findings: Integrating remote sensing mapping with a process-based biogeochemistry model, Denitrification and Decomposition (DNDC), was utilized to quantify the regional CH_4 emissions from the entire rice paddies in study region. Based on site validation and sensitivity tests, geographic information system (GIS) databases with the spatially differentiated input information were constructed to drive DNDC upscaling for its regional simulations. Results showed that (1) The large change in total methane emission that occurred in 2000 and 2010 compared to 1990 is distributed to the explosive growth in amounts of rice planted; (2) the spatial variations in CH_4 fluxes in this study are mainly attributed to the most sensitive factor soil properties, i.e., soil clay fraction and soil organic carbon (SOC) content, and (3) the warming climate could enhance CH_4 emission in the cool paddies.

Conclusions/Significance: The study concluded that the introduction of remote sensing analysis into the DNDC upscaling has a great capability in timely quantifying the methane emissions from cool paddies with fast land use and cover changes. And also, it confirmed that the northern wetland agroecosystems made great contributions to global greenhouse gas inventory.

Editor: Julian Clifton, University of Western Australia, Australia

Funding: 1. National Natural Science Foundation of China (No. 41001202). (http://www.nsfc.gov.cn/nsfc/cen/00/download.html) 2. National Basic Research Program of China (973) (No. 2010CB951603). (http://www.most.gov.cn/eng/index.htm) The two projects granted by the funders provided substantial financial support for the study, and took a role in soil data collection and analysis, agricultural management information survey for preparing the manuscript.

Competing Interests: The authors have declared that no competing interests exist.

* E-mail: yuan.zhang75@gmail.com

Introduction

Methane (CH_4) is a major greenhouse gases (GHG). According to the Intergovernmental Panel on Climate Change (IPCC) report, atmospheric CH_4 concentrations have risen to 1774 ppb in 2005 [1]. Many studies have proved that agricultural activities are responsible for approximately 50% of global atmospheric inputs of CH_4, wherein the rice paddies have been identified as a major source [2]. Over 10% of atmospheric CH_4 was attributed to the emissions from global rice paddies [3,4]. Thus, how to accurately estimate the CH_4 emissions from rice paddies has become an pressing issue for assessing the environment impacts of agroecosystems, and further making GHG mitigation strategies at regional or global levels.

As an important rice producing country, China possesses approximately 20% of the world's rice paddies which provides about 30% of the world's rice needs [5]. About 20% of all croplands in China were cultivated for rice production [6]. Such a huge CH_4 source could make a great contribution to global CH_4 inventory. To estimate the national inventory of CH_4 emission, a number of site-specific observations were conducted for measuring CH_4 flux at field sites in the major rice producing areas in Southern and Southeast of China [7–10]. These ground-based measurements were reliable for understanding the mechanics of CH_4 emission at local scale. Given the emergence of new frameworks for GHGs mitigation, however, it fails to respond to practicable requirements at national, regional, and global levels in the long run for high variations in spatial and temporal pattern of CH_4 emission with changing environmental conditions [11–15]. Consequently, based on the extrapolation of the understandings gained at site scale to a large spatial dimension, model simulations were required to meet the demands for spatiotemporal analysis of CH_4 emissions from rice fields.

Model estimation of CH_4 emissions from rice fields began with empirical models based on the regression relationships between

CH_4 emission rate and rice biomass or yield [16–18]. However, these "easy-to-use" approaches were unable to reasonably explain biogeochemical processes involved in CH_4 production, oxidation and emission, and also explicitly delineate the emissions variation at regional scale across a wide range of soil conditions and management practices. In this case, many physical models consequently were developed based on biogeochemical process to quantify the comprehensive effects of ecological drivers, soil and climate factors and management alternatives on agricultural production and environment [19–21]. For their capability in simulating CH_4 production and oxidation process in paddies, so were extensively utilized to regional or global CH_4 estimation [22–27].

Of among, the Denitrification and Decomposition (DNDC) model is a generic model that simulates the biogeochemical processes leading to GHG emissions from soil [19,28,29]. It has been adapted to simulations of GHG (e.g. CO_2, N_2O or CH_4) emissions from a wide range of systems such like dryland crop, pasture, rice paddy, and forest systems [30]. For rice-cropping system, DNDC has been substantially tested/validated against observed CO_2, N_2O or CH_4 fluxes during the past two decades [25,31–33]. Satisfactory results were achieved in a number of countries across the world like the U.S., China, Thailand, India, Japan, etc. [32–36]. Many studies proved that DNDC is most applicable for estimating CH_4 emissions from rice paddies at regional scale [26,27].

To implement the upscaling for multi-temporal, regional CH_4 estimation, accurate acquisition of spatial distribution of rice field was indispensable to advance the regional applications of DNDC. Remote sensing (RS) for mapping rice could provide more accurate spatial information of rice fields than conventional census data. Many researchers utilized remotely sensed data (optical or microwave) for mapping the spatial distribution of paddy rice at regional scale [37,38]. DNDC has been discussed for upsacling by integrating the RS technique to compile greenhouse gas inventories, identify spatial patterns in emission, or explore scenarios for GHG mitigation [39–41].

In this study, a representative region of paddy rice production, the Three-Rivers Plain (TRP) in Northern China, was selected for regional CH_4 estimation. This region possesses climate, soil and management conditions differing from that in the tropical or subtropical rice regions. With the introduction of RS analysis into the DNDC upscaling, this study aims at characterizing the spatiotemporal patterns of the CH_4 emissions from rice fields in the TRP over the past two decades (from 1990 to 2010), and further for quantifying the contribution of the rice paddies within the TRP to global methane.

Methods

Study area

The study area, the TRP, is located in northeast China (48.5°–43.8°N and 129.2°–135.1°E). Three major rivers, Songhua River, Wusuli River and Heilong River, whose watersheds cover almost the entire territory (10.93 million hectares) of the eastern part of Heilongjiang Province (Fig. 1). This region lies at 45 to 60 m geographic elevation above sea level with a gentle and flat topographic relief. Croplands is cultivated in this region from early May to early October for each cropping year with leaving fields fallow for a long period (approximately 7 months). Annual mean temperature is ranged from 2.6 to 5.2°C, and annual precipitation ranged from 330 to 850 mm during the period of 1980–2010. The soils are fertile and rich in organic matter. The flat topography, fertile soils and abundant water resources have made the alluvial plain favorable for crop production. In the 1950s, land reclamation campaigns in the Northeast China converted a majority of the natural swamplands into farmlands. Over the past six decades, the region has experienced drastic changes in the land use. Especially in the recent 20 year, over one million ha of lands has been cultivated as rice paddy.

The rice fields in the TRP, with the highest latitude in not only China but also the world, are one of the most important resources producing high quality rice for the region even the whole county. In this region, a small population of farmer owes a large amount of agro-lands, and modern cultivation managements are extensively practiced. Only single-season rice is planted in the region with growing season from late May to late September. Continuously deepwater flooding (with >10 cm water depth) is widely adapted in rice fields cross the TRP. Urea and synthetic fertilizer are predominantly applied without any organic matter amended. About 10% of the rice straw is normally left as stubble in the fields after harvest in October, and the stubble is incorporated into the soils with tillage before the beginning of the next rice season.

Model Validation

To validate the applicability of DNDC model for the rice fields in the TRP, field experiments were conducted at a paddy site in the Honghe Farm (at 47°35′N and 133°31′E) in 2004 and 2006 within the plain (Fig. 1). Ecological factors to drive DNDC model for simulating the CH_4 production and oxidation in rice paddy included three major factors, i.e. climate, soil properties and management practices. These factors (i.e. model inputs) were used to run DNDC for the experimental site. Daily meteorological data (air temperature and precipitation) were acquired from the local climate station, a part of the Ecological Experimental Station of Mire-Wetland in the TRP run by the Chinese Academy of Science. Soil physical and chemical properties in the field site were obtained from the ground-based measurements. Rice cultivation information was collected from the log documents of field measurements. Three treatments were conducted in a same paddy field with specific management practices (see details in [41]). Three treatments are: 60 kg N/ha of N-fertilizer application rate in 2004 (T1); 150 kg N/ha of N-fertilizer application rate in 2004 (T2); and 150 kg N/ha of N-fertilizer application rate in 2006 (T3). The measurements of CH_4 fluxes were conducted twice per week with static chamber method through the rice-growing period (from late May to early October). The measured CH_4 flux data were used to compare with the modeled CH_4 fluxes at daily time step. Statistical tools such as the root mean square error (RMSE), the coefficient of model efficiency (EF) and the coefficient of model determination (CD) were adopted to assess the "goodness of fit" of model predictions. Normally, value for EF is less than or equal to 1. A positive value indicates that the simulated values describe the trend in the measured data better than the mean of the observed values. The CD value is larger than or equal to 0. When a CD value of 1 or above, it indicates that the model describes the measured data better than the mean of the observations. Taken together, EF and CD allow RMSE to be further interpreted where standard error values of the measurements are unavailable. Detailed description on the calculation of the RMSE, CD and CD were listed in [42].

Construction of database for Regional Simulation

For the purpose of characterizing the spatiotemporal patterns of CH_4 emissions, it is necessary to construct geographic information system (GIS) database for regional simulation with DNDC upscaling. This database included multi-temporal rice field maps, soil properties, daily weather data, and farming management practices cover the study region.

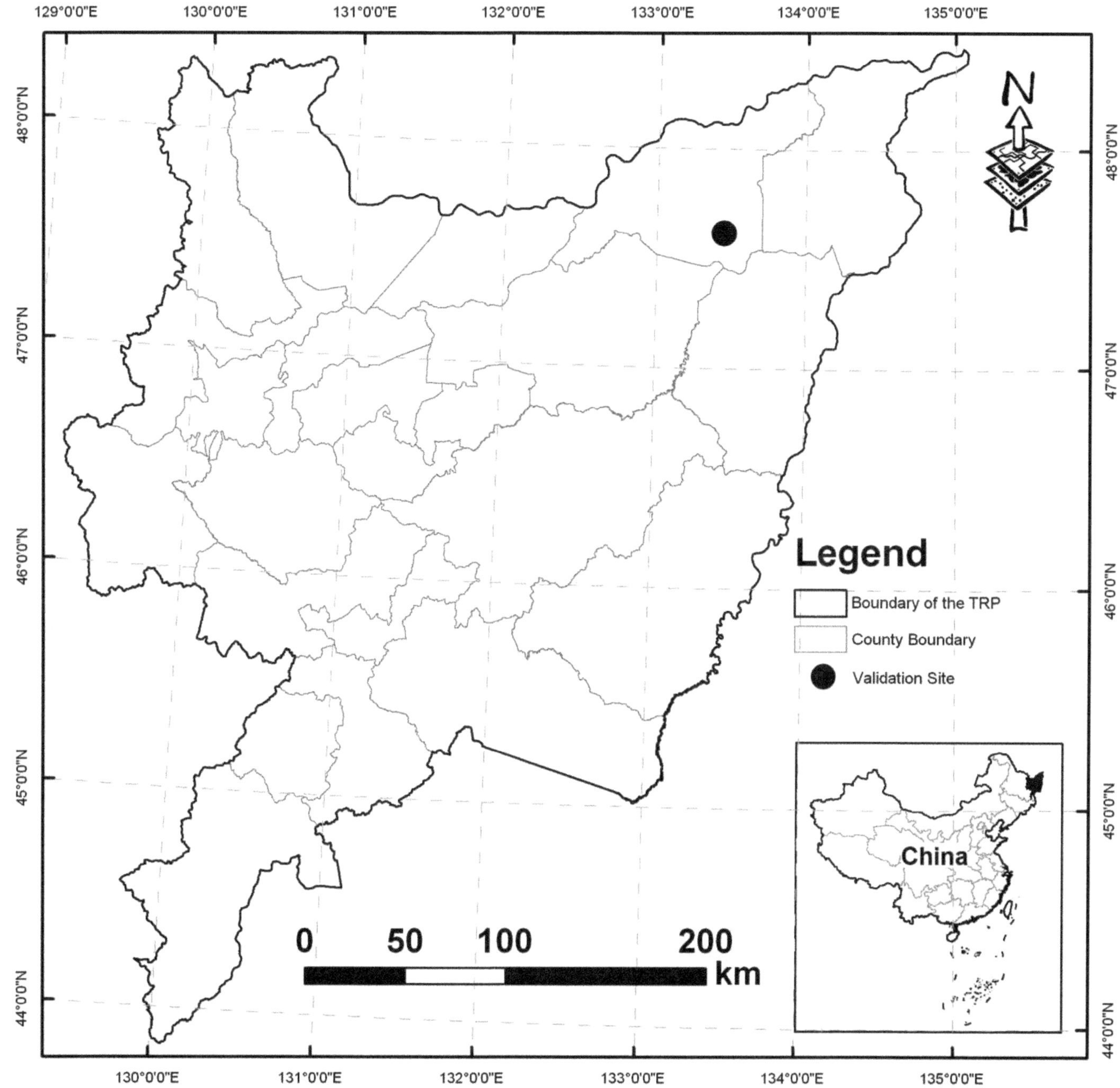

Figure 1. Location of the Three-Rivers Plain.

To delineate the spatial distribution of rice fields in the TRP, three rice maps were retrieved from Landsat thematic mapper (TM) RS imagery acquired in approximately 1990, 2000 and 2010 (Table 1). Total of 30 TM images with a high spatial resolution of 30 m were selected in the tillering or near-mature stage of rice in 1990 (9 images), 2000 (9 images) and 2010 (12 images). These Landsat TM images accessed from the EarthExplorer Interface (http://edcsns17.cr.usgs.gov/EarthExplorer/) were digitized by visual interpretation technology at the GIS software environment of ArcGIS 9.2, which were used to extract the detailed spatial distribution of paddies in the TRP. In terms of the unique phenology features of rice, three accurate rice maps were successfully extracted from the clear remote sensing images.

The climate dataset were composed of daily maximum and minimum air temperature, precipitation and mean wind speed observed in 1990, 2000 and 2010. They were acquired from 7 basic weather stations in China (China Meteorological Data Sharing Service System at http://data.cma.gov.cn/). The entire plain was segmented into 7 sub-regions at which one weather station was located. We assumed the rice growth was with the same climate condition within each sub-region.

Soil data were derived from the soil dataset developed by the Institute of Soil Science, Chinese Academy of Sciences, which was compiled based on the second national soil survey of China conducted in 1980–1990s [43,44]. The soil spatial dataset was a grid data with cell size of 10 km×10 km, which contains soil

Table 1. Remotely sensed Landsat TM imagery used for retrieving rice paddy.

	Acquisition date		
Path/Row No.	**1990**	**2000**	**2010**
113/26	-	-	19/09/2010
113/27	19/10/1992	05/09/2002	19/09/2010
113/28	-	-	19/09/2010
113/29	-	-	19/09/2010
114/27	12/06/1989	11/08/2002	06/06/2010
114/28	16/09/1989	25/09/2001	06/06/2010
114/29	29/09/1988	25/09/2001	-
115/27	25/06/1991	12/08/2000	14/09/2009, 17/09/2010
115/28	02/09/1993	12/08/2000	14/09/2009
115/29	17/10/1992	31/08/2001	14/09/2009
116/27	04/09/1991	07/09/2001	08/09/2010
116/28	23/05/1994	07/09/2001	08/00/2010

texture (sand, silt and clay percentage) and physical and chemical properties (e.g. organic matter, pH and bulk density) at multiple-layer profile. In the TRP, there were 605, 815 and 826 soil cells cover the rice paddies of the TRP in 1990, 2000 and 2010, respectively. Each cell containing soil properties in the top layer (0–10 cm) were used to drive the DNDC simulations. And then, spatial overlay analysis was performed to segment three rice thematic maps with the soil data with cell size of 10 km×10 km.

Generally, unlike the extensive farmland use mode in South China, the management practices are relatively identical cross the study region. The indistinctive differences in rice cultivation practices make less variation in CH_4 emissions from rice paddy in this study area. Therefore for the multiple temporal simulation of CH_4 emission, we assumed the general management practices in each simulated year was spatially identical for the entire rice paddies in TRP. Detailed information on paddy cultivation (e.g. planting/harvesting date, tillage/irrigation regime, fertilizer application, residue management, etc.) in 1990, 2000 and 2010 were investigated by communicating with a number of local agronomists and farmers (Table 2).

The spatially differentiated information above listed was compiled in the GIS database of DNDC for the regional simulation. DNDC was performed for characterizing the spatio-temporal patterns of CH_4 emission with three input datasets in 1990, 2000 and 2010, respectively. DNDC run twice for each grid cell with the maximum and minimum values of the soil properties, which formed a range of CH_4 emission that was later used for quantifying the uncertainty generated from the DNDC upscaling [25]. Based on the modeled CH_4 flux and the rice field acreage within each cell, the total yearly CH_4 emissions from the cell could be calculated. Three cell-level spatial patterns of CH_4 emissions were mapped for the whole plain, and then the regional emissions were accumulated for evaluating their contributions to global CH_4 inventory.

Results

Validation results and sensitivity factors analysis

The measured CH_4 fluxes at the three treatments mentioned above were compared with modeled results. Figure 2 shows the comparisons between the modeled CH_4 fluxes with observations. As a whole, the modeled results showed a fair agreement with observations although minor discrepancies exist across the three treatments. Results showed that the RMSE values were 0.190, 0.304 and 0.344 for treatment T1, T2 and T3, respectively. The EF values were positive (>0.8), and the CD values were greater than 1 for all the three treatments. Validation test has proved that the DNDC is capable of better capturing the seasonal behaviors of CH_4 fluxes from the experimental site within the study area [41,45].

Methane production and oxidation in rice fields are controlled by many factors such as climate variables, soil properties, or agricultural management practices [46]. The sensitivity test provided crucial information for finding out the most sensitive factors from all input parameters, which could affect the modeled results for regional estimations. In the sensitivity test, baseline scenario was first set based on the average climate, soil and management conditions cover the study region. The simulated result from baseline scenario was taken as a benchmark of CH_4 emissions for accessing those of other scenarios. And then, within a predefined range, DNDC were performed by varying single one of all input parameters while keeping all other input parameters constant (Table 3). The model responses to changes of these factors on CH_4 emission from rice paddies in the TRP were presented respectively in Fig. 3.

Specifically, the response of CH_4 emission to changes in climate factors was investigated by running DNDC using alternative climate scenarios. The modeled results indicated that precipitation changes (±20%) have no significant impacts on CH_4 emissions for the continuously flooded paddies, whereas the fluctuation of CH_4

Table 2. Management practices on the rice paddies in the TRP.

Items	1990	2000	2010
Tillage	5/15: Plow depth of 20 cm	5/15: Plow depth of 20 cm	5/14: Plow depth of 20 cm
Rice cultivation	5/25: transplanting; 9/25: harvesting, grain yield of 2400 kg C/ha	5/25: transplanting; 9/25: harvesting, grain yield of 2700 kg C/ha	5/25: transplanting; 9/25: harvesting, grain yield of 3000 kg C/ha
Flooding	5/15–8/25: Continuously flooding, water depth of 10 cm	5/15–8/25: Continuously flooding, water depth of 10 cm	5/14–8/25: Continuously flooding, water depth of 10 cm
Fertilization	6/1: Urea (24 kg N/ha); 7/1: Urea (36 kg N/ha)	6/1: Urea+Synthetic fertilizer (36 kg N/ha); 7/1: Urea (54 kg N/ha)	1/1: Urea+Synthetic fertilizer (48 kg N/ha); 7/1: Urea (72 kg N/ha)
Manure application	No	No	No
Residues incorporation	10%	10%	10%

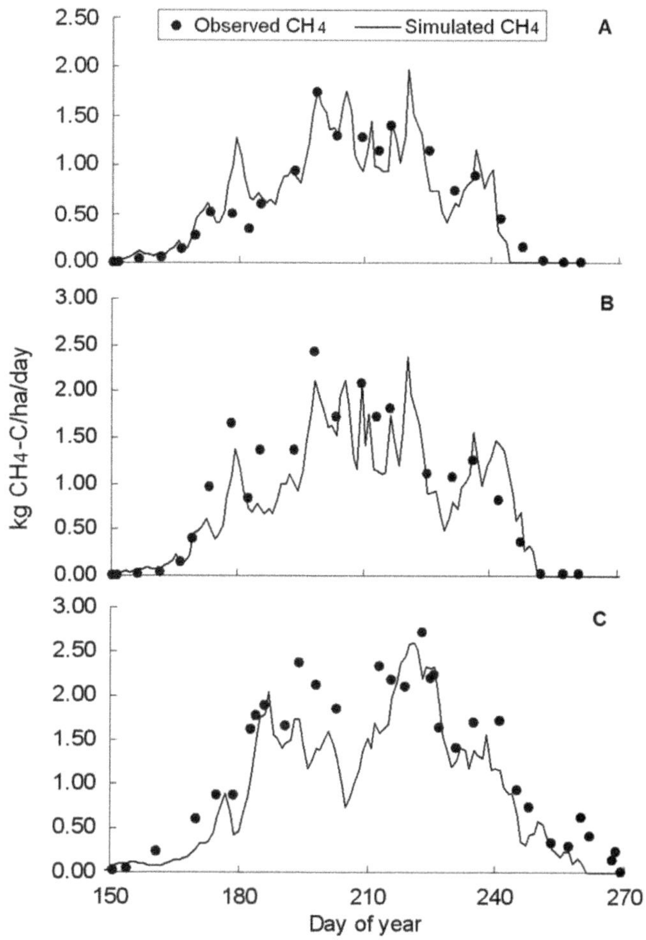

Figure 2. Simulated vs. observed CH_4 fluxes in validation site. (A) 60 kg N/ha of N-fertilizer application rate in 2004, (B) 150 kg N/ha of N-fertilizer application rate in 2004, (C) 150 kg N/ha of N-fertilizer application rate in 2006. (data from Zhang et al., 2011).

emissions incurred by temperature changes (±2°C) was remarkable. The higher is temperature, the more CH_4 emitted from rice paddy (Fig. 3). Temperature is therefore the most sensitive climate factor. This could be explained that the higher temperature accelerated the decomposition and fermentation process of soil organic matter (SOM) [47].

The spatial heterogeneity of soil property has a great impact on the pattern of CH_4 emission. In this test, four major soil parameters (soil organic carbon (SOC) content, clay fraction, pH and bulk density) were tested within specific changing range (Fig. 3). Test results indicated that soil clay content was the most sensitive factor, followed by the SOC content. For the anaerobic soil flooded by deepwater, the sandy loam soil was more likely to produce more CH_4 than the clay loam soil because large porosity existed in sandy soil could promote CH_4 transmission and release from flooded fields. The soil with high SOC content could provide more dissolved organic carbon to the methanogens, which is favorable to CH_4 production [25,48–50]. In contrast, CH_4 emission was less sensitive to both soil pH and bulk density.

Various scenarios on CH_4 emissions with three major management practices (flooding regime, residue incorporation and N-fertilizer application rate) were simulated by DNDC (Fig. 3). There was no significant impact of the incorporation rate of rice straw and residue on CH_4 emission from the incorporated fields. In the TRP, rice straw and residue were normally left in fields after rice harvest, no CH_4 emitted from the fields which have been already drained over one month. The nitrogen level meeting the need of the optimal rice-grain production is 120 kg N/ha. CH_4 emission increased with increasing N-fertilizer application when the applied amplitude was less than the optimal nitrogen demands for physical development of rice plant. When the application rate reached a level meeting the need for the optimal productivity, additional fertilizer application didn't make any more impact on CH_4 emission. Flooded rice paddy provided favorable environment for methanogenesis. Mid-season drainage changed the anaerobic status of CH_4 production, and thus reduced the CH_4 emission from rice fields [51–53]. Simulated results showed that for the rice paddy in cool climate region the CH_4 emission was reduced by 45% for 10 days draining duration in the peak tillering and early maturity stages. In the perspective of mitigating CH_4 emission from rice paddy fields, mid-season drainage would become a potential opportunity through adjusting present management practices.

Changes in rice paddies of the Three-Rivers Plain

Three accurate rice paddies maps cover the Three-Rivers Plain were delineated for each decade from 1990 to 2010 from TM images acquired in nominal 1990, 2000 and 2010 (Fig. 4). The rice paddies were mainly distributed in the lowland areas along with

Table 3. Environmental Factors for sensitivity tests.

Environmental Factors		Baseline value	Range tested
Climate	Annual mean temperature (°C)	3.8	1.8–5.8
	Total annual precipitation (mm)	550	440–660
Soil Property	Clay fraction (%)	0.32	0.1–0.6
	Initial soil C fraction (%)	2.78	1.0–6.0
	Bulk density (g/cm^3)	1.25	1.0–1.6
	Soil pH	6.6	5.0–8.0
Management Practices	Depth of continuous flooding water (cm)	10	Not varied
	Drained days (d)	0	2–12
	Residue incorporation (%)	10	0–100
	N-Fertilizer application (kg N/ha)	150	60.0–180.0

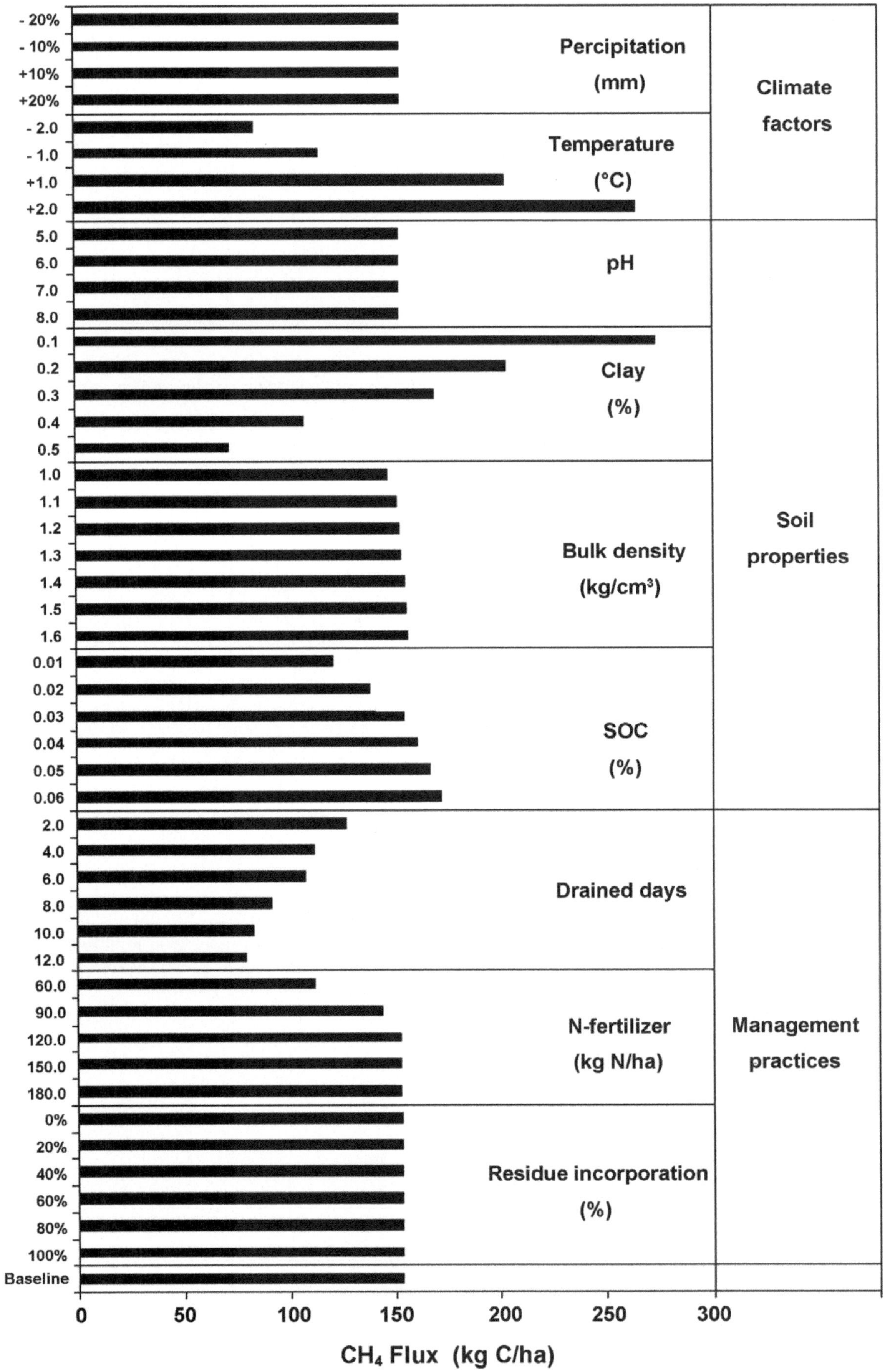

Figure 3. Sensitivity tests of environment factors driving CH_4 emissions from rice paddies.

the major rivers. The extracted total area of paddy fields was 0.23, 1.22 and 1.63 million ha in 1990, 2000 and 2010, respectively. Statistical results showed that the rice area has increased approximate one million ha for the first 10 years (from 1990 to 2000), whereas about 0.4 million ha for the last 10 years (from 2000 to 2010). Although the trend in changes of total area of rice paddies was constant within the TRP, the increment rate was slowing down gradually. The results indicated the change intensity of land use/covers in the last decade was evidently smaller than the first one, which could be attributed to the limited available lands (reclaimed or converted) for new rice cultivation. From the perspective of spatial distribution, a majority of rice was cultivated in the western and southern lowlands of the plain in 1990. Some small patches of rice fields were sparsely distributed in the eastern of the TRP where the complexity of natural wetlands and drylands existed (Fig. 4 (A)). In contrast, a large amount of drylands have

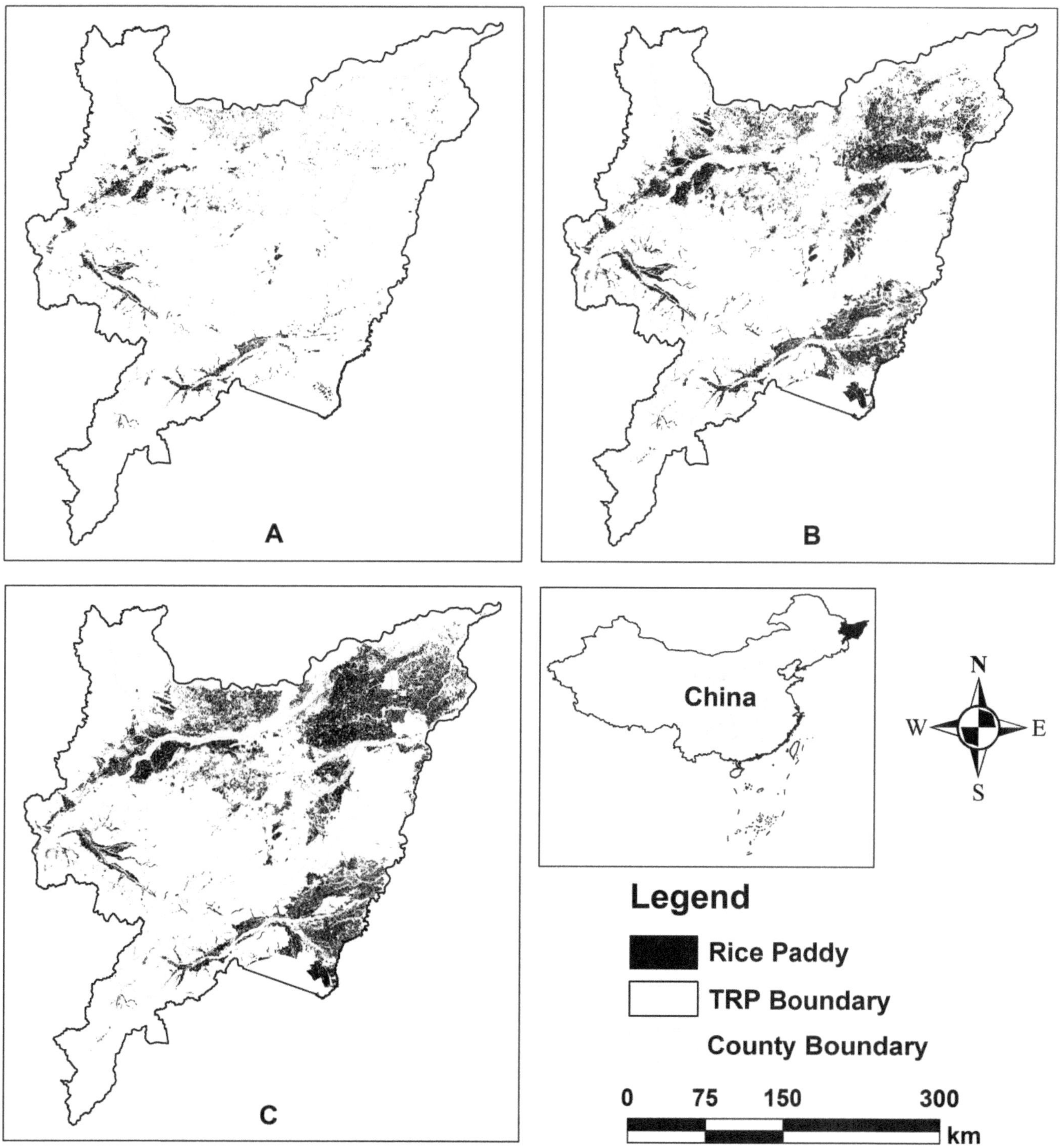

Figure 4. Rice paddies maps in the Three-Rivers Plain. (A) 1990, (B) 2000, (C) 2010.

been extensively cropped as rice paddy since 2000 (Fig. 4 (B and C)). At present, the northeast and southeast have become the major area of rice production for the Three-Rivers Plain. These areas have been playing a role in providing the high quality rice and ensuring food security for the Northern China, as well as the country.

Estimation of CH_4 emissions for the Three-Rivers Plain

Driven by the three GIS database separately constructed for 1990, 2000 and 2010, DNDC simulated the CH_4 flux cell-by-cell across the entire rice fields in the TRP. The total emissions were then calculated by multiplying the modeled flux by rice area in each cell to produce regional CH_4 emissions for each simulated year. The spatial and temporal patterns of CH_4 emissions were mapped cover the domain using GIS tools. Fig. 5 and Fig. 6 showed CH_4 emission rates and total emissions at the cell scale, respectively. The two maps showed clear spatial patterns in CH_4 emissions across the domain. Detailed descriptions were discussed as below.

In 1990, the simulated CH_4 emission rates were 20.46–175.92 kg C/ha, and the average of 71.05 kg CH_4-C were emitted from one hectare of rice field. Relatively higher CH_4 flux (>120 kg CH_4-C/ha) was located at a few of soil grid cells sparsely distributed in the southern tip and northern of the TRP, whereas lower flux (<30 kg CH_4-C/ha) were in the most northeastern (Fig. 5 (A); Table 4). The western and southeastern of the plain, the major rice production area, emitted more CH_4 (>200 tons (1 ton = 1000 kg)) than other areas. The northeastern was a lower emission area (<50 tons per cell) where a large number of farmlands were cropped as dryland crops at that time (Fig. 6 (A)).

Significant changes in CH_4 emission pattern happened in 2000. Higher CH_4 flux (>250 kg CH_4-C/ha) appeared in the central and the northern while the northeastern remained lower flux (<100 kg CH_4-C/ha) (Fig. 5 (B)). The mean flux reach to 180 kg CH_4-C/ha with large gradient of 34–457 kg CH_4-C/ha (Table 4). Yearly total emission showed that those cells with high CH_4 emission (>400 tons) accounted for over 40% of the domain (Fig. 6 (B)).

The recent CH_4 emission pattern in 2010 was presented in Fig. 5 (C) and Fig. 6 (C). Those cell with higher CH_4 flux (>250 kg CH_4-C/ha) sparsely distributed over the entire rice paddies, and lower flux (<100 kg CH_4-C/ha) in the northeastern kept constant. Although the simulated emission rates of CH_4 in 2010 was slightly less than that in 2000 (Table 4), their patterns in total emission was very comparable (Fig. 6 (C)).

In summary, statistical results indicated that the regional average fluxes were around 71, 137 and 180 kg CH_4-C/ha, in 1990, 2000 and 2010, respectively. The average of total CH_4 emitted from the entire rice paddies of the TRP was 0.025, 0.211 and 0.203 Tg CH_4-C (1 Tg = 10^9 kg) in the three years (Table 4).

Changes analysis for the spatiotemporal pattern in CH_4 emission

From the simulated results of the TRP, the highest emission rate was 8–16 times higher than the lowest one in each of the three simulated years (Table 4). Such a huge difference in CH_4 emission rate was due to the variations in the soil properties. Past study has demonstrated that those soil cell with lower CH_4 flux contained higher clay fraction, and the higher flux normally occurred in those cells with relatively higher SOC content [41]. Consequently, the spatial patterns in CH_4 fluxes in this study are mainly attributed to the Most Sensitive Factor (MSF) of soil properties, i.e. soil clay fraction and SOC content (Fig. 3).

As far as the yearly emissions from paddy cells were concerned, statistics analysis showed that the standard deviation (SD) of CH_4 emission was 36.44 tons CH_4-C in 1990, which was far less than that in 2000 and 2010 (258.85 and 245.54 tons CH_4-C). Otherwise, although the minimum of yearly emission were very similar (0.01~0.03 tons CH_4-C) for all three simulated years, the maximum in 2000 and 2010 (>1900 tons CH_4-C) were much more than that in 1990 (<680 tons CH_4-C). Therefore, there was an observable emission variation cross the entire study region in 2000 and 2010 while the character in spatial variation was not distinct in 1990 (Fig. 4). Obviously the clear spatial variation in yearly CH_4 emissions from cell data was attributed to the spatial distribution of rice fields.

In Table 4, average CH_4 emissions (0.025 Tg) in 1990 was far less than in 2000 and 2010 (>0.2 Tg). The huge difference could be attributed to several reasons. First of all, total area of rice paddies was the primary determinant of total emission. Only 0.23 million ha of rice were planted in 1990. However since 2000, the rice paddy areas have exceeded 1.2 million ha (Fig. 4). Such rapid increase in rice planted area would inevitably lead to the consequential increase in total CH_4 emissions. Additionally, the CH_4 emissions rate in 1990 was also smaller than in other simulated years. This could be firstly related to the fact that more biomass production with more N-fertilizer application. The sensitivity tests have indicated that although the effects of N-fertilizer application rate on CH_4 emissions seems not very evident, it did increase the crop biomass or yields that could indirectly enhance the CH_4 production. In 1990s, only 60 kg N/ha of N-fertilizer rate were applied to the rice fields by compared with 90 and 120 kg N/ha of N-fertilizer rate in 2000 and 2010, separately.

Otherwise, the temperature was a positive climate factor driving CH_4 production from rice fields. The emission fluxes increased with temperature because the higher temperature accelerated soil organic matter (SOM) decomposition and fermentation process, which has been proved in past research [44], and also presented in the sensitivity test (Fig. 3). Although the rice area in 2010 was 30% more than that in 2000, total CH_4 emissions (0.203 Tg CH_4-C) in 2010 was slightly less than that (0.211 Tg CH_4-C) in 2000. The results was largely related to this fact that higher flux (180 kg CH_4-C/ha) in 2000 than that in 2010 (137 kg CH_4-C/ha). By comparing the average of 10-day temperature from rice transplanting to harvesting stage (late May through September), the temperature during the rice-growing season in 2000 was apparently1.13–2.65°C higher than that in 2010 [Fig. 7]. Therefore, the temperature would be taken as a major factor for regional estimation of total CH_4 emission from rice paddies at long term and large regional scale.

Discussion

Rice paddy mapping with multi-temporal RS data can quantify the dynamics of cropland use. Since 1990, the agricultural lands have been experiencing significant changes in the TRP in the very northeastern part of China. In the major rice producing areas with the cooler climate in China, fast rise in rice cultivation was driven by the market demand for high yield/quality rice. More upland crops like corn or soybeans were converted to paddy rice in Northern China during past two decades. How to real-time estimate the GHG emission induced by this kind of rapid land-use transformation have been becoming an important objective for making the scientific GHG inventory for China. Consequently, characterizing and quantifying the spatiotemporal pattern of CH_4

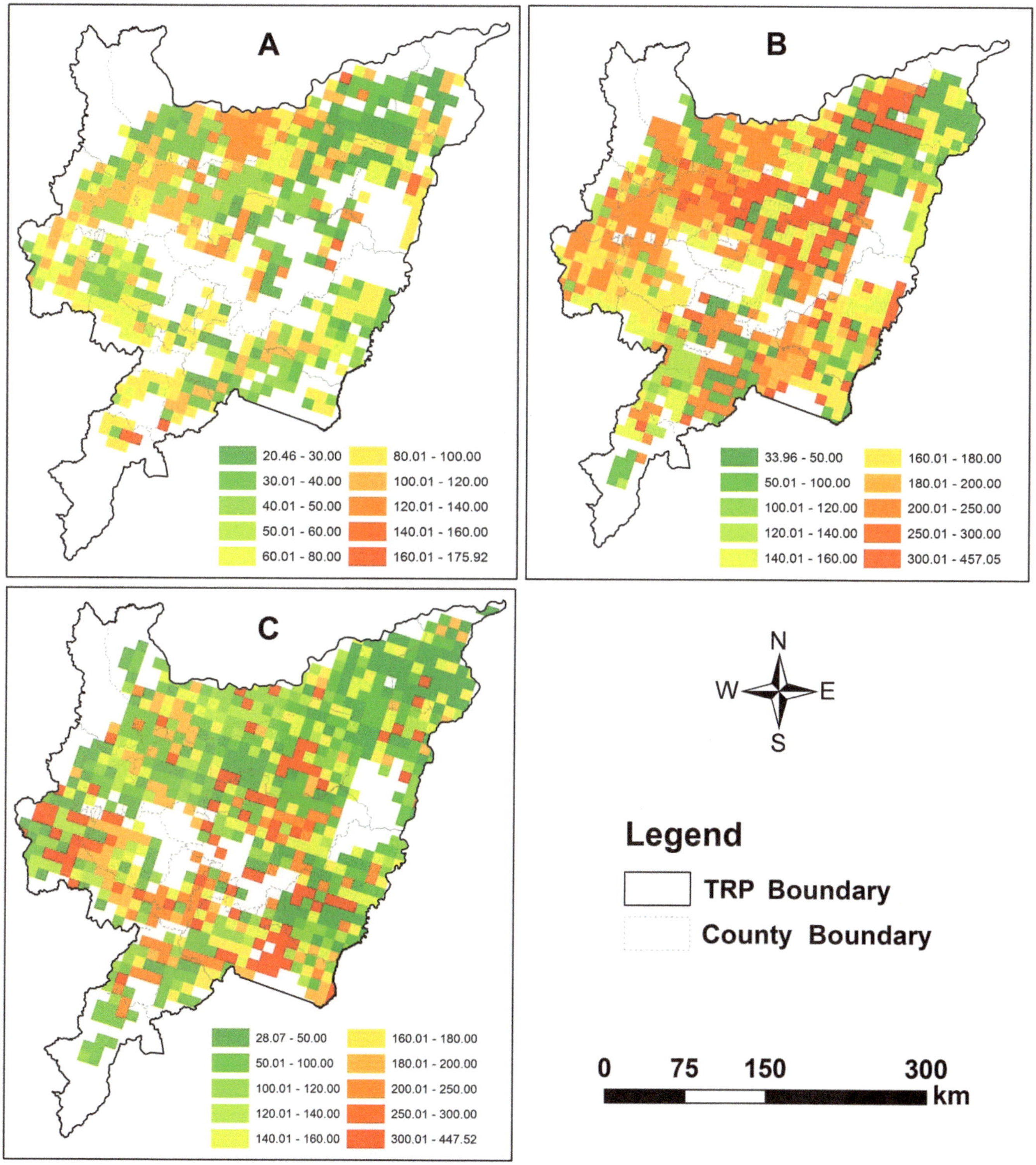

Figure 5. Mean CH_4 emission rates (kg CH_4-C/ha) of paddy fields. (A) 1990, (B) 2000, (C) 2010.

emissions from rice paddies could be an interested topic for researcher in the future.

In our precious study, DNDC upscaling has been utilized to quantify CH_4 emissions from rice fields of the study area in 2006 after the elaborate calibration and site validation [41]. This present study extended previous study to multi-temporal CH_4 estimation for characterizing spatial and temporal dynamics of CH_4 emissions over past two decades (from 1990 to 2010). The modeled average of CH_4 fluxes for the simulated year of 1990, 2000 and 2010 was 70, 180 and 137 kg CH_4-C/ha, respectively.

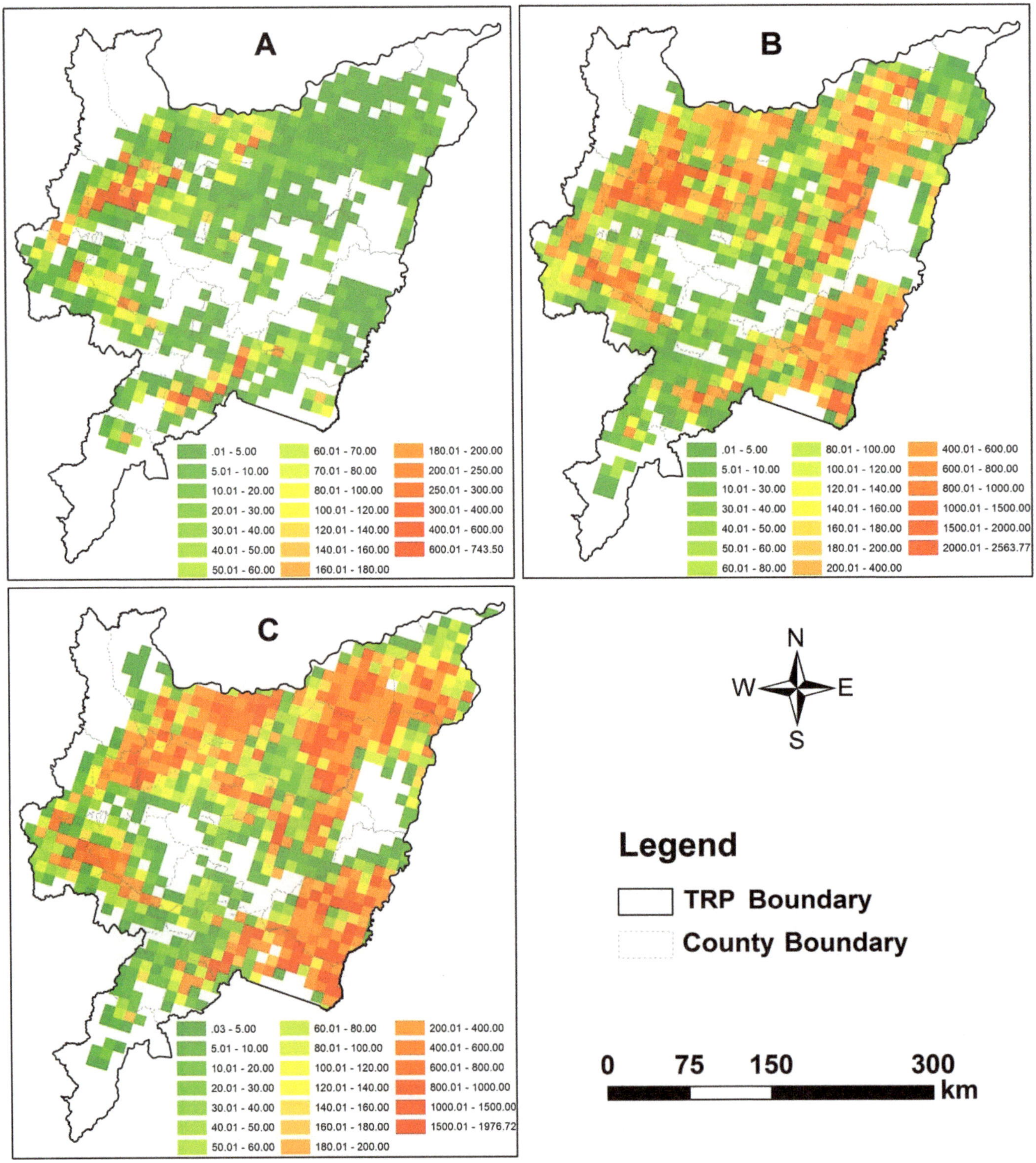

Figure 6. Yearly total CH_4 emissions (ton CH_4-C) from rice paddies. (A) 1990, (B) 2000, (C) 2010.

These results are comparative with the average emission rates observed and modeled in the Taihu Lake region of Southeast China (15–198 kg CH_4-C/ha/year) [27].

In the past studies, a baseline emission factor of 1.30 kg CH_4/ha/day in the 2006 IPCC Guidelines [54] was often recommended to estimate the regional or global CH_4 emissions from rice paddies. Thus, With the prevail management practices of 100-day continuously flooded without organic amendments for rice paddies in the TRP, the total CH_4 emissions based on the IPCC approach was 0.030, 0.159 and 0.212 Tg CH_4-C in the

Table 4. Simulated CH_4 emission rates and total emissions for three simulated year.

Items		1990	2000	2010
Emission rates (kg C/ha)	Max.	175.9	457.05	447.52
	Min.	20.46	33.96	28.07
	Avg.	71.05	180.42	136.85
Total emissions per cell (ton C)	Max.	679.2	2563.77	1976.72
	Min.	0.01	0.01	0.03
	Avg.	36.44	258.85	245.54
	SD	73.01	348.92	329.09
Total emissions (Tg C)	Max.	0.027	0.228	0.241
	Min.	0.022	0.193	0.165
	Avg.	0.025	0.211	0.203

1990, 2000 and 2010, respectively. Comparative analysis indicated that the simulated values in 1990 and 2010 (0.025 and 0.203 Tg) were slightly less than that of IPCC estimation, whereas the simulated values (0.211 Tg) in 2000 were significantly larger than the IPCC estimation. In this study, we took into account the integrated influence of various ecological drivers to CH_4 emissions from rice paddy. These drivers including climate, soil and management factor, were applied to drive DNDC model for regional CH_4 estimation. Thus this method introduced in this study would make great improvement for the CH_4 estimation compared to the IPCC method based on baseline emission factor.

Our modeled results further confirmed the high latitude wetland agroecosystems like rice paddy in Northern China was an important anthropogenic CH_4 source. With the increase in the rice-growing area, the rice paddies in the TRP could make more contribution to global CH_4 inventory. During the period of past decades in this region, natural swamp wetlands were first converted into dryland croplands and then, into anthropogenic wetlands (rice paddies). This kind of unique changes in land use/covers would consequentially resulted in huge environmental impacts. A number of ground-based observations on CH_4 emissions from natural wetlands have been substantially conducted in the high latitude plain of China [55]. These studies provided important supports for regional estimation of CH_4 emissions with spatial modeling technology. Thus quantifying the dynamics (i.e. the net increment) of CH_4 emissions in the process of land transformation in this region would be an interested topic for researchers in the future.

During the recent years, China authority has paid more attention to GHG inventory and mitigation. Developing an effective method towards assessing the magnitude of impacts from rice-cropping systems would be significant for meeting the social and research needs. The results in this study demonstrated huge potential of integrating biogeochemical models with RS mapping technology for meeting the environmental challenges rise in coming years in China.

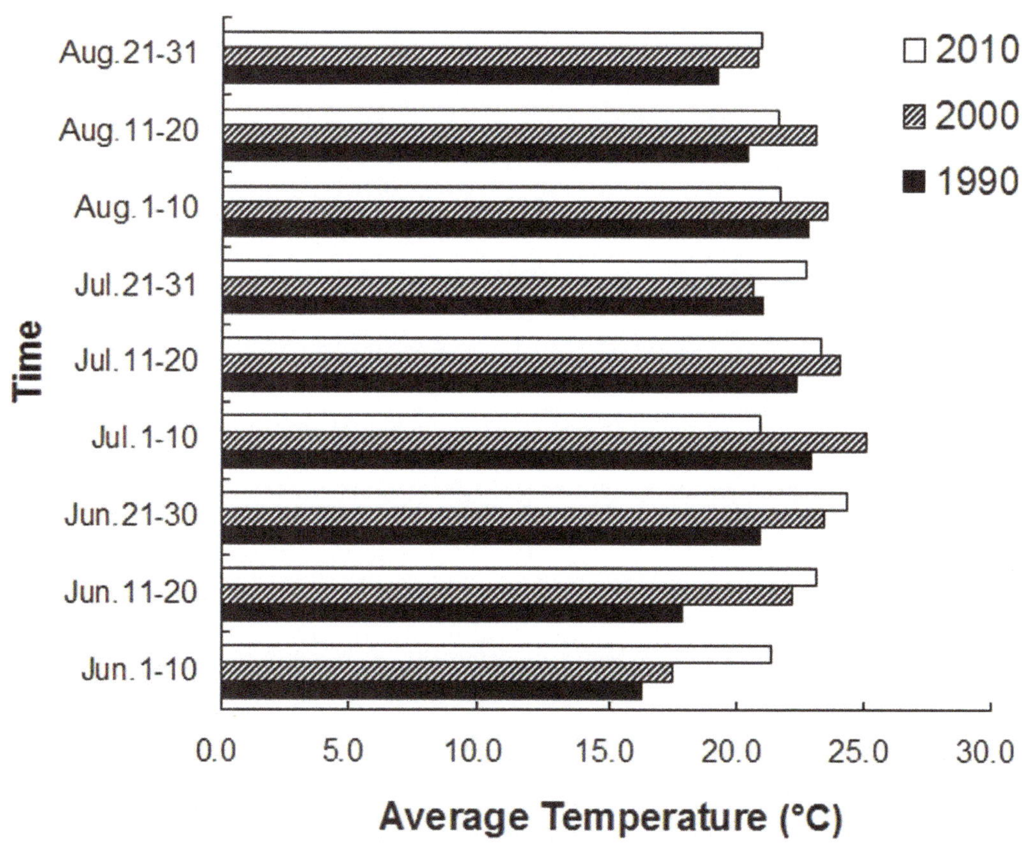

Figure 7. Average of 10-day temperature during the rice-growing season.

Author Contributions

Conceived and designed the experiments: YZ. Performed the experiments: YZ SS FZ. Analyzed the data: YZ SS. Contributed reagents/materials/analysis tools: YZ FZ. Wrote the paper: YZ SS. Provided the funding for the study: RS WG.

References

1. IPCC (2007) Climate Change 2007: The physical science basis. In: Solomon S, Qin D, Manning M, Chen Z, Marquis M, et al. (2007) Contribution of working group I to the fourth assessment report of the Intergovernmental Panel on Climate Change. Cambridge: Cambridge University Press.
2. USEPA (2006) Global anthropogenic non-CO2 greenhouse gas emissions: 1990–2020 (June 2006 Revised), Available: http://www.epa.gov/climatechange/economics/downloads/GlobalAnthroEmissionsReport.pdf via the Internet. Washington, DC: Office of Atmospheric Programs, USEPA.
3. Neue HU (1993) Methane emission from rice fields. Biosciences 43: 466–474.
4. Scheehle EA, Kruger D (2006) Global anthropogenic methane and nitrous oxide emissions. Energy J 22: 33–44.
5. FAO (2011) FAO statistical databases. Available at: http://faostatfaoorg/ (verified in 17 May 2011). Rome.
6. NBSC (2011) China Statistical Yearbook-2010. Beijing: China Statistics Press.
7. Wang MX, Shangguan XJ, Shen RX, Wassmann R, Seiler W (1993) Methane production, emission and possible control measures in the rice agriculture. Adv Atmos Sci 10: 307–314.
8. Wassmann R, Schuetz H, Papen H, Rennenberg H, Seller W, et al. (1993) Quantification of methane emissions from Chinese rice fields (Zhejiang Province) as influenced by fertilizer treatment. Biogeochemistry 20: 83–101.
9. Cai ZC, Xing GX, Shen GY, Xu H, Yan XY, et al. (1999) Measurements of CH4 and N2O emissions from rice paddies in Fengqiu, China. Soil Sci Plant Nutr 45: 1–13.
10. Cai ZC, Tsuruta H, Minami K (2000) Methane emission from rice fields in China: measurements and influencing factors. J Geophys Res 105: 17231–17242.
11. Khalil MAK, Rasmussen RA, Wang MX, Ren LX (1991) Methane emission from rice fields in China. Environ Sci Technol 25: 979–981.
12. Yao H, Zhuang Y, Chen ZL (1996) Estimation of methane emission from rice paddies in Mainland China. Global Biogeochem Cy 10: 641–649.
13. Van Bodegom PM, Scholten JCM (2001) Microbial processes of CH4 production in a rice paddy soil: model and experimental validation. Geochim Cosmochim Acta 65: 2055–2066.
14. Verburg PH, Van Der Gon HACD (2001) Spatial and temporal dynamics of methane emissions from agricultural sources in China. Glob Change Biol 7: 31–47.
15. Verburg PH, van Bodegom PM, van der Gon HACD, Bergsma A, van Breemen N (2006) Upscaling regional emissions of greenhouse gases from rice cultivation: Methods and sources of uncertainty. Plant Ecol 182: 89–106.
16. Anastasi C, Dowding M, Simpson VJ (1992) Future CH4 emission from rice production. J Geophys Res 97: 7521–7125.
17. Sinha SK (1995) Methane emission from rice paddies: excellent methodology but poor extrapolation. Curr Sci 68: 643–646.
18. Kern JS, Gong Z, Zhang G, Zhuo H, Luo G (1997) Spatial analysis of methane emissions from paddy soils in China and the potential for emissions reduction. Nutr Cycl Agroecosys 49: 181–195.
19. Li CS, Frolking S, Frolking TA (1992) A model of nitrous oxide evolution from soil driven by rainfall events. 2. Model applications. J Geophys Res 97: 9777–9783.
20. Cao M, Dent JB, Heal OW (1995) Modelling of methane emission from rice paddies. Global Biogeochem Cy 9: 183–195.
21. Matthews RB, Wassmann R, Arah J (2000) Using a crop/soil simulation model and GIS techniques to assess methane emissions from rice fields in Asia, I. Model development. Nutr Cycl Agroecosys 58: 141–159.
22. Cao M, Dent JB, Heal OW (1995) Methane emissions from China's paddyland. Agric Ecosys Environ 55: 129–137.
23. Cao M, Gregson K, Marshall S, Dent JB, Heal OW (1996) Global methane emissions from rice paddies. Chemosphere 33: 879–897.
24. Bachelet D, Neue HU (1993) Methane emission from wetland rice areas of Asia. Chemosphere 26: 219–237.
25. Li CS, Mosier A, Wassmann R, Cai Z, Zheng X, et al. (2004) Modeling greenhouse gas emissions from rice-based production systems: sensitivity and upscaling. Global Biogeochem Cy 18: GB1043.
26. Pathak H, Li CS, Wassmann R (2005) Greenhouse gas emissions from Indian ricefields: calibration and upscaling using the DNDC model. Biogeosciences 2: 113–123.
27. Zhang L, Yu D, Shi X, Weindorf D, Zhao L, et al. (2009) Quantifing methane emissions from rice fields in the Taihu Lake region, China by coupling a detailed soil database with biogeochemical model. Biogeosciences 6: 739–749.
28. Li CS, Frolking S, Frolking TA (1992) A model of nitrous oxide evolution from soil driven by rainfall events: I. Model structure and sensitivity. J Geophys Res 97: 9759–9776.
29. Li CS, Frolking S, Harriss RC (1994) Modeling carbon biogeochemistry in agricultural soils. Global Biogeochem Cy 8: 237–254.
30. Giltrap DL, Li CS, Saggar S (2010) DNDC: A process-based model of greenhouse gas fluxes from agricultural soils. Agric Ecosys Environ 136: 292–230.
31. Zhang Y, Li C, Zhou X, Moore III B (2002) A simulation model linking crop growth and soil biogeochemistry for sustainable agriculture. Ecol Model 151: 75–108.
32. Li CS, Qiu J, Frolking S, Xiao X, Salas W, et al. (2002) Reduced methane emissions from largescale changes in water management in China's rice paddies during 1980–2000. Geophys Res Lett 29: 1972.
33. Cai ZC, Sawamoto S, Li CS, Kang G, Boonjawat J, et al. (2003) Field validation of the DNDC model for greenhouse gas emissions in East Asian cropping systems. Global Biogeochem Cy 17: 1107.
34. Zheng XH, Wang MX, Wang YS, Shen RX, Shangguan XJ, et al. (1997) CH4 and N2O emissions from rice paddies in southeast China. Chin J Atmos Sci 21: 167–174.
35. Babu YJ, Li CS, Frolking S, Nayak DR, Datta A, et al. (2005) Modelling of methane emissions from rice-based production systems in India with the denitrification and decomposition model: field validation and sensitivity analysis. Curr Sci 89: 1904–1912.
36. Babu YJ, Li CS, Frolking S, Nayak DR, Adhya TK (2006) Field validation of DNDC model for methane and nitrous oxide emissions from rice-based production systems of India. Nutr Cycl Agroecosys 74: 157–174.
37. Frolking S, Qiu J, S. B, Xiao X, Liu J, et al. (2002) Combining remote sensing and ground census data to develop new maps of the distribution of rice agriculture in China. Global Biogeochem Cy 16: 1091.
38. Xiao X, Boles S, Liu J, Zhuang D, Frolking S, et al. (2005) Mapping paddy rice agriculture in southern China using multi-temporal MODIS images. Remote Sens Environ 95: 480–492.
39. Takeuchi W, Yasuoka Y, Tamura M (2001) Estimation of Mehtane Emission from Paddy Fields in Central Plain of Thailand by Scaling Technique. In: Proceedings of the 22nd Asian Conference on Remote Sensing, 5–9 November 2001. Singapore 1: 422–427.
40. Salas W, Boles S, Li CS, Babu Y, Xiao X, et al. (2007) Mapping and modeling of greenhouse gas emissions from paddies with satellite radar observations and the DNDC biogeochemical model. Aquat Conserv 17(3): 319–329.
41. Zhang Y, Wang YY, Su SL, Li CS (2011) Quantifying methane emissions from rice paddies in Northeast China by integrating remote sensing mapping with a biogeochemical model. Biogeosciences 8: 1225–1235.
42. Smith P, Smith JU, Powlson DS, McGill WB, Arah JRM, et al. (1997) A comparison of the performance of nine soil organic matter models using datasets from seven long-term experiments. Geoderma 81: 153–225.
43. Shi XZ, Yu DS, Warner ED, Pan XZ, Petersen GW, et al. (2004) Soil database of 1:1,000,000 digital soil survey and reference system of the Chinese genetic soil classification system. Soil Surv Horiz 45: 129–136.
44. Yu DS, Shi XZ, Wang HJ, Sun WX, Chen JM, et al. (2007) Regional patterns of soil organic carbon stocks in China. J Environ Manage 85: 680–689.
45. Wang Y, Chen W, Zhao Z, Gu J (2008) Characteristics and estimation of CH4, N2O emission from cold paddy field in the Sanjiang Plain. Trans CSAE 24: 170–176.
46. Yan XY, Yagi K, Akiyama H, Akimoto H (2005) Statistical analysis of the major variables controlling methane emission from rice fields. Glob Change Biol 11: 1131–1141.
47. Cao M, Gregson K, Marshall S (1998) Global methane emissions from wetlands and its sensitivity to climate change. Atmos Environ 32: 3293–3299.
48. Yagi K, Minami K (1990) Effect of organic matter application on methane emission from some Japanese rice fields. Soil Sci Plant Nutr 36: 599–610.
49. Wassmann R, Neue HU, Bueno C (1998) Methane production capacities of different rice soils derived from inherent and exogenous substrates. Plant Soil 203: 227–237.
50. Holzapfel-Pschorn A, Seiler W (1986) Methane emission during a cultivation period from an Italian rice paddy. J Geophys Res 91: 11803–11814.
51. Sass RL, Fisher FM, Wang YB, Turner FT, Jund MF (1992) Methane emission from rice fields: The effect of flood water management. Global Biogeochem Cy 6: 249–262.
52. Yagi K, Tsuruta H, Kanda K, Manami K (1996) Effect of water management on methane emission from a Japanese rice field: Automated methane monitoring. Global Biogeochem Cy 10: 255–267.
53. Li CS, Salas W, DeAngelo B, Rose S (2006) Assessing alternatives for mitigating net greenhouse gas emissions and increasing yields from rice production in China over the next 20 years. J Environ Qual 35: 1554–1565.
54. IPCC (2006) 2006 IPCC Guidelines for National Greenhouse Gas Inventories. In: Eggleston HS, Buendia L, Miwa K, Ngara T, Tanabe K, eds. Prepared by the National Greenhouse Gas Inventories Programme. Hayama: the Institute for Global Environmental Strategies (IGES).
55. Song C, Xu X, Tian H, Wang Y (2009) Ecosystem–atmosphere exchange of CH4 and N2O and ecosystem respiration in wetlands in the Sanjiang Plain, Northeastern China. Global Change Biology 15: 692–705.

Assessing the Impact of Copper and Zinc Oxide Nanoparticles on Soil

Daniel Collins[1], Todd Luxton[2], Niraj Kumar[3], Shreya Shah[1], Virginia K. Walker[3], Vishal Shah[1]*

1 Department of Biology, Dowling College, Oakdale, New York, United States of America, **2** National Risk Management Research Laboratory, U.S. Environmental Protection Agency, Cincinnati, Ohio, United States of America, **3** Department of Biology and School of Environmental Studies, Queen's University, Kingston, Ontario, Canada

Abstract

It is not known if the annual production of tonnes of industrial nanoparticles (NPs) has the potential to impact terrestrial microbial communities, which are so necessary for ecosystem functioning. Here, we have examined the consequences of adding zero valent copper and zinc oxide NPs to soil in pots that were then maintained under field conditions. The fate of these NPs, as well as changes in the microbial communities, was monitored over 162 days. Both NP types traveled through the soil matrix, albeit at differential rates, with Cu NPs retained in the soil matrix at a higher rate compared to ZnO NPs. Leaching of Cu and Zn ions from the parent NPs was also observed as a function of time. Analysis of microbial communities using culture-dependent and independent methods clearly indicated that Cu and ZnO NPs altered the microbial community structure. In particular, two orders of organisms found in rhizosphere, *Flavobacteriales* and *Sphingomonadales*, appeared to be particularly susceptible to the presence of NPs. Together, the migration of NPs through soil matrix and the ability of these potential pollutants to influence the composition of microbial community in this field study, cannot help but raise some environmental concerns.

Editor: Vipul Bansal, RMIT University, Australia

Funding: The work was funded by National Science Foundation grant # 966741, NSERC (Canada) grant, and a Queen's Research Chair award. The funders had no role in study design, data collection and analysis, decision to publish, or preparation of the manuscript.

Competing Interests: The authors have declared that no competing interests exist.

* E-mail: ShahV@dowling.edu

Introduction

Engineered metal nanoparticles (NPs) have numerous applications in a variety of consumer goods and industrial processes due to their unique physical, chemical and biological properties. As a consequence, the elevated production to meet this demand will result in an increase in NP release into the environment. To date, there is very little understanding on how such discharged NPs will influence microbial biodiversity. Microorganisms play important roles in geologic, hydrologic and ecological cycles, and any change in microbial diversity can potentially influence environmental quality and health, and even human development [1–3]. Previously, the effect of engineered metal NPs on terrestrial microbial communities has been tested under laboratory conditions [4–10], but here we have investigated the effect of zinc oxide nanoparticles (ZnO NPs) and zero valent copper nanoparticles (Cu NPs) on the soil microbial community in pots under field conditions.

ZnO NPs are used in electronics, personal care products, biosensors, food additives, pigments and rubber manufacture [11]. Similarly, zero valent Cu NPs are used in electronics, ceramics, films, polymers, inks, metallics, lubricant oil, coatings and health care products [12]. Their prevalence virtually ensures that these NPs may be considered pollutants and indeed, it has been reported that NPs have been detected in waste streams [13–14]. In these experiments, an effort has been made to correlate the observed toxicity with the speciation and migration of the NPs through the soil matrix. As a result, we believe that this is the first study to investigate the fate and effect of NPs under field conditions.

Results and Discussion

Soil and NP Characterization

The agricultural soil used was rich in total organic carbon (TOC), total Kjeldahl nitrogen (TKN) and phosphorus with values of 13100 mg/kg, 980 mg/kg and 357 mg/kg, respectively. The average soil pH was 7.5±0.2 (n = 5). Low levels of metals including Cu and Zn, as determined by acid digestion, were present in the soil (Table S1). K-edge X-ray Absorption Near Edge Structure (XANES) spectra indicated that Cu^{2+} orZn^{2+} in the untreated soil were adsorbed to a mineral or organic surface. To elucidate this further, reference spectra of Cu and Zn adsorbed to ferrihydrite, alumina γ-Al_2O_3, bentonite, and humic acid (Sigma Aldrich, St Louis, MO) were used as analogs for organic matter and mineral phase present in soils and to model the Cu and Zn XANES data using Linear Combination Fitting (LCF) (Figure S1). Results from the LCF analysis indicated that for Cu, 43% was adsorbed to alumina and 57% to humic acid. For Zn, the distribution was 27% ferrihydrite, 32% smectite, and 41% alumina. For both metals, the species present represent Cu and Zn phases that are present in both unperturbed and metal-contaminated soils [15–20].

NPs were analyzed using several techniques, with transmission and scanning electron microscope (TEM and SEM) images showing that both the Cu and ZnO NPs were aggregated. Attempts to de-aggregate the material were unsuccessful, and as a result, obtaining the true particle size distribution was not possible. Aggregation also prevented the use of dynamic light scattering as a means of estimating particle size. TEM data obtained for the Cu NPs revealed a wide range of particle sizes (Figure S2 and not shown) with particles generally spherical in shape and sized between <10 and 200 nm. SEM images revealed that the ZnO NPs were elongated (between 15–50 nm in width and 50–20 nm in length) and lacked well-defined crystal faces (Figure S2 and not shown).

Powdered X-ray diffraction (XRD) (Figure 1) identified three prominent crystalline phases associated with the Cu NPs: metallic Cu, cuprite (Cu_2O), and tenorite (CuO). Based on X-ray peak intensities the most abundant phase was metallic Cu followed by CuO and Cu_2O (Figure 1a). The presence of three phases was not unexpected due to the oxidation of Cu on the outer shell and is similar to previously reported results [21–22]. Diffraction patterns for the ZnO NPs revealed a single ZnO phase (Figure 1B). NP speciation was also evaluated by LCF analysis of the normalized, normalized derivative, and χ data. For Cu the reference materials included metallic Cu, Cu_2O, and CuO (Table 1; Figure S3). Consistent with the XRD results, the LCF analysis showed that metallic Cu was the most abundant phase followed by CuO and Cu_2O (Figure S3). Again, consistent with the XRD analysis, XANES spectra for Zn indicated that ZnO was the only phase present in the NP sample (Figure S4).

The point of zero charge (PZC) for the Cu NPs from alkalimetric titration was at a pH of 9.4, and close to previous values ([23] and references therein) (Figure S5). Titration of the Cu NP suspension between pH 6-5 resulted in the immediate oxidation of metallic copper in the NPs. XRD data from titration samples collected immediately after suspension showed a color change at pH 5, coincident with the disappearance of metallic Cu and a dramatic increase in the intensity and presence of diffraction peaks associated with cuprite (Figure S6). The PZC for ZnO NPs was 8.7, again close to previous values ([23] and references therein) (Figure S5). The elevated PZC values for both Cu and ZnO NPs indicated that the NPs would possess a positive surface charge in the soil used for the current study, based on the soil pH (7.5).

Fate of the Nanoparticles

Cu and ZnO NPs were deposited on the soil surface and allowed to freely migrate through the soil while exposed to environmental conditions over 160 days. Daily temperature and precipitation (total of 495 mm) were monitored over the field study period (Figure 2).

The first soil pots were sampled 24 h after NP addition and prior to any precipitation. There was no translocation of the Cu or Zn, and no detectable change in Cu and Zn soil concentrations as a function of depth or NP speciation based on XANES analysis (Figures 3, 4, 5). It should be noted that Cu and Zn K-edge XANES data (Figures 4 and 5, dashed lines) show spectral features unique to the NPs (black dashes) and the Cu and Zn species present in the untreated soil (red dashes). Figure 4 also shows a pre-edge feature indicative of zero valent Cu (Cu^0; shoulder), and a post-edge feature associated with adsorbed Cu (Cu^{2+}; red dashes). Figure 5 demonstrates the XANES spectra post-edge peak indicative of ZnO (black dashes), and a post-edge peak associated with adsorbed Zn (red dashes).

Prior to sampling on day 7, there was a snowfall (total of 2 mm of precipitation). XANES analysis at the surface of the day 7 samples showed that Cu and ZnO NPs were the only phases present, but there was a significant decrease in the Zn concentration and little change in the Cu concentration (Figure 3). Although no NPs were detected in the 2nd horizon at 24 h, by 7 days there was a significant increase in the Cu and Zn concentration at 3–8 cm, and detectable levels in the lower 3rd horizon. For Cu NP treatment pots, the emergence of the Cu^0 shoulder in the 2nd horizon (Figure 4B, black dashes) provides direct evidence for NP transport in the soil. Further, LCF analysis of the three horizon sections at this sampling period revealed evidence of Cu NPs in all three sections (Table 2). For ZnO NPs, the presence of the post-edge Zn peak at 9679 eV (Figure 5B, black dashed line) indicates that some of these particles had leached into the 2nd horizon. As with the Cu NPs, LCF analysis of all three horizons indicated that ZnO was also present in the 3rd-horizon section (Table 2). It is unlikely that the modest amount of

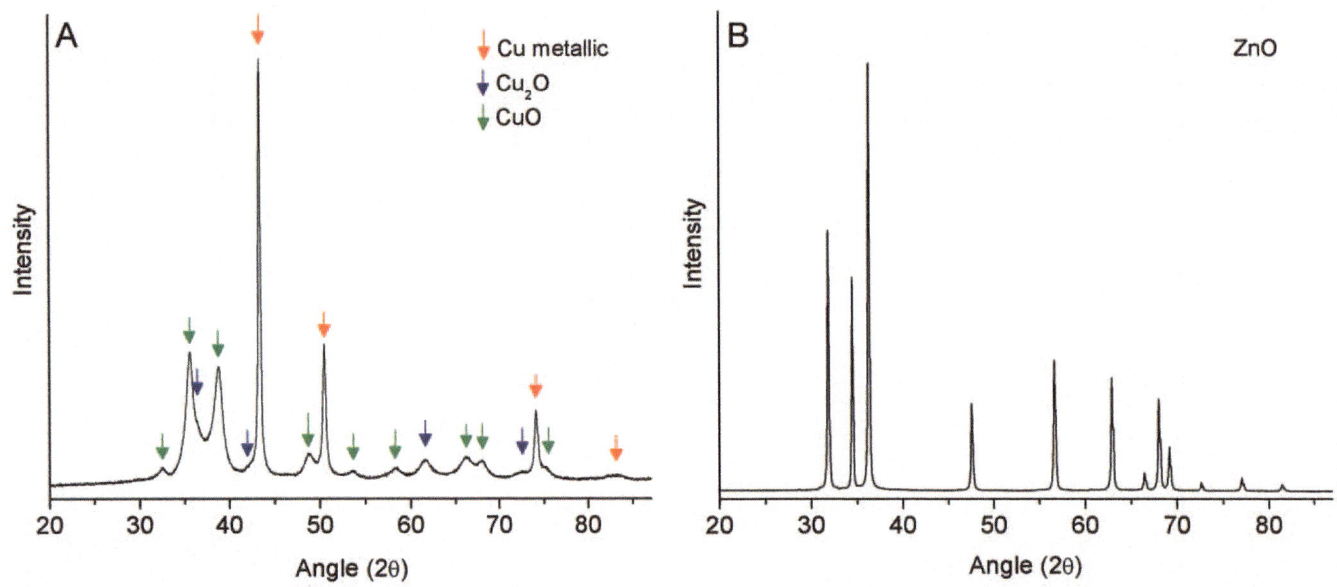

Figure 1. X-ray diffraction patterns for copper (A) and zinc oxide (B) nanoparticles. Different color arrows indicate specific copper phases. Speciation of the Cu phase was determined by comparison with the International Center for Diffraction data (ICDD) powder diffraction files with experimental data (Cu metal *01-085-1326*, Cu_2O *01-078-2076*, CuO *00-041-0254*). ICDD files for ZnO were *01-079-0206*.

Table 1. Powdered X-ray diffraction analysis of the Cu NPs, showing the linear combination fit (with R values referring to a general goodness of fit parameter).

Fitting Spectra	Cu metal %	+/− %	Cu_2O %	+/− %	CuO %	+/− %	Chi^2	Reduced Chi^2	R value
Norm xm(E)	41.11	0.20	14.17	0.50	44.72	0.20	0.015	$3.9*10^{-4}$	$7.9*10^{-4}$
Der Norm xm(E)	36.96	0.51	13.86	0.62	49.18	0.51	0.005	$1.4*10^{-5}$	0.0148
$\chi(k)k^2$	28.73	2.31	20.28	3.43	50.99	2.18	2.81	0.015	0.04
Average	35.60		16.10		48.30				

melting snow formed large pores in the soil facilitating NP transport in all the experimental pots. Indeed, several studies evaluating ZnO and Cu NP transport in saturated columns have shown that both NPs are mobile [24–26] and this mobility is likely enhanced by absorption to organic acids through steric interactions and electrostatic stabilization [24–28]. Thus, the high organic carbon content in the agricultural soil likely contributed to an organic surface coating and resulted in NP mobility through the soil. Under our experimental conditions it appears that ZnO is more mobile than Cu based on the decrease in Zn concentration in the surface layer and the coincident increase of Zn in the 2nd horizon (Figure 3). When these results are combined with those reported in the literature, it suggests that metallic Cu NPs can be considered the least mobile NPs when compared to Fe_3O_4, CuO, TiO_2 and ZnO NPs [24].

In addition to the migration of NPs, dissolution of Cu and ZnO NPs was also apparent after 7 days. The 9-fold increase in Cu and Zn concentration in the 2nd-horizon was partially due to translocation. However, since the Cu and ZnO NPs only accounted for 50% of the Cu and Zn species identified (Figures 3, 4B, 5B, Table 2), it appears that the NPs in the 1st horizon released Cu and Zn ions that then migrated downwards (Figures 3, 4B, 5B, Table 2). The same conclusion can be made for Cu and Zn in the 3rd horizon, based on the ~2.5-fold increase in metal concentration with ~40 and 25% of the Cu and Zn species, respectively, present in the NP form (Figures 3, 4B, 5B, Table 2).

The study site received significant precipitation in the form of rain and snow in the 7–30 day period (Figure 2). Nevertheless, there was no significant difference in the concentration or speciation of Cu and Zn in the surface layer and only modest differences in the 2nd horizon at the third sampling period, compared to the 7 day samples (Figure 3). In the 3rd horizon, however, the concentration of Zn and Cu doubled. This significant increase in Cu NPs in the 3rd horizon was further evidenced by the emergence of the Cu^0 pre-edge feature (Figure 4C, black dashes; Table 2), and clearly demonstrated the continued leaching of these NPs. Zinc speciation in the 3rd horizon indicated significant dissolution of the ZnO (Table 2) and was corroborated by the strong post-edge feature (Figure 5C, red dashes), associated with an adsorbed Zn phase. It has been reported that there is a rapid dissolution/transformation of ZnO to a Zn adsorbed phase (pH 7 kaolin suspension) within 24 h [29], and therefore, the dissolution of ZnO NP in soil over a 30 day period would not be unexpected. The 30-day samples also provided additional evidence for the increased mobility of ZnO NPs compared to Cu NPs since there was a marked increase in Zn concentrations throughout the soil profile as compared to Cu. The persistence and decreased mobility of Cu NPs may be a cause of concern with respect to their potential ecological impact on soil surfaces.

After 160 days with increased temperatures and much precipitation (Figure 2), including 10 mm of rain in a 24 h period, the concentration of Zn in the soil had returned to levels close to that of untreated soil (Figure 3). XANES analysis showed that some Cu NP remained but there was no significant presence of the original ZnO NPs. The observation of mobile Cu and Zn species at this time point (Figures 4D; 5D; Table 2), suggests that desorption of the adsorbed metals could be related to the increased temperature.

Less than 0.1% of the original NPs were recovered in the soils. There is no evidence in the literature suggesting that these NPs are resistant to microwave-assisted acid extraction and strict quality

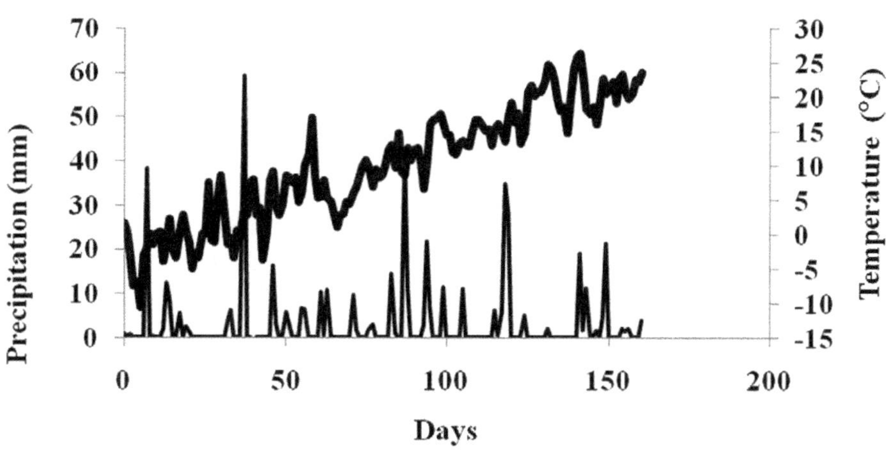

Figure 2. Temperature and precipitation over time at the field study location. Bold and thin lines represent temperature and precipitation, respectively. The parameters were measured using Davis-Vantage Pro2. Day 0 was January 19, 2011.

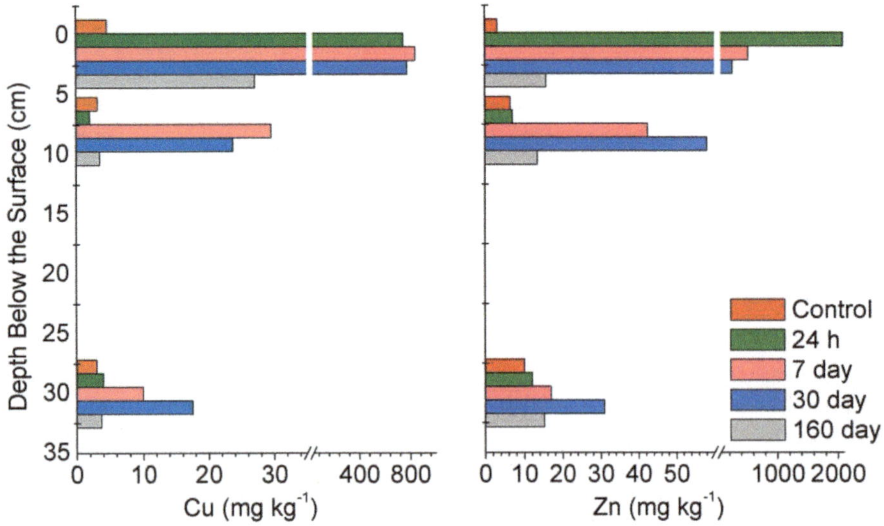

Figure 3. Concentration of Cu and Zn in the extracted soil cores as a function of depth and time. The break in the X-axis occurs at 35 and 60 mg kg^{-1} for Cu and Zn, respectively.

assurance and control procedures (including, and not limited to blank, spiked, and numerous controls), gave no indication that recovered NP mass was associated with instrument error. Our soil-nanoparticle experimental pots were open systems, and no attempt was made to collect the leachate or prevent the erosion of nanoparticles caused by wind, which would be necessary for determining the relationship between NPs and soil on a mass basis. In the future, consideration must be given to controlling wind erosion as well as the collection of soil leachate. Not withstanding these caveats, the analyses presented here clearly demonstrate that both Cu and ZnO NPs are mobile in agricultural soils and have the potential to escape beyond an initial 'spill' site.

Microbial Community Analysis

The impact of Cu and ZnO NPs on the soil microbial community was measured using culture-idependent (Biolog® ecoplates) and culture-independent methods (FAME analysis and pyrosequencing, respectively). Physiological profiles obtained

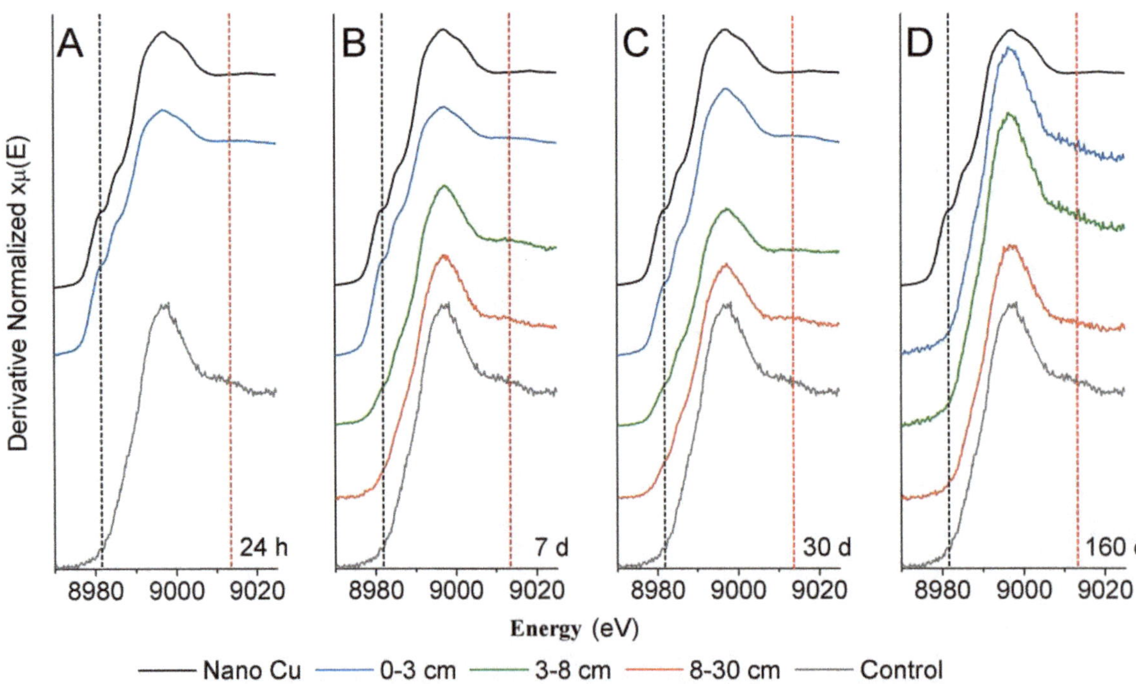

Figure 4. Cu K-edge powdered X-ray diffraction spectra of soil cores at different depths at the four time points. All panels (A–D) show spectra for a Cu NP reference (black line) and the untreated, control soil (grey line). Spectral features are indicated by dashed lines:the Cu pre-edge feature associated with zero valent Cu^0 (black dashed line) and the post edge feature in the untreated soil,a shoulder associated with adsorbed Cu^{2+} (red dashed line).

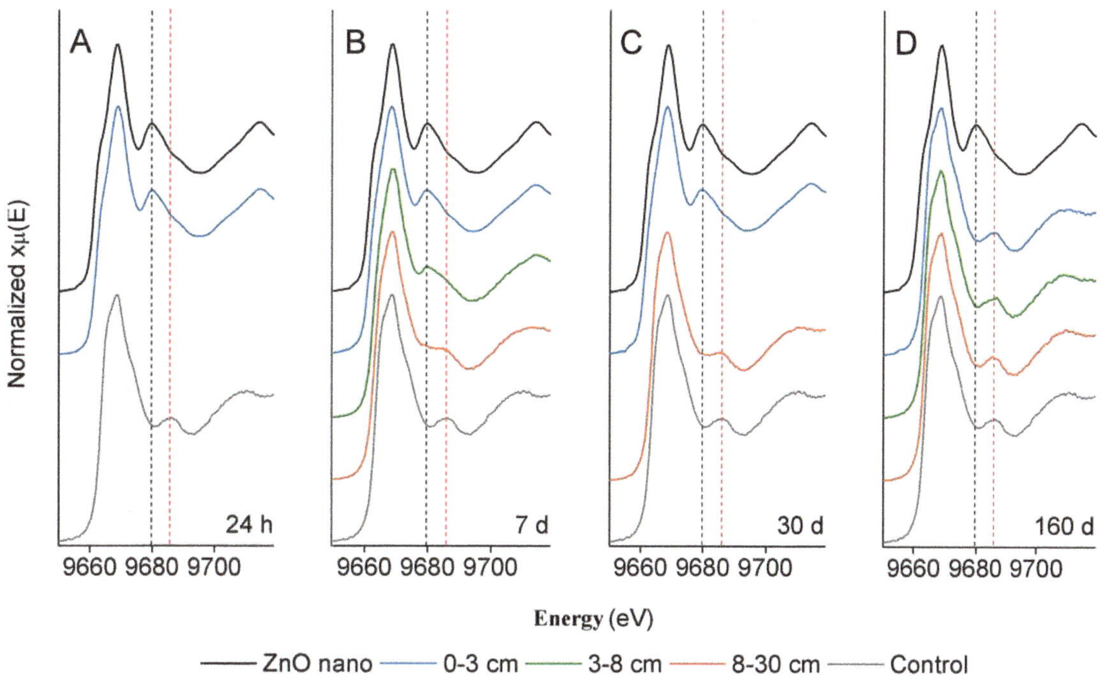

Figure 5. Zinc K-edge powdered X-ray diffraction spectra of soil cores at different depths at the four time points. All panels (A-D) contain a reference spectrum for ZnO NP (black line) and untreated soil (grey line). Spectral features associated either with the ZnO NPs (black dashed lines) or the untreated soil (red dashed lines) are indicated.

using Biolog® ecoplates showed that NP additions had an immediate impact on the culturable microorganisms (Table 3). Substrate richness (S) signifies the number of substrates used appreciably (optical density>25) by the community and it is a good indicator of the functional diversity within soil [30]. In the top horizon, the control soil had an S value of 22 but this value dropped to 2 (representing only pyruvic acid methyl ester and L-aspargine substrates) right after the addition of Cu NPs. Likewise, the S value dropped to 11 immediately following the addition of ZnO NPs to the soil. Parallel to these decreases in substrate utlization, a decrease in the Shannon diversity index (H) was observed for both the NPs. Since these physiological profiles were done *in vitro*, the added NPs would have been carried over with the soil samples explaining the observed high toxicity, as has been previously reported for Cu and ZnO NPs [12,28–30]. Indeed, the S and H values (Table 3) are inversely related the NP concentration (Figure 3). We therefore suggest that direct contact [31–34] between NPs and microbes, such

Table 2. K-edge X-ray absorption near-edge structure (XANES) spectral results (linear combination fit analysis) showing normalized Cu and Zn XANES data with Cu and Zn speciation are presented as a function of time and depth (corresponding to the spectral results in Figures 3 and 4).

		Cu species and NPs			Zn species and NPs			
Day	**Depth**	**Cu alumina %**	**Cu humic %**	**Cu NP %**	**Zn alumina %**	**Zn ferrihydrite %**	**Zn smectite %**	**ZnO NP %**
	0–3	0	0	100	0	0	0	100
1	3–8	–	–	–	–	–	–	–
	8–30	–	–	–	–	–	–	–
	0–3	0	0	100	0	0	0	100
7	3–8	31	12	57	7	0	40	53
	8–30	10	54	36	18	0	59	24
	0–3	0	0	100	0	0	0	100
30	3–8	10	25	66	–	–	–	–
	8–30	5	43	52	25	11	57	7
	0–3	22	69	9	36	4	60	0
160	3–8	8	74	18	38	2	61	0
	8–30	32	51	17	36	7	57	0

that is achieved in the *in vitro* physiological assays, may be necessary for acute bacterial toxicity. A previous study showed that when Cu NPs were homogenously mixed with the NPs and incubated under laboratory conditions only marginal changes in the microbial community were observed [4]. In this case, we speculate that very few of the microbes were in direct contact with the NPs.

Culture-independent FAME analysis showed no significant impact of the addition of either NP on the overall microbial soil profile across all three horizons and over time. Gram-negative bacteria were more prevalent (65%) than Gram-positive bacteria in control soil and this proportion remained the same throughout the horizons at all time points with minor changes irrespective of the NP treatment. Likewise, diatom and actinomycete fatty acid signatures showed no impact due to neither the NPs nor the metallic ion species.

While physiological profiles clearly showed microbial toxicity, likely due to the direct contact with the NPs, FAME analysis

Table 3. Microbial community distribution in the soil exposed to copper or zinc oxide nanoparticles under field conditions as measured using FAME analysis and microbial community diversity indicators as measured using community level physiological profile analysis.

LAYER	EXPERIMENTAL SAMPLE	FAME ANALYSIS								BIOLOG ANALYSIS		
		G+	G−	Diatoms	Eukaryote	Fungi	Methanobacter	Actinomycetes	Anaerobic	H	S	E
TOP	Control- 0 day	30	36	16	2	4	1	11	0	3.0	22	2.2
	Copper- 0 day	29	36	18	2	4	1	10	0	2.2	2	7.3
	Zinc- 0 day	30	37	17	2	4	1	9	0	2.5	11	2.4
	Control- 7 days	30	37	17	2	3	1	11	0	2.5	14	2.2
	Copper- 7 days	29	37	17	2	4	1	10	0	2.1	3	4.4
	Zinc- 7 days	29	37	17	3	4	1	8	0	2.1	5	3.0
	Control- 30 days	29	37	17	2	4	0	11	0	2.8	17	2.2
	Copper- 30 days	29	38	16	1	3	1	12	0	0.0	0	0.0
	Zinc- 30 days	28	37	16	2	4	2	11	0	2.7	15	2.3
	Control- 160 days	24	39	19	2	3	1	12	0	2.7	16	2.2
	Copper- 160 days	25	39	17	2	3	1	12	0	2.3	8	2.5
	Zinc- 160 days*									2.4	12	2.3
MIDDLE	Control- 0 day	31	35	16	2	4	1	10	0	3.0	24	2.1
	Copper- 0 day	31	35	16	2	3	1	11	0	3.1	24	2.4
	Zinc- 0 day	30	35	17	2	5	1	10	0	2.9	20	2.2
	Control- 7 days	31	37	16	2	3	1	10	0	2.8	15	2.3
	Copper- 7 days	31	35	16	2	4	1	10	0	2.9	20	2.2
	Zinc- 7 days	29	36	17	2	4	1	11	0	2.9	18	2.3
	Control- 30 days	28	36	17	3	4	1	10	0	2.8	18	2.3
	Copper- 30 days	28	38	16	1	3	1	13	0	2.6	14	2.2
	Zinc- 30 days	28	38	17	2	4	0	12	0	2.7	15	2.2
	Control- 160 days	25	40	18	2	3	2	12	0	2.4	11	2.3
	Copper- 160 days	25	40	16	2	3	2	13	0	2.4	12	2.3
	Zinc- 160 days	24	40	18	2	3	2	12	0	2.4	11	2.3
BOTTOM	Control- 0 day	32	35	16	2	5	1	10	0	2.9	17	2.3
	Copper- 0 day	30	36	16	2	4	1	11	0	3.0	21	2.4
	Zinc- 0 day	30	36	17	2	4	1	10	0	3.0	23	2.3
	Control- 7 days	29	37	16	2	4	1	10	0	2.7	14	2.2
	Copper- 7 days	28	36	17	3	6	0	10	0	2.8	18	2.3
	Zinc- 7 days	28	37	17	2	5	2	10	0	2.8	16	2.2
	Control- 30 days	29	36	17	2	4	1	11	0	2.7	15	2.5
	Copper- 30 days	29	37	15	1	3	1	13	0	2.4	9	2.2
	Zinc- 30 days	29	37	17	2	4	1	11	0	2.8	19	2.2
	Control- 160 days	26	40	17	2	3	2	11	0	2.3	12	2.3
	Copper- 160 days	25	40	16	2	3	2	13	0	2.3	11	2.2
	Zinc- 160 days	25	41	16	2	3	2	11	0	2.3	12	2.2

*FAME data for top layer zinc 160 days not obtained due to contamination of sample.

indicated that the proportion of major bacterial types did not change after NP exposure. Thus, it was still not clear if the overall community composition significantly altered under these field conditions. Therefore, we additionally surveyed the bacterial community through pyrosequencing having first treated the soil so as to decrease the proportion of DNA derived from dead or dying cells, a method that is not applicable prior to FAME analysis. Although DNA community analysis is the best means of evaluating soil changes, caution needs to be exercised when using such a responsive technique in field experiments since microbial flora can be unevenly distributed, with variation introduced by eukaryotes, including invertebrates, birds, plants and mammals. Indeed, the higher percentages of *Enterobacteriales* and *Lactobacillales* in some of the samples (Figure 6, Table S2) can likely be attributed to the bird feces [35–38]. Despite such perturbations, however, an overall impact of the microbial community by the introduced NPs was evident (Table 4).

Control soil samples showed that members of the *Rhizobiales*, *Flavobacteriales* and *Sphingomonadales* orders formed the majority within the bacterial community (Figure 6, Table S2), as would be expected in agricultural soils [39]. All of these organisms are found in high numbers in the rhizosphere [40]. Viable cells are unlikely to be introduced by vertebrate excreta and thus any shift in the ratio of members of these orders should be suitable for the evaluation of NP toxicity. The ratio of these bacterial orders in the three horizons as a function of time and treatment is shown in Table 4. When comparing the ratio in control *vs* NP-treated samples for the top horizon at time 0, one can infer that representatives of *Rhizobiales* are not immediately impacted by Cu or ZnO NPs. In contrast, bacteria belonging to *Flavobacteriales* appeared to be highly susceptible to Cu NPs, if not to ZnO NPs (Table 4). As the NPs were translocated towards the lower horizons as a function of time (Figure 3), the concordant decrease in *Flavobacteriales* is evident in these horizons. The majority of the identified genera representing this order are mesophiles and thus more metabolically active at moderate temperatures. As a result, it would have been active bacteria that were exposed to the NPs/metallic ion species as they travelled to lower horizons while the soil warmed to summer conditions, and active microbes appear to be more susceptible to NP toxicity [4,41]. In fact, there was almost a complete absence of bacteria belonging to both *Flavobacteriales* and *Sphingomonadales* in the 3rd horizon at 160 days. Taken together, the pyrosequencing data validates the hypothesis that exposure to NPs significantly alters the microbial community structure, similar to the results obtained by laboratory studies using metal NPs [4,8]. It should be recalled that at 160 days, almost no original NPs were detected in this horizon (Figure 3), suggesting that toxicity had either occurred prior to that time or that leachate containing ion species affected the consortia. The low sensitivity of *Rhizobiales* to Cu and ZnO NPs is distinct from a previous study [4] that showed susceptibility to 0.067% Ag NPs. It is possible that the differential sensitivity of *Rhizobiales* to types of metal NPs could be attributed to different mechanism of action for individual NPs. Further studies need to be carried out to test the hypothesis. Based on this first field study of the effects of Cu and ZnO NPs on soils, we can conclude that they migrated through the soil matrix as NPs, and the leaching of ions from the NPs occured parallel to that of particle transport. We have been careful not only to investigate the effects of metallic NPs on soil chemistry and biota but have also considered the combined effect of NPs and ions when evaluating the results. The microbial community is indeed altered by the NPs/metal ions as they migrate through the matrix. While it is too soon to correlate these structural changes in the community to functional changes in the ecosystem, considering that we are manufacturing tonnes of NPs yearly [42], we believe that immediate attention is warranted to understand this correlation and to study the overall impact of NPs on bio-geochemical cycles.

Materials and Methods

Nanoparticle Characterization

Zero valent copper and zinc oxide NPs were purchased from Sun Innovations (Fremont, CA) and extensively characterized prior to use. NP crystal structure was determined by powder XRD using a PANalyticalXpert pro MPD (Westborough, MA) with CuKα radiation and a scan rate of 0.02° θ, from 10 to 85° 2θ. Six scans were collected and averaged prior to analysis. The PZC for Cu and ZnO was determined by measuring the point of zero net proton charge in a NaCl background electrolyte [43]. Titrations were carried out from pH 10−4.5 and 10−6.5 for Cu and ZnO, respectively. The oxidation state and local bonding environment of Cu and ZnO NPs were examined using X-ray absorption fine structure (XAFS) spectroscopy. The K-edge spectra were collected at beam line 10-BM (Materials Research Collaborative Access Team, Advanced Photon Source, Argonne National Laboratory, Argonne IL). NPs were diluted with polyvinylpyrrolidone (PVP) to a final concentration of 0.05%. The diluted samples were then compressed into pellets using a hand press and sealed between two strips of Kapton tape prior to analysis. Adsorption spectra were collected at the K-edge energies of 8979 and 9659 eV, respectively. Data collection was done in fluorescence mode using a 4-element solid-state Si-detector. The synchrotron was operated at 7.0 GeV at a nominal 100 mA fill current. The energy of a Si (111) double crystal monochromator was calibrated using an elemental Cu and Zn foil. Scans were collected from 8779–9979 eV and 9459–10659 eV for Cu and Zn, respectively. All spectra were collected at room temperature at ambient pressure with a minimum of three scans (and up to 5) collected for each sample. Particle size and shape were determined by field emission scanning electron microscopy (FESEM) (JEOL JSM-7000Scanning Electron Microscope, Peabody, MA), and scanning transmission electron microscopy (STEM)(JEOL JEM-2100F Transmission Electron Microscope, Peabody, MA). Samples for TEM and SEM analysis were prepared by evaporation of a dilute ethanol suspension. A 0.1 or 3 µL droplet was deposited onto a nickel mesh TEM grid or carbon tape, respectively and dried at 80°C for 1 h. The SEM operated at operated at 15 kV with a 2 nm resolution and the TEM was operated at 200 kV with a resolution of 0.1 nm. Nanoparticle characterization experiments were carried out in triplicate and the results represent an average of three measurements.

Soil Characterization

The soil used for the field study was obtained from an agricultural soil-distribution center (New York State; 40.76° N, 73.27° W), from a single mound of topsoil. When the experiments were initiated, the first 15–16 cm of the soil was frozen, and thus soil was taken from deeper in the mound. The soil was transported in plastic bags and any visible debris (plant matter, rocks, and wood chips) was removed manually before soil was added to the pots. A portion of the soil was sampled to determine the pH by measuring the supernatant of a 1:1 soil to Milli-Q water mixture. Similarly, the chemical composition of the soil was determined by inductively coupled plasma optical emission spectroscopy (ICP-OES) after carrying out microwave assisted nitric/hydrochloric acid digestion [44]. Briefly, a 0.2–0.5 g sample of the homogenized, dried soil was extracted via microwave-assisted digestion

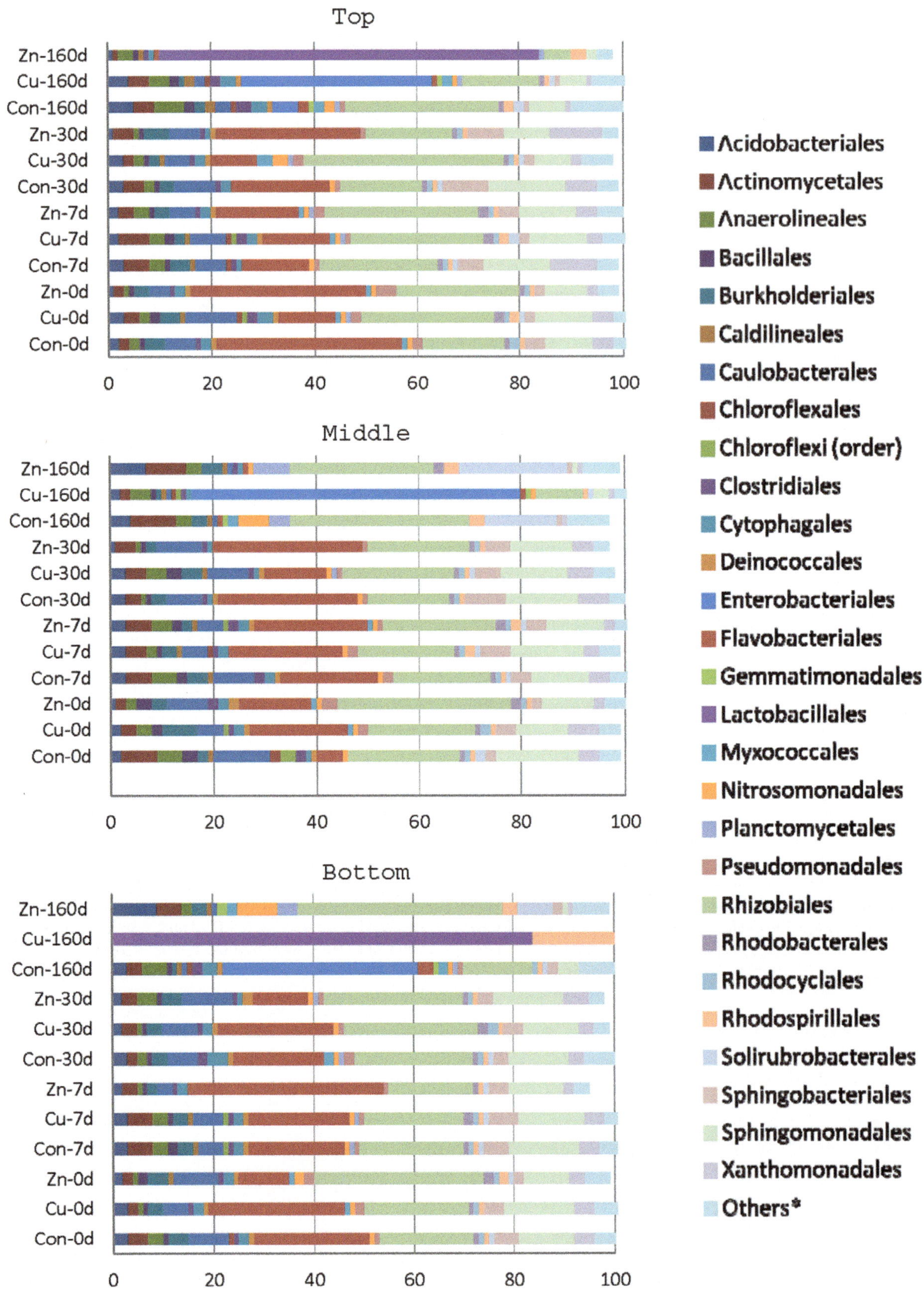

Figure 6. Relative abundance of bacterial orders (%) determined by pyrosequencingin the top, middle and bottom horizons of unexposed and nanoparticle-exposed soils. Horizontal colored bars represent different bacterial orders identified in soils from control (Con), Cu NP (Cu) and Zn NP (Zn) experimental pots treated for 160, 30, 7 and 0 days. *Others represents all of the orders in which the percentage was <1.

Table 4. % of Flavobacteriales, Rhizobiales and Sphingomonadales in control soil and that exposed to nanoparticles. (Con, Control; Cu, Cu nanoparticles; Zn, ZnO nanoparticles).

	Con-0d	Cu-0d	Zn-0d	Con-7d	Cu-7d	Zn-7d	Con-30d	Cu-30d	Zn-30d	Con-160d	Cu-160d	Zn-160d
Top horizon												
Flavobacteriales	60%	23%	52%	26%	25%	28%	38%	16%	52%	6%	5%	9%
Rhizobiales	26%	54%	36%	47%	53%	53%	33%	71%	32%	77%	70%	70%
Sphingomonadales	14%	23%	12%	27%	22%	19%	29%	14%	17%	18%	25%	21%
Mid horizon												
Flavobacteriales	11%	38%	24%	39%	41%	40%	47%	25%	48%	2%	6%	2%
Rhizobiales	51%	42%	58%	39%	35%	40%	29%	47%	32%	97%	72%	94%
Sphingomonadales	38%	21%	18%	22%	25%	20%	24%	28%	20%	1%	22%	4%
Bottom horizon												
Flavobacteriales	43%	44%	19%	35%	38%	58%	33%	38%	20%	14%	0%	1%
Rhizobiales	36%	34%	64%	38%	38%	25%	45%	44%	53%	65%	100%	98%
Sphingomonadales	21%	23%	18%	26%	24%	17%	22%	18%	26%	21%	0%	1%

and chemical extraction using boiling nitric and hydrochloric acid. The resulting extracts were analyzed by inductively coupled plasma optical emission spectroscopy (ICP-OES; Thermo Elemental IRIS Intrepid; Madison, WI). The accuracy ($\pm$5–7.5%) and precision ($\pm$3–12% depending on the element) of the instrument were verified before and after analysis. TOC, TKN and total phosphorus in the soil were analyzed commercially (Long Island Analytical Laboratories, New York) using EPA 9060, ASTM D3590-89 & 02(A) and SM 18–21 4500 – PE methods, respectively. pH and chemical composition of the soil were determined three times and averaged.

Soil Treatments

Plastic planting pots (23.6 cm×26.7 cm, Ames True Temper) containing a rolled rim and a saucer at the bottom were filled with soil to a height of 20.3 cm and transported to the field site (located close to the soil originating point at 40.72° N, 73.09° W). The area was secured to prevent unauthorized access. No specific permits were required from any agencies as the contaminated soil was never directly introduced into the environment. In addition, as Dowling College houses The Center for Estuarine, Environmental, and Coastal Oceans Monitoring (CEECOM) at this location, no specific permission was required to place the pots at this location. The field study did not involve endangered or protected species. After transport, the soil was allowed to stabilize for 7 days prior to adding the Cu or ZnO NPs (550 mg; starting on Jan. 19, 2011) by sprinkling over a 5 cm diameter at the center of the soil surface. Control pots received no NPs. At specified times (see results) sample pots were transported to the lab, and a 'core sample' was taken from the center by inserting a poly vinyl chloride (PVC) pipe (5 cm diameter). The cored soil sample was separated based on the depth with the top horizon (0−2.5 cm), 2nd horizon (2.5–7.5 cm) and 3rd horizon (7.5–20 cm). The soil from each horizon was mixed for 5 min to decrease soil heterogeneity. One pot each of control, Cu NPs, and ZnO NPs were removed at each time point. Time points were chosen as follows: short term incubation within a month of NP inoculation for the analysis of acute, local effects (1, 7 and 30 days), and long-term incubation (160 days) for long-term transport investigation. Weather data were recorded using Davis-Vantage Pro2 (Product #6312, Model #6152). The instrument was located no more than 2 m from the pots, ensuring it was not directly over them.

Solid Phase Analysis

After sampling, samples were freeze dried (Millrock Technology, Bench Top Model, Kingston, NY), fractionated based on particle size (2 mm−250 μm, 250–75 μm, and <75 μm) and stored in airtight containers prior to analysis. The concentration of Zn and Cu in each of the soil particle size fractions was determined by microwave-assisted acid digestion using the previously described methods.

The oxidation state and local bonding environment of the NPs in the soil were examined using XAFS spectroscopy as described except that PVP was not added. K-edge spectra were collected for the <75 μm NP fraction. In addition to the soil samples, mineral reference compounds were analyzed. Reference compounds included: metallic Cu, Cu_2O, CuO, $CuCO_3$, aqueous Cu^{2+}, $CuCl_2$, metallic Zn, ZnO, $ZnCO_3$, aqueous Zn^{2+}, and $ZnCl_2$. Reference scans were also collected for Cu and Zn adsorbed to ferrihydrite, γAl_2O_3, smectite, birnesite, and humic acid. Initial Cu and Zn concentrations were 500 mg L^{-1} and conducted at pH 6, in 0.01 M NaCl, and a suspension density of 2 g L^{-1}.

XAFS data processing and analysis were conducted using the Athena software package in the IFEFFIT computer program [45]. The collected fluorescence data were averaged and normalized to the edge jump height. The K-edge inflection point was determined using the energy at the maximum in the first derivative of the normalized spectra. Cu and Zn speciation in the soils was determined by linear combination fits utilizing a least squares procedure to fit the reference compounds to the sample spectra. During the fitting procedure, the energy shift of the reference compounds was constrained to $\pm$0.5 eV of the K-edge inflection point. The number of reference compounds to describe a soil samples was kept to a minimum and any reference compound returned with $<\sim$10% or negative concentrations was omitted from the fitting process.

Microbial CLPP and FAME Analysis

Microbial community physiological profiles were obtained in triplicate using Biolog® Ecoplates (Biolog®, Hayward, CA) using previously established methods [8]. The average of three absorption readings was employed and the values of control wells were subtracted. Occasional negative values were set to zero. Evenness (E) as well as S and H values were calculated as described [5].

Fatty acids were extracted from soil samples and analyzed by Microbial ID (Newark, DE).

DNA Isolation and 454-pyrosequencing

Soil samples (0.5 g) were first treated with ethidiummonoazide (EMA) as described [46] in order to preferentially amplify DNA from viable cells. DNA was extracted from soil (0.25 g) using Power Soil™ DNA isolation kits (MO BIO Laboratories Inc, Carlsbad, CA). DNA extraction was carried out in duplicate and the samples pooled prior to pyrosequencing. Bacterial tag-encoded FLX ampliconpyrosequencing (bTEFAP) was performed as described previously using the following bacterial primers: Gray28F 5′TTTGATCNTGGCTCAG and Gray519r 5′ GTNTTACNGCGGCKGCTG [47]. One-step PCR (30 cycles) was conducted using a mixture of Hot Start and HotStar (QIAGEN, CA, USA) high fidelity Taq polymerases. Amplicons originating and extending from the bacterial 28F primer were used for initial generation of the sequencing library. Tag-encoded FLX amplicon pyrosequencing analyses utilized a Roche 454 FLX instrument with titanium reagents and procedures were performed at the Research and Testing Laboratory (Lubbock, TX) based upon their protocols (www.researchandtesting.com). Subsequently, all failed sequence reads, low quality sequence ends, tags and primers were removed and any non-bacterial ribosomal RNA (rRNA) sequences and chimeras were removed using B2C2 [48] as previously described [47]. To curate the short reads (<150bp), sequences with ambiguous base calls and sequences with homopolymers>6bp>6bp were removed. To determine the identity of bacteria in the curated data, sequences were denoised, assembled into operational taxonomic units (OTUs) at 96.5% identity, and queried using a distributed. NET algorithm via Blastn+ (KrakenBLAST www.krakenblast.com) against the database of high quality 16S rRNA bacterial sequences. Using a. NET and C# analysis pipeline, the resulting outputs were compiled and data reduction analysis performed as described previously [49].

Based upon the sequence identity determined as the percent of the total length query sequence, aligned with a known sequence in the database, each sequence was assigned to the appropriate taxonomic level based upon the following criteria. Sequences with identity scores, to known or well characterized 16S rRNA sequences, greater than 97% identity were resolved at the species level, between 95–97% at the genus level, between 90–95% to the family and between 85–90% to the order level 80–85% to class and 77–80% to phyla. The percentage of sequences at each bacterial level were then used to determine the relative abundance within and among the individual samples based upon relative numbers of reads within each. Evaluations presented at each taxonomic level, including percentage compilations represent all sequences resolved to their primary identification or their closest relative [48].

Supporting Information

Figure S1 Cu (A) and Zn (B) K-edge XANES spectra for the untreated soil and Cu and Zn adsorbed to reference materials. The dashed red line represents the LCF best fit. For Cu the LCF results were 43% Cu adsorbed to alumina and 57% Cu adsorbed to humic Acid. For Zn the LCF results were 27% Zn adsorbed to ferrihydrite, 32% Zn adsorbed to smectite, and 41% Zn adsorbed to alumina.

Figure S2 Visualization of the NPs showing A) a representative TEM micrograph of Cu NPs B) a representative SEM image of ZnO NPs.

Figure S3 Linear combination fitting results for Cu K-edge A) normalized XANES spectra, B) normalized 1st derivative XANES spectra, and C) extended X-ray Adsorption Fine Structure (EXAFS) $\chi(k)k^2$ data. Relative percentage values are presented in Table 1.

Figure S4 Normalized and the first derivative of the normalized XANES spectra for ZnO nanoparticles, compared to ZnO and $ZnCO_3$ reference materials. Dashed grey lines in the figures highlight the energies corresponding to spectral features present in the ZnO NPs.

Figure S5 Alkalimetric titration curves for Cu and ZnO nanoparticles. The dashed lines indicate the point of zero charge. The grey-circled area indicates the pH region where metallic Cu is oxidized to Cu^{1+} and Cu^{2+}.

Figure S6 X-ray diffraction pattern of pre- and post-titration Cu nanoparticles. Different colored arrows indicate specific Cu phases. A break in the Y-axis occurs at an arbitrary value and was used to illustrate changes in the finer features of the XRD pattern.

Table S1 Elemental composition of the untreated soil determined by EPA Method 3051, at horizons 1–3 (0–3, 3–8 and 8–30 cm, respectively), expressed as mg/kg showing the relative standard deviation (RSD).

Table S2 Bacterial population present in the soil classified at order level. (Con, Control; Cu, Cu nanoparticles; Zn, ZnO nanoparticles).

Acknowledgments

We thank Mr. Sixto Portilla for his help in collecting weather data and Kristy Moniz for assistance with the DNA amplification. Any opinions expressed in this paper are those of the author(s) and do not, necessarily, reflect the official positions and policies of the USEPA. Any mention of products or trade names does not constitute recommendation for use by the USEPA.

Author Contributions

Conceived and designed the experiments: VS TL VW. Performed the experiments: DC NK SS VS TL VW. Analyzed the data: VS TL VW. Contributed reagents/materials/analysis tools: VS TL VW. Wrote the paper: DC NK SS VS TL VW.

References

1. Hill GT, Mitkowski NA, Aldrich-Wolfe L, Emele LR, Jurkonie DD (2000) Methods for assessing the composition and diversity of soil microbial communities. Appl Soil Ecol 15: 25–36.
2. Pampulha ME, Ferreira Mass, Oliveira A (2007) Effects of a phosphinothricin based herbicide on selected groups of soil microorganisms. J Basic Microbiol 47: 325–331.
3. Colwell RR (1997) Microbial diversity: the importance of exploration and conservation. J Industrial Micro Biotechnol 18: 302–307.
4. Kumar N, Shah V, Walker VK (2011) Perturbation of an arctic soil microbial community by metal nanoparticles. J Hazard Mater 190: 816–822.
5. Kumar N, Shah V, Walker VK (2012) Influence of a nanoparticle mixture on an arctic soil community. Environ ToxicolChem 31: 131–135.
6. Shah V, Belozerova I (2008) Influence of metal nanoparticles on the soil microbial community and germination of lettuce seeds. Water Air Soil Pollut 197: 143–148.
7. Hansch M, Emmerling C (2010) Effects of silver nanoparticles on the microbiota and enzyme activity in soil. J Plant Nutr Soil Sci 173: 554–558.
8. Ge Y, Schimel JP, Holden PA (2011) Evidence for negative effects of TiO_2 and ZnO nanoparticles on soil bacterial communities. Environ SciTechnol 45: 1659–1664.
9. Cullen LG, Tilston EL, Mitchell GR, Collins CD, Shaw LJ (2011) Assessing the impact of nano- and micro-scale zerovalent iron particles on soil microbial activities: particle reactivity interferes with assay conditions and interpretation of genuine microbial effects. Chemosphere 82: 1675–1682.
10. Kato S, Nakamura R, Kai F, Watanabe K, Hashimoto K (2010) Respiratory interactions of soil bacteria with (semi)conductive iron-oxide minerals. Environ Microbiol 12: 3114–3123.
11. Bai XD, Gao PX, Wang ZL, Wang EG (2003) Dual-mode mechanical resonance of individual ZnOnanobelts. ApplPhysLett82:4806–4808.
12. Rispoli F, Angelov A, Badia D, Kumar A, Seal S, et al. (2010) Understanding the toxicity of zero valent copper nanoparticle against *Escherichia coli*. J Hazard Mater 180:212–216.
13. Gystrzejewska-Piotrowska G, Golimowski J, Urban PL (2009) Nanoparticles: Their potential toxicity, waste and environmental management. Waste Manage 29: 2587–2595.
14. Biswas P, Wu C-Y (2005) Nanoparticles and the environment. J Air Waste Manage Assoc 55: 708–746.
15. Jacquat O, Voegelin A, Juillot F, Kretzschmar R (2009) Changes in Zn speciation during soil formation from Zn-rich limestones. Geochim Cosmochim Acta73: 5554–5571.
16. Jacquat O, Voegelin A, Kretzschmar R (2009) Soil properties controlling Zn speciation and fractionation in contaminated soils. Geochim Cosmochim Acta 73: 5256–5272.
17. Jacquat O, Voegelin A, Kretzschmar R (2009) Local coordination of Zn in hydroxy-interlayered minerals and implications for Zn retention in soils. Geochim Cosmochim Acta 73: 348–363.
18. Strawn DG, Baker LL (2008) Speciation of Cu in a contaminated agricultural soil measured by XAFS, mu-XAFS, and mu-XRF. Environ Science Technol 42: 37–42.
19. Strawn DG, Baker LL (2009) Molecular characterization of copper in soils using X-ray absorption spectroscopy. Environ Pollution 157: 2813–2821.
20. Voegelin A, Jacquat O, Pfister S, Barmettler K, Scheinost AC, et al. (2011) Time-dependent changes of Zinc speciation in four soils contaminated with Zincite or Sphalerite. Environ Science Technol 45: 255–261.
21. Midander M, Cronholm P, Karlsson HL, Elihn K, Moller L, et al. (2009) Surface characteristics, copper release, and toxicity ofnano- and micrometer-sized Copper and Copper(II) Oxide particles: A Cross-Disciplinary Study. Small 5: 389–399.
22. Nishimura S, Takagaki A, Maenosono S, Ebitani K (2010) In situ time-resolved XAFS study on the formation mechanism of Cu nanoparticles using Poly (N-vinyl-2-pyrrolidone) as a capping agent. Langmuir 26: 4473–4479.
23. Kosmulski M (2009) Compilation of PZC and IEP of sparingly soluble metal oxides and hydroxides from literature. Advances Colloid Interface Science152: 14–25.
24. Ben-Moshe T, Dror I, Berkowitz B (2010) Transport of metal oxide nanoparticles in saturated porous media. Chemosphere81:387–393.
25. Jeong S-W, Kim S-D (2009) Aggregation and transport of copper oxide nanoparticles in porous media. J. Environ. Monit.11: 1595–1600.
26. Kanel SR, Al-Abed SR (2011) Influence of pH on the transport of nanoscale zinc oxide in saturated porous media. J. Nanopart. Res. 13: 4035–4047.
27. Petosa AR, Jaisi DP, Quevedo IR, Elimelech M, Tufenkji N (2010) Aggregation and deposition of engineered nanomaterials in aquatic environments: role of physicochemical interactions. Environ Sci Technol 44: 6532–6549.
28. Zhou D, Keller AA (2010) Role of morphology in the aggregation kinetics of ZnO nanoparticles. Water Res 44: 2948–2956.
29. Scheckel KG, Luxton TP, El Badawy AM, Impellitteri CA, Tolaymat TM (2010) Synchrotron speciation of Silver and Zinc Oxide nanoparticles aged in a kaolin suspension. Environ Sci Technol44:1307–1312.
30. Yan F, McBratney AB, Copeland L (2000) Functional substrate biodiversity of cultivated and uncultivated A horizons of vertisols in NW New South Wales. Geoderma 96: 321–343.
31. Ramyadevi J, Jeyasubramanian K, Marikani A, Rajakumar G, Rahuman AA (2012) Synthesis and antimicrobial activity of copper nanoparticles. Materials Lett 71: 114–116.
32. Yoon K-Y, Byeon JH, Park J-H (2007) Susceptibility constants of *Escherichia coli* and *Bacillus subtilis* to silver and copper nanoparticles. Sci Total Env 373: 572–575.
33. Thill A, Zeyons O, Spalla O, Chauvat F, Rose J, et al. (2006) Cytotoxicity of CeO_2 nanoparticles for *Escherichia coli*. Physico-chemical insight of the cytotoxicity mechanism. Environ Sci Technol 40: 6151–6156.
34. Liu Y, He L, Mustapha A, Li H, Hu ZQ, et al. (2009) Antibacterial activities of zinc oxide nanoparticles against *Escherichia coli* 0157: H7. J Appl Microbiol 107: 1193–1201.
35. Lu J, Domingo JWS, Lamendella R, Edge T, Hill S (2008) Phylogenetic diversity and molecular detection of bacteria in gull feces. Appl Env Microbiol 74: 3969–3976.
36. Angelakis E, Raoult D (2010) The increase of *Lactobacillus* species in the gut flora of newborn broiler chicks and ducks is associated with weight gain. PLoS One 5: e10463.
37. Feare CJ, Sanders MF, BlascoF, Bishop JD (1999) Canada goose (*Brantacana-densis*) droppings as a potential source of pathogenic bacteria. J Royal Society for the Promotion of Health 119: 146–155.
38. Glunder G (1981) Occurrence of Enterobacteriaceae in feces of Garnivorous Passeriform birds. Avian Diseases 25: 195–198.
39. Lauber CL, Hamady M, Knight R, Fierer N (2009) Pyrosequencing-based assessment of soil pH as a predictor of soil bacterial community structure at the continental scale. Appl Environ Microbiol 75: 5111–5120.
40. Peterson SB (2008) Interactions among rhizosphere bacteria. PhD Thesis. The University of Wisconsin-Madison.
41. Haggstrom JA, Klabunde KJ, Marchin GL (2010) Biocidal properties of metal oxide nanoparticles and their halogen adducts. Nanoscale 2: 399–405.
42. Simonet BM, Valcarcel M (2009) Monitoring nanoparticles in the environment. Anal Bioanal Chem 393: 17–21.
43. Zelazny LW, He L, Vawormhoudt A (1996) Charge analysis of soils and anion exchange. In Sparks DL editor. Methods of Soil Analysis: Chemical Methods of Soil Analysis, Part 3, Madison, WI: Soil Sci. Soc. Of Am. Book Series. 1231–1253.
44. USEPA (1986) Metalitc analytes. 3.2, Sample preparation methods. Method 3050/3051. In Acid digestion of sediments, sludges, and soils. USEPA: Washington, DC. Vol. Volume 1., Section A. Part I. Chapter 3.
45. Ravel B, Newville M (2005) ATHENA, ARTEMIS, HEPHAESTUS: data analysis for X-ray absorption spectroscopy using IFEFFIT. JSynchrotron Radiat12: 537–541.
46. Pisz JM, Lawrence JR, Schafer AN, Siciliano SD (2007) Diffrentiation of genes extracted from non-viable versus viable microorganisms in environmental samples using ethidium monoazide bromide. J Microbiol Methods 71: 312–318.
47. Callaway TR, Dowd SE, Wolcott RD, Sun Y, McReynolds JL, et al. (2009) Evaluation of the bacterial diversity in fecal contents of laying hens fed various molting diets by using bacterial tag-encoded FLX ampliconpyrosequencing. PoultSci 88:298–302.
48. Gontcharova V, Youn E, Wolcott RD, Hollister EB, Gentry TJ (2010) A Windows-based doftware for batch fepletion of chimeras from Bacterial 16S rRNA gene datasets. Open Microbiol J 4: 47–52.
49. Dowd SE, Callaway TR, Wolcott RD, Sun Y, McKeehan T, et al. (2008) Evaluation of bacterial diversity in the feces of cattle using 16S rDNA bacterial tag-encoded FLX ampliconpyrosequencing (bTEFAP). BMC Microbiol 8: 125.

5

Convergent Adaptations: Bitter Manioc Cultivation Systems in Fertile Anthropogenic Dark Earths and Floodplain Soils in Central Amazonia

James Angus Fraser[1,2]*, Alessandro Alves-Pereira[3,4], André Braga Junqueira[5,6], Nivaldo Peroni[7], Charles Roland Clement[4,5]

1 Departamentos de Antropología e Historia, Universidad de los Andes, Bogotá, Cundinamarca, Colombia, **2** Department of Anthropology, University of Sussex, East Sussex, United Kingdom, **3** Post-graduate Program in Genetics and Plant Breeding, Universidade de São Paulo, Piracicaba, Estado de São Paulo, Brazil, **4** Laboratório de Evolução Aplicada, Universidade Federal do Amazonas, Manaus, Amazonas, Brazil, **5** Coordenação de Tecnologia e Inovação, Instituto Nacional de Pesquisas da Amazônia, Manaus, Amazonas, Brazil, **6** Centre for Crop Systems Analysis/Technology and Agrarian Development Group Wageningen University, Wageningen, Gelderland, The Netherlands, **7** Departamento de Ecologia e Zoologia, Universidade Federal de Santa Catarina, Florianópolis, Santa Catarina, Brazil

Abstract

Shifting cultivation in the humid tropics is incredibly diverse, yet research tends to focus on one type: long-fallow shifting cultivation. While it is a typical adaptation to the highly-weathered nutrient-poor soils of the Amazonian *terra firme*, fertile environments in the region offer opportunities for agricultural intensification. We hypothesized that Amazonian people have developed divergent bitter manioc cultivation systems as adaptations to the properties of different soils. We compared bitter manioc cultivation in two nutrient-rich and two nutrient-poor soils, along the middle Madeira River in Central Amazonia. We interviewed 249 farmers in 6 localities, sampled their manioc fields, and carried out genetic analysis of bitter manioc landraces. While cultivation in the two richer soils at different localities was characterized by fast-maturing, low-starch manioc landraces, with shorter cropping periods and shorter fallows, the predominant manioc landraces in these soils were generally not genetically similar. Rather, predominant landraces in each of these two fertile soils have emerged from separate selective trajectories which produced landraces that converged for fast-maturing low-starch traits adapted to intensified swidden systems in fertile soils. This contrasts with the more extensive cultivation systems found in the two poorer soils at different localities, characterized by the prevalence of slow-maturing high-starch landraces, longer cropping periods and longer fallows, typical of previous studies. Farmers plant different assemblages of bitter manioc landraces in different soils and the most popular landraces were shown to exhibit significantly different yields when planted in different soils. Farmers have selected different sets of landraces with different perceived agronomic characteristics, along with different fallow lengths, as adaptations to the specific properties of each agroecological micro-environment. These findings open up new avenues for research and debate concerning the origins, evolution, history and contemporary cultivation of bitter manioc in Amazonia and beyond.

Editor: Lee A. Newsom, The Pennsylvania State University, United States of America

Funding: The research was funded by Leverhulme Trust (Grant F/00 230/W) and the CNPq (Brazilian National Research Council) (CT Amazonia 575588/08-0). The funders had no role in study design, data collection and analysis, decision to publish, or preparation of the manuscript.

Competing Interests: The authors have declared that no competing interests exist.

* E-mail: james.angus.fraser@gmail.com

Introduction

Shifting cultivation has been a predominant mode of traditional agriculture in the rainforests of the lowland Neotropics [1], humid Africa [2], the Indian subcontinent [3] and parts of South East Asia [4] and Oceania [5] for thousands of years. Shifting cultivation in the humid tropics is incredibly diverse [6], yet both empirical studies and theoretical discussions tend to focus on one type: long-fallow shifting cultivation (LFSC). LFSC is an *extensive* form of crop production, which entails a relatively short cropping period of 1–3 years, followed by a longer fallow period (normally 10–20 years), with land cleared by slashing and burning fallow vegetation (or sometimes mature forest) [7]. The long fallow period is necessary because soils of the *terra firme* (non-flooded upland plateaus) in the tropical world tend to be highly-weathered infertile Oxisols and Ultisols (US classification system). Hence, this form of land-use is spatially (fields must shift frequently) and temporally (fields must be fallowed for years before they can be cultivated again) *extensive*. Extensive systems are widely held to be well adapted to such soils because a nutrient flush provided by the burning of secondary or mature forest is used for a single cropping period, before swidden fields are left to fallow in order to restore fertility via the growth of secondary forest [8]. It has long been assumed that any intensification of LFSC (i.e., a reduction of fallow periods) in the humid tropics will cause crop yields to decline, as infertile Oxisols and Ultisols are not capable of withstanding a sustained reduction of fallow periods, leading to the eventual collapse of the system. These assumptions about the prevalence and precariousness of extensive shifting cultivation have influenced theories of cultural development in Amazonia, where Steward [9] and Meggers [10] held that the small scattered

Amerindian settlements practicing LFSC found in the interfluvial *terra firme* today represent an optimal adaptation to a low productivity environment. While their theories have long since gone out of fashion [11], studies of the cultivation of the Amazonian staple bitter manioc (*Manihot esculenta* Crantz) continue to focus almost exclusively on LFSC practiced by current Native Amazonian populations [12] and other traditional populations [13] in the poor soils of marginal interfluvial *terra firme* environments.

In Amazonia, however, the most populated areas – both today and in the late pre-Columbian period – are along major white-water rivers, such as the Solimões, Amazon and Madeira, all of which have nutrient-rich floodplain soils used for agriculture and adjacent paleo-floodplains that are somewhat more nutrient-rich than the typical Oxisols and Ultisols [14], as well as areas of Anthropogenic Dark Earths [15]. These regions are among the most suitable for human habitation and population growth, since they are characterized by an abundance of fish as well as fertile soils. Students of pre-Columbian Amazonia have long emphasized the agricultural potential of these regions. Lathrap [16] argued that the productivity of the white-water floodplains underwrote the growth of large and dense human settlements, sparking competition and warfare between groups for these rich but spatially and temporally limited areas. However, although the flood pulse provides a yearly flush of nutrients, extreme floods once every decade or so can destroy crops, settlements and even the high floodplain itself, which makes these environments relatively unpredictable. Denevan [17] argued that agriculture began as multi-cropping in floodplain environments and developed later in forested uplands as intensive multi-cropping systems with short fallow. For him, long fallow, extensive cultivation is a late development associated with the introduction of efficient forest clearance tools, especially metal axes and machetes.

Denevan recognized that large scale settlements could not have been supported exclusively by seasonal floodplain cultivation, given the riskiness of producing food in areas subjected to floods. He proposed that settlements would preferentially be located on bluffs along the river margins, allowing the exploitation of both the fertile floodplain zones and the *terra firme* areas that were safe from flooding but located on less fertile soils [17,18]. This occupation pattern would ultimately result in the transformation of upland soils: many bluffs in central Amazonia feature fertile anthropogenic soils known as Anthropogenic Dark Earths (ADE) [19–21]. These soils are associated with Native Amazonian settlements of the late pre-Columbian period (2000–500 years before present) [22,23], and are most abundant and largest in whitewater regions, because these are where the largest populations were located in pre-Columbian times [24]. Amazonian ADE form through human inputs of organic and inorganic matter [e.g., biomass wastes, manure, bones, ash, charcoal and ceramics] [15]. Amazonian ADE are enduringly fertile in part due to their historical enrichment in highly stable black carbon, which has a half-life of 1000 years [25]. ADE exhibit much higher levels of chemical elements essential for plant growth, such as phosphorous, calcium, magnesium, zinc and manganese, than the weathered and infertile soils in which they were formed. ADE sites exhibit a highly fertile 'core area', which grades into more subtly modified soils, with a continuum of fertility between them and surrounding soils [26,27]. It is likely that ADE were appreciated for plant cultivation by pre-Columbian Native Amazonian peoples, just as they are by Amazonian people today [28].

These historical arguments point to the possibility that local people may also intensify shifting cultivation when circumstances permit today. Fertile soils, such as ADE and floodplain soils, provide farmers with the opportunity to intensify shifting cultivation, increasing both crop productivity and frequency of cultivation [29], raising the carrying capacity of the landscape [30]. Therefore, the study of swidden systems on fertile soils allows us to examine the extent to which fallow lengths, bitter manioc landraces, and other aspects of swidden systems vary on different soils. While numerous studies have addressed many dimensions of shifting cultivation, few have compared swidden systems on different soils and in different ecological contexts within a circumscribed geographical area [31].

The Amazonian Staple: Bitter Manioc

Manioc landraces are classified as either bitter or sweet depending on cyanogenic glucoside (CG) content. In Central Amazonia, bitter manioc is the staple crop today, whilst sweet manioc is only of secondary importance [32]. Sweet manioc has low CG content in their tuberous roots ($<$100 ppm fresh weight), while bitter manioc has larger amounts of cyanogenic glycosides ($>$100 ppm fresh weight) [33]. There are an estimated 7000 landraces of manioc worldwide [34], but this is surely an underestimate given farmers' continual selection and propagation of new landraces. Differences in the color, form and size of leaves and stems, in the size, number, color and cyanide content of roots, and in the rate of growth of tuberous roots are determined by genetic differences, although the environmental conditions in which plants grow also affect manioc phenotypes, including CG content [33]. Farmers comprehend a "landrace" as a set of individuals sharing particular morphological characteristics that differentiate them from other landraces; they also distinguish them by giving them a particular name [35,36]. These morphological characteristics are shaped by genotype-environment interactions that can result in different phenotypic expressions of the same genotype, and create morphotypes that are identified as distinct landraces [37]. Landraces are the result of generations of farmer selection in local environments, and are therefore well adapted to local growing conditions, which has been demonstrated in South America [13,38,39], in humid Africa [40] and Asia [41]. Several students of manioc have asserted that bitter manioc does not yield well in fertile soils [42–44], perhaps because it has been so thoroughly studied on nutrient-poor soils. However, about 30% of Central Amazonian manioc, both bitter and sweet, is produced in the floodplain [45]. It was observed that on highly fertile soils manioc plants tend to invest more in aboveground biomass than in the tubers [29,43]. However, there are bitter manioc landraces that yield well in the floodplain and in ADE, possibly even better than in Oxisols and Ultisols [46].

Manioc is vegetatively propagated via stem cuttings, which grow into plants that are genetically identical to the mature plants from which they were cut. However, manioc retains its ability to reproduce sexually and produces seeds that lie dormant in fallow vegetation [47]. When fallows are cleared for cultivation, seeds are stimulated to germinate by increasing temperatures caused by both soil exposure and the heat of the burn phase in swidden-fallow systems [48], and seedlings appear [49]. When cuttings from seedlings are incorporated into a landrace, its genetic diversity is increased because it becomes a polyclonal landrace [36,48,50–52]. This increased genetic diversity provides the raw material for adaptation to new conditions, such as ADE and the floodplain, as different clones in the landrace have different possibilities for adaptation to different agro-ecosystems. The extent to which volunteer seedlings are incorporated is variable in modern bitter manioc cultivation in the Neotropics. Studies in Guyana [53], French Guiana [54] and the Atlantic Forest in Brazil [55] found a relatively high level of incorporation. However,

Stocker found only a small amount of seedling incorporation amongst farmers in Pará, in eastern Brazilian Amazonia ([56]:162–63). Nonetheless, this is a recurring pattern in many traditional farming systems in Amazonia and beyond [13].

This article examines the hypothesis that contemporary Amazonian people have developed divergent bitter manioc cultivation systems as adaptations to the properties of different soils in the landscapes that they inhabit. We test this hypothesis by comparing swidden systems in fertile (ADE and floodplain) and infertile (Oxisols and Ultisols) soils with respect to their assemblages of manioc landraces, the performance (from the farmer's point of view) and productivity of the most common landraces in different soils, along with the genetic relationships between the most common landraces, and length of fallow periods. Hence, we will examine both the adaptations of manioc landraces and of swidden-fallow systems. In earlier publications [46,57,58], we hypothesized, based on initial interpretations of ethnobotanical data, that "weak" bitter manioc landraces grown in ADE originated in the floodplain, given farmers' knowledge and similar adaptations (fast maturation and low starch content).

Results and Discussion

Farmer theory and practice

In the preliminary stage of research we conducted multi-sited participant observation in numerous communities of the study region (Figure 1). Subsequent quantitative work was driven by hypotheses generated during this stage. While qualitative data have been presented at length elsewhere [57], here we summarize major findings and their interplay with quantitative data. We found that, contrary to what was expected from the literature [29,31,43,59–61], bitter manioc was widely cultivated in ADE. More specifically, farmers stated that not only did they *i*) perceive that certain landraces yield better in certain soils (the basis for our *pri* index and yield observations, see below), but that this informed their *ii*) selection of different landraces for planting in different soils (the basis for the *la* index), and *iii*) that fertile soils (ADE/ floodplain) could be farmed with much shorter fallow periods than infertile soils (the basis for our fallow length data).

Most intriguingly, however, was the finding that farmers expressed these perceptions and behaviors in the form of a local theory of strength and weakness (the categories strong and weak are present in traditional agriculture in other areas of the world ([62]:148)). Along the middle Madeira River, the categories weak and strong express the suitability of different landraces for planting in different soil-successional scenarios. Briefly, weak landraces are fast yielding (5 months–1 year), but rot if left too long in the ground, and produce less starch than stronger landraces. Strong landraces, on the other hand, are slow yielding (1–3 years), but produce more starch than weaker landraces. Farmers claim that weak landraces are best suited to planting in "weak" land (land with young fallow), whilst strong landraces are said to be suited to "strong" land (land with older fallow). Elsewhere, we reported that landraces described as weak were planted more frequently in ADE and the floodplain, whilst those described as strong were planted more frequently in Oxisols and Ultisols ([63]:400). The specifics of this local theory have been elaborated at length [46,57,58], and here we would just emphasize two points. This local theory would appear to provide evidence of the cognitive aspects of adaptation – local people have theorized the emergent properties of their adaptive knowledge and practices – and express these using the simple yet powerful metaphors of strength and weakness. Secondly, this theory is not limited to the middle Madeira; an independent study recently found an identical theory of weak and strong manioc amongst floodplain and *terra firme* manioc farmers in and around the Brazilian Sustainable Development Reserves of Mamirauá and Amanã, close to the town of Tefé on the middle Solimões River [64]. The fact that the same local theory exists in two localities hundreds of kilometers apart suggests that this theory is likely to inform adaptive aspects of manioc farming in various locations throughout Amazonia.

Diversity of manioc landraces

A total of 50 landraces were found, with 29 in ADE, 20 in Oxisols, 20 in Ultisols and 23 in the floodplain. Most of the landraces cultivated on the *terra firme* are planted in more than one type of soil: among the 29 landraces cultivated in ADE, 18 (62%) are also planted in Ultisols and 19 (65.5%) in Oxisols. Ultisols and Oxisols shared 75% (15) of their landraces. On the other hand, most of the landraces that occurred in the floodplains (19, or 82.6%) only occurred in this environment, with only four landraces shared with other types of soil (one shared with Ultisols and ADE, and three that occurred in all four types of soil). The number of landraces cultivated in each village varied between 5 (Verdum) and 21 (Vista Alegre), and tended to be lower in floodplain villages ($N = 5$; mean $\pm$ standard deviation: 8.2 ± 3.1) than villages located on the *terra firme* ($N = 4$; 15.7 ± 5.0), principally because villages on the *terra firme* have access to more than one type of soil (and, in the case of Água Azul, even cultivate in the floodplain). The average number of landraces cultivated in each manioc field was similar among the different types of soil (ADE: 2.8 ± 1.9; Oxisols: 3.4 ± 1.5; Ultisols; 3.2 ± 2.0; Floodplain: 3.1 ± 1.3) (Dataset S1).

Along the Upper Negro River, communities commonly have 60 to 89 landraces, somewhat higher when compared to other local communities in Amazonia and in the Atlantic Forest in Brazil (53–58; [13]). The diversity of bitter manioc landraces on the *terra firme* along the middle Madeira is quite low, although higher than what Grenand reported along the Cuieiras River near Manaus (six) [65]. Perhaps this is due to a greater market orientation in the Madeira River communities, and higher population pressure. By contrast, the number of landraces we registered on the floodplain is much higher than reports for Careiro Island (three; [65]) at the confluence of the Negro and Solimões Rivers, and for communities between Santarém and Óbidos (maximum of four), along the middle Amazon River [66], although Pereira [67] reported 43 bitter manioc landraces at the confluence of the Solimões and Japurá Rivers. The generally low numbers in the literature most likely reflect the lack of attention to manioc in the floodplain compared to the *terra firme*, as pointed out by Denevan [68] for studies of agriculture in the floodplain in general. Various studies mention fast maturing manioc in the floodplain, without reporting the number of landraces [69–71].

Landrace composition of manioc fields in different soils

The composition of manioc landraces in fields on different types of soil is significantly different, which is shown both in the general model (NPMANOVA; $F = 16.46$, $p = 0.01$) and also in all possible pairwise comparisons (six) between the four different soil types (all with p values lower than 0.05, corrected for multiple comparisons) (Figure 2). In general, manioc fields in Oxisols and Ultisols cluster together, while in Floodplains they form an almost completely separate group, with fields in ADE occupying an intermediate area in the ordination space (Figure 2). Also, fields in Oxisols and Ultisols are more homogeneous, while in Floodplains and ADE they are more heterogeneous, especially in the latter case. The areas of overlap and the dispersion of some points through the figure indicate that the composition of manioc landraces is very

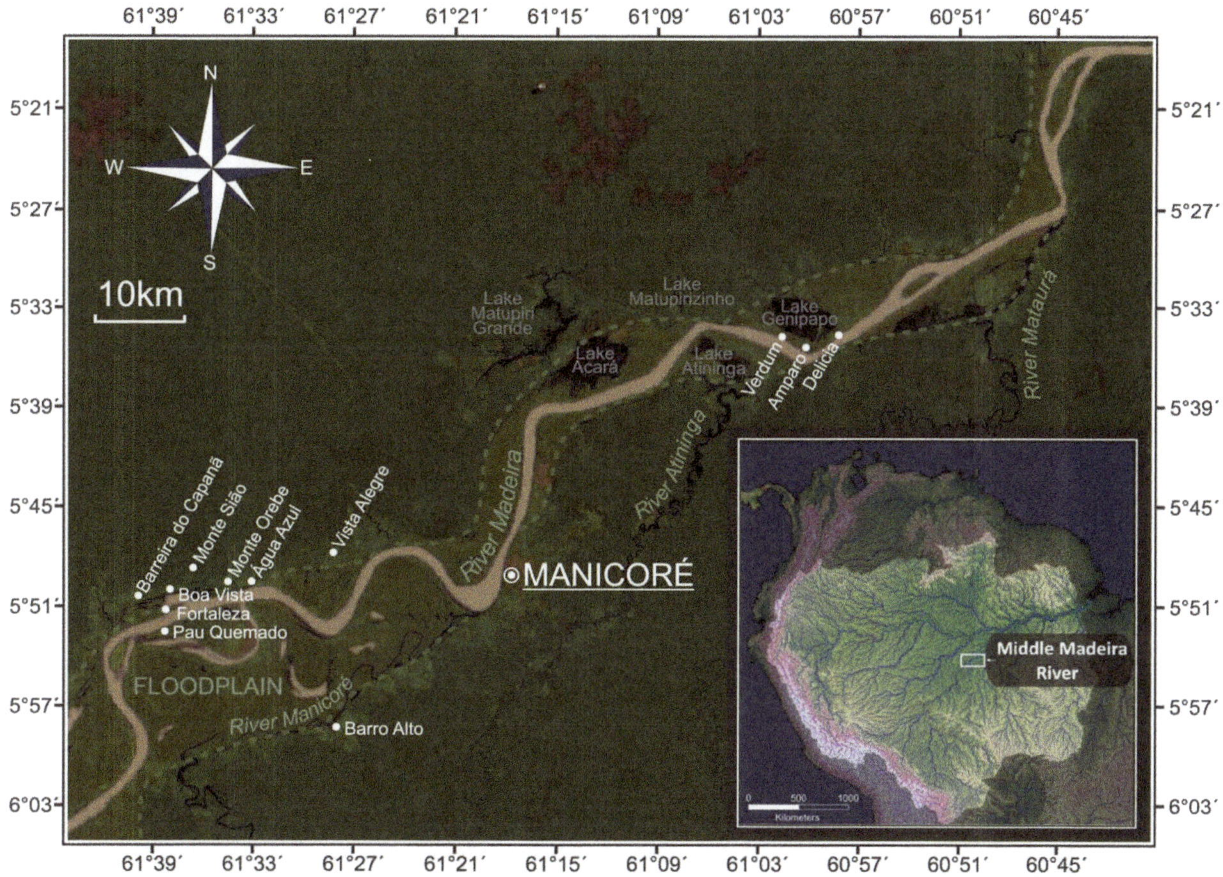

Figure 1. Map of the middle Madeira River region, Amazonas State, Brazil, showing communities where research was carried out. The inset map shows the location of the middle Madeira River in Northern South America. Map by Victoria Frausin.

heterogeneous and highlights the practice of landrace exchange between soils. Within-field heterogeneity is commonly reported [12,13,38,72], as is exchange among farmers [35,36,38–40,48,73], but exchange among farmers on different soil types has not been reported previously, perhaps because the heterogeneity of soils has been less carefully studied than that of bitter manioc landraces.

Farmer perceptions of landrace performance

A Performance Ranking Index (*pri*) for each landrace in each type of soil on the *terra firme* was calculated. Most of the landraces mentioned by the farmers in the interviews in all three soils on the *terra firme* (ADE, Oxisol and Ultisol) have a very low *pri* score, either because they are rarely cited in that specific soil or because farmers do not rank them among the best performers in that soil (Figure 3) (Dataset S1). *Roxinha*, *Tartaruga*, *Arroz* and *Aruari* are among the five landraces with higher *pri* in all three soils. However, while the landraces *Jabuti* and *Arroz* are considered the best performers in Oxisols and Ultisols, in ADE this place is occupied by *Tartaruga*, *Pirarucu Branco* and *Roxinha Branca* (Figure 3).

There were strong positive correlations between the area occupied by each landrace in the fields (*la*) and their performance (*pri*) in the soils (Pearson's Correlation - Ultisol: $\rho = 0.901$; Oxisol: $\rho = 0.912$; ADE: $\rho = 0.898$). This high degree of correlation between *pri* and *la* supports the notion that farmers' planting behavior is shaped by their perceptions of the performance of different landraces in different types of soil. While this pattern is common and has been mentioned elsewhere (e.g. [74]:50), the correlations between this pair of indices permits a quantitative demonstration of this behavior.

Farmer estimates of manioc yields

Our farmer yield observations reveal that there are significant differences between yields in five of the six most popular landraces when planted in different soils along the middle Madeira River (Table 1). These findings support our inference of a high degree of adaptation in bitter manioc systems found on different soils, since landraces that perform better in particular soils tend to be more predominant in those soils, as farmers respond positively to good yields by planting more of specific landraces. This is reflected in the *la* and *pri* indices: *Jabuti* and *Roxinha* are significantly more predominant in Oxisols and Ultisols than in ADE, while *Tartaruga* and *Pirarucu Branco* are more predominant in ADE than in Oxisols or Ultisols.

Fallow length and the intensification of swidden systems

Intensification in swidden-fallow systems refers, among other factors, to a shortening of fallow length [75]. Intensification on infertile soils can result in exhaustion of fertility and breakdown of the system. Fertile soils provide opportunities for sustainable reduction of fallow lengths. The average fallow length for 55 fields on Ultisols was 13.2 ± 8.1 years and for 64 Oxisol fields the average length was 21.3 ± 15.4 years, while for 71 ADE fields it was 6.5 ± 6.8 years and for 59 floodplain fields it was 1.6 ± 4.0. There were significant differences between fallow lengths in different soils

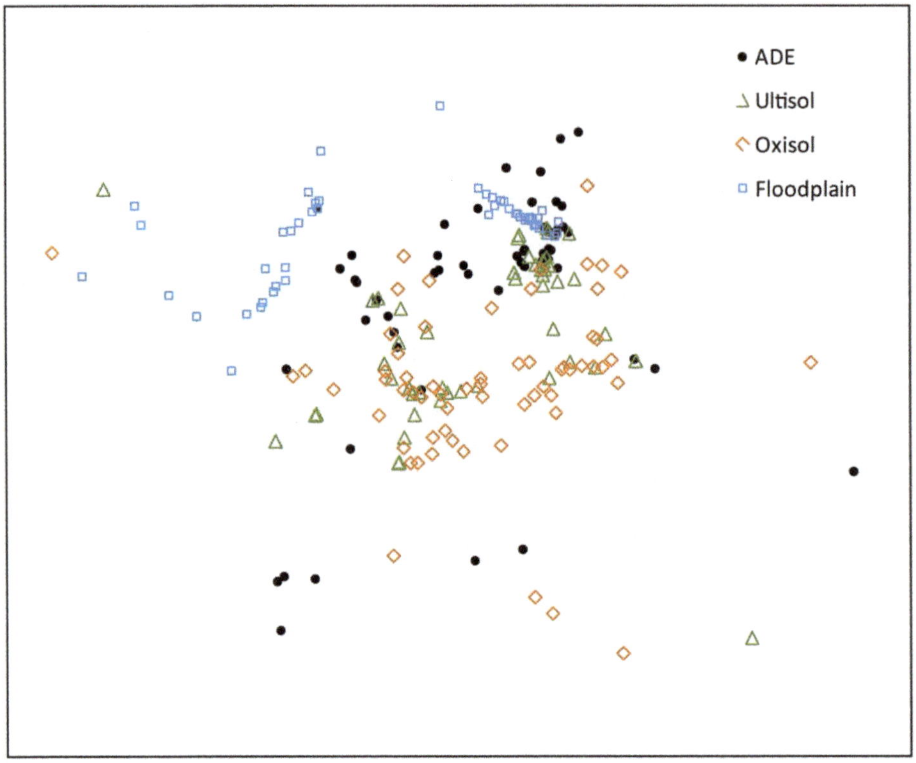

Figure 2. Non-metric multidimensional scaling (NMDS) showing the composition of bitter manioc landraces sampled in fields on Ultisols, Oxisols, Anthropogenic Dark Earths (ADE) and on floodplains along the middle Madeira River (percentage of explanation of the bidimensional model: 35.8%). Each point represents a bitter manioc field, and its position in the graph is a bidimensional representation of the Bray-Curtis dissimilarity between the fields.

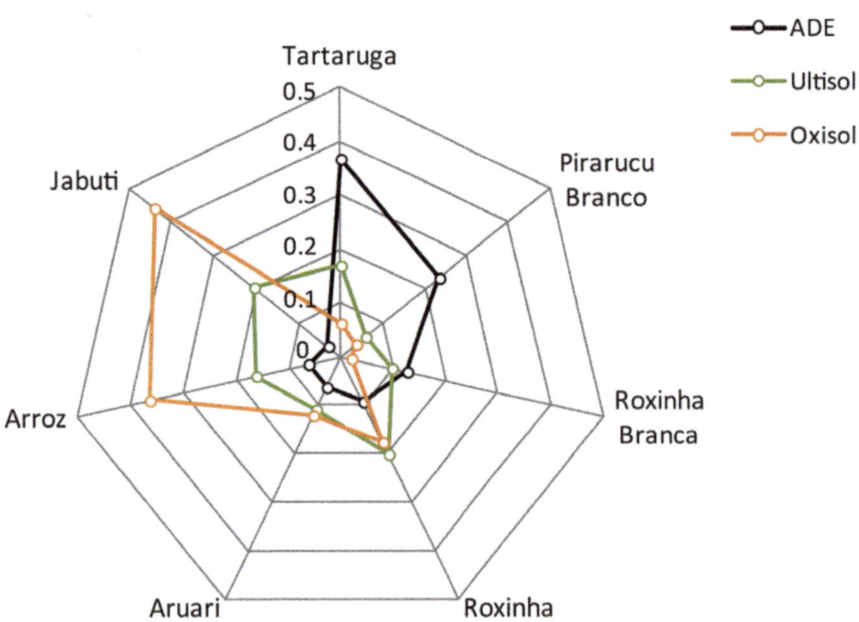

Figure 3. Farmer perceptions of performance for the six most frequent manioc landraces cultivated in Oxisols, Ultisols and Anthropogenic Dark Earths (ADE) along the middle Madeira River. Numbers in the axis indicate the Performance-Rank Index (*pri*) of each landrace in each type of soil, calculated based on the perception of 162 farmers interviewed on the Middle Madeira. Dots indicate the value of *pri* for each variety in the three types of soil.

Table 1. Farmer-recorded production of 50 kg sacks of manioc flour that would be obtained from approximately 0.5 ha of swidden plot at various localities along the Middle Madeira River, Central Amazonia, Brazil, in 2007–2008.

Manioc landrace	ADE		Oxisol/Ultisol		Floodplain		ANOVA	
	n	(avg ± SD)	n	(avg ± SD)	n	(avg ± SD)	F	p
Jabuti	5	24.2±5.1	6	44.2±8.6			20.6	0.001
Arroz	4	31±2.6	4	30.5±4.2			0.04	ns
Tartaruga	6	44.4±6.1	7	26.7±3.5			43.0	<0.001
Roxinha	5	25.2±3.11	7	40.1±4.3			43.5	<0.001
Pirarucu Branco	4	53.2±7.0^{a}	4	25.7±4.6^{b}	5	55±11.2^{a}	15.8	0.001
Pirarucu Amarelo	4	66.7±24.9^{a}	4	30.5±4.4^{b}	4	70.5±11.3^{a}	7.6	0.012

Superscript characters indicate significant differences in pairwise comparisons using Tukey's HSD test ($p<0.05$).

(ANOVA, $F=44.51$, $p<0.001$, while Tukey's post hoc test revealed significant differences between all possible pairwise comparisons except ADE vs Floodplain). It is important to emphasize that local factors, such as population density, also play important roles in shaping fallow lengths, and these are discussed at length for each locality in the Supporting Information, along with other factors (Text S1).

Generation, selection and exchange of landrace diversity

Local farmers identify and exchange genetic diversity of bitter manioc landraces [38,72]. Along the middle Madeira River, we observed that certain people that we categorized as "key individuals" play an important role in the exchange of new landraces originating from seedlings, conservation of existing landraces and knowledge associated with them. Key individuals were identified as being the individuals who most informants mentioned during open interviews as being responsible for introducing new landraces and keeping a stock of all landraces present in each community. Interviews with these key individuals revealed that they consciously recognize new phenotypes and are always intentionally experimenting with new landraces from seedlings and with clonal material from their kin in other communities [13]. These key individuals perform a vital role in the identification and distribution of new manioc genetic diversity, because they try out new landraces in local micro-environments and, if they prove to be exceptional, are responsible for their distribution within communities, and to kin in other communities. The fact that each locality only has a few key individuals points to the critical role that they play in the circulation of exceptional landraces throughout the region, but some new genotypes can circulate unconsciously mixed with other individuals that are morphologically similar.

At the six localities examined in this study, both the incorporation of seedlings into pre-existing landraces and the creation of new landraces from seedlings are relatively common practices. Farmers recognize that seedlings often appear in the newly burnt fields *before* cuttings have been planted and are morphologically different from the planted landraces, as they tend to grow taller and only have a single tuberous root growing straight downwards. Most farmers (53–66% across communities) interviewed simply ignore or weed out the seedlings, while others (11–32%, including, but not restricted to key individuals) take cuttings from mature seedlings and plant them separately (Table 2). When mature, these volunteers are either: i) incorporated into an existing landrace, e.g., *Jabuti*; ii) incorporated as a sub-landrace of an existing landrace (such as *Jabuti-Preto*); iii) named as a combination of two landraces (such as *Jabuti-Arroz*); or iv) established as a new landrace with a different name.

All the landraces analyzed with microsatellite markers had observed heterozygosities (H_O) higher than expected heterozygosities (Table 3), which appears to be related to both the incorporation of seedlings and the selection of heterozygous individuals [76]. For example, the landrace *Arroz* from ADE at Barreira do Capanã had an observed heterozygosity of 0.495 and 5 different multi-locus genotypes (MLGs) in the 20 individuals examined. The landraces *Jabuti* from Oxisols and *Tartaruga* from floodplain had 7 and 6 different MLGs, and H_O of 0.567 and 0.707, respectively. The number of MLGs reflects either the incorporation of seedlings or the unintentional "mixing" of landraces (which may happen, for example, when they are morphologically similar), both of which contribute to increasing heterozygosity. Incorporation of seedlings into recognized landraces increases intra-landrace heterogeneity, because farmers tend to select the largest and healthiest volunteers, which tend to be the most heterozygous [53,54]. These practices are critical in maintaining and amplifying the genetic diversity of manioc landraces, and in adapting landraces to new environmental conditions, such as different soils.

Along the middle Madeira River the performance in different soils is a major factor in varietal selection by farmers (Figure 4). The selection of distinct traits in different environments where the landraces are cultivated may be correlated to the genetic differentiation found among the bitter manioc landraces grown in different soil types (Figure 5; [77]). Evidence of genetic divergence among landraces cultivated in floodplain from those cultivated in ADE and infertile soils was also found when analyzing intra-varietal genetic diversity. In all examples the landraces grown in ADE and infertile soils had the same most common MLGs. On the other hand, landraces grown in the floodplain, except *Pirarucu Amarelo*, had distinct MLGs from those with the same landrace name but grown in ADE or infertile soils. Additionally, in the floodplain some farmers observed that volunteers occur more frequently in higher areas of the floodplain. These areas are not flooded every year and the seeds have time to be acted upon by agro-ecological management. The planting of manioc in different zones, the flood regime, farmer perceptions and the combination of landraces they select drive the generation of new landraces in the floodplain. This is consistent with the somewhat higher number of landraces from the floodplain observed in this study, and may be related to the genetic differentiation among the landraces grown in the floodplain from those grown in upland soils (Figure 4).

Table 2. Number of informants, both total and key individuals for manioc management, number of landraces cultivated, and the ways that farmers manage seedlings at six communities along the middle Madeira River, municipality of Manicoré, Amazonas, Brazil.

	Informants		Number of landraces	Seedling Management		
	Total	Key		Intentionally incorporate	Incorporate at random	Remove
Barro Alto	37	3	12	0.32	0.14	0.54
Barreira do Capanã/Boa Vista	29	6	19	0.17	0.14	0.66
Água Azul	13	4	9	0.23	0.31	0.46
Vista Alegre	11	4	21	0.27	0.09	0.55
Água Azul floodplain	9	3	13	0.11	0.33	0.55
Genipapo floodplain	15	4	14	0.26	0.20	0.53

Farmer selection for agronomic performance? Comparative perspectives

Our findings call for re-evaluation of the widespread notion that farmer identification and selection of manioc landraces is weakly associated with agronomic performance. This notion is based on Boster's research on the perceptual distinctiveness of manioc amongst the Aguaruna Jivaro of the Peruvian Amazon. Boster concluded that "there are significant differences between the [manioc] cultivars in their responses to environmental factors…", but "the Aguaruna are generally unaware of or unconcerned with these differences" ([72]:40). Boster's own garden experiments did, however, show that "some cultivars grow well in sandy alluvial soils on river islands and do poorly on other sites, while other cultivars yield about the same regardless of soil conditions." Boster also noted that the Cashinahua, another Native Amazonian group that resides in Peru and Brazil, do appear to select particular landraces for planting in certain kinds of soil, according to their well-known ethnographer, Ken Kensinger [78]. The Aguaruna Jivaro reside in hill country, where there is certainly a lot of soil variation and there will be some high quality soils, and the Cashinauhua live along the fertile whitewater Juruá River (Tarauacá, Jordão, Breu, Muru, Envira) and in the Purus basin (Upper Purus and its affluent the Curanja) [79,80]. Each of these areas offers a variety of very different soils to the manioc farmer. However, even in the Upper Negro Region, an area notorious for infertile soils, according to Wilson [81], there appears to be non-random planting of sweet manioc in more fertile soils by Tukanoan manioc farmers. Heckler and Zent, on the other hand, describing research amongst the Piaroa in Venezuela, found that "specific genotype-micro-condition interactions were demonstrably not a significant factor in determining planting patterns" ([82]:684). Similarly, Salick and colleagues, reporting on a study amongst the Amuesha in Peruvian Amazonia, found that "soil seemed to have little predictive power for cassava phenotype occurrence." ([38]:11).

Table 3. Indices of intra-varietal genetic diversity [Observed (H_O) and expected (H_E) heterozygosities, and number of multi-locus genotypes (No. MLGs)] for five bitter manioc landraces cultivated in different soil types in the middle Madeira River region, based on variation detected with 10 microsatellite markers.

Landrace	Soil	Locality	H_O	H_E	No. MLGs	MLGs
Pirarucu Branco	ADE	Barreira do Capanã	0.495	0.250	2	**A**, g
	FP	Pau Queimado	0.505	0.255	2	**B**, h
Tartaruga	ADE	Barro Alto	0.503	0.268	2	**B**,F
	OX	Barro Alto	0.507	0.276	4	**B**, F, i, j
	FP	Verdum	0.707	0.361	5	**C**, k, l, m, n
Arroz	ADE	Barreira do Capanã	0.495	0.460	5	A, **D**, o, p, q
	OX	Água Azul	0.503	0.257	2	**D**, r
Jabuti	OX	Barreira do Capanã	0.567	0.445	6	D, **E**, s, t, u, v
Pirarucu Amarelo	ADE	Água Azul	0.607	0.306	2	**F**, x
	FP	Água Azul	0.607	0.307	1	**F**

Capital letters indicate the MLGs that were present in more than one landrace. Boldfaced letters indicate the most common MLG for a given landrace, while small letters indicate MLGs present at low frequencies.

The findings of Boster, Heckler and Zent, and Salick et al. explain the persistence of the notion that farmer identification and selection of manioc landraces is weakly associated with agronomic performance. Our findings, however, when combined with those of Kensinger and Wilson, support the conclusion that the way in which *at least some* Amazonian farmers consciously identify and select landraces *is* shaped by agronomic performance. Indeed, the perceptual weak-strong theory of manioc detailed above and elsewhere appears to be built upon and indeed express in schematic/metaphoric form local knowledge of selection for agronomic performance. This raises questions regarding the distribution of practices of selection for agronomic performance, and which factors (such as the presence/absence of fertile soils, indigenous/non-indigenous farmers etc.) contribute to the presence of these practices and associated knowledge in different parts of Amazonia. We conjecture that selection for agronomic performance would be more likely to emerge amongst people inhabiting areas with greater heterogeneity of soil types.

Conclusions

We examined manioc cultivation in four different soil types at six localities, and found that farmers plant different assemblages of bitter manioc landraces in different soils and that shorter fallow

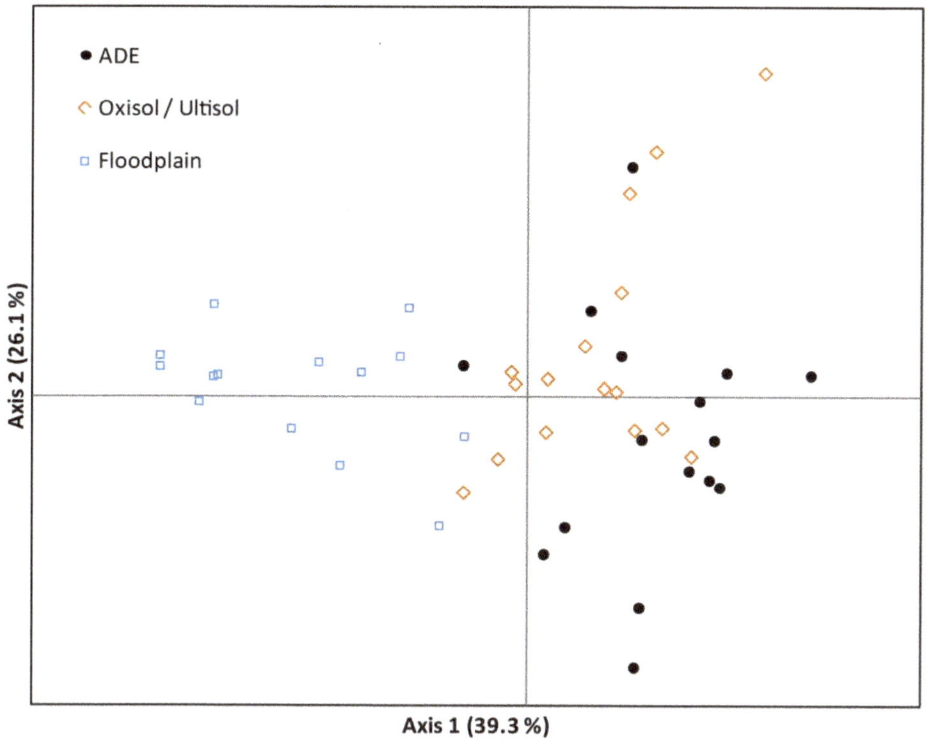

Figure 4. Principal Coordinates Analysis based on diversity revealed by 10 microsatellites markers showing the dispersion of 48 swiddens from three soil types (17 in ADE soils, 14 in floodplain soils and 17 in Oxisols/Utilsols) in six communities along the middle Madeira River. The two coordinates together explain 65.4% of the variation in the matrix.

lengths were strongly associated with more fertile soils. Most popular landraces were shown to exhibit significantly different yields when planted in different soils. From this we can infer that farmers have selected different sets of landraces with different perceived agronomic characteristics, along with different fallow lengths, as adaptations to the specific properties of each agroecological micro-environment. On the *terra firme*, intentional selection of certain landraces for cultivation in certain soils was demonstrated by the close relationship between *Performance Ranking Index* and *Landrace Area* for landraces in different soil types. These findings suggest that the local theory revolving around metaphors of strength and weakness may represent a cognitive manifestation of adaptive farming behavior: a way to express in simple terms local knowledge of the relationship between different sets of bitter manioc landrace traits and soil properties and fallow stages. Although landraces grown in ADE and the floodplain share similar phenotypical characteristics, specifically fast maturation and low starch content, and are both described as "weak" by

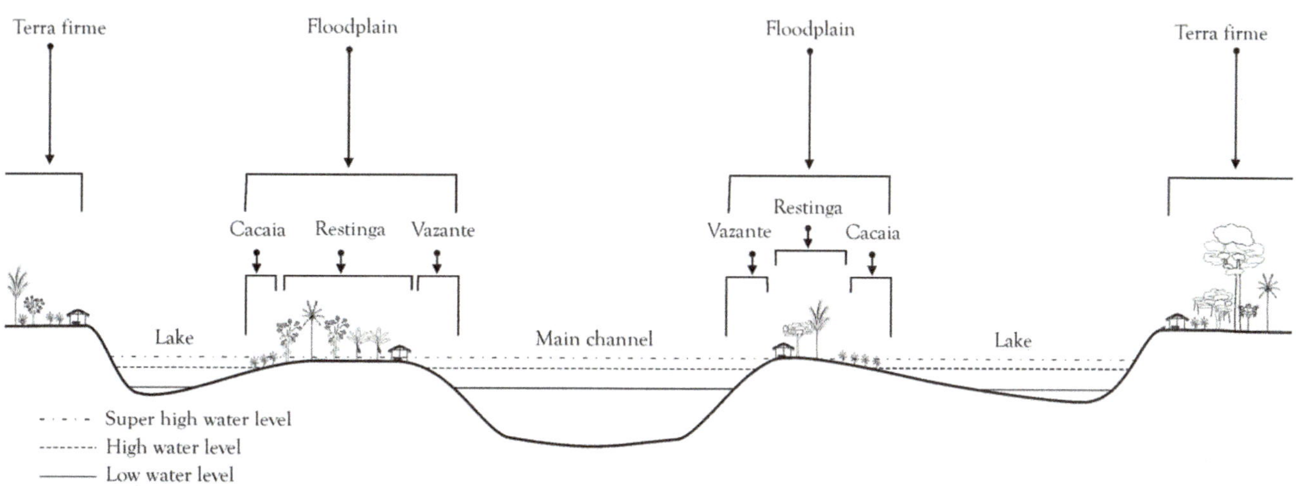

Figure 5. Schematic representation of floodplain zones relative to the main channel, lakes and *terra firme* (non-flooded upland plateaus). *Vazante* is the local term on the middle Madeira for the banks of the main channel, *Restinga* for the high levee floodplain and *Cacaia* for the back-swamp area. Drawn by Victoria Frausin.

farmers, these groups of landraces were not genetically more related to each other, as we had predicted from ethnobotanical observation. Rather, the landraces grown in upland soils (ADE and Oxisols/Ultisols) are more related to each other and genetically differentiated from floodplain landraces. For landraces cultivated in the floodplain and in ADE, the selection for convergent adaptive traits appears to be associated with the similar ecological adaptations to nutrient-rich soils and short periods for rapid growth and yield before stress. The stresses are quite different, however: floods in the floodplain and enhanced weed pressure in ADE. Nonetheless, they yield convergent adaptive results.

If the similar ecological adaptations observed for ADE and floodplain landraces are outcomes of adaptive convergence of traits directed by farmer's selection, what drives that selection? On ADE, farmer selection for fast maturing landraces coupled with a reduction in fallow periods is likely to be driven by population pressure and permitted by ADE's high nutrient availability. When ADE is cultivated at Barro Alto and Água Azul, where population pressure exerts more of an influence, the most intensive swidden-fallow systems were found (Text S1). These intensive systems were associated with a reduction in the diversity of landraces in general, suggesting that when farmers reduce fallow lengths they may select only the fastest maturing landrace that is widely available for planting. In the floodplain, selection for fast maturing landraces is certainly a strategy to cope with the high seasonality of the floodplain determined by the flood pulse. Hence two different swidden-fallow systems have developed as convergent adaptations to intensification in nutrient-rich anthropogenic and floodplain soils. This intensification may be driven by population pressure (*terra firme*) and the flood pulse (floodplain); alternatively, owing to their greater fertility, it may be that the floodplain and ADE are the only places that offer the opportunity to produce food and/or money in a shorter time period – and this could also be a driver for the selection of these common traits.

Our findings broaden our understanding of the diversity of bitter manioc swidden-fallow systems in contemporary Amazonia, and allow some ethnographic projection. Firstly, the intensive short-fallow shifting cultivation in the floodplain and on bluffs with ADE that we reported along the middle Madeira are clearly modern analogs of Denevan's [17,18] hypothesis that pre-conquest food production systems were much more intensive than the extensive long-fallow shifting cultivation common today, which he argues is a post-conquest adaptation. Secondly, Arroyo-Kalin [32] recently hypothesized that during the pre-Columbian period manioc was selected for high toxicity in swiddens in the infertile agricultural hinterland further from settlements, but farmers were selecting for lower toxicity in fertile ADE soils forming in and around homegardens. The phenomenon of farmer selection being driven by landrace agronomic performance in different soils would support the possibility that farmer selection in the pre-Columbian period for planting in different environments was a key driver in the emergence of "bitter" and "sweet" manioc and the continuum of toxicity that underlie them. Given the importance of bitter manioc cultivation to our understanding of pre-Columbian and contemporary populations in Central Amazonia, we hope that this study will encourage further research into the under-investigated area of intensified bitter manioc cultivation in fertile soils.

Materials and Methods

Study area and sampling design

We focus on the middle Madeira River, Amazonas State, Brazil (Figure 1), where traditional farmers plant their staple crop, bitter manioc, in infertile Oxisols and Ultisols of the *terra firme*, and in fertile ADE and floodplain soils. Amazonian Dark Earths are easily recognizable by their very dark brown or black coloring, high fertility, and pottery shards. The other *terra firme* soils were classified in the field according to local ethnopedological knowledge and physical properties (color, presence of potsherds, etc.). It was found that the local ethnopedological category "*barro*" (clay), recognized by its red/yellow coloring and high clay content in the A horizon, is broadly coterminous with Oxisols, while the local category "*areia*" (sand), recognized by dark brown, grey or black coloring and highly friable and "sandy" (although probably pseudosands) A horizon is broadly coterminous with Ultisols [26]. Along the Madeira River, farmers recognize three major zones in the floodplain: i) the highly fertile sides of the main channel, locally known as the *vazante*; ii) the high floodplain, known as the *restinga*; iii) and the back-swamp area, known as the *cacaia* (Figure 5). These terms are local and identify fewer categories than Denevan [68] observed along the floodplain of the Ucayali River, in eastern Peru, between the towns of Pucallpa and Panaillo.

Both the collection of plant material and the interviews took place after prior informed consent was obtained at each community. Anthropological fieldwork was authorized by a scientific expedition (EXC 022/05) granted by the Brazilian National Research Council (CNPq). Four *terra firme* localities (the second and third comprised of several communities) were selected for semi-structured quantitative interviews on bitter manioc cultivation on ADE, Oxisols and Ultisols: Barro Alto, Barreira do Capanã/Boa Vista, the Água Azul Coast, and Vista Alegre (Figure 1). These localities were selected because they had the greatest numbers of farmers cultivating this crop on ADE [83]. All farmers cultivating bitter manioc on ADE in these localities at the time of research were interviewed. We selected an equivalent number of families farming Oxisols and Ultisols using the snowball method whereby new families were enrolled through those who had already been interviewed ([84]:184–85). Two floodplain localities, each comprising several communities, were selected. Upstream from Manicoré families resident in the floodplain communities Fortaleza and Pau Queimado, and a few families at Água Azul and Monte Sião who had fields in the floodplain were selected. Downstream from Manicoré are the communities Verdum, Amparo and Delícia. These communities are among the longest established floodplain communities on the middle Madeira River. All farmers present at these floodplain communities during visits were interviewed. Key individuals who maintained the highest number of bitter manioc landraces were identified through open-interviews with different community members. In order to minimize gender/age biases, and to verify information from multiple sources, interviews were conducted in the household, normally with all of the family present, as family members would often intervene to correct the orator or provide additional information.

Participant observation and open interviews

JAF conducted extensive participant observation in all communities of the study area prior to and during subsequent quantitative work (from September 2006 to March 2008), in order to build rapport and open up areas of interest that cannot be gleaned from close reading of the literature [85]. This entailed engaging in daily activities, including planting, harvesting and processing manioc, and conducting open, unstructured interviews with local people. The advantage of this type of initial qualitative approach is that it places no limitations on and is non-reductive in relation to reality. This open-ended approach led to the generation of the hypothesis that this paper addresses (that there are different adaptive manioc systems in different soils) and facilitated the discovery of the local theory of weakness and strength. The disadvantage of such an unstructured qualitative approach is that it cannot test the hypotheses it generates, and this is why we generated various novel quantitative indicators to measure aspects of bitter manioc cultivation systems from which adaptability could be inferred (planting behaviour, perceptions of landrace performance in different soils, yields and fallow lengths).

Ethnobotanical Data, Indices and Analyses

In total, we conducted 249 semi-structured interviews at the six localities between May 2007 and March 2008. Quantitative ethnobotanical data were gathered on a) the area occupied by each manioc landrace (landrace area – *la* – see below) in fields on four soil types (ADE, Ultisols, Oxisols and Floodplain), b) perceptions of relative performance (performance ranking index – *pri* – see below) of the same landraces in the three different *terra firme* soils, c) fallow lengths, and d) incorporation of seedlings. In order to test if fields on different soils exhibit different compositions of landraces, data on *la* were ordered through a Non-Metric Multidimensional Scaling (NMDS) based on the Bray-Curtis dissimilarity, followed by non-parametric multivariate analyses of variance (MANOVA; [86]). Non-parametric MANOVAs were performed for the general model (i.e., including all soil types simultaneously) and also for the six possible pairwise comparisons between the soil types. Critical p-values were adjusted by the Holm correction for multiple comparisons [87]. The relationship between *la* and *pri* was investigated using simple correlations. Composite soil samples were collected in each field and the results from chemical analysis were used for an *a posteriori* classification of soil types [26].

Landrace Area Index and Fallow Lengths

In order to identify the different landraces cultivated and quantify their abundance in the fields, we relied on the local nomenclature for the landraces and on numerical estimates provided by each farmer, using a local unit commonly used for quantifying manioc stem cuttings. In general, farmers along the middle Madeira recall the amount of each landrace they plant in terms of the number of bundles ("*feixes*") of each landrace they planted in their fields. We asked each farmer (a) which landraces he/she cultivated, (b) how many bundles of each landrace he had planted in his/her current manioc field and (c) how big was his/her field. Using this information we calculated the area occupied by each landrace in each field (*la*) through the formula:

$$la = \frac{nbv}{nbT} \times arf$$

where *la* is the *landrace area* index, *nbv* is the number of bundles of a given landrace planted in a given field, *nbT* is the total number of bundles planted in that field, and *arf* is the area of the field. Since we did not observe differences in the spacing used to plant the landraces, we regard the area occupied by each landrace (*la*) as a good estimator of the abundance of each landrace in the field. When all landraces are taken into account, we interpret this data as the varietal composition of the fields. We collected data on fallow lengths by asking each farmer how old the fallow was that he/she had cleared in order to establish the field. Data for the calculation of *la* and on fallow lengths were collected with 249 farmers at all six localities, of which 190 farmers on the *terra firme* and 59 on the floodplain.

Performance Ranking Index and seedling management

During participant observation it emerged that farmers claimed that certain landraces performed better in particular soils [83]. In order to quantify this, we asked all 190 *terra firme* farmers (a) to name all landraces that they cultivated in each type of soil and then (b) to rank the landraces mentioned according to their performance in that specific soil. Based on Sutrop's "Cognitive Salience Index" [88], we combined data on the frequency with which a given landrace was cited, its mean position in the performance ranking and calculated a Performance Ranking Index (*pri*) of each landrace in each type of soil, using the formula:

$$pri = \frac{F}{(N \times mP)}$$

where *pri* is the Performance Ranking index, *F* is the frequency of citations of the landrace, *N* is the number of farmers interviewed who cultivate in this specific type of soil, and *mP* is the mean position of the landrace in the rankings.

Participant observation also revealed that some farmers purposefully separate cuttings from volunteer seedlings for later planting, others simply harvest them together with other landraces (resulting in their random incorporation into pre-existing landraces), and others purposefully exclude volunteers [83]. Once these three categories were established, the incorporation of seedlings was measured by asking 114 farming families if they: a) *separated* seedlings, that is took cuttings from seedlings and planted them apart to see how they did; b) *randomly* incorporated seedlings, where they do plant cuttings of seedlings volunteers, but randomly mixed up with the other landraces; or c) *removed* them, cutting the volunteers out as weeds when they appeared in the field.

Farmer landrace yield estimates

Thirty one farmers were selected to record the production of different bitter manioc landraces in mono-varietal patches within 43 fields on the four different soil types. They were selected on the basis of literacy, their expressed interest in participating in the study (i.e., willingness to commit to recording production), and, crucially, the fact that they had mono-varietal patches within manioc fields during the study period. Bitter manioc fields are harvested bit by bit depending on labor availability and demand for manioc flour for subsistence consumption and sale. Amazonian farmers also do not tend to have accurate scales. They take a great interest in the amount of manioc flour produced, however, since it is equivalent to, as one woman put it, their "daily bread" (e.g., carbohydrate staple) and additionally an important product for sale in markets. In order to measure production therefore we asked farmers to record how many sacks of manioc flour (each sack is 50 kg) they produced from each mono-varietal-landrace patch. Fields were visited and we did not record significant differences in spacing, so we consider them to be all planted at similar density (around 1 meter between plants – c. 5000 plants per 0.5 ha). The size of mono-varietal patches varied (mean 0.286 ha±0.123 SD), so we corrected each one up to 0.5 ha in order to make the data commensurate and comparable using ANOVA (Table 1).

Collection of genetic data

The collection of plant material for genetic analyses was carried out in 2009. Authorization for interviewing farmers was obtained from *Instituto Nacional de Pesquisas da Amazônia*'s Committee for Research Ethics (protocol 235/09) and our collecting was registered in the System for Authorization and Information on Biodiversity, coordinated by the Chico Mendes Institute for

Biodiversity, of the Ministry of the Environment (number of register: 10020-5). No proprietary traditional knowledge was accessed, allowing us to meet Resolution 21 requirements for basic research that does not require authorization from Brazil's Council for Genetic Patrimony (CGEN in the Brazilian acronym), which was consulted before field work.

To investigate how the genetic diversity of bitter manioc was organized in different environments of cultivation we collected leaf samples of each landrace present in each of a total of 48 swiddens in all the localities except Vista Alegre: 17 in ADE soils, 14 in floodplain soils and 17 in Oxisols/Utilsols. A total of 184 individuals were sampled, representing 43 different bitter manioc landraces. The genetic variation of landraces was accessed using ten microsatellite loci [89,90], which are short sequences of repetitive DNA. Seven of them (GA21, GA126, GA131, GA134, GA136, GA140, GAGG5) were described by Chavarriaga-Aguirre *et al.* [89] and three (SSRY13, SSRY89, SSRY 164) by Mba *et al.* [90]. Each microsatellite locus refers to a unique genomic region and the genetic variation results from differences in the number of repetitive units among individuals.

To investigate the extent of genetic variation within landraces and the genetic identity of landraces with the same name but grown in different soil types, some of the most commonly cultivated bitter manioc landraces in the region (*Pirarucu Branco*, *Tartaruga*, *Arroz*, *Jabuti* and *Pirarucu Amarelo*) had 20 or 30 individuals sampled (Dataset S2). To evaluate the dispersion of bitter manioc swiddens along the genetic diversity revealed by microsatellites, a Principal Coordinate Analysis (PCoA), based on Euclidian distances, was carried out with GenAlEx v.6.4 [91]. Among the parameters used to describe intra-varietal genetic diversity, we estimated the observed (H_O) and expected (H_E) heterozygotes with GenAlEx v.6.4, while the number of multi-locus genotypes (MLGs) was analysed with GenClone v.2.0 [92]. H_O varies from 0 to 1, and corresponds to the probability that a given microsatellite locus has two different numbers of repetitions in a given individual. H_E also varies from 0 to 1, and corresponds to the proportion of heterozygote individuals that were expected under the Hardy-Weinberg Equilibrium based on the number of different microsatellite forms found for a given landrace. The MLGs describe the number of different microsatellite combinations present in a given group of individuals (in this case, the landraces). The detailed sampling strategy, material and methods and other genetic results are described elsewhere [76,77].

Supporting Information

Dataset S1 Landrace area and performance ranking index.

Dataset S2 Intra-varietal genotypes and soil differentiation genotypes.

Text S1 Manioc cultivation at the community level.

Acknowledgments

We extend our heartfelt thanks to the people of the Middle Madeira River for participating in our research. JAF thanks the CNPq (Brazilian National Research Council) for granting research permission (EXC 022/05). AAP thanks the *Coordenação de Aperfeiçoamento de Pessoal de Nível Superior*—CAPES for a master's scholarship. CRC thanks CNPq for a research fellowship (*bolsas de produtividade*). NP thanks CAPES for a research fellowship (PRODOC).

Author Contributions

Conceived and designed the experiments: JAF AAP CRC NP. Performed the experiments: JAF AAP NP ABJ. Analyzed the data: JAF AAP NP ABJ CRC. Wrote the paper: JAF AAP NP ABJ CRC.

References

1. Arroyo-Kalin M (2011) Slash-burn-and-churn: Landscape history and crop cultivation in pre-Columbian Amazonia. Quaternary International 249: 4–18.
2. Neumann K, Bostoen K, Höhn A, Kahlheber S, Ngomanda A, et al. (2011) First farmers in the Central African rainforest: A view from southern Cameroon. Quaternary International 249: 53–62.
3. Kingwell-Banham E, Fuller DQ (2011) Shifting cultivators in South Asia: Expansion, marginalisation and specialisation over the long term. Quaternary International 249: 84–95.
4. Barker G, Gilbertson D, Reynolds T, editors (2012) Rainforest Foraging and Farming in Island Southeast Asia: The Archaeology and Environmental History of the Niah Caves, Sarawak. Cambridge: Cambridge University Press.
5. Denham T, Haberle S (2008) Agricultural emergence and transformation in the Upper Wahgi valley, Papua New Guinea, during the Holocene: theory, method and practice. Holocene 18: 481–496.
6. Brookfield H, Padoch C (1994) Appreciating agrodiversity: A look at the dynamism and diversity of indigenous farming practices. Environment 36: 6–21.
7. Conklin H (1961) The study of shifting cultivation. Current Anthropology 2: 27–61.
8. Nye PH, Greenland DJ (1960) The Soil under Shifting Cultivation. Harpenden, UK: Technical Communication 51, Commonwealth Bureaux of Soils. 156 p.
9. Steward JH, editor (1948) Handbook of South American Indians. Volume 3: The Tropical Forest Tribes. Washington, D.C: Bulletin 143, Bureau of American Ethnology, Smithsonian Institution. 567 p.
10. Meggers BJ (1971) Amazonia: Man and Culture in a Counterfeit Paradise. Washington: Smithsonian Institution Press. 194 p.
11. Stahl PW (2002) Paradigms in paradise: revising standard Amazonian prehistory. The Review of Archaeology 23: 39–51.
12. Rival L, McKey D (2008) Domestication and diversity in manioc (Manihot esculenta Crantz ssp. esculenta, Euphorbiaceae). Current Anthropology 49: 1119–1128.
13. Emperaire L, Peroni N (2007) Traditional management of agrobiodiversity in Brazil: A case study of manioc. Human Ecology 35: 761–768.
14. Irion G (1984) Sedimentation and sediments of Amazon rivers and evolution of the Amazon landscape since Pliocene times. In: Sioli H, editor. The Amazon - Limnology and Landscape Ecology of a Mighty Tropical River and its Basin. The Hague: Springer. pp. 201–204.
15. Glaser B, Birk JJ (2012) State of the scientific knowledge on properties and genesis of Anthropogenic Dark Earths in Central Amazonia (terra preta de Indio). Geochimica et Cosmochimica Acta 82: 39–51.
16. Lathrap DW (1970) The Upper Amazon. Southampton: Thames & Hudson. 256 p.
17. Denevan WM (2001) Cultivated Landscapes of Native Amazonia and the Andes; Clark G, Goudie A, Peach C, editors. Oxford: Oxford University Press. 400 p.
18. Denevan WM (1996) A bluff model of riverine settlement in prehistoric Amazonia. Annals of the Association of American Geographers 86: 654–681.
19. Glaser B, Woods WI, editors (2004) Amazonian Dark Earths: Explorations in Space and Time. Berlin: Springer. 216 p.
20. Lehmann J, Kern DC, Glaser B, Woods WI, editors (2003) Amazonian Dark Earths. Origins, Properties and Management. Dordrecht: Kluwer Press. 523 p.
21. Woods WI, Teixeira WG, Lehmann J, Steiner C, WinklerPrins AMGA, et al. (2009) Amazonian Dark Earths: Wim Sombroek's Vision. Berlin: Springer. 504 p.
22. Neves EG (2008) Ecology, ceramic chronology and distribution, long-term history, and political change in the Amazonian floodplain. In: Silverman H, Isbell WH, editors. Handbook of South American Archaeology: Springer. pp. 359–403.
23. Heckenberger M, Neves EG (2009) Amazonian archaeology. Annual Review of Anthropology 38: 251–266.
24. Petersen JB, Neves EG, Heckenberger MJ (2001) Gift from the past: *terra preta* and prehistoric occupation in Amazonia. In: McEwan C, Barreto C, Neves EG, editors. Unknown Amazon Culture in Nature in Ancient Brazil. 1st ed. London: The British Museum Press. pp. 86–107.
25. Kuzyakov Y, Subbotina I, Bogomolova I, Xu XL (2009) Black carbon decomposition and incorporation into soil microbial biomass estimated by ^{14}C labeling. Soil Biology & Biochemistry 41: 210–219.
26. Fraser J, Teixeira W, Falcão N, Woods W, Lehmann J, et al. (2011) Anthropogenic soils in the Central Amazon: From categories to a continuum. Area 43: 264–273.

27. Schmidt MJ (2010) Reconstructing tropical nature: Prehistoric and modern anthrosols (*terra preta*) in the Amazon rainforest, upper Xingu river, Brazil [PhD Dissertation]: Department of Geography, University of Florida. 1594 p.
28. Erickson CL (2008) Amazonia: The Historical Ecology of a Domesticated Landscape. In: Silverman H, Isbell WH, editors. The Handbook of South American Archaeology. New York: Springer. pp. 157–183.
29. German LA (2003) Historical contingencies in the coevolution of environment and livelihood: contributions to the debate on Amazonian Black Earth. Geoderma 111: 307–331.
30. Smith NJH (1980) Anthrosols and human carrying capacity in Amazonia. Annals of the Association of American Geographers 70: 553–566.
31. German LA (2004) Ecological praxis and blackwater ecosystems: A case study from the Brazilian Amazon. Human Ecology 32: 653–683.
32. Arroyo-Kalin M (2010) The Amazonian formative: Crop domestication and anthropogenic soils. Diversity 2: 473–504.
33. McKey D, Cavagnaro TR, Cliff J, Gleadow R (2010) Chemical ecology in coupled human and natural systems: people, manioc, multitrophic interactions and global change. Chemoecology 20: 109–133.
34. Hershey CH (1994) Manihot Genetic Diversity: International network for cassava genetic resources. International Crop Network Series, IPGRI 10.
35. Boster JS (1985) Selection for perceptual distinctiveness: evidence from Aguaruna cultivars of Manihot esculenta. Economic Botany 39: 310–325.
36. Elias M, McKey D (2000) The unmanaged reproductive ecology of domesticated plants in traditional agroecosystems: An example involving cassava and a call for data. Acta Oecologica-International Journal of Ecology 21: 223–230.
37. Emperaire L, Pinton F, Second G (1998) Gestion dynamique de la diversité varie tale du manioc en Amazonie du nord-ouest. Nature, Science et Societé 6: 27–42.
38. Salick J, Cellinese N, Knapp S (1997) Indigenous diversity of cassava: Generation, maintenance, use and loss among the Amuesha, Peruvian Upper Amazon. Economic Botany 51: 6–19.
39. Elias M, Rival L, McKey D (2000) Perception and management of cassava (Manihot esculenta Crantz) diversity among Makushi Amerindians of Guyana (South America). Journal of Ethnobiology 20: 239–265.
40. Delêtre M, McKey D, Hodkinson T (2011) Marriage exchanges, seed exchanges, and the dynamics of manioc diversity. Proceedings of the National Academy of Sciences of the United States of America 108: 18249–18254.
41. Ellen RF, Soselisa HL (2012) A Comparative Study of the Socio-ecological Concomitants of Cassava (Manihot esculenta Crantz) Diversity, Local Knowledge and Management in Eastern Indonesia. Ethnobotany Research & Applications 10: 15–35.
42. Jones W (1959) Manioc in Africa: Stanford University Press. 315 p.
43. Roosevelt AC (1980) Parmana: Prehistoric Maize and Manioc Susbistence along the Amazon and Orinoco. New York: Academic Press. 320 p.
44. Ohly J, Junk W (1999) Multiple Use of Central Amazon Floodplains: Reconciling Ecological Conditions, Requirements for Environmental Protection, and Socioeconomic Needs. In: Padoch C, Ayres JM, Pinedo-Vasquez M, Henderson A, editors. Varzea: Diversity, Development and Conservation of Amazonias's whitewater floodplains. New York: New York Botanical Garden. pp. 283–299.
45. Gutjahr E (2000) Prospects for arable farming in the floodplains of the Central Amazon. In: W. J . Junk, J. J . Ohly, Piedade MTF, Soares MGM, editors. The Central Amazon Floodplain: Actual Use and Options for a Sustainable Management. Leiden: Backhuys. pp. 141–170.
46. Fraser JA (2010) The diversity of bitter manioc (Manihot esculenta Crantz) cultivation in a whitewater Amazonian landscape. Diversity 2: 586–609.
47. McKey D, Emperaire L, Elias M, Pinton F, Robert T, et al. (2001) Local management and regional dynamics of varietal diversity of cassava in Amazonia. Genetics Selection Evolution 33: S465–S490.
48. Pujol B, Gigot G, Laurent G, Pinheiro-Kluppel M, Elias M, et al. (2002) Germination ecology of cassava, Manihot esculenta Crantz, Euphorbiaceae, in traditional agroecosystems: Seed and seedling biology of a vegetatively propagated domesticated plant. Economic Botany 56: 366–379.
49. Elias M, Penet L, Vindry P, McKey D, Panaud O, et al. (2001) Unmanaged sexual reproduction and the dynamics of genetic diversity of a vegetatively propagated crop plant, cassava (Manihot esculenta Crantz), in a traditional farming system. Molecular Ecology 10: 1895–1907.
50. Elias M, Muhlen GS, McKey D, Roa AC, Tohme J (2004) Genetic diversity of traditional south american landraces of cassava (Manihot esculeatta Crantz): An analysis using microsatellites. Economic Botany 58: 242–256.
51. Pujol B, Renoux F, Elias M, Rival L, McKey D (2007) The unappreciated ecology of landrace populations: Conservation consequences of soil seed banks in Cassava. Biological Conservation 136: 541–551.
52. Peroni N (1998) Taxonomia folk e diversidade intraespecífica de mandioca (Manihot esculenta Crantz) em roças de agricultura tradicional em áreas de Mata Atlântica, São Paulo. [Masters Dissertation]. São Paulo: University of São Paulo.
53. Elias M, McKey D, Panaud O, Anstett MC, Robert T (2001) Traditional management of cassava morphological and genetic diversity by the Makushi Amerindians (Guyana, South America): Perspectives for on-farm conservation of crop genetic resources. Euphytica 120: 143–157.
54. Pujol B, David P, McKey D (2005) Microevolution in agricultural environments: how a traditional Amerindian farming practice favours heterozygosity in cassava (Manihot esculenta Crantz, Euphorbiaceae). Ecology Letters 8: 138–147.
55. Peroni N, Kageyama PY, Begossi A (2007) Molecular differentiation, diversity, and folk classification of "sweet" and "bitter" cassava (Manihot esculenta) in Caicara and Caboclo management systems (Brazil). Genetic Resources and Crop Evolution 54: 1333–1349.
56. Stocker P (2006) Family farmers and manioc in contemporary Brazil: The management of agrobiodiversity and change [PhD Dissertation]: University of Edinburgh. 359 p.
57. Fraser JA (2010) Caboclo horticulture and Amazonian dark earths along the middle Madeira River, Brazil. Human Ecology 38: 651–662.
58. Fraser JA, Clement CR (2008) Dark earths and manioc cultivation in Central Amazonia: A window on pre-Colombian agricultural systems? Boletim do Museu Paraense Emílio Goeldi Ciências Humanas 3: 175–194.
59. Hiraoka M, Yamamoto S, Matsumoto E, Nakamura S, Falesi I, et al. (2003) Contemporary use and management of Amazonian Dark Earths. In: Lehmann J, Kern D, Glaser B, Woods W, editors. Amazonian Dark Earths: Origin, properties, and management. Dordrecht: Kluwer Academic Publishers. pp. 387–406.
60. German LA (2001) The dynamics of *Terra Preta*: an integrated study of human-environmental interaction in a nutrient-poor Amazonian ecosystem [Unpublished Ph.D Dissertation]: University of Georgia. 312 p.
61. German LA (2003) Ethnoscientific understandings of Amazonian Dark Earths. In: Lehmann J, Kern D, Glaser B, Woods WI, editors. Amazonian Dark Earths: Origin, Properties and Management. Dordrecht: Kluwer. pp. 179–201.
62. Harlan J (1992) Crops and Man. Madison, Wisconsin.: American Society of Agronomy Inc. Crop Science Society of America Inc. 284 p.
63. Fraser JA, Junqueira AB, Kawa NC, Moraes CP, Clement CR (2011) Crop Diversity on Anthropogenic Dark Earths in Central Amazonia. Human Ecology 39: 395–406.
64. Lima D, Steward A, Richers BT (2012) Trocas, experimentações e preferências: um estudo sobre a dinâmica da diversidade da mandioca no Médio Solimões, Amazonas. Boletim do Museu Paraense Emílio Goeldi: Ciências Humanas.
65. Grenand F (1993) Bitter manioc in the lowlands of Tropical America: from myth to commercialization. In: C. M. Hladik, A. Hladik, O. F. Linares, H. Pagezy, Semple A et al., editors. Tropical Forests and Food: Biocultural Interactions and Applications to Development. Paris: UNESCO and Parthenon.
66. Castro F (2009) Patterns of Resource Use by Caboclo Communities in the Middle-Lower Amazon. In: C. R. Adams, R. Murrieta, W. Neves, Harris M, editors. Amazon peasant societies in a changing environment: Political ecology, invisibility and modernity in the rainforest. New York: Springer. pp. 157–177.
67. Pereira KJC (2008) Agricultura tradicional e manejo da agrobiodiversidade na Amazônia Central: Um estudo de caso nos roçados de mandioca das Reservas de Desenvolvimento Sustentável Amanã e Mamirauá, Amazonas [PhD Dissertation]. São Paulo: University of São Paulo. 223 p.
68. Denevan W (1984) Ecological heterogeneity and horizontal zonation of agriculture in the Amazon floodplain. In: M . Schmink and C . Wood, editor. Frontier Expansion in Amazonia. Gainsville: University of Florida Press. pp. 311–336.
69. Smith NJH (1999) The Amazon River Forest: A Natural History of Plants, Animals, and People. Oxford: Oxford University Press. 208 p.
70. Noda SN, Noda H, Santos HP (2000) Family farming systems in the floodplains of the State of Amazonas. In The Central Amazon Floodplain: Actual Use and Options for a Sustainable Management. In: W. J. Junk, J. J. Ohly, Piedade MTF, Soares MGM, editors. The Central Amazon Floodplain: Actual Use and Options for a Sustainable Management. Leiden: Backhuys. pp. 215–242.
71. Lima RMB, Saragoussi M (2000) Floodplain home gardens on the Central Amazon in Brazil. In: Junk WJ, J. J. Ohly, Piedade MTF, Soares MGM, editors. The Central Amazon Floodplain: Actual Use and Options for a Sustainable Management. Leiden: Backhuys. pp. 243–268.
72. Boster JS (1984) Classification, cultivation and selection of Aguaruna cultivars of Manihot esculenta (Euphorbiaceae). Advances in Economic Botany 1: 34–47.
73. Pujol B, Gigot G, Laurent G, Pinheiro-Kluppel M, Elias M, et al. (2002) Germination ecology of cassava (Manihot esculenta Crantz, Euphorriaceae) in traditional agroecosystems: Seed and seedling biology of a vegetatively propagated domesticated plant. Economic Botany 56: 366–379.
74. Lebot V (2009) Tropical Root and Tuber Crops: Cassava, Sweet Potato, Yams and Aroids. Oxford, UK: CAB International.
75. Boserup E (1993 (1965)) The conditions of agricultural growth. The economics of agrarian change under population pressure. London: Earthscan Publications Ltd.
76. Alves Pereira A (2011) Dinâmica evolutiva de mandioca (Manihot esculenta Crantz) em três tipos de solos manejados por Caboclos na região do Médio Rio Madiera, Amazonas. Manaus, Brazil: Instituto Nacional de Pesquisas da Amazônia. 87 p.
77. Alves-Pereira A, Peroni N, Abreu AG, Gribel R, Clement CR (2012) Genetic structure of traditional varieties of bitter manioc in three soils in Central Amazonia. Genetica 139: 1259–1271.
78. Kensinger KM (1975) Studying the Cashinahua. In: Kensinger KM, editor. The Cashinahua of Eastern Peru Studies in Anthropology and Material Culture, vol 1. Providence, Rhode Island: Brown University. pp. 9–85.
79. Aquino TVd, Iglesias MP (2002) Habitantes: Os Kaxinawá. In: Carneiro da Cunha M, Almeida MBd, editors. Enciclopédia da floresta O alto juruá: práticas e conhecimentos das populações. São Paulo: Companhia das Letras. pp. 147–160.

80. Ricardo CA, editor (2000) Povos Indígenas no Brasil, 1996–2000. São Paulo: Instituto Socioambiental.
81. Wilson W (2002) Soils Utilized for Gardens by Tukanoans in Northwestern Amazonia and their Impact on Cassava (Manihot esculenta Crantz) Cultivar Selection. Culture & Agriculture 24: 20–30.
82. Heckler S, Zent S (2008) Piaroa Manioc Varietals: Hyperdiversity or Social Currency? Human Ecology 36: 679–697.
83. Fraser JA (2009) Amazonian dark earths and *Caboclo* subsistence on the Middle Madeira River, Brazil [PhD Dissertation]: University of Sussex.
84. Bryman A (2008) Social Research Methods. Oxford, UK: Oxford University Press. 800 p.
85. Bernard HR (2006) Research methods in anthropology: qualitative and quantitative approaches. Oxford: Altamira Press. 803 p.
86. McCune B, Grace JB (2002) Analysis of Ecological Communities: MjM Software Design. 304 p.
87. Wright SP (1992) Adjusted p-values for simultaneous inference. Biometrics 48: 1005–1013.
88. Sutrop U (2001) List task and a cognitive salience index. Field Methods 13: 263–276.
89. Chavarriaga-Aguirre P, Maya MM, Bonierbale MV, Kresovich S, Fregene MA, et al. (1998) Microsatellites in cassava (Manihot esculenta Crantz): discovery, inheritance and variability. Theoretical and Applied Genetics 97: 493–501.
90. Mba REC, Stephenson P, Edwards K, Melzer S, Nkumbira J, et al. (2001) Simple sequence repeat (SSR) markers survey of the cassava (Manihot esculenta Crantz) genome: towards an SSR-based molecular genetic map of cassava. Theoretical and Applied Genetics 102: 21–31.
91. Peakall R, Smouse PE (2006) Genalex 6: Genetic analysis in Excel. Population genetic software for teaching and research. Molecular Ecology Notes 6: 288–295.
92. Arnaud-Haond S, Belkhir K (2007) GENCLONE: a computer program to analyse genotypic data, test for clonality and describe spatial clonal organization. Molecular Ecology Notes 7: 15–17.

6

Soil Microbial Responses to Elevated CO_2 and O_3 in a Nitrogen-Aggrading Agroecosystem

Lei Cheng[1¤], Fitzgerald L. Booker[2,3], Kent O. Burkey[2,3], Cong Tu[1], H. David Shew[1], Thomas W. Rufty[3], Edwin L. Fiscus[2,3], Jared L. Deforest[4], Shuijin Hu[1]*

1 Department of Plant Pathology, North Carolina State University, Raleigh, North Carolina, United States of America, **2** Plant Science Research Unit, United States Department of Agriculture, Agricultural Research Service, Raleigh, North Carolina, United States of America, **3** Department of Crop Science, North Carolina State University, Raleigh, North Carolina, United States of America, **4** Department of Environmental and Plant Biology, Ohio University, Athens, Ohio, United States of America

Abstract

Climate change factors such as elevated atmospheric carbon dioxide (CO_2) and ozone (O_3) can exert significant impacts on soil microbes and the ecosystem level processes they mediate. However, the underlying mechanisms by which soil microbes respond to these environmental changes remain poorly understood. The prevailing hypothesis, which states that CO_2- or O_3-induced changes in carbon (C) availability dominate microbial responses, is primarily based on results from nitrogen (N)-limiting forests and grasslands. It remains largely unexplored how soil microbes respond to elevated CO_2 and O_3 in N-rich or N-aggrading systems, which severely hinders our ability to predict the long-term soil C dynamics in agroecosystems. Using a long-term field study conducted in a no-till wheat-soybean rotation system with open-top chambers, we showed that elevated CO_2 but not O_3 had a potent influence on soil microbes. Elevated CO_2 (1.5×ambient) significantly increased, while O_3 (1.4×ambient) reduced, aboveground (and presumably belowground) plant residue C and N inputs to soil. However, only elevated CO_2 significantly affected soil microbial biomass, activities (namely heterotrophic respiration) and community composition. The enhancement of microbial biomass and activities by elevated CO_2 largely occurred in the third and fourth years of the experiment and coincided with increased soil N availability, likely due to CO_2-stimulation of symbiotic N_2 fixation in soybean. Fungal biomass and the fungi:bacteria ratio decreased under both ambient and elevated CO_2 by the third year and also coincided with increased soil N availability; but they were significantly higher under elevated than ambient CO_2. These results suggest that more attention should be directed towards assessing the impact of N availability on microbial activities and decomposition in projections of soil organic C balance in N-rich systems under future CO_2 scenarios.

Editor: Jonathan H. Badger, J. Craig Venter Institute, United States of America

Funding: Lei Cheng was supported by a fellowship from the USDA-ARS Plant Science Research Unit (Raleigh, NC) and was in part by an USDA grant to SH (2009-35101-05351). The funders had no role in study design, data collection and analysis, decision to publish, or preparation of the manuscript.

Competing Interests: The authors have declared that no competing interests exist.

* E-mail: shuijin_hu@ncsu.edu

¤ Current address: Institute for Environmental Genomics, Department of Botany and Microbiology, University of Oklahoma, Norman, Oklahoma, United States of America

Introduction

Soil microbes critically affect plant and ecosystem responses to climate change by modulating organic C decomposition and nutrient availability for plants. Experimental evidence accumulated over the last several decades has clearly shown that climate change factors such as CO_2 enrichment in the atmosphere can significantly alter plant growth [1,2] and the availability of organic C, N and cation nutrients for microbes [3,4,5]. Ozone is a greenhouse gas with demonstrated inhibitory effects on plant growth and resource allocation belowground [6,7]. Although less well-studied, O_3 is considered to have an impact on soil microbial processes [6]. Alterations in soil microbes can, in turn, profoundly influence soil C processes and the long-term potential of terrestrial ecosystems as a C sink to mitigate anthropogenic sources of atmospheric CO_2. However, predicting what these changes will be is hampered by our limited understanding of the underlying mechanisms by which soil microbes respond to altered resource availability.

The current prevailing hypothesis, building on the assumption that soil microbes are generally C limited [8], predicts that elevated CO_2 increases soil microbial biomass and activities due to enhanced soil C availability [9,10,11], whereas O_3 reduces them due to lower C allocation belowground [6,12,13]. This broad hypothesis has been extensively tested over the past two decades for CO_2 but less so with O_3 [4,10,12,13,14,15,16]. Though C availability to microbes has been commonly reported to increase under elevated CO_2 [5,17,18] and to decrease under elevated O_3 [6,13,19], results of soil microbial responses to elevated CO_2 and O_3 have been inconsistent [6,11,20,21]. In a meta-analysis study, de Graaff *et al.* (2006) found that elevated CO_2 increased microbial biomass C and microbial respiration by 7.7% and 17.1%, respectively, across 40 studies that mainly included herbaceous species. In the meantime, Hu *et al.* (2006) reviewed 135 studies examining elevated CO_2 effects on a suite of soil microbial parameters such as microbial biomass and respiration and found that microbial biomass C and microbial respiration increased under elevated CO_2 in 19 of 40 studies and 20 of 38 studies, respectively, but remained unchanged or even decreased in the remainder. Despite considerable efforts in the past two decades, there is a lack of conceptual understanding of why and how these inconsistencies in CO_2 and O_3 effects on microbes occur.

Soil microbial responses to elevated CO_2 can also be influenced by CO_2-induced alterations in soil moisture [9,22], grazing activity of soil animals [23], and soil nutrient availability [4,9]. Such mechanisms could operate either singly or in combination with changes in soil C availability. In particular, CO_2-induced alteration in the stoichiometery of available C and N has been proposed to be a primary control over microbial responses to elevated CO_2 [4,9,24]. Soil N availability may influence microbial responses to elevated CO_2 by affecting both physiological activities and the community structure composition of microbes [9,25]. When soil N was limiting, competitive plant N uptake can significantly reduce soil N availability for microbes under elevated CO_2, limiting microbial decomposition over the short-term [4]. Conversely, high N availability in soil often increases microbial activities [26,27] and favors bacteria over fungi [28,29]. Yet most studies that examined microbial responses to elevated CO_2 were conducted in N-limiting forest and grassland ecosystems [20,21]. It remains unclear how soil microbes respond to elevated CO_2 and O_3 in N-rich or N-aggrading agroecosystems.

Many crop plants, particularly C_3 crops, are usually responsive to elevated CO_2 and O_3 [2,30]. For instance, it has been estimated that elevated CO_2 alone increased the shoot biomass of soybean and wheat by 48% and 16%, respectively [2,31], but elevated O_3 reduced them by 21% and 18% [32,33]. Also, elevated CO_2 has been shown to ameliorate O_3 effects on plants by reducing O_3 uptake and increasing C assimilation rates [30,32,34]. However, whether the CO_2- and O_3-induced changes in plant biomass translate into alterations in soil C sequestration depends largely on the responses of soil microbial processes. Additionally, elevated CO_2 significantly increased symbiotic N_2 fixation in legumes such as soybean and peanut [35,36], whereas elevated O_3 tended to reduce it [36]. It has been suggested that high N availability in agro- and grassland ecosystems can sustain plant responses to rising CO_2 over a long time frame and provide an opportunity for soil C sequestration in soil in a higher CO_2 world [2,37,38,39]. Convincing evidence is still lacking, but soil microbial responses may be indicative for understanding the long-term soil C dynamics in high N or N-aggrading ecosystems [27,40].

In a long-term study examining climate change effects on soil C dynamics in a wheat-soybean agroecosystem with no-till practice, we continually monitored a suite of soil microbial parameters in response to elevated CO_2 and O_3 for more than four years. Because soybean and its symbiotic N_2 fixation are sensitive to elevated CO_2 and O_3 [35,36], we expected that elevated CO_2 would enhance both C and N inputs belowground through increasing residue returns, while elevated O_3 would offset this CO_2 effect. Also, we expected that the stoichiometry of available C and N for microbes might change over time as a portion of residue C was mineralized and released back to the atmosphere as CO_2 while a large proportion of residue N was retained in the system. Consequently, alterations in C and N availability for microbes induced by elevated CO_2 and O_3 may further affect microbial biomass and activities over time, and possibly induce a shift in the microbial community structure. Therefore, our specific objectives were to: 1) document the time-course of CO_2 and O_3 effects on microbial biomass, activities and community structure; and, 2) examine how changes in microbial parameters were related to residue inputs and soil C and N availability.

Materials and Methods

Site description

We initiated a long-term field experiment in May 2005 to investigate the response of a wheat-soybean rotation agroecosystem to elevated atmospheric CO_2 and O_3 using open-top field chambers (OTC). The experimental site is located at the Lake Wheeler Experimental Station, 5 km south of North Carolina State University, Raleigh, North Carolina, USA (35°43′N, 78°40′W; elevation 120 m). Annual mean temperature is 15.2°C and annual mean precipitation is 1050 mm. The field had been left fallow for eight years prior to this study. Before CO_2 and O_3 treatments were initiated, the soil was repeatedly turned-over using a disc implement and rotovator. The soil is an Appling sandy loam (fine, kaolinitic, thermic Typic Kanhapludult), well drained with a pH of 5.5, and contained 9.0 g C and 0.86 g N kg^{-1} soil when the experiment started.

This experiment was a 2×2 factorial design with four treatments randomly assigned into four blocks. Four different trace-gas treatments were: (*a*) charcoal-filtered air and ambient CO_2 (CF); (*b*) charcoal-filtered air plus ambient CO_2 and 1.4 times ambient O_3 (+O_3); (*c*) charcoal-filtered air plus 180 μl l^{-1} CO_2 (+CO_2); and (*d*) charcoal-filtered air plus 180 μl l^{-1} CO_2 and 1.4 times ambient O_3 (+CO_2+O_3). The seasonal daily average concentrations of CO_2 and O_3 over the experimental duration are shown in Table 1. The purpose of filtration of ambient air with activated charcoal was to reduce the concentrations of ambient O_3 to levels considered nonphytotoxic to soybean and wheat plants. Ozone was deemed as a major air pollutant in this area, while other air pollutants such as NO_2 and SO_2 were below the phytotoxic levels at the experimental location [41].

Soybean [cv. CL54 RR (Year 1), Asgrow 5605 RR (Years 2 and 3) and SS RT5160N RR (Year 4)] was planted each spring followed by soft red winter wheat (Coker 9486) in the fall using no-till practices. Plants were exposed to reciprocal combinations of CO_2 and O_3 within cylindrical OTCs (3.0 m diameter×2.4 m tall) from emergence to physiological maturity. Carbon dioxide was released from a 14-ton liquid-receiving tank 24 h daily and monitored at canopy height using an infrared CO_2 analyzer (model 6252, Li-Cor Inc. Lincoln, NE, USA). Ozone was generated by electrostatic discharge in dry O_2 (model GTC-1A, Ozonia North America, Elmwood Park, NJ, USA) and dispensed 12 h daily (08:00–20:00 hours Eastern Standard Time) in proportion to concentrations of ambient O_3. The O_3 concentration in the chambers was monitored at canopy height with a UV photometric O_3 analyzer (model 49, Thermo Environmental Instruments Co., Franklin, MA, USA). During wheat growing seasons, each plot initially received 48 g NH_4NO_3 (equivalent to 24 kg N ha^{-1}) in November each year, followed by an additional input of 192 g NH_4NO_3 (equivalent to 96 kg N ha^{-1}) in March. Plots were treated with lime, K and P in November during the

Table 1. The seasonal daily average (12 h) CO_2 and O_3 concentrations at canopy height during the 4-year period.

Crop	CO_2 (μl l^{-1})		O_3 (nl l^{-1})	
	Ambient	Elevated	CF	Elevated
Soybean	376.0±0.4	555.0±0.7	19.9±0.3	65.7±0.4
Wheat	388.0±0.4	547.0±0.5	20.7±0.2	49.8±0.3

CF: charcoal filtered air. Values are mean ± s.e.m.

experiment according to soil test recommendations. During soybean growing seasons, plants were irrigated with drip lines to prevent visible signs of water stress, but no additional N fertilizers were applied. Upon senescence of the plants, all aboveground plant biomass in each chamber was harvested. Soybean plants were divided into leaves, stems, husks and seeds, while wheat plants were separated as straw, chaff and seeds, then dried and quantified. Afterward, residues other than seeds were uniformly returned to their corresponding treatment plots and evenly distributed on the soil surface.

Soil sampling

The chamber plot was divided into two parts: the sampling area (an inner circular area with a diameter of 2.4 m) and the border area (for purpose of reducing chamber effects, 0.3 m in width). To avoid taking soil samples from the same location, the sampling area was divided into 448 small subplots (10×10 cm). Soil sampling locations were determined using a random number generator and each subplot was sampled only once. In June and November of each year (Year 1–Year 4), corresponding to harvest time for each crop, we used a 5-cm diameter soil corer to take three soil cores to 20 cm depth in the center of three pre-determined subplots from each chamber. Three additional soil cores were immediately taken from the border areas to fill holes in the sampling areas. Sample holes in border areas were refilled by soil cores taken just outside of each chamber. Soil cores were separated into 0–5 cm, 5–10 cm and 10–20 cm depth fractions. Core sections were then pooled by the depth fraction into three soil samples per chamber. Soil samples were also collected at the mid-growing season to check whether microbial parameters were significantly different from those obtained at the end of the growing season. Soil samples (0–5 and 5–10 cm) were collected in each April of the first two years and each August of the last two years, corresponding to the maximal physiological activity of wheat and soybean plants, respectively. All samples were sealed in plastic bags, stored in a cooler and transported to the laboratory.

Field moist soils were mixed thoroughly and sieved through a 4-mm mesh within 24 hours of the field sampling and all visible residues and plant roots were carefully removed. Subsamples (~20 g) were then taken immediately; frozen and stored at -20°C for the phospholipid fatty acid (PLFA) analysis and the rest of soils were stored at 4°C for other microbial and chemical analyses. A 10-g subsample was oven-dried at 105°C for 48 h and weighed for the determination of the water content. All the soil and microbial data were calculated on the dry weight basis of soils.

Sample analyses

C and N contents in plant residues and soils. Air-dried subsamples of aboveground plant components (stems, leaves, husks and seeds of soybean; straw, chaff and seeds of wheat) were ground in a Tecator Cyclotec mill fitted with a 1-mm screen (FOSS, Eden Prairie, MN, USA). Soil samples were ground into fine powder using an 8000-D Mixer Mill (SPEX CertiPrep Inc. Metuchen, NJ, USA). The C and N concentrations in various plant components and in soil were determined with a CHN elemental analyzer (Carla Erba and model 2400, Perkin Elmer Co., Norwalk, CT, USA). Aboveground residue C and N inputs to soil were calculated by adding up the C and N, respectively, in all aboveground plant components except for seeds.

Symbiotic N_2 fixation in soybean. Soybean N_2 fixation was estimated using the conventional N accumulation method [42]. To estimate total N_2 fixation by soybean in each season, we first estimated total biomass N in wheat and soybean, respectively. Total aboveground plant N was calculated by directly adding up the N in all plant components. Root biomass was estimated by using the fixed root:shoot ratios of wheat (0.07) and soybean (0.22) according to the literature [2,31,43,44,45]. We also assumed that the C:N ratio of roots was the same as that of shoots [46]. Then, we used wheat in the following season as the nonfixing plant to estimate total N fixed by soybean in each season by subtracting total N in wheat plants from total N in soybean plants on a per chamber basis. Further, the CO_2 effect on N_2 fixation was estimated by subtracting total N in soybean in ambient CO_2 from elevated CO_2. Although wheat has been often used as a non-fixing control plant [42], we realized that this method does not provide an exact estimate of N_2 fixation by soybean plants in the field. Using wheat plants as the non-fixing control in our system should provide a conservative underestimate of soybean N_2 fixation because: 1) N inputs to soil through soybean root exudates and fine root turnover were not considered; 2) inorganic N fertilizers (120 kg N ha^{-1}) were applied for wheat; and 3) wheat should have also obtained significantly higher amounts of N from the mineralization of soybean residues. In addition to estimating N_2 fixation, changes in total soil N over time were documented by comparing the soil N content (0–5 cm soil layer) at the end of the fourth year to the pretreatment soil N content (0–5 cm soil layer).

Soil microbial biomass C and N. Soil microbial biomass C (MBC) and biomass N (MBN) were determined using the fumigation-extraction method [47]. Twenty-g dry weight equivalent soil samples were fumigated with ethanol-free chloroform for 48 h and then extracted with 50 mL of 0.5 M K_2SO_4 by shaking for 30 min. Another 20-g sample of non-fumigated soil was also extracted with 50 mL of 0.5 M K_2SO_4. Soil extractable organic C in both fumigated and non-fumigated K_2SO_4 extracts was measured using a TOC analyzer (Shimadzu TOC-5050A, Shimadzu Co., Kyoto, Japan). Soluble inorganic N in the extracts of fumigated and non-fumigated soils was quantified on the Lachat flow injection analyzer (Lachat Instruments, Milwaukee, WI, USA) after digestion with alkaline persulfate [48]. The differences in extractable organic C and inorganic N between fumigated and non-fumigated soils were assumed to be from lysed soil microbes. The released C and N were used to calculate MBC and MBN using a conversion factor of 0.45 (k_{EC}) and 0.45 (k_{EN}), respectively [47,49].

Soil extractable C and N. The concentration of organic C in non-fumigated soil extracts was used to represent soil extractable C. The extractable inorganic N referred to the sum of NH_4^+-N and NO_3^--N in non-fumigated soil extracts.

Soil microbial respiration. We determined soil heterotrophic respiration using an incubation-alkaline absorption method [50]. In brief, 20-g dry mass equivalent soil samples were adjusted to moisture levels of around 60% water holding capacity, placed in 1-L Mason jars, and then incubated at 25°C in the dark for 2 weeks. Respired CO_2 was trapped in 5 mL of 0.25 M NaOH contained in a beaker suspended in the jar. After the first week incubation, NaOH solutions were replaced with fresh solutions. The CO_2 captured in the NaOH solution was titrated with 0.125 M HCl to determine the amount of CO_2 evolved from the soil. Soil microbial respiration (SMR) rate was expressed as mg CO_2 kg^{-1} soil d^{-1} by averaging the data across two 1-wk incubations.

Net soil N mineralization. Potential N mineralization was determined following a 4-wk incubation at 25°C in the dark. Soil NH_4^+ and NO_3^- in un-incubated and incubated subsamples (20-g each) were extracted with 50 mL of 0.5 M K_2SO_4 by shaking for 30 min. The concentrations of inorganic N were then measured on the Lachat flow injection analyzer. Net mineralized N (NMN) was determined by the difference in extractable total inorganic N

(NH_4^+-N + NO_3^--N) between incubated and un-incubated soil samples.

Phospholipid fatty acids. PLFAs were extracted following a procedure described by Bossio *et al.* (1998). Briefly, 10 g of freeze-dried soils (0–5 cm soil layer) were extracted using a solution containing $CH_3OH:CH_3Cl:PO_4^{3-}$ (vol/vol/vol 2:1:0.8). Solid phase extraction columns (Thermo Scientific, Vernon Hills, IL, USA) were used to separate phospholipids from neutral and glycol-lipids. The phospholipids were then subjected to an alkaline methanolysis to form fatty acid methyl esters (FAMEs). The resulting FAMEs were separated and measured using gas chromatography on a HP GC-FID (HP6890 series, Agilent Technologies, Inc. Santa Clara, CA, USA); peaks were identified using the Sherlock Microbial Identification System (v. 6.1, MIDI, Inc., Newark, DE, USA). We chose the following fatty acids, i14:0, i15:0, a15:0, 15:0, i16:0, 16:1ω7c, i17:0, a17:0, 17:0cy, 17:0, 18:1ω7c, and 18:1ω5c, to represent the bacterial PLFAs [51,52,53], and the other three fatty acids (16:1ω5c, 18:2ω6.9c and 18:1ω9c) as the fungal PLFAs [52,53,54]. We used the ratio of signature fungal and bacterial PLFAs as an indicator of soil microbial community structure [51,53,55]. The fatty acid profile of soil microbes was examined in soils collected in years 1, 3, 4, and 5.

Statistical analysis

We examined results for the entire experimental duration from 2005–2009 (Year 1–Year 4), and used the linear mixed model [56] to test the main effects of CO_2, O_3 and the interaction of CO_2 and O_3, and whether these changed over time. We employed a set of covariance structures including compound symmetric model (CS), the first-order autoregressive model [AR (1)], and autoregressive with random effect to reduce autocorrelation. The *P* values for treatments and interaction terms were reported based on the covariance structure that minimized Akaike information criterion (AIC) and Bayesian information criterion (BIC) [56]. Data for soil and microbial parameters from mid-seasons (Appendix S1), plant and soil N contents, fungal and bacterial PLFAs and the fungi:bacteria ratios were subjected to the analysis of variance using the mixed model. To test for relationships between variables, we conducted a correlation analysis using all the data generated over the 4-year period. A Chi-square (χ^2) test was also conducted to examine whether the CO_2 effect on microbial biomass, respiration and the community structure were correlated with the CO_2 effect on N availability. We thus developed four contingency tables for SMR, MBC, MBN, and fungi:bacteria ratio, respectively. All statistical analyses were performed using the SAS 9.1 (SAS Institute, Inc., Cary, NC, USA). For all tests, $P \leq 0.05$ was considered to indicate a statistically significant difference.

Results

Soybean N_2 fixation and plant residue C and N inputs to soil

Elevated CO_2 significantly increased symbiotic N_2 fixation by soybean plants, while O_3 decreased it. Over the 4-year period, total N derived from symbiotic N_2 fixation in soybean was estimated at 92.5, 68.4, 119.1 and 109.7 g N m^{-2} in the CF, +O_3, +CO_2 and +CO_2+O_3 treatments, respectively. Compared to the CF treatment, the +CO_2 treatment significantly increased the net N inputs to soil from symbiotic N_2 fixation (excluding seed harvests) on average by 43%, while the +O_3 treatment decreased it by 23% over the 4-year period (Table 2). However, there was no significant effect of $CO_2 \times O_3$ on the net N inputs derived from symbiotic N_2 fixation in any year (Table 2).

Table 2. Effects of elevated CO_2 and O_3 on soybean N_2 fixation and total N in the surface soil.

	Year 1	Year 2	Year 3	Year 4
N inputs to soil derived from soybean N_2 fixation (g N m^{-2})				
Treatment				
CF	11.4±0.8	2.9±0.4	6.0±0.5	7.5±0.3
+O_3	10.6±0.5	1.8±0.4	5.0±0.5	4.5±0.5
+CO_2	15.1±0.4	4.8±0.2	8.3±0.7	10.0±0.5
+CO_2+O_3	14.7±1.4	3.8±0.2	7.8±0.3	8.6±0.7
Source				
O_3	NS	**	NS	**
CO_2	***	***	***	***
$CO_2 \times O_3$	NS	NS	NS	NS
Total soil N in the surface soil (0–5 cm) (g N m^{-2})				
Treatment				
CF	85.7±8.2	ND	ND	93.7±11.7
+O_3	74.1±16.0	ND	ND	90.0±3.1
CO_2	76.4±10.7	ND	ND	95.6±6.4
+CO_2+O_3	75.1±4.7	ND	ND	97.5±8.1

Values shown for N inputs to soil from N_2 fixation exclude seeds.
Values are mean ± s.e.m. *** ($P<0.001$) and ** ($P<0.01$) denote statistically significant main treatment effects, ANOVA mixed models. ND, not determined. NS, not significant. CF, charcoal-filtered ambient air. +O_3, elevated O_3.+CO_2, elevated CO_2.+CO_2+O_3, elevated CO_2+O_3. The main treatment effects of CO_2, O_3 and the $CO_2 \times O_3$ interaction on soil N were not statistically significant for any years.

Elevated CO_2 significantly increased both soybean aboveground residue C and N inputs to the chambers in all years ($P \leq 0.001$ for each year; Fig. 1), leading to an average increase by 38% and 30%, respectively, over the experimental period. Elevated CO_2 also significantly increased wheat residue C inputs by 15%, but did not affect wheat residue N inputs (Fig. 1). Elevated O_3 had no significant effects on wheat residue C and N inputs, but reduced soybean C and N inputs by 12% (Fig. 1). No significant $CO_2 \times O_3$ interaction was observed on soybean and wheat residue C and N inputs ($P>0.1$). Additionally, the total amounts of C and N in soybean residues were significantly different ($P<0.01$) among four years, which primarily resulted from the differences in biomass production of three different cultivars as well as the variability among years.

Soil organic C and N, and extractable C and N

Over the experimental period, no significant CO_2 or O_3 effects on total soil organic C (data not shown) or total soil N (Table 2) were observed. However, there was a significant increase in total soil N over the experimental period in all treatments compared to the soil N before the treatments were applied in 2005. Although the magnitude of increase in soil N over time tended to be higher under elevated (26%) than ambient (17%) CO_2 plots, this effect was not statistically significant.

Neither CO_2 nor O_3 treatments had any significant effects on soil extractable C in the whole soil profile or on the interactions between time and gas treatments over the 4-year period (Appendix S2). In general, elevated CO_2 tended to increase concentrations of total extractable inorganic N (Appendix S3). Soil extractable N in

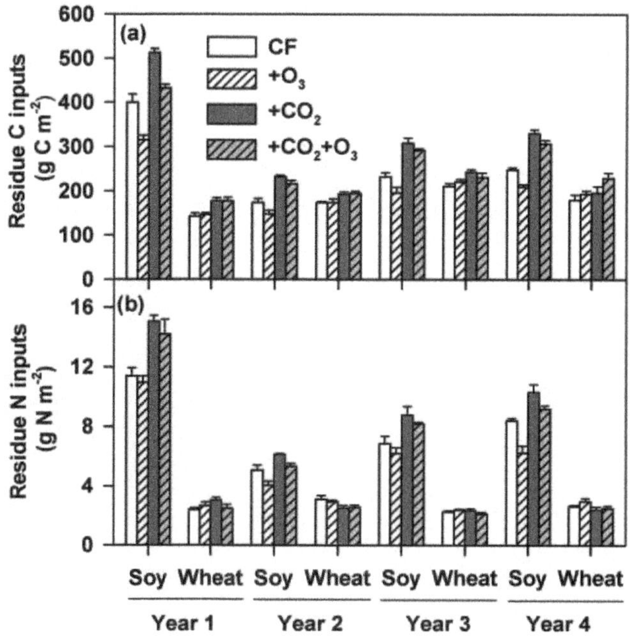

Figure 1. Aboveground residue C and N inputs under elevated CO_2 and O_3. Soy, soybean. CF, charcoal-filtered ambient air. $+O_3$, elevated O_3.$+CO_2$, elevated CO_2.$+CO_2+O_3$, elevated CO_2+O_3. Data represent means (n = 4) ± s.e.m. (a) Residue C inputs. Soybean residue C inputs: CO_2 effect, $P \leq 0.001$ for every year; O_3 effect, $P<0.05$ for every year; $CO_2 \times O_3$, $P>0.1$ for every year. Wheat residue C inputs: CO_2 effect, $P<0.05$ for every year; O_3 effect, $P>0.1$ for every year; $CO_2 \times O_3$, $P>0.1$ for every year, ANOVA mixed model. (b) Residue N inputs. Soybean residue N inputs: CO_2 effect, $P \leq 0.001$ for every year; O_3 effect, $P<0.01$ only for year 2 and 4; $CO_2 \times O_3$, $P>0.1$ for every year. Wheat residue N inputs: CO_2 effect, $P<0.05$ for year 2 (significantly decreased) but >0.1 for year 1, 3 and 4; O_3 effect, $P>0.1$ for every year; $CO_2 \times O_3$, $P>0.1$ for every year, ANOVA mixed model.

elevated CO_2 plots increased by, on average, 17% ($P>0.1$), 18% ($P<0.01$) and 8% ($P>0.1$), respectively, in 0–5, 5–10, and 10–20 cm soil layers over the 4-year period. There were no significant effects of O_3 and the $CO_2 \times O_3$ interaction on soil extractable N (Appendix S2).

Microbial biomass C and biomass N

Elevated CO_2 significantly enhanced both MBC and MBN in the 0–5 cm soil layer, leading to an average increase of 8% ($P<0.05$; Fig. 2a) and 14% ($P<0.05$; Fig. 2d), respectively, over the 4-year period. However, these increases resulted primarily from CO_2-induced enhancement in the third and fourth years of the experiment (Fig. 2). Both MBC and MBN at elevated CO_2 remained unchanged in the top soil layer during year 1 and 2, but increased on average by 14% and 26%, respectively, within year 3 and 4 of the experiment. The CO_2 effects were also significant for MBN in the 10–20 cm soil layer ($P<0.05$), but not significant for MBC (Appendix S2). However, neither O_3 nor the $CO_2 \times O_3$ interaction had any significant impacts on MBC or MBN along the soil profile (Appendix S2).

Soil microbial respiration (SMR)

Over the 4-year period, atmospheric CO_2 enrichment increased SMR rates (Fig. 3). Compared to ambient CO_2, SMR under elevated CO_2 was 26% ($P<0.05$), 17% ($P>0.1$) and 31% ($P<0.05$) higher in 0–5, 5–10 and 10–20 cm soil layers, respectively. Similar to microbial biomass, the observed increases in SMR were largely due to the CO_2 stimulation effects in the third and fourth years of the experiment (Fig. 3). In the 0–5 cm soil layer, for example, elevated CO_2 only increased SMR by 9% in the first two years, but by 43% over the subsequent two years. Neither the O_3 effect nor the $CO_2 \times O_3$ interaction resulted in significant effects on SMR in any soil layer (Appendix S2).

Metabolic quotient of soil microbes (the respiration rate per unit of microbial biomass C) under elevated CO_2 increased on average by 16% (repeated measures mixed models; CO_2 effect: $P=0.003$), 9% ($P=0.2$), and 20% ($P=0.02$) in 0–5, 5–10, and 10–20 cm soil layers, respectively, over the 4-year period (Appendix S4). The CO_2 effect on the metabolic quotient also changed considerably over time. In the 0–5 cm soil layer, CO_2 enrichment slightly increased the metabolic quotient by 7% in the first two years, but significantly increased it by 25% within the following two years. Neither the O_3 effect nor the $CO_2 \times O_3$ interaction resulted in significant impacts on metabolic quotient of soil microbes in any soil layer (Appendix S2).

Net soil N mineralization

Similar to the effects on SMR and MBN, CO_2 enrichment significantly stimulated the rate of net soil N mineralization at both 0–5 ($P<0.01$; Fig. 4a) and 10–20 ($P<0.05$; Fig. 4c) cm soil layers. On average, net mineralizable N (NMN) in elevated CO_2 plots was 13%, 5%, and 26% higher than those in ambient CO_2 plots, respectively, in the 0–5, 5–10, and 10–20 cm soil layers. Again, these effects were mainly due to the CO_2-induced increases within the year 3 and 4 of the experiment (Fig. 4). Elevated CO_2 showed no impacts on net soil N mineralization in the first two years, but caused an average increase by 22%, 12% and 49%, respectively, in the 0–5, 5–10, and 10–20 cm soil layers during the third and fourth years. In contrast, neither the O_3 treatment effect nor the $CO_2 \times O_3$ interaction were statistically significant (Appendix S2).

Stratification of soil microbial parameters under elevated CO_2

The time course of CO_2-effects on various parameters along the soil profile was significantly different. In the top 5-cm soil samples, MBC fluctuated over the whole period (Fig. 2a), but MBN, SMR and NMN started to increase by the third year (Fig. 2d, 3a and 4a). In the deeper soils (5–10 and 10–20 cm), MBC significantly decreased (Fig. 2b and 2c), MBN remained unchanged (Fig. 2e and 2f), but SMR and NMN increased (Fig. 3b, 3c, 4b and 4c) in years 3 and 4. Over the first two years of the experiment, all these parameters remained largely unaffected by CO_2 enrichment in the 5–10 and 10–20 cm soil layers (Figs. 2, 3, 4). By the third and fourth years, elevated CO_2 had no impacts on MBC in the deeper soil depths (Fig. 2b and 2c), but still increased SMR and NMN (Fig. 3b, 3c, 4b and 4c).

The microbial parameters from soil samples collected at the mid-seasons were similar with those at the harvest of the corresponding growing season and those results were shown in Appendix S1.

PLFAs of soil microbes and the microbial community structure

Two trends in fungal and bacterial PLFAs emerged. First, the abundance of fungal and bacterial PLFAs and fungi:bacteria ratios remained largely unchanged in the first two years but decreased significantly in years 4 and 5 of the experiment (Fig. 5). On average, fungal and bacterial PLFAs decreased by 31% and 13%, respectively, from the years 1–3 to years 4–5. Second, elevated CO_2 significantly increased microbial PLFA biomass which was due only to increased fungal PLFA biomarkers starting from year 3

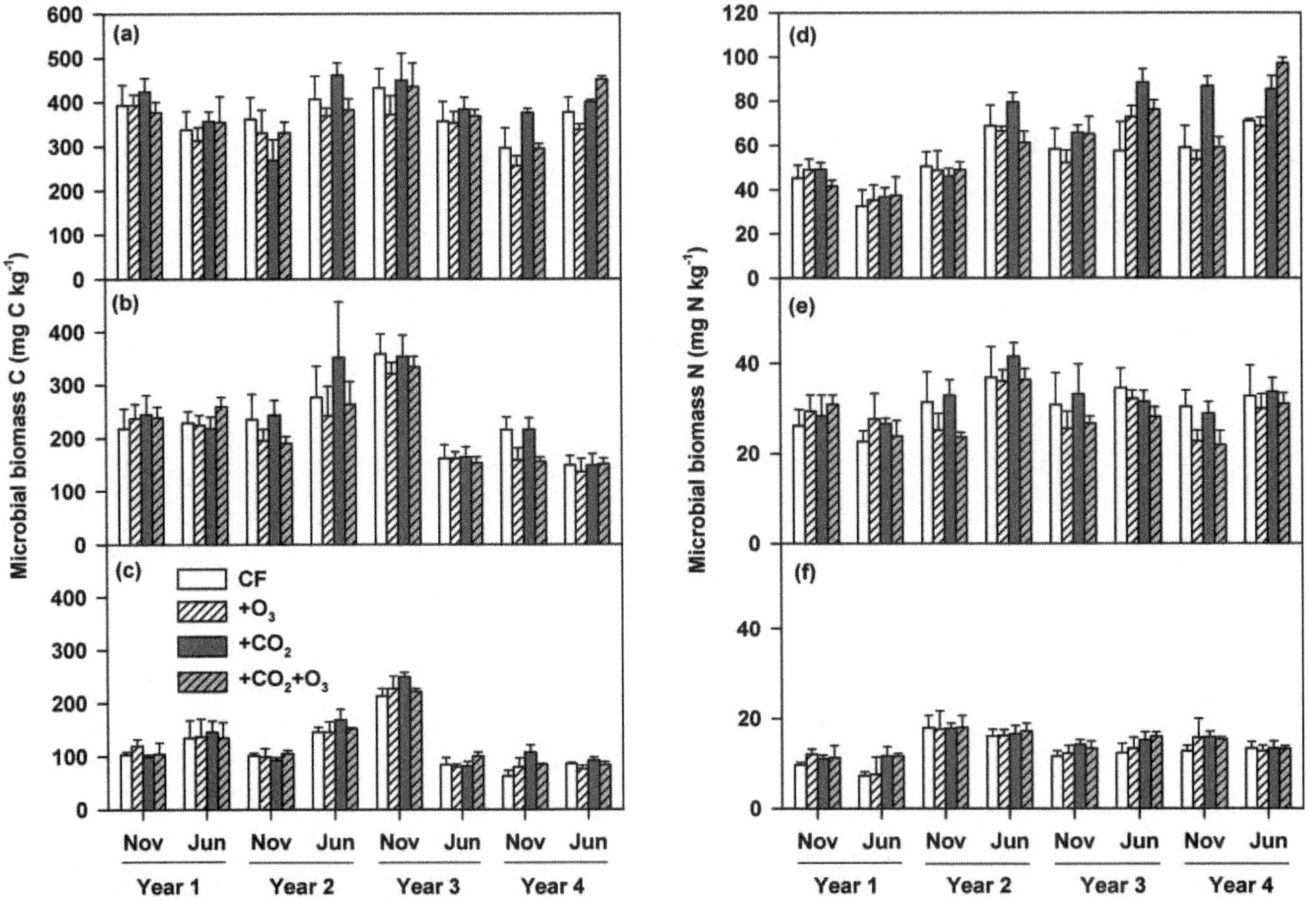

Figure 2. Effects of elevated CO_2 and O_3 on soil microbial biomass C and N. CF, charcoal-filtered ambient air. +O_3, elevated O_3.+CO_2, elevated CO_2.+CO_2+O_3, elevated CO_2+O_3. Microbial biomass C: (a) 0–5 cm soil layer (Repeated measures mixed model; CO_2 effect: $P=0.026$; $CO_2 \times$Time: $P>0.1$), (b) 5–10 cm soil layer (Repeated measures mixed model; CO_2 effect: $P>0.1$; $CO_2 \times$Time: $P>0.1$) and (c) 10–20 cm soil layer (Repeated measures mixed model; CO_2 effect: $P>0.1$; $CO_2 \times$Time: $P>0.1$). Microbial biomass N: (d) 0–5 cm soil layer (Repeated measures mixed model; CO_2 effect: $P=0.025$; $CO_2 \times$Time: $P=0.018$), (e) 5–10 cm soil layer (Repeated measures mixed model; CO_2 effect: $P>0.1$; $CO_2 \times$Time: $P>0.1$) and (f) 10–20 cm soil layer (Repeated measures mixed model; CO_2 effect: $P=0.040$; $CO_2 \times$Time: $P>0.1$). The O_3 and $CO_2 \times O_3$ effects were not significant for all soil layers. Data represent means (n = 4) ± s.e.m.

($P<0.05$; Fig. 5a). As such, the fungi:bacteria ratio increased significantly due to elevated CO_2 ($P<0.05$; Fig. 5c). Neither the O_3 effect nor the $CO_2 \times O_3$ interaction had significant effects on fungal and bacterial PLFAs at any time points (Fig. 5a, 5b).

Correlation analysis

Correlation analysis, conducted among extractable C, extractable N, MBC, MBN, SMR and NMN, showed that all six parameters were significantly correlated with each other, though the coefficients varied considerably (Appendix S5). Net soil N mineralization can best explain the variation of MBC ($R^2=0.44$), MBN ($R^2=0.72$) and SMR ($R^2=0.72$). The χ^2 values for MBC vs N, MBN vs N, SMR vs N and the fungi:bacteria ratio vs N were 14.7 ($P<0.001$), 14.5 ($P<0.001$), 10.8 ($P<0.01$) and 4 ($P<0.05$) (Appendix S6), respectively, indicating that the CO_2 effects on microbial biomass, activity and the community structure were closely related to CO_2-induced alterations in N availability.

Discussion

The effect of elevated CO_2 on soil microbes

Results obtained in this study showed that elevated CO_2 influenced microbial processes over time likely through its impacts on C and N availability (Figs. 2, 3, 4, 5, Appendix S5 and F). Microbial responses to elevated CO_2 have so far largely been considered in the context that soil microbes are C-limited [8,11] while plant growth is N-limited [39,57]. Many experiments have provided evidence that increased C availability induced by elevated CO_2 enhanced soil microbial biomass and/or activities [5,10,11,20,26,58] and can alter the structure of soil microbial communities in favor of fungal growth [16,25,55,59]. In the current study, elevated CO_2 significantly increased C and N availability for microbes by enhancing both aboveground (and presumably belowground) soybean and wheat residue C and N inputs in all four years (Fig. 1). However, microbial biomass, respiration and the community structure did not respond significantly until the third year (Fig. 2a, 2d, 3a, and 5c). Likely, it took a period of time for the increased residue inputs to accumulate to levels that affected soil microbial processes. It should be noted that the CO_2-stimulation in soybean residue inputs was lower in year 2 compared with in other years (Fig. 1), which may have contributed to the time-lag in elevated CO_2 effects on soil microbial processes observed in the present study.

Along with available C in soil, other factors such as soil moisture, soil food-web interactions and nutrient availability have also been suggested to affect microbial responses to elevated CO_2,

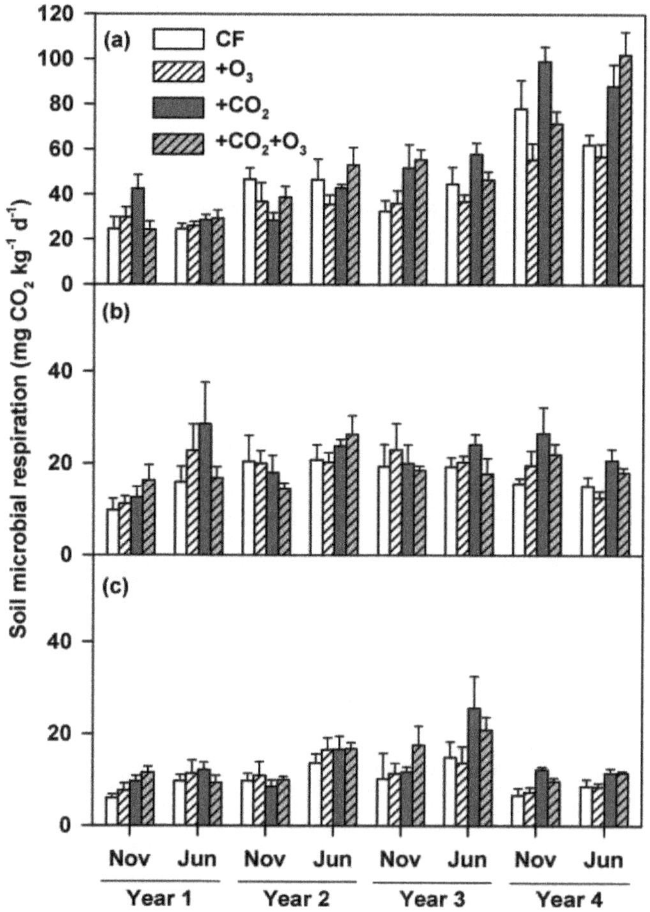

Figure 3. Effects of elevated CO_2 and O_3 on soil microbial respiration. CF, charcoal-filtered ambient air. +O_3, elevated O_3.+CO_2, elevated CO_2.+CO_2+O_3, elevated CO_2+O_3. (a) 0–5 cm soil layer (Repeated measures mixed model; CO_2 effect: $P=0.012$; $CO_2\times$Time: $P=0.003$). (b) 5–10 cm soil layer (Repeated measures mixed model; CO_2 effect: $P>0.1$; $CO_2\times$Time: $P>0.1$). (c) 10–20 cm soil layer (Repeated measures mixed model; CO_2 effect: $P=0.044$; $CO_2\times$Time: $P>0.1$). The O_3 and $CO_2\times O_3$ effects were not significant for all soil layers. Data represent means (n = 4) ± s.e.m.

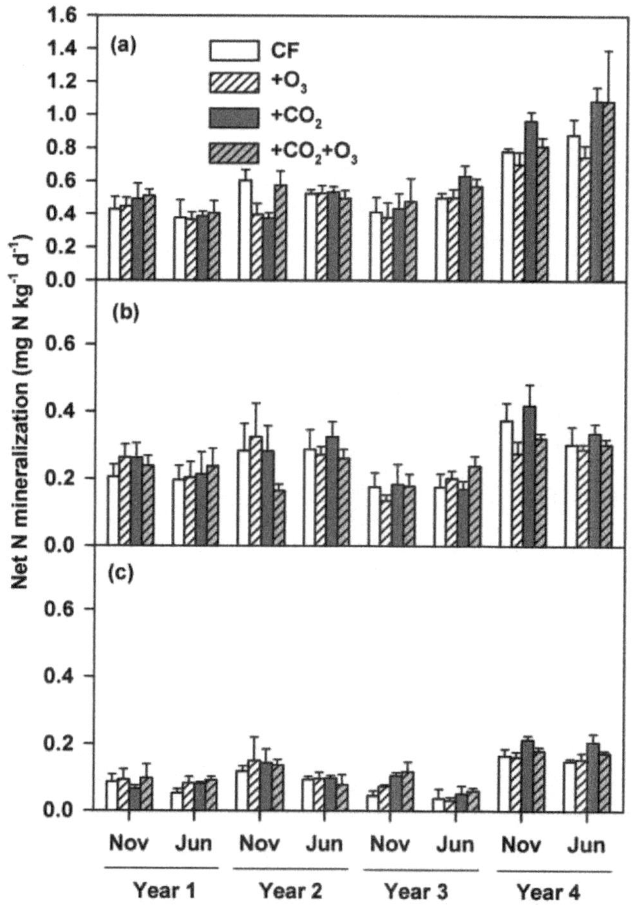

Figure 4. Effects of elevated CO_2 and O_3 on net soil N mineralization. CF, charcoal-filtered ambient air. +O_3, elevated O_3.+CO_2, elevated CO_2.+CO_2+O_3, elevated CO_2+O_3. (a) 0–5 cm soil layer (Repeated measures mixed model; CO_2 effect: $P=0.002$; $CO_2\times$Time: $P=0.011$), (b) 5–10 cm soil layer (Repeated measures mixed model; CO_2 effect: $P>0.1$; $CO_2\times$Time: $P>0.1$) and (c) 10–20 cm soil layer (Repeated measures mixed model; CO_2 effect: $P=0.019$; $CO_2\times$Time: $P>0.1$). The O_3 and $CO_2\times O_3$ effects were not significant for all soil layers. Data represent means (n = 4) ± s.e.m.

either singly or in combination [9,11]. The availability of soil N has so far received the most attention in studies of elevated CO_2 effects on soil microbes [9,20,21] because N is the most abundant nutrient element required for microbial growth [60]. The coincidence of higher microbial activities with increasing soil N availability and microbial biomass N in the CO_2 treatments during the third and fourth years of the experiment suggests a link between soil N availability and microbial responses to elevated CO_2 in this N-aggrading system. With the surface placement of residues in no-till systems, N existing in plant residues (mainly soybean) gradually moves into the soil profile, particularly the top soil layer, through leaching and decomposition processes. Higher N inputs from residues of plants grown under elevated CO_2 (Fig. 1b), which stemmed from both CO_2-stimulation of N_2 fixation (Table 2) [35,36] and possibly plant N retention [4,61], can in turn increase soil N availability for microbes. In a recent meta-analysis of 131 manipulation studies with tree species, Dieleman *et al.* (2010) found that the CO_2-enhancement of microbial activity and decomposition was positively correlated with increasing soil N availability. These results are similar to the CO_2-stimulation of microbial growth and activities along with increasing available soil N observed in our study. Also, it has been well documented that mineral N additions can stimulate decomposition of plant residues, particularly the non-lignin components [28,62]; thus increased soil N availability can significantly facilitate decomposition of non-lignin components of crop residues. All exoenzymes responsible for disintegrating organic materials are N-rich proteins, and sufficient supplies of N for microbes may facilitate enzyme production [28,63], though addition of inorganic N could also suppress lignin-degrading enzymes [64]. Other studies have also recently showed that N addition stimulated microbial respiration [26] and decomposition activities [25].

Changes in fungal and bacterial PLFAs, and their ratios observed in our study provide new insights into how alterations in the relative availability of C and N can modulate microbial activities and their responses to elevated CO_2. First, the coincidence between decreased fungi:bacteria ratios in years 4 and 5 in comparison with previous years and increased N availability and MBN over time is consistent with the general concept that high N availability favors bacteria over fungi [28,29].

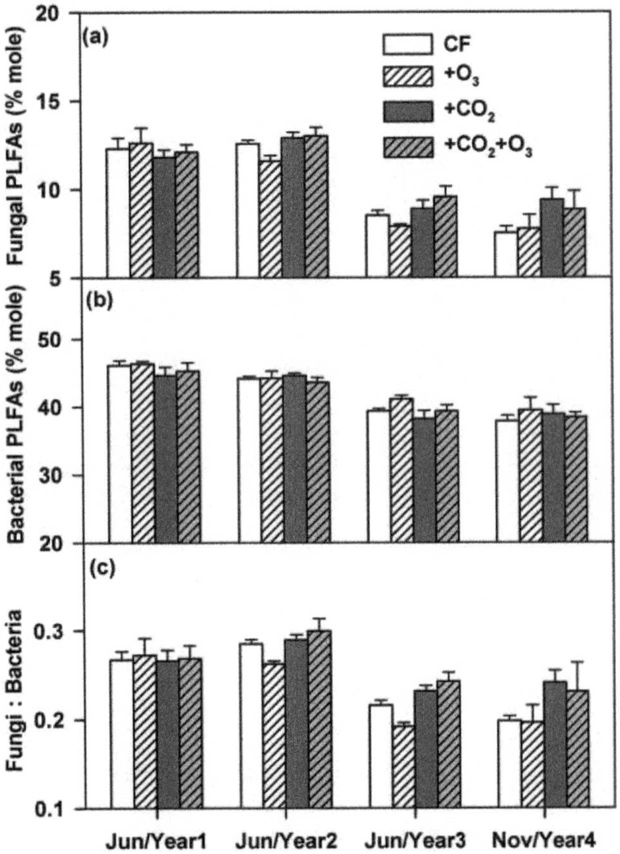

Figure 5. Effects of elevated CO_2 and O_3 on microbial community composition. CF, charcoal-filtered ambient air. +O_3, elevated O_3.+CO_2, elevated CO_2.+CO_2+O_3, elevated CO_2+O_3. (a) the abundance of fungal phospholipid fatty acids (PLFAs) [ANOVA mixed model; CO_2 effect: Jun/Year 1 (P>0.1), Jun/Year 2 (P=0.01), Jun/Year 3 (P=0.029), Nov/Year 4 (P=0.015); O_3 effect: P>0.1 for all four time points; $CO_2 \times O_3$: P>0.1 for all four time points)], (b) the abundance of bacterial PLFAs (ANOVA mixed model; CO_2 effect: P>0.1 for all four time points, O_3 effect: P>0.1 for all four time points; $CO_2 \times O_3$: P>0.1 for all four time points) and (c) the ratio of fungal to bacterial PLFAs in top (0–5 cm) soils [ANOVA mixed model; CO_2 effect: Jun/Year 1 (P>0.1), Jun/Year 2 (P=0.006), Jun/Year 3 (P<0.001), Nov/Year 4 (P=0.03); O_3 effect: P>0.1 for all four time points; $CO_2 \times O_3$: Jun/Year 1 (P>0.1), Jun/Year 2 (P=0.019), Jun/Year 3 (P=0.032), Nov/Year 4 (P>0.1))]. Data represent means (n=4) ± s.e.m.

Evidently, high N inputs due to N fertilization of wheat and soybean N_2 fixation in our system gradually increased soil N availability and altered the soil microbial community composition over time. In a loblolly pine system, Feng *et al.* (2010) also observed that N fertilization reduced the fungi:bacteria ratio. Second, significantly higher fungi:bacteria ratios under elevated than ambient CO_2 indicate that CO_2-enhancement of C inputs may still play a major role in shaping the community structure in N-rich agroecosysystems, as shown in many forests and grasslands [4,16,25,55,59,65]. What was surprising is that the significant increases in microbial respiration (Fig. 3a) concurred with decreased microbial PLFA biomass (Fig. 5a, 5b), leading to an increase in metabolic quotient over time as well as under elevated CO_2 (Appendix S4). Since both fungi and bacteria identified by PLFAs represent the most active part of soil microorganisms [52,59], these results indicate that high N inputs may have stimulated microbial physiological activities and/or microbial biomass turnover. Taken together, our results suggest that CO_2-induced changes in soil N availability might be an important factor that concurrently mediated elevated CO_2 effects on soil microbes and microbial feedbacks in this N-aggrading agroecosystem.

The findings that the stimulation of soil microbes under elevated CO_2 over the course of the experiment may have significant implications for understanding residue turnover and soil C sequestration in agroecosystems under future climate change scenarios. In many natural and semi-natural ecosystems, the CO_2-induced stimulation of plant growth may not persist because of nutrient limitation [9,39]. In agricultural ecosystems, however, N is typically not a limiting factor for plant growth due to the application of chemical N fertilizers and/or the incorporation of legume plants, and CO_2-stimulation of biomass production is expected to be sustained [2,31,66]. Therefore, it has been suggested that elevated CO_2 can increase long-term C storage in agroecosystems, particularly in combination with no-tillage management [38,67,68]. However, this assumption does not fully consider the C output from agroecosystems: unlike forest ecosystems where the standing biomass constitutes a major C pool, most agroecosystems must accumulate C in the soil for ecosystem C sequestration to occur. Consequently, the fate of returning residues will largely determine the potential of agroecosystem C sequestration. The close correlations between N availability and both microbial respiration and metabolic quotient under elevated CO_2 in our study (Fig. 3, Appendix S4 and Appendix S6) indicate that soil microbes became more active with CO_2 enrichment. Our results, along with other previous findings [27], suggest that high N availability may significantly increase soil organic C turnover in agroecosystems through stimulating residue decomposition under future CO_2 scenarios, highlighting the need to examine the interactive effect of soil N availability and atmospheric CO_2 on soil organic C dynamics.

It is also interesting to note that microbial parameters along the soil profile exhibited different patterns under elevated CO_2 (Figs. 2, 3, 4). No-till systems are characterized by vertical stratification of soil organic C and microbial biomass because of continuous residue surface placement [37,69]. Rapid deceases in MBC and diminished CO_2 effects on MBC starting in the third year in deeper soil layers (Fig. 2b and 2c) seems to suggest that alteration in C availability caused by residue placement may dominate microbial responses. However, other parameters [MBN (Fig. 2f), SMR (Fig. 3c) and NMN (Fig. 4c)] did not decrease correspondingly with MBC and continued to significantly respond to elevated CO_2 (Appendix S2), suggesting that other factors may significantly exert control. High correlations between MBN and SMR, and NMN in deeper soil depths (Appendix S5) suggest that N availability critically modulated microbial activities. In no-till systems, root-derived C is the primary source for deep soil C and CO_2-stimulation of both fine and deep roots has been proposed as a potential mechanism that facilitates C sequestration there [37]. However, higher metabolic quotient (Appendix S4), SMR (Fig. 3) and NMN (Fig. 4) under elevated CO_2 indicate that not only were microbes more active but also organic C turnover was more rapid in the deeper soil layers. Consequently, high root production under elevated CO_2 might stimulate C losses from deep soil layers by priming decomposition of indigenous organic matter [15,58,66,70,71]. Long-term experiments are critically needed to examine whether the stimulation of SMR and NMN in our study is transient or will be sustained over time.

Effects of elevated O_3 and $CO_2 \times O_3$ on soil microbes

Elevated O_3 often leads to a substantial decline in the aboveground biomass of O_3-sensitive plants [7,30,72] and subsequent C allocation belowground [6,19]. In the current study, the statistically significant decline in plant residue C primarily stemmed from O_3-reduction of soybean residue C (by 12% on average; $P<0.05$). The unresponsiveness of wheat to O_3 was likely due to the relatively low O_3 concentrations during the wheat growing season (Table 1), use of a relatively O_3-tolerant cultivar, and possibly other environmental conditions (for example, temperature and light levels). Ambient O_3 concentrations during winter wheat growing seasons are usually low due to the lower concentrations of precursors of O_3 formation and the lower temperatures during the winter and the early spring. The decrease in soybean residue N inputs under elevated O_3 (Fig. 1b) resulted from O_3-induced reduction in residue biomass and possibly symbiotic N_2 fixation in soybean plants (Table 2) [36]. However, no significant O_3 effects were detected on any soil microbial parameters in this study (Figs. 2, 3, 4, 5, Appendix S2). These results suggest that N inputs through both fertilization and N_2 fixation in our system might overtake O_3-induced reduction of residue N in affecting soil microbes. Alternatively, these results also suggest that the magnitude of reductions in both C and N under elevated O_3 were insufficient to substantially affect soil microbial activity in our experiment. In an OTC experiment under conventional tillage practice, Islam *et al.* (2000) also found that elevated O_3 had no significant impacts on soil microbial respiration.

Our results showed that elevated O_3 tended to reduce soybean residue C and N inputs under elevated CO_2 (Fig. 1). This indicated that added O_3 prevented a portion of the CO_2-induced stimulation in biomass production from occurring. Such a pattern, however, was not observed for microbial parameters over the course of the experiment (Figs. 2, 3, 4, 5). The lack of microbial responses to O_3 under elevated CO_2 suggests that, as noted above, the magnitude of the combination of elevated CO_2 and O_3 effect on residue C and N inputs was not enough to influence soil microbes in the current study. It is also possible that O_3 might not necessarily diminish the stimulation effect of elevated CO_2 on C allocation belowground through fine root biomass, root exudation and turnover during plant growth, as observed in the Rhinelander free-air CO_2 and O_3 enrichment study using tree species [73]. Regardless of the underlying causes, our results suggest that O_3 may have limited impact on soil microbial processes in agricultural systems under future CO_2 scenarios and that its effect will be dependent on the sensitivity of crop cultivars to O_3.

Conclusions

In summary, results obtained from this study showed that the responses of soil microbes and their community structure to elevated CO_2 significantly changed through time in the N-aggrading wheat-soybean rotation system, and that these may be largely related to CO_2-induced alterations in soil C and N availability. While soil microbial biomass, activities and the community structure composition were little affected by elevated CO_2 in the first two years, they significantly responded to CO_2 enrichment in the third and fourth years of the experiment as N availability increased. However, O_3 effects on soil C and N availability were likely insufficient in magnitude to produce detectable changes in the soil microbial parameters measured. Together, these results highlight the urgent need for considering the interactive impact of C and N availability on microbial activities and decomposition when projecting soil C balance in N-rich systems under future CO_2 scenarios.

Supporting Information

Appendix S1 Effects of CO_2 enrichment on soil microbial parameters during mid-growing seasons.

Appendix S2 *P* values of analyses of repeated measures linear mixed models of CO_2, O_3 and time effects, and all interactions over 4 years.

Appendix S3 Effects of elevated CO_2 and O_3 on soil extractable N.

Appendix S4 Effects of elevated CO_2 and O_3 on metabolic quotient of soil microbes.

Appendix S5 Linear correlations among microbial respiration, microbial biomass C and N, extractable C and N, net N mineralization of soils over the 4-year period.

Appendix S6 Chi-square test of relationship between the CO_2 effect on N availability and the CO_2 effect on microbial biomass, respiration and the community structure.

Acknowledgments

We thank Walter Pursley and Erin Silva of USDA-ARS Plant Science Research Unit for field assistance and maintaining CO_2 and O_3 facilities. We are grateful to Guillermo Ramirez and Lisa Lentz of Soil Science Department of North Carolina State University for their help in nitrogen analyses. We are also grateful to Dr. Yiqi Luo and two anonymous reviewers for their constructive suggestions.

Author Contributions

Conceived and designed the experiments: KOB FLB ELF SH LC CT. Performed the experiments: LC FLB CT. Analyzed the data: LC SH. Contributed reagents/materials/analysis tools: HDS TWR JLD. Wrote the paper: LC SH.

References

1. Drake BG, GonzalezMeler MA, Long SP (1997) More efficient plants: A consequence of rising atmospheric CO_2? Annual Review of Plant Physiology and Plant Molecular Biology 48: 609–639.
2. Kimball BA, Kobayashi K, Bindi M (2002) Responses of agricultural crops to free-air CO_2 enrichment. In: Sparks DL, ed. Advances in Agronomy, Vol 77. San Diego: Academic Press Inc. pp 293–368.
3. Cheng L, Zhu J, Chen G, Zheng X, Oh NH, et al. (2010) Atmospheric CO_2 enrichment facilitates cation release from soil. Ecology Letters 13: 284–291.
4. Hu S, Chapin FS, Firestone MK, Field CB, Chiariello NR (2001) Nitrogen limitation of microbial decomposition in a grassland under elevated CO_2. Nature 409: 188–191.
5. Hungate BA, Holland EA, Jackson RB, Chapin FS, Mooney HA, et al. (1997) The fate of carbon in grasslands under carbon dioxide enrichment. Nature 388: 576–579.
6. Andersen CP (2003) Source-sink balance and carbon allocation below ground in plants exposed to ozone. New Phytologist 157: 213–228.
7. USEPA (2006) Air Quality Criteria for Ozone and Related Photochemical Oxidants. Washinton, D.C.: U.S. Environmental Protection Agency. pp 705–734.
8. Smith JL, Paul EA (1990) The significance of soil microbial biomass estimations. In: Bollag JM, Stotzky G, eds. Soil Biochemistry. New York: Marcel Dekker, Inc. pp 357–396.
9. Hu S, Firestone MK, Chapin FS (1999) Soil microbial feedbacks to atmospheric CO_2 enrichment. Trends in Ecology & Evolution 14: 433–437.
10. Zak DR, Pregitzer KS, Curtis PS, Teeri JA, Fogel R, et al. (1993) Elevated atmospheric CO_2 and feedback between carbon and nitrogen cycles. Plant and Soil 151: 105–117.
11. Zak DR, Pregitzer KS, King JS, Holmes WE (2000) Elevated atmospheric CO_2, fine roots and the response of soil microorganisms: a review and hypothesis. New Phytologist 147: 201–222.
12. Islam KR, Mulchi CL, Ali AA (2000) Interactions of tropospheric CO_2 and O_3 enrichments and moisture variations on microbial biomass and respiration in soil. Global Change Biology 6: 255–265.

13. Phillips RL, Zak DR, Holmes WE, White DC (2002) Microbial community composition and function beneath temperate trees exposed to elevated atmospheric carbon dioxide and ozone. Oecologia 131: 236–244.
14. Blagodatskaya E, Blagodatsky S, Dorodnikov M, Kuzyakov Y (2010) Elevated atmospheric CO_2 increases microbial growth rates in soil: results of three CO_2 enrichment experiments. Global Change Biology 16: 836–848.
15. Langley JA, McKinley DC, Wolf AA, Hungate BA, Drake BG, et al. (2009) Priming depletes soil carbon and releases nitrogen in a scrub-oak ecosystem exposed to elevated CO_2. Soil Biology & Biochemistry 41: 54–60.
16. Rillig MC, Field CB, Allen MF (1999) Soil biota responses to long-term atmospheric CO_2 enrichment in two California annual grasslands. Oecologia 119: 572–577.
17. Pendall E, Mosier AR, Morgan JA (2004) Rhizodeposition stimulated by elevated CO_2 in a semiarid grassland. New Phytologist 162: 447–458.
18. Talhelm AF, Pregitzer KS, Zak DR (2009) Species-specific responses to atmospheric carbon dioxide and tropospheric ozone mediate changes in soil carbon. Ecology Letters 12: 1219–1228.
19. Grantz DA, Gunn S, Vu HB (2006) O_3 impacts on plant development: a meta-analysis of root/shoot allocation and growth. Plant Cell and Environment 29: 1193–1209.
20. Hu S, Tu C, Chen X, Gruver JB (2006) Progressive N limitation of plant response to elevated CO_2: a microbiological perspective. Plant and Soil 289: 47–58.
21. de Graaff MA, van Groenigen KJ, Six J, Hungate B, van Kessel C (2006) Interactions between plant growth and soil nutrient cycling under elevated CO_2: a meta-analysis. Global Change Biology 12: 2077–2091.
22. Rice CW, Garcia FO, Hampton CO, Owensby CE (1994) Soil microbial responses in tall grass prairie to elevated CO_2. Plant and Soil 165: 67–74.
23. Jones TH, Thompson LJ, Lawton JH, Bezemer TM, Bardgett RD, et al. (1998) Impacts of rising atmospheric carbon dioxide on model terrestrial ecosystems. Science 280: 441–443.
24. Barnard R, Le Roux X, Hungate BA, Cleland EE, Blankinship JC, et al. (2006) Several components of global change alter nitrifying and denitrifying activities in an annual grassland. Functional Ecology 20: 557–564.
25. Feng XJ, Simpson AJ, Schlesinger WH, Simpson MJ (2010) Altered microbial community structure and organic matter composition under elevated CO_2 and N fertilization in the duke forest. Global Change Biology 16: 2104–2116.
26. West JB, Hobbie SE, Reich PB (2006) Effects of plant species diversity, atmospheric [CO_2], and N addition on gross rates of inorganic N release from soil organic matter. Global Change Biology 12: 1400–1408.
27. Dieleman WIJ, Luyssaert S, Rey A, De Angelis P, Barton CVM, et al. (2010) Soil [N] modulates soil C cycling in CO_2-fumigated tree stands: a meta-analysis. Plant, Cell & Environment 33: 2001–2011.
28. Fog K (1988) The effect of added nitrogen on the rate of decomposition of organic matter. Biological Reviews of the Cambridge Philosophical Society 63: 433–462.
29. Kaye JP, Hart SC (1997) Competition for nitrogen between plants and soil microorganisms. Trends in Ecology & Evolution 12: 139–143.
30. Fiscus EL, Booker FL, Burkey KO (2005) Crop responses to ozone: uptake, modes of action, carbon assimilation and partitioning. Plant Cell and Environment 28: 997–1011.
31. Ainsworth EA, Davey PA, Bernacchi CJ, Dermody OC, Heaton EA, et al. (2002) A meta-analysis of elevated [CO_2] effects on soybean (*Glycine max*) physiology, growth and yield. Global Change Biology 8: 695–709.
32. Feng ZZ, Kobayashi K, Ainsworth EA (2008) Impact of elevated ozone concentration on growth, physiology, and yield of wheat (*Triticum aestivum* L.): a meta-analysis. Global Change Biology 14: 2696–2708.
33. Morgan PB, Ainsworth EA, Long SP (2003) How does elevated ozone impact soybean? A meta-analysis of photosynthesis, growth and yield. Plant Cell and Environment 26: 1317–1328.
34. Booker FL, Fiscus EL (2005) The role of ozone flux and antioxidants in the suppression of ozone injury by elevated CO_2 in soybean. Journal of Experimental Botany 56: 2139–2151.
35. Rogers A, Ainsworth EA, Leakey ADB (2009) Will elevated carbon dioxide concentration amplify the benefits of nitrogen fixation in legumes? Plant Physiology 151: 1009–1016.
36. Tu C, Booker FL, Burkey KO, Hu S (2009) Elevated atmospheric carbon dioxide and O_3 differentially alter nitrogen acquisition in peanut. Crop Science 49: 1827–1836.
37. Prior SA, Runion GB, Rogers HH, Torbert HA, Reeves DW (2005) Elevated atmospheric CO_2 effects on biomass production and soil carbon in conventional and conservation cropping systems. Global Change Biology 11: 657–665.
38. Lal R (2004) Soil carbon sequestration impacts on global climate change and food security. Science 304: 1623–1627.
39. Reich PB, Hobbie SE, Lee T, Ellsworth DS, West JB, et al. (2006) Nitrogen limitation constrains sustainability of ecosystem response to CO_2. Nature 440: 922–925.
40. Liu LL, Greaver TL (2010) A global perspective on belowground carbon dynamics under nitrogen enrichment. Ecology Letters 13: 819–828.
41. Booker FL, Prior SA, Torbert HA, Fiscus EL, Pursley WA, et al. (2005) Decomposition of soybean grown under elevated concentrations of CO_2 and O_3. Global Change Biology 11: 685–698.
42. Warembourg FR (1993) Nitrogen fixation in soil and plant systems. In: Knowles R, Blackburn TH, eds. Nitrogen Isotope Techniques. San Diego: Academic Press. pp 127–157.
43. Kimball BA, Pinter PJ, Garcia RL, LaMorte RL, Wall GW, et al. (1995) Productivity and water use of wheat under free-air CO_2 enrichment. Global Change Biology 1: 429–442.
44. McMaster GS, LeCain DR, Morgan JA, Aiguo L, Hendrix DL (1999) Elevated CO_2 increases wheat CER, leaf and tiller development, and shoot and root growth. Journal of Agronomy and Crop Science 183: 119–128.
45. Nissen T, Rodriguez V, Wander M (2008) Sampling soybean roots: A comparison of excavation and coring methods. Communications in Soil Science and Plant Analysis 39: 1875–1883.
46. Hu S, Wu J, Burkey KO, Firestone MK (2005) Plant and microbial N acquisition under elevated atmospheric CO_2 in two mesocosm experiments with annual grasses. Global Change Biology 11: 213–223.
47. Vance ED, Brookes PC, Jenkinson DS (1987) An extraction method for measuring soil microbial biomass-C. Soil Biology & Biochemistry 19: 703–707.
48. Cabrera ML, Beare MH (1993) Alkaline persulfate oxidation for determining total nitrogen in microbial biomass extracts. Soil Science Society of America Journal 57: 1007–1012.
49. Jenkinson DS, Brookes PC, Powlson DS (2004) Measuring soil microbial biomass. Soil Biology & Biochemistry 36: 5–7.
50. Coleman DC, Anderson RV, Cole CV, Elliott ET, Woods L, et al. (1978) Trophic interactions in soils as they affect energy and nutrient dynamics .4. flows of metabolic and biomass carbon. Microbial Ecology 4: 373–380.
51. Bossio DA, Scow KM, Gunapala N, Graham KJ (1998) Determinants of soil microbial communities: Effects of agricultural management, season, and soil type on phospholipid fatty acid profiles. Microbial Ecology 36: 1–12.
52. Frostegard A, Baath E (1996) The use of phospholipid fatty acid analysis to estimate bacterial and fungal biomass in soil. Biology and Fertility of Soils 22: 59–65.
53. Zhang W, Parker KM, Luo Y, Wan S, Wallace LL, et al. (2005) Soil microbial responses to experimental warming and clipping in a tallgrass prairie. Global Change Biology 11: 266–277.
54. Balser TC, Treseder KK, Ekenler M (2005) Using lipid analysis and hyphal length to quantify AM and saprotrophic fungal abundance along a soil chronosequence. Soil Biology & Biochemistry 37: 601–604.
55. Carney KM, Hungate BA, Drake BG, Megonigal JP (2007) Altered soil microbial community at elevated CO_2 leads to loss of soil carbon. Proceedings of the National Academy of Sciences of the United States of America 104: 4990–4995.
56. Littell RC, Milliken GA, Strooup WW, Wolfinger RD, Schabenberger O (2006) SAS for Mixed Models. Cary, NC, USA: SAS Institute Inc.
57. Oren R, Ellsworth DS, Johnsen KH, Phillips N, Ewers BE, et al. (2001) Soil fertility limits carbon sequestration by forest ecosystems in a CO_2-enriched atmosphere. Nature 411: 469–472.
58. Phillips RP, Finzi AC, Bernhardt ES (2011) Enhanced root exudation induces microbial feedbacks to N cycling in a pine forest under long-term CO_2 fumigation. Ecology Letters 14: 187–194.
59. Jin VL, Evans RD (2010) Microbial ^{13}C utilization patterns via stable isotope probing of phospholipid biomarkers in Mojave Desert soils exposed to ambient and elevated atmospheric CO_2. Global Change Biology 16: 2334–2344.
60. Duboc P, Schill N, Menoud L, Vangulik W, Vonstockar U (1995) Measurements of sulfur, phosphorus and other ions in microbial biomass: influence on correct determination of elemental composition and degree of reduction. Journal of Biotechnology 43: 145–158.
61. Finzi AC, DeLucia EH, Hamilton JG, Richter DD, Schlesinger WH (2002) The nitrogen budget of a pine forest under free air CO_2 enrichment. Oecologia 132: 567–578.
62. Knorr M, Frey SD, Curtis PS (2005) Nitrogen additions and litter decomposition: A meta-analysis. Ecology 86: 3252–3257.
63. Schimel JP, Weintraub MN (2003) The implications of exoenzyme activity on microbial carbon and nitrogen limitation in soil: a theoretical model. Soil Biology & Biochemistry 35: 549–563.
64. DeForest JL, Zak DR, Pregitzer KS, Burton AJ (2004) Atmospheric nitrate deposition, microbial community composition, and enzyme activity in northern hardwood forests. Soil Science Society of America Journal 68: 132–138.
65. Zak DR, Ringelberg DB, Pregitzer KS, Randlett DL, White DC, et al. (1996) Soil microbial communities beneath *Populus grandidentata* grown under elevated atmospheric CO_2. Ecological Applications 6: 257–262.
66. Peralta AL, Wander MM (2008) Soil organic matter dynamics under soybean exposed to elevated [CO_2]. Plant and Soil 303: 69–81.
67. Paustian K, Andren O, Janzen HH, Lal R, Smith P, et al. (1997) Agricultural soils as a sink to mitigate CO_2 emissions. Soil Use and Management 13: 230–244.
68. West TO, Post WM (2002) Soil organic carbon sequestration rates by tillage and crop rotation: A global data analysis. Soil Science Society of America Journal 66: 1930–1946.
69. Beare MH, Parmelee RW, Hendrix PF, Cheng WX, Coleman DC, et al. (1992) Microbial and faunal interactions and effects on litter nitrogen and decomposition in agroecosystems. Ecological Monographs 62: 569–591.
70. Fontaine S, Barot S, Barre P, Bdioui N, Mary B, et al. (2007) Stability of organic carbon in deep soil layers controlled by fresh carbon supply. Nature 450: 277–280.

71. de Graaff M-A, Classen AT, Castro HF, Schadt CW (2010) Labile soil carbon inputs mediate the soil microbial community composition and plant residue decomposition rates. New Phytologist 188: 1055–1064.
72. Booker F, Muntifering R, McGrath M, Burkey K, Decoteau D, et al. (2009) The Ozone Component of Global Change: Potential Effects on Agricultural and Horticultural Plant Yield, Product Quality and Interactions with Invasive Species. Journal of Integrative Plant Biology 51: 337–351.
73. Pregitzer KS, Burton AJ, King JS, Zak DR (2008) Soil respiration, root biomass, and root turnover following long-term exposure of northern forests to elevated atmospheric CO_2 and tropospheric O_3. New Phytologist 180: 153–161.

7

Arbuscular Mycorrhizal Colonization Alters Subcellular Distribution and Chemical Forms of Cadmium in *Medicago sativa* L. and Resists Cadmium Toxicity

Yuanpeng Wang[1,2]⁹, Jing Huang[1]⁹, Yanzheng Gao[1]*

1 Institute of Organic Contaminant Control and Soil Remediation, College of Resource and Environmental Sciences, Nanjing Agricultural University, Nanjing, Jiangsu Province, P.R. China, **2** Department of Chemical and Biochemical Engineering, College of Chemistry and Chemical Engineering, Xiamen University, Xiamen, Fujian Province, P.R. China

Abstract

Some plants can tolerate and even detoxify soils contaminated with heavy metals. This detoxification ability may depend on what chemical forms of metals are taken up by plants and how the plants distribute the toxins in their tissues. This, in turn, may have an important impact on phytoremediation. We investigated the impact of arbuscular mycorrhizal (AM) fungus, *Glomus intraradices*, on the subcellular distribution and chemical forms of cadmium (Cd) in alfalfa (*Medicago sativa* L.) that were grown in Cd-added soils. The fungus significantly colonized alfalfa roots by day 25 after planting. Colonization of alfalfa by *G. intraradices* in soils contaminated with Cd ranged from 17% to 69% after 25–60 days and then decreased to 43%. The biomass of plant shoots with AM fungi showed significant 1.7-fold increases compared to no AM fungi addition under the treatment of 20 mg·kg^{-1} Cd. Concentrations of Cd in the shoots of alfalfa under 0.5, 5, and 20 mg·kg^{-1} Cd without AM fungal inoculation are 1.87, 2.92, and 2.38 times higher, respectively, than those of fungi-inoculated plants. Fungal inoculation increased Cd (37.2–80.5%) in the cell walls of roots and shoots and decreased in membranes after 80 days of incubation compared to untreated plants. The proportion of the inactive forms of Cd in roots was higher in fungi-treated plants than in controls. Furthermore, although fungi-treated plants had less overall Cd in subcellular fragments in shoots, they had more inactive Cd in shoots than did control plants. These results provide a basis for further research on plant-microbe symbioses in soils contaminated with heavy metals, which may potentially help us develop management regimes for phytoremediation.

Editor: Vishal Shah, Dowling College, United States of America

Funding: This work was supported by the National Natural Science Foundation of China (21077056, 41071212, 41071302), the Natural Science Foundation of Jiangsu Province (BE2011780), the Fundamental Research Funds for the Central Universities of China (KYZ201109), and the Fok Ying Tung Education Foundation (122045). The funders had no role in study design, data collection and analysis, decision to publish, or preparation of the manuscript.

Competing Interests: The authors have declared that no competing interests exist.

* E-mail: gaoyanzheng@njau.edu.cn

⁹ These authors contributed equally to this work.

Introduction

Cadmium (Cd) is a widespread hazardous heavy metal. Many agricultural soils have elevated concentrations of Cd resulting from management practices such the application of sewage sludge or animal manure. Mining activities can also lead to high concentrations of Cd in surrounding lands. These and other practices may threaten environmental quality and sustainable food production [1,2]. Excessive concentrations of Cd are toxic to plants and profoundly interfere with a series of physiological processes such as enzyme activity, respiration, photosynthesis, and nutrient assimilation [3].

However, some plants that can grow in Cd-contaminated soils have evolved mechanisms for tolerating heavy metals inside plant cells [4]. There is some evidence that metal tolerance and detoxification in plants can be achieved by confining toxins to a subcellular distribution or by changing their chemical structure. For example, zinc (Zn) and Cd are preferentially stored in vacuoles of the epidermal and mesophyll cells of the Zn/Cd hyperaccumulator *Thlaspi caerulescens* [5,6]. Similarly, in *Brassica juncea* and *Arabidopsis thaliana*, leaf trichomes appear to be preferential storage and detoxification sites for Cd [7]. In Cd-tolerant *Salix viminalis*, pectin-rich layers of the collenchyma cell walls of leaf veins are an important Cd sink [8].

Although some recent studies have shown that several plants can tolerate and detoxify Cd, few studies have examined the effects of soil microbes that form associations with these plants, especially arbuscular mycorrhizal (AM) fungi. AM fungi play significant roles in the recycling of plant nutrients, maintenance of soil structure, detoxification of noxious chemicals, control of plant pests, and regulation of plant growth and its interactions with the soil environment [9,10]. Plant-microbe symbioses are ubiquitous in both natural and most anthropogenically influenced soils [11–13]. In addition, AM fungi have been shown to transport and immobilize Cd in the root system, which reduces the detrimental effects of the metal on plant physiology [14]. Therefore, the use of mycorrhizal plants for land remediation has been proposed. However, how AM fungi might affect other aspects of heavy metal accumulation and distribution in plants remains uncertain.

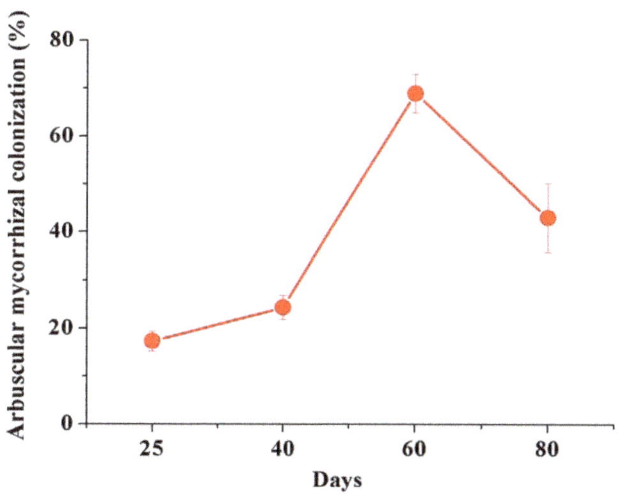

Figure 1. Arbuscular mycorrhizal colonization (%) of alfalfa (*Medicago sativa* L.) exposed to 20 mg kg^{-1} Cd in soil.

Alfalfa (*Medicago sativa* L.) is a deep-rooting perennial plant with high biomass production. It grows quickly, is tolerant to drought, and does not have any reported environmental hazards. In this sense, it is an ideal natural resource for the remediation of contaminated soils [15,16]. Batch laboratory experiments have determined that alfalfa can bind various heavy metal ions [17]. In addition, alfalfa has been shown to tolerate and take up heavy metals from soil; however, relatively few studies have investigated the effects of AM fungi on the uptake and distribution of metals (and the forms of metals) in alfalfa [18,19].

In the present study, we investigated the effects of AM fungi on the growth and subcellular uptake of Cd of alfalfa planted in soil contaminated with Cd. We also examined forms of Cd taken up by the plants and quantified the amounts of each form of Cd in plant tissues. Our results elucidate the interaction among AM fungi, alfalfa, and Cd-contaminated soil, and provide new insights into how AM fungi may affect plant uptake and distribution of heavy metals.

Materials and Methods

Soil Characterization

Uncontaminated loamy soil (13.4% sand, 61.9% silt, 24.7% clay, pH 6.0) was collected from the top 20 cm of a field site in Jiangning, Nanjing, China, and the total Cd concentration in soil was very low at 0.014 $mg{\cdot}kg^{-1}$. According to the United States Department of Agriculture soil taxonomy, the collected soil was Eutric Planosol [20]. The soil was air-dried and sieved (<0.45 mm) to remove plant materials, soil macrofauna, and stones. The samples were sieved using a 10-mm mesh sieve and wrapped in aluminum foil packages of approximately 1 kg. Each package was autoclaved at 121°C for 30 min to inactivate native AM fungi. A representative uncontaminated soil sample was analyzed for organic matter (2.41%), total nitrogen (1.02 $g{\cdot}kg^{-1}$), available phosphorous (7.70 $mg{\cdot}kg^{-1}$), and available potassium (60.38 $mg{\cdot}kg^{-1}$).

Cd Treatments

Subsamples contaminated with Cd were prepared from bulk soil. Briefly, Cd concentrations of 0.5, 5, and 20 $mg{\cdot}kg^{-1}$ soil were used, added as $CdCl_2{\cdot}2.5H_2O$. The Cd salt was dissolved in distilled water, sprayed onto the soil samples, and mixed thoroughly. Then the treated soil samples were incubated at 70% maximum field water-holding capacity for 3 months. After incubation, the subsamples were air-dried naturally, fully homogenized, and stored for use. Three months pre-treatment of the soil samples may have an impact on soil heavy metals availability. However, this pre-treatment made the soil samples close to real contaminated soil. Although Cd had different speciation in the soil and the availability of Cd were not presented in the study, the total concentration of Cd is a good representation of the contamination.

Seedling Preparation and Incubation

Seeds of alfalfa were soaked in distilled water for 30 min and then germinated in commercial potting mixture. Seedlings were watered daily with a nutrient solution for 2 weeks. Then uniform seedlings were selected and transplanted to porcelain pots, each containing 0.4 kg Cd-contaminated soil. *Glomus intraradices* was selected as the AM fungi, based on our previous study [21]. Mycorrhizal pots were inoculated with 15 g *G. intraradices* spores. Non-mycorrhizal controls received an equivalent amount of sterilized inoculum to provide similar conditions, except for the absence of the active mycorrhizal fungi. All pots were equilibrated in a glass greenhouse to 50% water-holding capacity. All treatments were carried out in triplicate.

The plants were grown in the greenhouse with average day and night temperatures of 32°C and 16°C, respectively. For plant biomass and colonization by the AM fungi *G. intraradices,* the plants were harvested at different time (25, 40, 60 and 80 days after transplantation). For subcellular and chemical forms of Cd, the plants were harvested after 80 days after transplantation. They were removed from the pots, and their roots were washed in tap water to remove soil and then washed with deionized water.

Mycorrhizal Analysis

Root samples of the alfalfa plants were cleared and stained as described by Phillips and Hayman [22], and then a 1-g subsample of fresh root was randomly taken and cut into approximately 1 cm pieces to estimate the proportion of total root length colonized by the AM fungi. Mycorrhizal colonization was determined using the gridline intersect method [23]. Briefly, the stained root segments were arranged lengthwise on a thin layer of PVA mounted on a microscope slide. A hairline graticule inserted into the eyepiece of a compound microscope acted as a line of intersection with the roots. Fungal structures at each intersection were calculated by observation at 200× magnification (Nikon, TE2000).

Subcellular Fractions

Frozen materials were homogenized in cold extract buffer containing 50 mM Hanks' balanced salt solution (HEPES), 500 mM sucrose, 1.0 mM dithiothreitol (DTT), 5.0 mM ascorbic acid, and 1.0% (w/v) polyvinyl polypyrrolidone (PVPP), and adjusted to pH 7.5 with NaOH. Cells were separated into the four fractions of cell wall, soluble fraction, organelle, and cell membranes using the differential centrifugation technique reported by Lozano-Rodriguez et al. [24] with some modifications [4]. The homogenate was sieved through a nylon cloth (100 µm mesh size) and washed with extraction buffer. This residue, together with the pellet retained after centrifugation of the filtrate at 100 g for 5 min, constituted the cell wall-bound metals. The supernatant contained the remaining metals, which were further separated into three fractions. First, the supernatant was centrifuged at 10,000 g for 30 min, producing a pellet containing the organelles. Then it was centrifuged at 100,000 g for 30 min, resulting in a pellet containing the cell membranes. The remaining

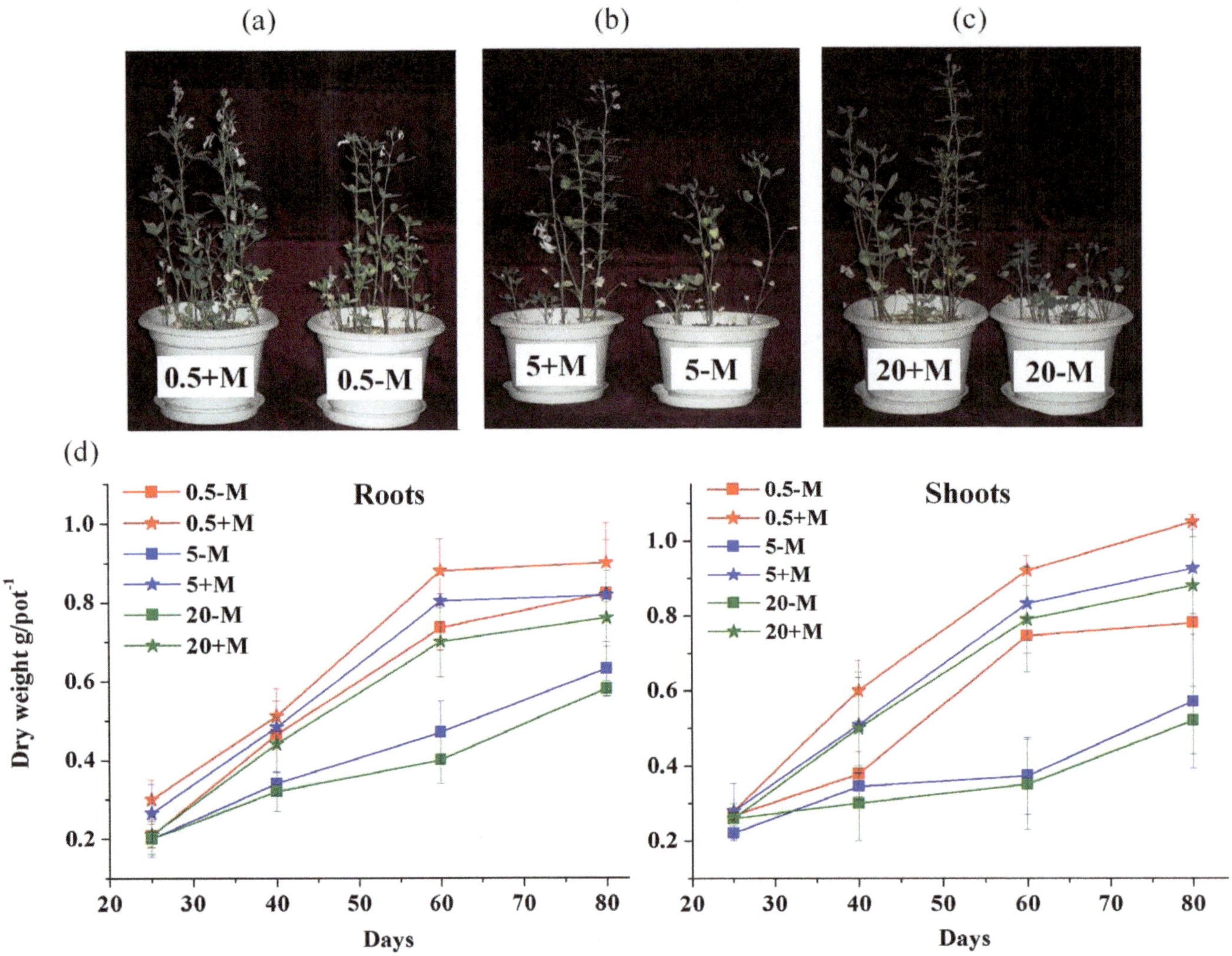

Figure 2. Effect of Cd^{2+} toxicity on plant growth in plants inoculated with the AM fungus after 80 days (a: 0.5 mg kg^{-1} Cd treatment; b: 5 mg kg^{-1} treatment; c: 20 mg kg^{-1} treatment) and the biomass of alfalfa for untreated plants (No AMF) and treated plants (AMF) after 25–80 days of growth in 0.5, 5 and 20 mg kg^{-1} Cd soil (d) (0.5-M means 0.5 mg kg^{-1} Cd treatment without AM fungal inoculation; 0.5+M means 0.5 mg kg^{-1} Cd treatment with AM fungal inoculation).

supernatant contained the soluble fraction. All steps were performed at 4°C.

Chemical Forms

To determine the chemical forms of Cd in alfalfa, the following extractions were performed in the following order: (1) 80% ethanol, extracting inorganic Cd giving priority to nitrate/nitrite, chloride, and aminophenol cadmium; (2) deionized water (d-H_2O), extracting water soluble Cd-organic acid complexes and $Cd(H_2PO_4)_2$; (3) 1 M NaCl, extracting pectates and protein-integrated Cd; (4) 2% acetic acid (HAc), extracting undissolved cadmium phosphate including $CdHPO_4$ and $Cd_3(PO_4)_2$ and other Cd-phosphate complexes; and (5) 0.6 M HCl, extracting cadmium oxalate [4].

Frozen tissues were homogenized in extraction solution with a mortar and pestle, diluted at a ratio of 1:100 (w/v), and shaken for 22 h at 25°C. Then the homogenate was centrifuged at 5000×g for 10 min, obtaining the first supernatant solution in a conical beaker. The sedimentation was re-suspended twice in extraction solution, shaken for 2 h at 25°C, and centrifuged at 5000×g for 10 min. Then the supernatants of the three suspensions and centrifuge steps for each of the five extraction solutions were pooled. Each pooled solution was evaporated on an electric plate at 70°C to constant weight.

Analysis of Heavy Metals

Before metal analysis, 5 μL concentrated HNO_3 was added to each 1 mL sample solution. Both plant materials and the centrifuged fractions were dried at 105°C for 24 h. Plant materials were weighed to determine the dry weight (DW). Then all plant materials and centrifuged fractions were wet-digested in concentrated HNO_3:$HClO_4$ (7:3, v:v). Metal concentrations of all plant materials and all centrifuged fractions were determined using inductively coupled plasma optical emission spectrometry (ICP-OES, Optima 2000DV, Perkin-Elmer Co., USA).

Statistical Analysis

All data were analyzed using Microsoft Excel, Origin and SPSS. The treatment effects were carried out with one-way ANOVA and

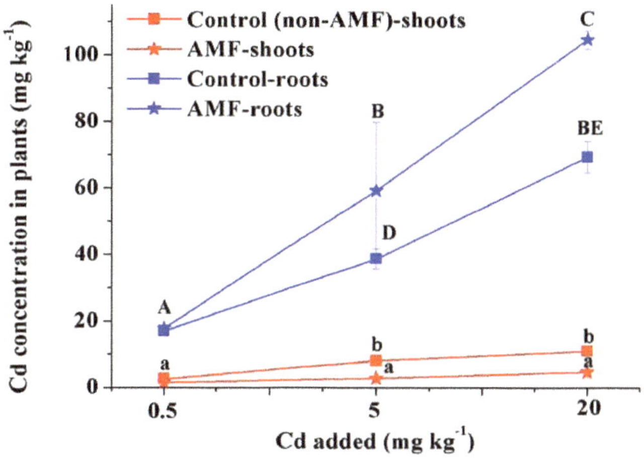

Figure 3. Accumulated Cd in alfalfa inoculated with *G. intraradices* after 80 days of seedling growth. Sharing a common lowercase are not significantly different in the Cd concentration in plant shoots and the same capital are not significantly different in the Cd concentration in plant roots ($P<0.05$).

the LSD multiple range test was used to determine the statistical significance ($P<0.05$) between pairs with SPSS.

Results

Colonization by the AM Fungi *G. intraradices*

Because the soils in our greenhouse experiments were sterilized, there was no colonization by the AM fungi or any indigenous fungi in untreated plants. In plants inoculated with *G. intraradices*, mycelia in endodermis of root cells were observed after 25 days of growth. They first penetrated root cells, and then slowly branched from the outer cortical cells to the inner cortical cells. Some of intracellular hyphae were straight and slowly began to form arbuscule. Vesicles could be detected at 60 days after inoculation. At 80 days, spores formed and filled the entire cell (Fig. S1). Colonization of alfalfa by *G. intraradices* in soils contaminated with Cd ranged from 17% to 69% after 25–60 days and then decreased to 43% (Fig. 1).

Plant Growth

The biomass of all plants increased with incubation time, reaching maximum values at 80 days. The shoots and roots biomass of untreated plants decreased with increasing Cd concentration, however, AM fungal inoculation significantly increased the biomass of roots and shoots compared to untreated plants. When Cd addition concentration was 20 mg kg^{-1}, the biomass of plant shoots with AM fungi showed significant 1.7-fold increases compared to no AM fungal addition ($P<0.05$). On the other hand, there were no significant differences on the biomass with AM fungal inoculation under 0.5 mg kg^{-1} and 5 mg kg^{-1} Cd treatments (Fig. 2).

Plant Uptake of Cd

Concentrations of Cd in the shoots of alfalfa were 2.56, 8.02, and 11.02 mg·kg^{-1} under 0.5, 5, and 20 mg·kg^{-1} Cd without AM fungal inoculation; these values are 1.87, 2.92, and 2.38 times higher, respectively, than those of fungi-inoculated plants. However, Cd concentrations were higher in the roots than in the shoots, especially in fungi-inoculated plants (Fig. 3).

Subcellular Distribution of Cd

Table 1 summarizes the subcellular distribution of Cd in shoots and roots. The recovery rate of Cd uptake was more than 85% for most treatments, and Cd concentrations in subcellular fractions increased with increasing heavy metal concentrations with or without AM inoculation. Most Cd was present in the soluble and cell wall fractions, with very little in the membrane.

Fungi-inoculated plants had less Cd in membranes after 80 days of incubation compared to untreated plants. In addition, fungal inoculation led to less Cd accumulation in subcellular fractions of shoots, increased levels (37.2–80.5%) in the cell walls of roots and

Table 1. Subcellular distribution of Cd in alfalfa.

Plant	Cd (mg/kg)	AM fungi	The amount of heavy metals in each cell fraction per unit (kg) (mg/kg)				
			Cell wall	Soluble fraction	Organelles	Membranes	Recovery[b] (%)
Root	0.5	+M[a]	10.77±0.52[b]	4.09±0.08	3.34±0.25	0.37±0.01	85.2[c]
		−M[d]	5.42±0.16	0.75±0.38	3.84±0.55	0.51±0.01	80.5
	5	+M	44.57±11.32	14.59±5.70	3.98±3.53	0.17±0.04	94.7
		−M	21.26±0.76	12.75±0.89	8.79±1.26	1.06±0.09	86.0
	20	+M	57.27±1.80	26.06±0.82	16.06±0.09	2.30±0.03	97.2
		−M	23.44±1.98	17.09±1.49	21.92±1.24	0.62±0.01	93.4
Shoot	0.5	+M	0.47±0.02	0.48±0.06	0.06±0.00	–	85.1
		−M	2.00±0.02	0.32±0.01	0.25±0.02	–	90.2
	5	+M	0.91±0.03	0.94±0.01	0.22±0.03	0.38±0.02	91.9
		−M	5.24±0.01	0.63±0.31	0.76±0.01	0.47±0.01	87.3
	20	+M	2.02±0.01	0.97±0.01	0.20±0.00	0.51±0.01	87.3
		−M	7.64±0.61	0.54±0.01	0.73±0.01	0.58±0.01	90.6

[a]With AM fungal inoculation.
[b]Mean±standard deviation(n = 3).
[c]Percentage recovery (%) = (cell wall +organelle+ soluble fraction+ membrane)×100%/total.
[d]Without AM fungal inoculation.

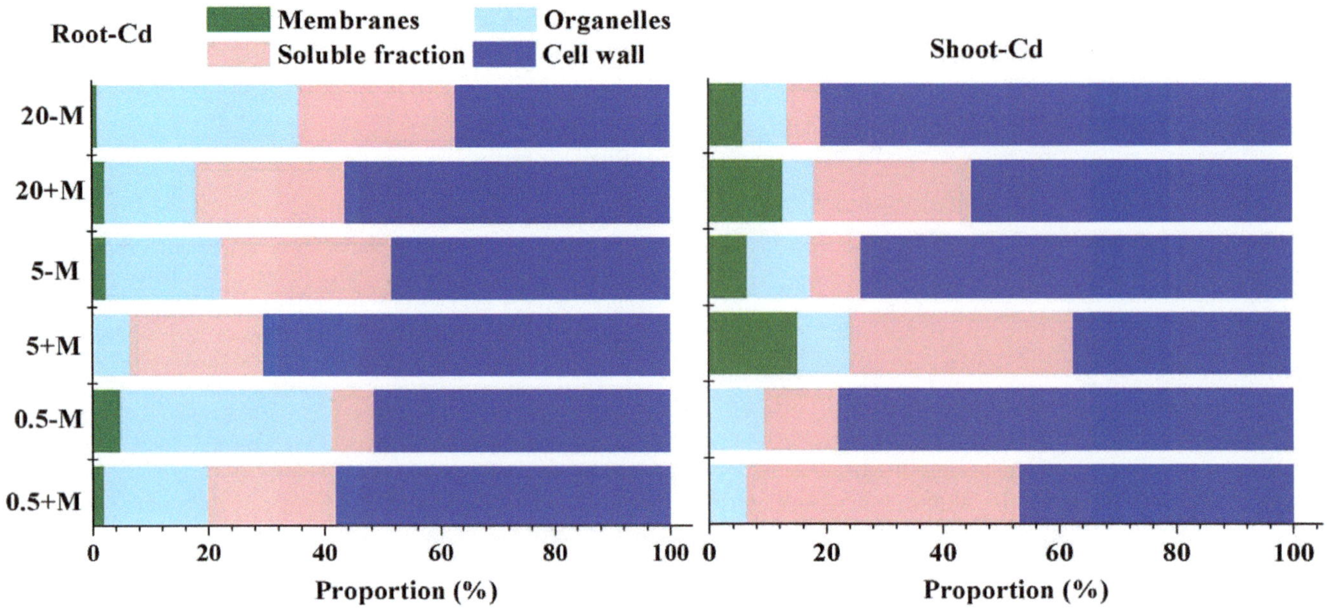

Figure 4. The proportion of Cd with a subcellular distribution in alfalfa.

shoots (Fig. 4), and decreased levels in organelles and membranes compared to controls.

Chemical Forms of Cd

Cd of different forms increased with increasing Cd exposure (Table 2). In roots, undissolved and pectates and protein-integrated Cd were predominant in both untreated and fungi-inoculated plants, whereas other forms were rare. In shoots, pectates and protein-integrated Cd and water soluble Cd, and Cd residues, made up the largest proportion of Cd, followed by inorganic and oxalate Cd.

The undissolved and oxalate Cd, and Cd residues, generally showed no or little toxicity to plants. Hence, these three forms were considered to be inactive. The proportion of these inactive forms of Cd in roots was higher in fungi-treated plants than in controls. Furthermore, although fungi-treated plants had less overall Cd in subcellular fragments in shoots, they had more inactive Cd in shoots than did control plants (Fig. 5).

Discussion

Certain plants can be used to clean up contaminated soils and waters. This detoxification ability may depend on what chemical forms of metals are taken up by plants and how the plants distribute the toxins in their tissues, and hence these factors may impact phytoremediation efforts [25–27]. One such plant is alfalfa, which is widely distributed in both contaminated and uncontaminated areas of China [15]. We investigated the subcellular distribution and chemical forms of Cd in alfalfa (*Medicago sativa* L.) inoculated with the arbuscular mycorrhizal (AM) fungi *Glomus*

Table 2. Distribution of each chemical form of Cd in alfalfa.

			The amount of Cd in each fraction per unit (kg) (mg/kg)					
Plant	Cd (mg/kg)	AM fungi	$F_{ethanol}$	Fd-H_2O	F_{Nacl}	F_{HAc}	F_{HCl}	$F_{residue}$
Root	0.5	+M	2.91±0.06	4.15±0.04	6.73±0.03	6.09±0.06	2.58±0.05	1.50±0.08
		−M	2.57±0.04	4.59±0.06	5.57±0.06	4.12±0.42	2.07±0.05	1.06±0.06
	5	+M	8.20±1.78	8.99±1.99	14.79±3.17	13.80±3.01	7.19±1.58	3.06±0.66
		−M	7.24±0.09	8.07±0.05	14.32±0.33	10.11±0.04	4.52±0.04	1.38±0.04
	20	+M	14.6±0.28	19.41±0.27	32.62±0.26	29.80±0.02	12.15±0.39	8.44±0.01
		−M	8.11±0.02	13.03±0.25	23.84±0.18	16.93±0.19	5.98±0.03	3.58±0.04
Shoot	0.5	+M	0.32±0.00	2.86±1.26	1.21±0.01	0.22±0.01	0.30±0.00	1.04±0.03
		−M	0.64±0.01	2.35±0.02	3.85±0.00	0.07±0.00	0.58±0.02	1.30±0.04
	5	+M	0.61±0.04	0.99±0.27	2.25±0.02	0.38±0.02	0.61±0.05	2.10±0.08
		−M	1.84±0.01	6.79±0.12	0.89±0.39	0.90±0.39	1.33±0.00	3.65±0.00
	20	+M	1.07±0.00	2.37±0.01	4.33±0.01	0.70±0.01	0.81±0.00	3.95±0.01
		−M	1.07±0.01	6.40±0.03	15.50±0.02	0.27±0.02	1.40±0.01	3.27±0.00

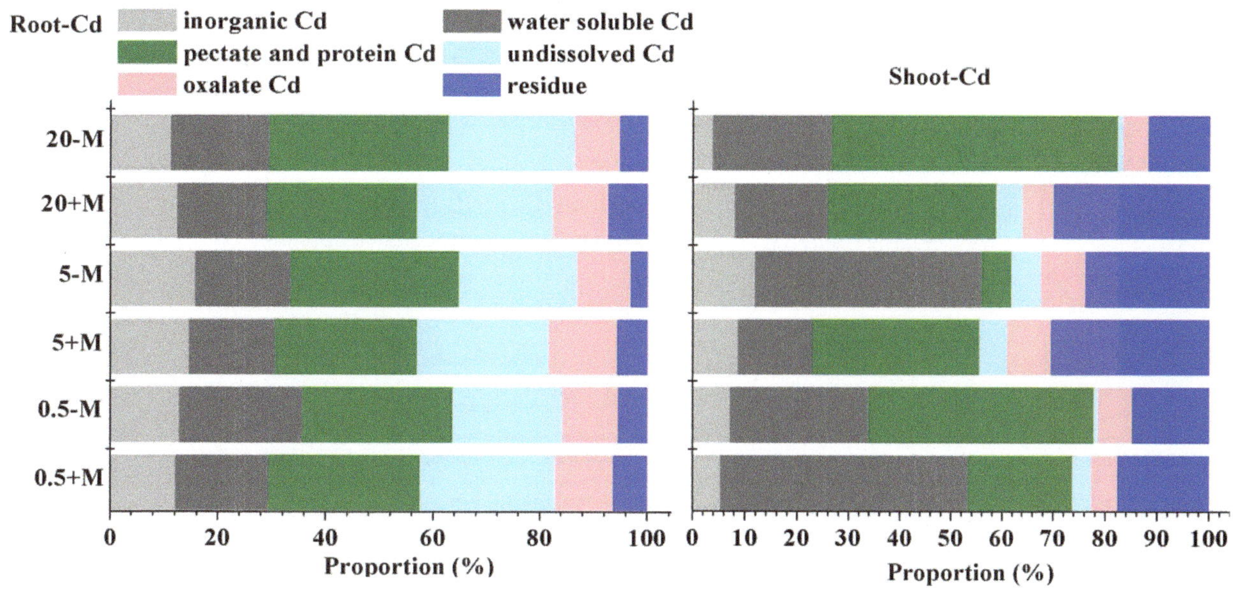

Figure 5. The proportion of each chemical form of Cd in alfalfa.

intraradices and planted in contaminated soil. Our results indicate that AM fungi have the potential ability of reducing Cd accumulation and beneficial consequences of mycorrhizal associations on plant growth in heavy metals contaminated soil.

High concentrations of Cd were toxic to alfalfa plants lacking AM fungi, which significantly decreased biomass. However, AM fungal inoculation significantly increased the biomass of roots and shoots compared to untreated plants and there were no significant differences on the biomass with AM fungal inoculation under 0.5 mg kg^{-1} and 5 mg kg^{-1} Cd treatments (Fig. 2). In present study, AM fungi alter Cd transformation in alfalfa and resist Cd toxicity by reducing Cd uptake from roots to shoots, distributing Cd in the subcellular fractions of cell walls, restricting Cd accumulation in membranes, and isolating Cd in an inactive state in roots and shoots.

Plants that accumulate and/or tolerate Cd have evolved mechanisms to resist taking up metals in the soil and/or to tolerate metals inside cells [28]. Some mycorrhizal fungi that grow in association with plants have been shown to reduce the accumulation of certain heavy metals in the shoots of host plants, presumably by increasing metal retention within roots [11,29]. Indeed, in the present study, plants inoculated with AM fungi had larger amounts of Cd in their roots and smaller amounts in their shoots (Fig. 3), improving the plants' resistance to Cd toxicity compared to controls. In addition, AM-treated plants distributed more Cd in subcellular fractions of tissues, further reducing Cd accumulation in membranes (Table 1, Fig 4). Excessive accumulation in plant membranes would destroy cell activity and thus inhibit plant growth [25,26]. Hence, selective distribution of toxins might be an important strategy for heavy metal tolerance and detoxification in plants.

The fungi-treated plants also sequestered more Cd in cell walls. Cell walls are mainly composed of cellulose, hemicellulose, pectin, and protein, whose surfaces tend to be negatively charged [30], serving as efficient sites for sequestering toxins.

Heavy metals that crossed cell walls eventually dissolved and became incorporated into organelles. However, we found that plants with AM fungal inoculation accumulated more Cd in the soluble fractions of roots and shoots. That is, more Cd remained in solution and never entered organelles, suggesting that the soluble fraction serves as another storage compartment for toxins in both roots and shoots. Therefore, the cell matrix between cells and organelles could be considered an intracellular buffer [31]. Soluble cellular components not only store heavy metals but also contain organo-ligands, which are mainly sulfur-rich peptides, organic alkali, and organic acids. Complexation of metals with organo-ligands within such storage sites decreases free ion activity and thus reduces toxicity [32].

Finally, AM fungi also improved tolerance to Cd by seemingly converting Cd into inactive forms, as treated plants had significantly more inactive forms of Cd in roots and shoots than did untreated plants. Different chemical species of heavy metals have different biological functions, with distinct toxicities and migration patterns [4,33]. For example, water-soluble Cd has a high capacity to migrate and is more deleterious to plant cells, whereas Cd in forms such as metal-phosphate complexes shows no or little toxicity to plants. In the present study, there was a significant increase in the proportion of inactive forms of Cd in the subcellular fractions of both roots and shoots of fungi-treated plants compared to controls (Fig. 5). Among these fractions, a large proportion of Cd was integrated with pectates and protein, suggesting that Cd may have been chelated by a polar material such as a hydroxyl or carboxyl group, forming a non-toxic complex.

Although heavy metal pollutants had significant effects on AM fungi in soil [34,35], numerous studies have demonstrated that symbiotic interactions between plants and AM fungi may improve plant tolerance to and uptake of heavy metals [36–38]. So inoculation of AM fungi would be important in phytoremediation of soils contaminated with heavy metals. Even some propagules of indigenous fungi already exist in soil, there may be competition among AM fungi with different ability of colonization. Therefore, inoculation of target AM fungal species that have been demonstrated to be effective in detoxification would be helpful to improve the efficiency of phytoremediation.

In summary, inoculation of AM fungi significantly enhanced the growth of alfalfa plants in a Cd-contaminated soil, increased total Cd in roots but decreased Cd concentrations in shoots. Our results also showed that AM fungi increased the proportion of Cd in cell wall while reducing the proportion in organelles and membranes. Moreover, AM fungi increased the proportion of inactive forms of Cd in roots and shoots. Together, our results illustrate that AM fungi enhanced plant resistance or tolerance to Cd through altering both the forms of Cd and the distribution of Cd among different plant tissues. These findings provide new insights into the roles of plant-microbe symbioses in mediating the impact of heavy metals on plants and highlight the need of further research on the contribution of AMF to phytoremediation.

Supporting Information

Figure S1 The status of roots inoculated with AM fungus *G. intraradices*. (a) 25 days, (b) 40 days, (c) 60 days (d) 80days.

Author Contributions

Conceived and designed the experiments: YG YW. Performed the experiments: YW JH. Analyzed the data: JH. Contributed reagents/ materials/analysis tools: JH. Wrote the paper: YW YG.

References

1. Bååth E, Diaz-Ravina M, Frostegard A, Campbell CD (1998) Effect of metal-rich sludge amendments on the soil microbial community. App Environ Microbio 64: 238–245.
2. Li TQ, Yang XE, Yang JY (2006) Zn Accumulation and subcellular distribution in the Zn hyperaccumulator *Sedum alfredii* Hance. Pedosphere 16: 616–623.
3. Hall JL. (2002) Cellular mechanisms for heavy metal detoxification and tolerance. J Exp Bot 53: 1–11.
4. Fu XP, Dou CM, Chen YX, Chen XC, Shi JY, et al. (2011) Subcellular distribution and chemical forms of cadmium in *Phytolacca americana* L. J Hazard Mater 186: 103–107.
5. Küpper H, Zhao FJ, McGrath SP (1999) Cellular Compartmentation of Zinc in Leaves of the Hyperaccumulator *Thlaspi caerulescens*. Plant Physiol 119(1): 305–312.
6. Vogel-Mikuš K, Simčič J, Pelicon P, Budnar M, Kump P, et al. (2008) Comparison of essential and non-essential element distribution in leaves of the Cd/Zn hyperaccumulator *Thlaspi praecox* as revealed by micro-PIXE. Plant Cell Environ 31: 1484–1496.
7. Ager FJ, Ynsa MD, Dominguez-Solis JR, Gotor C, Respaldiza MA, et al (2002) Cadmium localization and quantification in the plant Arabidopsis thaliana using micro-PIXE. Nucl Instrum Meth B 189: 494–498.
8. Vollenweider P, Cosio C, Günthard-Goerg S, Keller C (2006) Localization and effects of cadmium in leaves of cadmiumtolerant willow (*Salix viminalis L.*). Part II - Microlocalization and cellular effects of cadmium. Environ Exp Bot 58: 25–40.
9. Rojas LA, Yanez C, Gonzalez M, Lobos S, Smalla K, et al. (2011) Characterization of the metabolically modified heavy metal-resistant *Cupriavidus metallidurans* strain MSR33 generated for mercury bioremediation. PloS one 6(3): 0017555.
10. Zhang XH, Zhu YG, Lin AJ, Chen BD, Smith SE, et al. (2006) Arbuscular mycorrhizal fungi can alleviate the adverse effects of chlorothalonil on *Oryza sativa* L. Chemosphere 64: 627–1632.
11. Zhang XH, Lin AJ, Gao YL, Reidd RJ, Wong MH, et al. (2009) Arbuscular mycorrhizal colonisation increases copper binding capacity of root cell walls of *Oryza sativa* L. and reduces copper uptake. Soil Biol Biochem 41: 930–935.
12. Ismail Y, McCormick S, Hijri M (2011) Fungal symbiont of plant-roots modulates mycotoxin gene expression in the pathogen *Fusarium sambucinum*. PloS One 6(3): e17990.
13. Baslam M, Garmendia I, Goicoechea N (2011) Arbuscular mycorrhizal fungi (AMF) improved growth and nutritional quality of greenhouse-grown lettuce. J Agric Food Chem 59: 5504–5515.
14. Aloui A, Recorbet G, Robert F, Schoefs B, Bertrand M, et al. (2011) Arbuscular mycorrhizal symbiosis elicits shoot proteome changes that are modified during cadmium stress alleviation in *Medicago truncatula*. BMC Plant Biol 11: 75.
15. Su DC, Wong JWC, Jagadeesan H (2004) Implications of rhizospheric heavy metals and nutrients for the growth of alfalfa in sludge amended soil. Chemosphere 56(10): 957–965.
16. Kumar PBAN, Dushenkov V, Motto H, Raskin I (1995) Phytoextraction the use of plants to remove heavy metals from soils. Environ Sci Technol 29: 1232–1238.
17. Parsons JG, Peralta-Videa JR, Tiemann KJ, Saupe GB, Gardea-Torresdey JL (2005) Use of chemical modification and spectroscopic techniques to determine the binding and coordination of gadolinium(III) and neodymium(III) ions by alfalfa biomass. Talanta 67(1): 34–45.
18. Leung HM, Wu FY, Cheung KC, Ye ZH, Wong MH (2010) Synergistic effects of arbuscular mycorrhizal fungi and phosphate rock on heavy metal uptake and accumulation by an arsenic hyperaccumulator. J Hazard Mater 181: 497–507.
19. Latef AAHA (2011) Influence of arbuscular mycorrhizal fungi and copper on growth, accumulation of osmolyte, mineral nutrition and antioxidant enzyme activity of pepper (Capsicum annuum L.). Mycorrhiza 21: 495–503.
20. United States Department of Agriculture (USDA) (1988) Soil Taxomony: A Basic System of Soil Classification for Making and Interpreting Soil Surveys. Robert E. Krieger Publication, Florida.
21. Gao YZ, Li QL, LingWT, Zhu XZ (2011) Arbuscular mycorrhizal phytoremediation of soils contaminated with phenanthrene and pyrene. J Hazard Mater 185: 703–709.
22. Phillips JM, Hayman DS (1970) Improved procedures for clearing and staining parasitic and vesicular-arbuscular mycorrhizal fungi for rapid assessment of infection, Trans Br Mycol Soc 55: 158–161.
23. Brundrett MC, Piché Y, Peterson RL (1984) A new method for observing the morphology of vesicular-arbuscular mycorrhizae. Can J Bot 62: 2128–2134.
24. Lozano-Rodriguez E, Hernandez LE, Bonay P (1997) Distribution of cadmium in shoot and root tissues of maize and pea plants. J Exp Bot 48: 123–128.
25. Liu DH, Kottke I (2004) Subcellular localization of copper in the root cells of Allium sativum by electron energy loss spectroscopy(EELS). Bioresour Technol 94: 153–158.
26. Muñoz AHS, Kubachka K, Wrobel K, Corona JFG, Yathavakilla SKV, et al. (2006) Se-Enriched mycelia of *Pleurotus ostreatus*: distribution of selenium in cell walls and cell membranes/cytosol. J Agric Food Chem 54: 3440–3444.
27. He M, Ke CY, Wang WX (2010) Effects of cooking and subcellular distribution on the bioaccessibility of trace elements in two marine fish species. J Agric Food Chem 58: 3517–3523.
28. Miransari M (2011) Hyperaccumulators, arbuscular mycorrhizal fungi and stress of heavy metals. Biotechnol Adv 29; 645–653.
29. Redon PO, Béguiristain T, Leyval C (2008) Influence of *Glomus intraradices* on Cd partitioning in a pot experiment with *Medicago truncatula* in four contaminated soils. Soil Biol Biochem 40: 2710–2712.
30. Genre A, Ortu G, Bertoldo C, Martino E, Bonfante P (2009) Biotic and abiotic stimulation of root epidermal cells reveals common and specific responses to arbuscular mycorrhizal fungi. Plant Physiol 149: 1424–1434.
31. Aldrich MV, Gardea-Torresdey JL, Peralta-Videa JR, Parsons JG (2010) Plant growth and metal distribution in tissues of prosopis juliflora-velutina grown on chromium contaminated soil in the presence of *Glomus deserticola*. Environ Sci Technol 44: 7272–7279.
32. Bhatia NP, Walsh KB, Baker AJM (2005) Detection and quantification of ligands involved in nickel detoxification in a herbaceous Ni hyper-accumulator *Stack-housia tryonii Bailey*. J Exp Bot 56: 1343–1349.
33. Salt DE, Prince RC, Pickering IJ (2002) Chemical speciation of accumulated metals in plants: evidence from X-ray absorption spectroscopy, Microchem J 71: 255–259.
34. Ipsilantis I, Karpouzas DG, Papadopoulou KK, Ehaliotis C (2009) Effects of soil application of olive mill wastewaters on the structure and function of the community of arbuscular mycorrhizal fungi. Soil Biol Biochem 41: 2466–2476.
35. Bedini S, Turrini A, Rigo C, Argese E, Giovannetti M (2010) Molecular characterization and glomalin production of arbuscular mycorrhizal fungi colonizing a heavy metal polluted ash disposal island, downtown Venice. Soil Biol Biochem 42: 758–765.
36. Chen X, Wu CH, Tang JJ, Hu SJ (2005) Arbuscular mycorrhizae enhance metal lead uptake and growth of host plants under a sand culture experiment. Chemosphere 60: 665–671.
37. Wu FY, Ye ZH, Wu SC, Wong MH (2007) Metal accumulation and arbuscular mycorrhizal status in metallicolous and nonmetallicolous populations of *Pteris vittata* L. and *Sedum alfredii* Hance. Planta 226: 1363–1378.
38. Hammer EC, Rillig MC (2011) The influence of different stresses on *Glomalin* levels in an arbuscular mycorrhizal fungus-salinity increases *Glomalin* Content. PloS One 6(12): e28426.

8

Phylogenetic Diversity and Environment-Specific Distributions of Glycosyl Hydrolase Family 10 Xylanases in Geographically Distant Soils

Guozeng Wang[1,2], Kun Meng[1], Huiying Luo[1], Yaru Wang[1], Huoqing Huang[1], Pengjun Shi[1], Peilong Yang[1], Zhifang Zhang[2]*, Bin Yao[1]*

1 Key Laboratory for Feed Biotechnology of the Ministry of Agriculture, Feed Research Institute, Chinese Academy of Agricultural Sciences, Beijing, People's Republic of China, **2** Biotechnology Research Institute, Chinese Academy of Agricultural Sciences, Beijing, People's Republic of China

Abstract

Background: Xylan is one of the most abundant biopolymers on Earth. Its degradation is mediated primarily by microbial xylanase in nature. To explore the diversity and distribution patterns of xylanase genes in soils, samples of five soil types with different physicochemical characters were analyzed.

Methodology/Principal Findings: Partial xylanase genes of glycoside hydrolase (GH) family 10 were recovered following direct DNA extraction from soil, PCR amplification and cloning. Combined with our previous study, a total of 1084 gene fragments were obtained, representing 366 OTUs. More than half of the OTUs were novel (identities of <65% with known xylanases) and had no close relatives based on phylogenetic analyses. Xylanase genes from all the soil environments were mainly distributed in Bacteroidetes, Proteobacteria, Acidobacteria, Firmicutes, Actinobacteria, Dictyoglomi and some fungi. Although identical sequences were found in several sites, habitat-specific patterns appeared to be important, and geochemical factors such as pH and oxygen content significantly influenced the compositions of xylan-degrading microbial communities.

Conclusion/Significance: These results provide insight into the GH 10 xylanases in various soil environments and reveal that xylan-degrading microbial communities are environment specific with diverse and abundant populations.

Editor: Haixu Tang, Indiana University, United States of America

Funding: This work was supported by the key program of Transgenic Plant Breeding (2008ZX08003-002) and the China Modern Agriculture Research System (CARS-42). The funders had no role in study design, data collection and analysis, decision to publish, or preparation of the manuscript.

Competing Interests: The authors have declared that no competing interests exist.

* E-mail: binyao@caas.net.cn (BY); zhifangzhang@yahoo.com (ZZ)

These authors contributed equally to this work.

Introduction

Xylan is the second most abundant polysaccharide in nature, accounting for approximately one-third of all renewable organic carbon on earth [1–3]. It is made up of homopolymeric backbone chain of β-1,4-linked xylopyranose units with substitution of side chains at different positions [4]. Complete hydrolysis of xylan requires a large variety of cooperatively acting enzymes [3,5]; among them, endo-1,4-β-D-xylanase (EC 3.2.1.8) is a crucial component that cleaves the backbone of xylan. Xylanases belonging to glycosyl hydrolase (GH) families 10 and 11 are the most abundant and have been studied extensively, and enzymes with xylanase activity have also been found in GH 5, 7, 8 and 43 (http://www.cazy.org/fam/acc_GH.html) [3,6] and GH 30 [7,8]. GH 10 and 11 xylanases are truly distinct; they share no similarity at the amino acid level and have different three-dimensional structures [9] and mechanisms of action [10]. In the Pfam database (http://pfam.sanger.ac.uk/), more GH 10 xylanases (1093 sequences) are identified in more microbial sources than GH 11 xylanases (527 sequences). In addition, most of the xylanase genes from rumen microbiome [11], termite hindgut microbiome [12] and tundra soil [13] belong to GH 10. Therefore we specified on GH 10 xylanase for further studies.

Soil harbors abundant and diverse microorganisms which are critical to maintain soil functions such as soil structure formation, decomposition of organic matter and cycling of carbon, nitrogen, phosphorus, and sulphur [14,15]. Plant-litter input constitutes the main source of energy and matter for the extraordinarily diverse community of soil microorganisms connected by highly complex interactions in terrestrial ecosystem [16,17]. Xylan as one major structural polysaccharide in plant cells, its degradation is a key step in carbon cycle in the soil environment.

Soil microorganisms represent rich resources of novel natural products like antibiotics and biocatalysts [18]. Compared with the traditional isolation and pure cultivation approach that only reflects limited information of soil microbial communities, culture-independent molecular methods are more powerful to explore the functional genes of the uncultured majority in the soil [19]. Two strategies are usually used to obtain xylanase genes from soil genomic DNA. One is metagenomic library construction and screening, e.g. Hu *et al.* [20] screened several novel xylanase genes from the soil metagenomic library. The other is PCR based molecular methods that are used to obtain xylanase gene fragments and full-length genes by using inverse PCR or genome walking. Sunna and Bergquist [21] cloned a novel extremely thermostable xylanase from a hot pool environmental DNA sample. Both strategies have its advantages and disadvantages [22]. Our previous studies sought to gain an insight into the diversity of GH 10 and 11 xylanases in alpine tundra soil and rumen using degenerate PCR, gene cloning from environmental DNA and functional verification [13,23]. By using the same methods, the work presented in this report further addresses two main research goals: (i) to explore the similarities and differences of GH 10 xylanases genes from more soil samples with distinct environmental characteristics; (ii) to identity the factors that influence the diversity and distribution of xylanases in nature.

Results and Discussion

Physicochemical characters of five soil samples

Five vegetated and fell-field soil habitats spanning 22° of latitude and 34° of longitude were sampled to investigate the diversity of GH 10 xylanase genes of different terrestrial habitats. They were two soils (SS and GS) from Tianshan Mountain that represented low temperature soil environment with different vegetation types, a common agricultural soil from a maize field (FS), a hot spring sediment (HS) to represent high temperature soil environment, and a pond sediment (PS) and a muddy mangrove soil (MS) that represented freshwater and seawater environments, respectively. Details of their locations and physicochemical properties are given in Table 1.

Sequence analysis of GH 10 xylanase gene fragments

By using the total soil genomic DNA of each environmental sample as the template and the CODEHOP primers X_{10}-F and X_{10}-R specific for GH 10 xylanases [13], PCR products of the expected size, about 260 bp, were amplified and used to construct five clone libraries. After random confirmation of 300 clones from each library by PCR with primers M13F and M13R, a total of 1157 clones were sequenced; of them, 937 sequences were identified to be GH 10 xylanase gene fragments based on BLASTx analysis and contained the conserved residue Asn of GH 10 xylanases [24]. After removing the redundant fragments of each soil environment with Cd-hit program [25], 270 sequences were defined as distinct fragments with less than 95% identity (see Table S1, S2, S3, S4, S5).

Our previous study had identified 96 distinct GH 10 xylanase fragments in the tundra soil (GS) [13] that had different vegetation from SS. To compare the effects of vegetation on xylanase diversity and distribution, we combined the GS data with other five soils' to give a total of 366 distinct GH 10 xylanase fragments. Based on the BLASTp analysis, most sequences had low identities (< 65%) with known xylanases in GenBank (Figure 1), implying that these xylanases may be as yet undiscovered. Xylanase sequences from different soil environments varied in sequence identity distributions. About 18% of the distinct sequences from soil HS had the identity of more than 80% with known GH 10 sequences while only 3% in soil SS. Soil MS harbored the most novel sequences (identities <50%), representing 19% of the distinct sequences, and FS had the least, only 3% of the distinct sequences. Moreover, insertion or deletion of amino acids frequently occurred in this region. The lengths of GH 10 gene fragments displayed a wide range of variation (from 80 to 106 amino acids), and 57% of the distinct sequences had a length of 84–86 amino acids (Figure S1). The shortest sequence was PS33 (80 amino acids) that had 42% protein identity with a GH 10 protein from *Clavibacter michiganensis* [26]. The longest one was PS18 that contained 106 amino acids and had a maximum protein identity of 70% with *Prevotella ruminicola* xylanase [27]. These results suggested that our culture-independent molecular methods are efficient and reliable to retrieve xylanase genes of various lengths from different soil environments.

High genetic diversity of GH 10 xylanases in the soils

Using Distance-based OTU and Richness determination (DOTUR) software [28], rarefaction curves for GH 10 xylanases in each sample were calculated based on 6% cut-offs (Figure 2). The flat rarefaction curves in Figure 2 suggested that we have sequenced enough clones to account for the overall genes. Based on the UPGMA (average neighbor) clustering algorithm implemented in DOTUR, a total of 270 OTUs were identified (Table 1), 63 in soil SS, 34 in soil HS, 48 in soil MS, 64 in soil PS and 61 in

Table 1. Physicochemical characters of six soils and GH 10 xylanase fragment sequences obtained.

Soil	Location	Habitat	Temperature (°C)	pH	Total organic carbon (mg/g)	Total nitrogen (mg/g)	C/N ratio	Clones sequenced	Sequences recovered	OTUs
MS	21° 29′ N, 109° 45′ E	Mangrove swamp	19	5.52	33.3		2.2	231	190	48
HS	23° 26′ N, 103° 08′ E	Hot spring	74	8.27	1.2	0.07	17.5	218	156	34
PS	30° 42′ N, 120° 55′ E	Pond	15	7.36	9.7	1.3	7.6	246	210	64
FS	27° 55′ N, 116° 22′ E	Farmland	6	7.12	13.2	1.5	8.8	227	188	61
SS	43° 06′ N, 86° 50′ E	Snow lotus	2	8.22	56.5	4.1	13.8	235	193	63
GS	43° 06′ N, 86° 50′ E	Glacier	2	6.89	33.8	2.6	12.9	216	147	96
Total								1373	1084	366

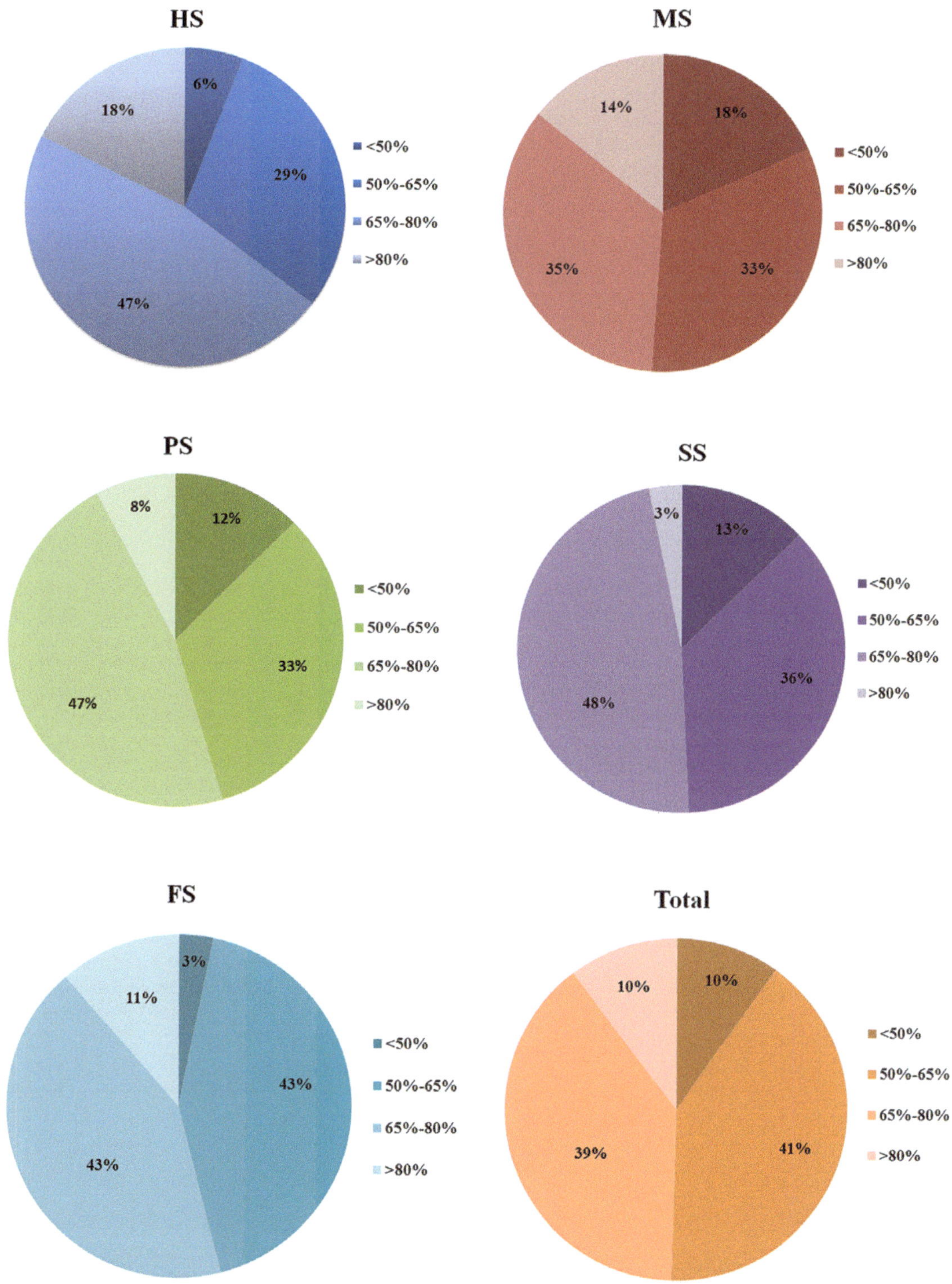

Figure 1. Amino acid sequence identities of GH 10 xylanase gene fragments from six soil environments to the known xylanases. Each sequence was analyzed with NCBI BLASTp (version 2.2.13) against the GenBank nr database. An E-score (expect value) cutoff of 10^{-10} (default) was applied and the top BLASTp hit to the known xylanases was collected. PS: pond sediment; MS: mangrove soil; HS: hot spring soil; FS: farmland soil; SS: snow lotus soil; GS: glacier soil.

soil FS. It meant that more GH 10 xylanase genes were present in the soil PS and much less in the soil HS. The most abundant OTU in each soil environment was different. HS3, the most abundant OTU in soil HS, represented about 20% (31/156) of all the sequences (Table S1) and had the highest identity (94%) with the xylanase sequence from the extreme thermophile *Dictyoglomus thermophilum* Rt46B.1 [29]. The most abundant OTUs in soil MS and PS were MS16 (17/190; Table S2) and PS159 (30/210; Table S3), respectively, both of which had the highest identities with the xylanase of *Prevotella bergensis* DSM 17361. The most abundant OTUs in soil SS and FS were SS153 (17/193; Table S4) and FS97 (13/188; Table S5), respectively; they had the highest identities with the xylanases from *Verrucomicrobiae bacterium* DG1235 and *Ktedonobacter racemifer* DSM 44963, respectively. The distinction of the most abundant OTUs in each soil implied that their major xylan-degrading microorganisms are different from each other. Moreover, the most abundant OTUs of each soil were found to be environment specific. For example, the most dominant OTU in the HS was closely related to the xylanase of the extreme thermophile *D. thermophilum* Rt46B.1 [29]. The most dominant OTUs in the PS and MS were related to the xylanase from anaerobic *P. bergensis*, a bacterium usually found in gastroenteric environments [30]. In the SS, the most dominant OTU was found to be related to the xylanase from Verrucomicrobiae bacterium, which belongs to the phylum Verrucomicrobia that has been reported to be dominant in cold environments [31].

To understand the environmental impacts on xylanase diversity and distribution, goat rumen that represents a strict anaerobic environment together with the tundra soil [13,23] were included for comparison analysis. A total of 418 sequences from seven samples and 16 representative reference sequences were used to construct phylogenetic tree (Figure 3) using ARB software [32]. Seventeen groups (A–Q) were separated based on the high bootstrap values; of them, only six groups A, E, G, K, L and O had reference sequences. Group G was the largest group containing 123 sequences from all seven environments along with the reference sequence of *P. ruminicola* [33]. Combining the sequences in group D, 38.8% of the distinct sequences (162/418) were closely related with GH 10 sequences of Bacteroidetes. Actinobacteria represented another main environmental xylanase producer, accounting for 22.0% of the sequences (92/418; group H, J, K, L, M, N and O). Firmicutes and Proteobacteria were the third and fourth xylanase producers, respectively. Xylanase genes have been found in other microbial sources, such as Verrucomicrobiae, Acidobacteria, Dictyoglomi, Thermotogae, fungi and unclassified bacteria in different soil environments. The phylogenetic diversity of xylanase genes in this study might not be very exact because horizontal gene transfer between family 10 xylanases is a common phenomenon in environments [34,35].

Our study suggested that GH 10 xylanase genes are mainly distributed in bacteria in soil environments. Based on the collection of GH 10 sequences in Pfam database, Firmicutes are the main microbial source of xylanase production, accounting for 23% of the sequences, and Bacteroidetes constitute only 11%. These data are significantly different from our results. The reason might be that most of the sequences collected by the Pfam database are from cultured microorganisms while uncultured microorganisms are the main xylanase producer in soil environments. Moreover, when we compared the soil xylanase genes with those from rumen microbiome [11] and termite hindgut microbiome [12], we found that xylanases from soils are more diverse. GH 10 xylanase genes from the soil environments under study were widely distributed in the phyla of Bacteroidetes, Proteobacteria, Acidobacteria, Firmicutes, Actinobacteria, Dictyoglomi and the fungi of Ascomycota, while GH 10 xylanase genes from rumen and termite hindgut were mainly distributed in the phyla of Bacteroidetes and Firmicutes, respectively. The results suggested that soil environment harbors a more complex microflora than gastrointestinal tract environments.

Bacteria and fungi are the key agents of xylan degradation, in which fungi have been supposed to play an important role in the initiation of plant fiber degradation during fermentative digestion [1,2,36]. However, fewer fungal GH 10 xylanase gene fragments were identified in this study. Only eight xylanase fragments from snow lotus soil, hot spring sediment and pond sediment were found to be related to the xylanases from Ascomycota, a phylum dominant in soils and important in xylan degradation [36]. This phenomenon has also been found in other environments, such as no fungal GH 10 xylanases in the rumen microbiome [11] and termite hindgut microbiome [12]. However, our unpublished data showed that more GH 11 xylanase fragment sequences (78/122) from the same soil environments were related to fungal instead of bacterial xylanases. According to the Pfam database (http://pfam.sanger.ac.uk/), more GH 11 sequences are from fungi (296/527), and more GH 10 sequences are from bacteria (870/1093). These facts suggest that bacteria and fungi are the main producers of GH 10 and 11 xylanases, respectively. Considering that GH 10 and 11 xylanases are distinct at structures and action modes [10], they must work in xylan degradation in a compatible instead of competitive way.

Three sequences from mangrove soil and farmland soil were closely related to the xylanase of extremely halophilic archaeon *Halorhabdus utahensis*, suggesting the existence of xylanase-producing Archaea in these soil environments.

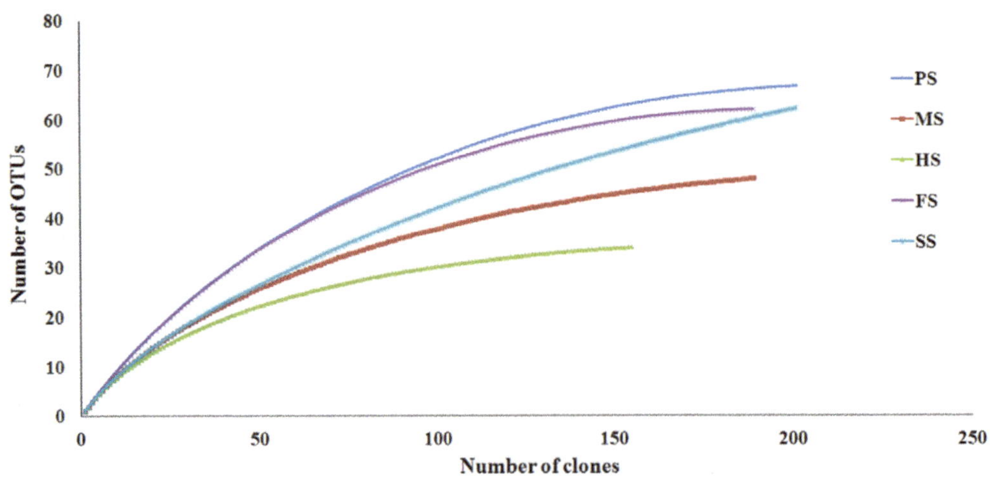

Figure 2. Rarefaction curves of GH10 xylanase fragments from six soil environments. The gene types were defined at the 6% cut-off. The curves represent the number of gene types as a function of total clones sequenced in libraries constructed from six soil genomic DNAs.

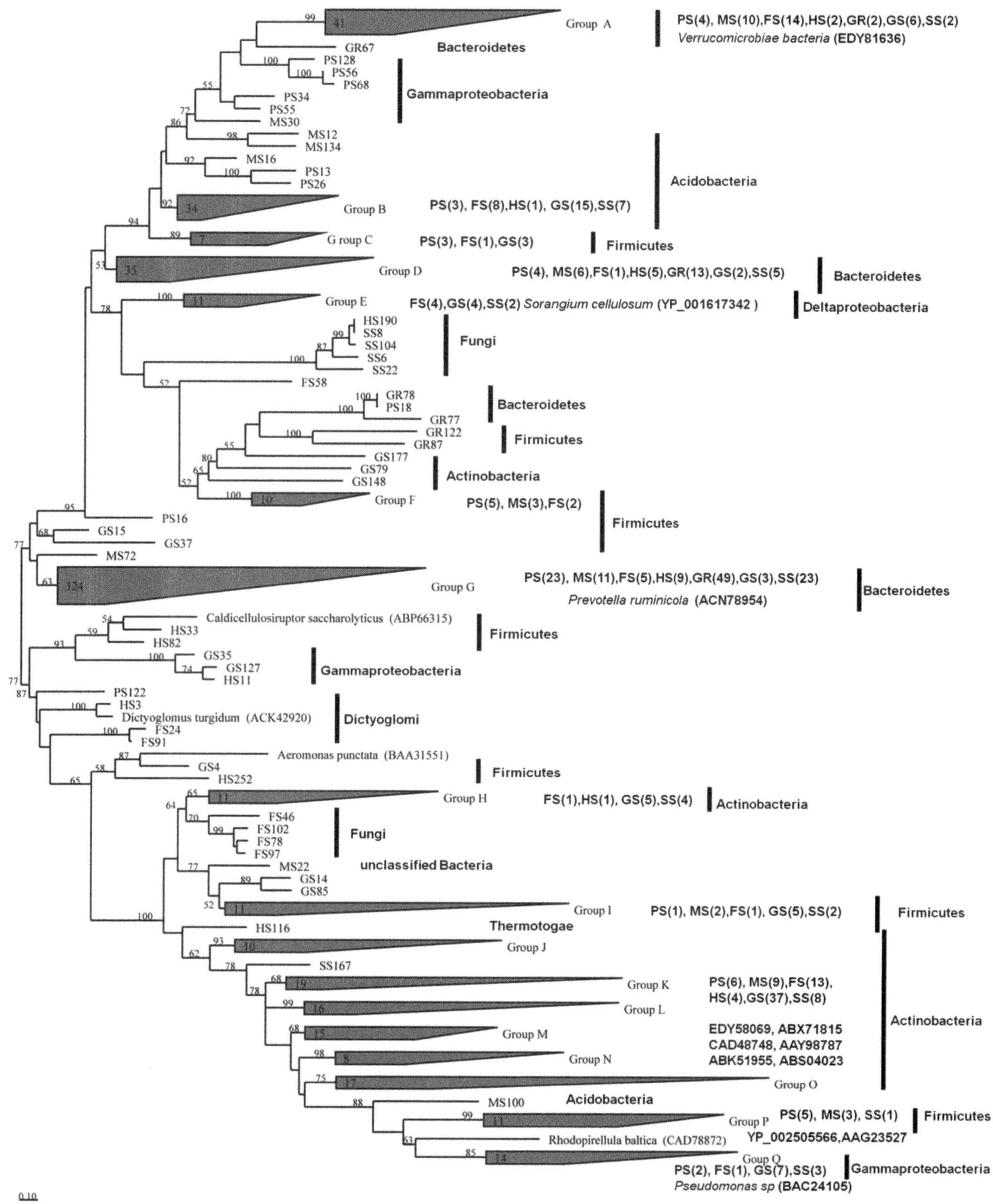

Figure 3. Phylogenetic tree constructed using ARB software using neighbor-joining method. A total of 19 representative sequences were randomly selected and used as references for phylogenetic tree construction. The lengths of the branches indicate the relative divergence among the amino acid sequences. Sequences with the same source were clustered together and the numbers of sequences in each soil were listed. The numbers at the nodes indicate bootstrap values based on 1000 replications and bootstrap values (>50) are displayed. The scale bar represents 0.1 amino acid substitution per position.

Effects of environmental factors on GH 10 xylanase in the soils

The diversity, abundance and distribution of GH 10 xylanase genes in soils varied a lot from each other. The GH 10 xylanase genes in different soil samples were compared by UniFrac analysis [37] (Figure 4). High Jackknife Count (100) separated seven soil environments into two big clusters: cluster α containing soil GS, FS, HS and SS, and cluster β of other samples (PS, MS and GR). Soil SS and soil HS were clustered together although their soil characters differed a lot except for soil pH (Table 1). The GH 10 xylanase genes in soil GS were significantly different from soil SS although they were sampled from the same area but with different vegetation. The GH 10 xylanase gene fragments from soil PS, MS and GR were grouped together in cluster β. These three environments are all anaerobic. Studies have shown that environmental factors like pH, oxygen content, temperature, organic matter and vegetation might have great effect on the distribution and diversity of functional genes [38–40]. In this study, we found that pH and oxygen content are the most important geochemical factors influencing the xylanase diversity and distribution in soils. Microbial xylanases are often secreted as extracellular enzymes or are associated with the outer membrane [2,3]. The extracellular location makes xylanases adapt to physicochemical conditions of the surrounding environment for efficient function.

Although the environmental factors differed a lot, 16 identical sequences were found between two or more soil samples (Figure S2). The same phenomenon has been reported by LeCleir *et al.* [41] in the study of chitinase genes in various aquatic habitats. For example, a fragment showing the highest similarity with a bifunctional GH 10 xylanase/CE 1 ferulic acid esterase of *P. ruminicola* [30] was found in six samples (MS150, GS122, HS7, SS121, FS144 and PS111); fragments MS36 and PS158, MS128 and PS36, MS13 and PS96, HS187 and SS32, and HS190 and SS8 were found to be identical. Another two pairs of sequences (HS262 and PS88, HS108 and PS21) were found both in soil HS and PS. Moreover, eight xylanase sequences from goat rumen were identical with that from soil environments (data not show). These results suggested that these xylanase genes are well conserved and broadly distributed in environment.

In conclusion, we identified large numbers of GH 10 xylanase fragments in five soil environments using culture-independent molecular methods. Most of the sequences have low identities with known xylanases and have no close relatives, suggesting the phylogenetic diversity and novelty of these xylanase genes. Compared to the xylanases from cultured bacteria, xylanases directly retrieved from soil environments have some specific distributions. The various diversity and distribution patterns of this functional gene in different soil environments suggest that xylan-degrading bacterial community is environment specific. Environmental factors such as pH and oxygen content may be the most significant geochemical factors that affected the distribution and diversity of microbial xylanase genes. Although we identified so many xylanase genes in soil environments, how and when these xylanase genes are expressed *in situ* is unknown. New approaches such as metatranscriptome analysis [42–44] will be powerful to solve these questions.

Materials and Methods

Ethics statement

No specific permits were required for the described field studies. All the soil samples were collected from the location that is not privately-owned or protected in any way. The field studies did not involve endangered or protected species because this study only concentrated on the soil samples.

Collection of soil samples

Five soil types with different physicochemical characteristics were sampled in China (Table 1). Rhizosphere soil of snow lotus was sampled from tundra soil of Tianshan Mountain at an elevation of 3525 m. Hot spring sediment with temperature of 74°C was sampled from Jinping, Yunnan. Pond sediment was sampled from the pond which was used to fish culturing in Zhejiang. Field soil was sampled from the farmland in Jiangxi which had been used to grow maize. Mangrove soil was sampled from Guangxi. For all the soil samples, three soil cores (at a depth of 0–10 cm, 5 cm in diameter) were collected from each soil site and stored at −70°C before use. Soil parameters, including total organic carbon and total nitrogen, were determined by using

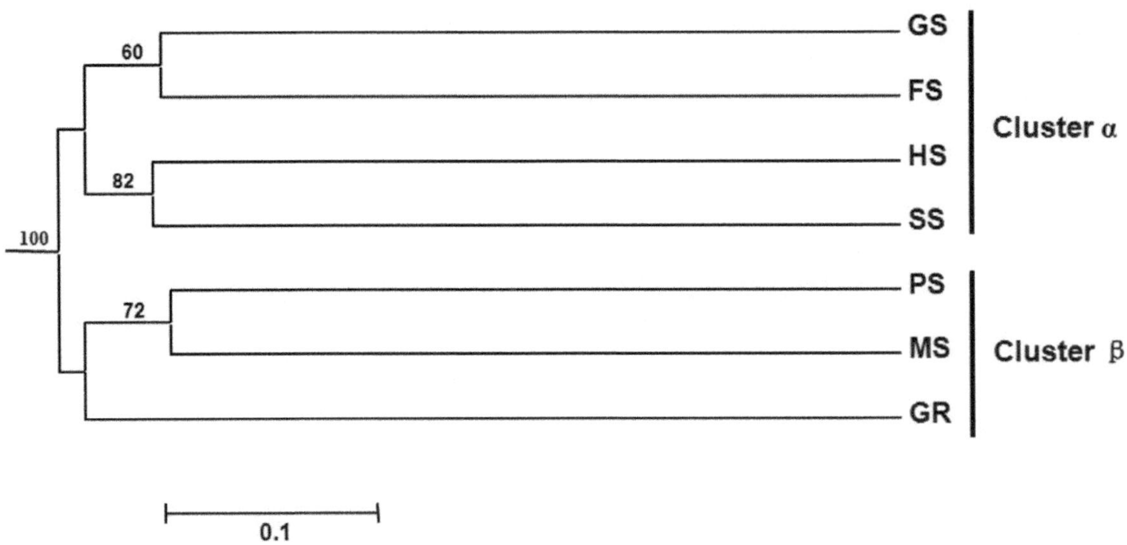

Figure 4. Cluster analysis of six soil samples with unweighted UniFrac algorithm.

Jackson's methods [45]. The pH of the soil sample was measured in a 1:2 soil-water suspension.

DNA extraction

Total soil genomic DNA was extracted using a modified sodium dodecyl sulphate (SDS)–based protocol specific for high molecular weight DNA from environmental samples [46]. Root material and small stones (if any) in the soil samples were removed by hand. One gram of each soil was added to a 2 ml screw-capped tube, followed by addition of 1 ml of lysis buffer (100 mM Tris–HCl, 100 mM EDTA, 1.5 M NaCl, 2% CTAB and 1% SDS) and complete mixing. Glass beads (0.2 ml of 0.1 mm in diameter and 0.1 ml of 0.5 mm in diameter) were added to the tubes, and agitated on MiniBead Beater-1 (Biospec, Bartlesville, OK) at the speed of 2500 rpm, 20 sec for three times. Then all the contents were removed to a 5 ml tube, another 3 ml of lysis buffer was added, mixed completely and incubated at 70°C for 2 h with gentle shaking every 20 min. After centrifugation at 16,000 $\times g$ and 4°C for 10 min, genomic DNA was recovered by precipitation with 0.7 volume of isopropanol and dissolved in TE buffer. RNase was added to the crude soil DNA to digest RNAs (37°C, 1 h). The presence and size of the soil DNAs were determined by agarose gel electrophoresis with ethidium bromide staining. DNA samples were recovered, purified with the Omega Gel Extraction Kit (Norcross, GA) and stored at −70°C before use.

PCR amplification, library construction and sequencing

The CODEHOP primers X_{10}-F (5′-CTACGACTGG-GAYGTNIBSAAYGA-3′) and X_{10}-R (5′-GTGACTCTG-GAWRCCIABNCCRT-3′) specific for GH 10 xylanase [13] were used to amplify GH 10 xylanase gene fragments by touchdown PCR from the purified metagenomic DNA of each soil sample. PCR products were visualized on an agarose gel, bands of the expected size (about 260 bp) were excised and purified using the Qiaquick gel extraction kit (Qiagen, Valencia, CA). Purified PCR products were ligated into vector pGEM-T Easy (Promega, Madison, WI), and then electroporated into *Escherichia coli* DH5α (TaKaRa, Tokyo, Japan) following the procedure recommended by the manufacturer. Three hundred positive transformants (white colonies) from each library were randomly picked for further confirmed by PCR with M13F (GTAAAACGACGGCCAGT) and M13R (GGATAACAATTT-CACACAGGA) and sequenced by Sunbiotech (Beijing, China) using Sanger method with the ABI-3730 automatic sequencer (Life Technologies, Carlsbad, CA).

Phylogenetic tree construction

Vector sequences introduced by automated Sanger sequencing machines were removed using Figaro software [47] (http://sourceforge.net/apps/mediawiki/amos/index.php?title = Figaro). Then the sequences were analyzed with NCBI BLASTx (version 2.2.13) against the GenBank nr database. An E-score (expect value) cutoff of 10^{-10} was applied. Nucleotide sequences identified as xylanase gene fragments were translated into amino acids by EMBOSS Transeq (http://www.ebi.ac.uk/emboss/transeq) and aligned with the known sequences in GenBank database at the protein level using ClustalW. Redundant amino acid sequences were removed using Cd-hit [25] with a 95% sequence identity cut-off.

ARB software [32] was used for construction of the sequence databases and phylogenetic trees (neighbor joining, Jukes-Cantor correction factor for protein with 1000 bootstrap replications). A total of 19 representative sequences were randomly selected and used as references for phylogenetic tree constructions. To test clustering, representative sequences and at least one environmental sequence from each group were used for phylogenetic analysis using minimum evolution (Jukes-Cantor correction) and maximum parsimony criteria with MEGA 4.0 [48].

Rarefaction and Unifrac analysis

Rarefaction curves for GH 10 xylanases in each sample were calculated, respectively, using DOTUR software [28]. Distance matrices of the clones were calculated at the protein level using default parameters of protdist in PHYLIP (http://evolution.genetics.washington.edu/phylip.html). Sequences were then assigned to OTUs based on the UPGMA (average neighbor) clustering algorithm implemented in DOTUR with default parameters for precision (0.01) and bootstrap (1000), and rarefaction results at 6% protdist were calculated.

UniFrac was applied to examine the differences in the composition of GH 10 xylanase among samples [37]. First, phylogenetic tree was constructed for the xylanase gene sequences using ARB software as described above. The phylogenetic tree and the environment file were subsequently uploaded to UniFrac (http://bmf2.colorado.edu/unifrac/index.psp). In the present study, "P test significance" and "cluster environments" options were used.

Nucleotide sequence accession numbers

The GH 10 xylanase gene fragments have been deposited into GenBank database under accession numbers FJ527914–FJ527975 for snow lotus soil, FJ918954–FJ919017 for pond sediment, FJ919018–FJ919065 for mangrove soil, FJ919066–FJ919094 for hot spring sediment, and FJ919095–FJ919155 for farmland soil.

Supporting Information

Figure S1 Length variation of GH 10 xylanase fragments from six soil environments.

Figure S2 Phylogenetic tree of identical xylanase fragments from two or more soil environments. The phylogenetic tree was constructed using the neighbor-joining method (MEGA 4.0). The lengths of the branches indicate the relative divergence among the amino acid sequences. The numbers at the nodes indicate bootstrap values based on 1000 replications and bootstrap values (>50) are displayed. The scale bar represents 0.1 amino acid substitution per position.

Table S1 The GH 10 xylanase gene fragments detected in the hot spring sediment (HS) and their closest relative based on amino acid sequence identity and similarity.

Table S2 The GH 10 xylanase gene fragments detected in the mangrove soil (MS) and their closest relative based on amino acid sequence identity and similarity.

Table S3 The GH 10 xylanase gene fragments detected in the pond sediment (PS) and their closest relative based on amino acid sequence identity and similarity.

Table S4 The GH 10 xylanase gene fragments detected in the snow lotus soil (SS) and their closest relative based on amino acid sequence identity and similarity.

Table S5 The GH 10 xylanase gene fragments detected in the farmland soil (FS) and their closest relative based on amino acid sequence identity and similarity.

Acknowledgments

The authors would like to thank Profs. Jiang Li in East China Institute of Technology and Ningfeng Wu in Biotechnology Research Institute of Chinese Academy of Agricultural Sciences for their assistance in soil sample collection.

Author Contributions

Conceived and designed the experiments: GW KM ZZ BY. Performed the experiments: GW KM YW PS. Analyzed the data: GW HL YW HH. Contributed reagents/materials/analysis tools: ZZ PY. Wrote the paper: GW ZZ BY.

References

1. Sunna A, Antranikian G (1997) Xylanolytic enzymes from fungi and bacteria. Crit Rev Biotechnol 17: 39–67.
2. Kulkarni N, Shendye A, Rao M (1999) Molecular and biotechnological aspects of xylanases. FEMS Microbiol Rev 23: 411–456.
3. Collins T, Gerday C, Feller G (2005) Xylanases, xylanase families and extremophilic xylanases. FEMS Microbiol Rev 29: 3–23.
4. Bastawde KB (1992) Xylan structure, microbial xylanases, and their mode of action. World J Microbiol Biotechnol 8: 353–368.
5. Thomson JA (1993) Molecular biology of xylan degradation. FEMS Microbiol Rev 10: 65–82.
6. Henrissat B, Claeyssens M, Tomme P, Lemesle L, Mornon JP (1989) Cellulase families revealed by hydrophobic cluster analysis. Gene 81: 83–95.
7. Haegeman A, Vanholme B, Gheysen G (2009) Characterization of a putative endoxylanase in the migratory plant-parasitic nematode *Radopholus similis*. Mol Plant Pathol 10: 389–401.
8. Luo H, Yang J, Li J, Shi P, Huang H, et al. (2010) Molecular cloning and characterization of the novel acidic xylanase XYLD from *Bispora* sp. MEY-1 that is homologous to family 30 glycosyl hydrolases. Appl Microbiol Biotechnol 86: 1829–1839.
9. Biely P, Vrsanska M, Tenkanen M, Kluepfel D (1997) Endo-β-1,4-xylanase families: differences in catalytic properties. J Biotechnol 57: 151–166.
10. Jeffries TW (1996) Biochemistry and genetics of microbial xylanases. Curr Opin Biotechnol 7: 337–342.
11. Hess M, Sczyrba A, Egan R, Kim TW, Chokhawala H, et al. (2011) Metagenomic discovery of biomass-degrading genes and genomes from cow rumen. Science 331: 463–467.
12. Warnecke F, Luginbuhl P, Ivanova N, Ghassemian M, Richardson TH, et al. (2007) Metagenomic and functional analysis of hindgut microbiota of a wood-feeding higher termite. Nature 450: 560–565.
13. Wang G, Wang Y, Yang P, Luo H, Huang H, et al. (2010) Molecular detection and diversity of xylanase genes in alpine tundra soil. Appl Microbiol Biotechnol 87: 1383–1393.
14. Hopkins DW, Dungait JAJ (2010) Soil microbiology and nutrient cycling. In: Dixon GR, Tilston EL, editors. Soil Microbiology and Sustainable Crop Production. Springer. 59–80.
15. Nielsen UN, Ayres E, Wall DH, Bardgett RD (2011) Soil biodiversity and carbon cycling: a review and synthesis of studies examining diversity-function relationships. Eur J Soil Sci 62: 105–116.
16. Hättenschwiler S, Tiunov AV, Scheu S (2005) Biodiversity and litter decomposition in terrestrial ecosystems. Annu Rev Ecol Evol Syst 36: 191–218.
17. Hattenschwiler S, Fromin N, Barantal S (2011) Functional diversity of terrestrial microbial decomposers and their substrates. C R Biol 334: 393–402.
18. Daniel R (2004) The soil metagenome_a rich resource for the discovery of novel natural products. Curr Opin Biotechnol 15: 199–204.
19. Daniel R (2005) The metagenomics of soil. Nat Rev Microbiol 3: 470–478.
20. Hu Y, Zhang G, Li A, Chen J, Ma L (2008) Cloning and enzymatic characterization of a xylanase gene from a soil-derived metagenomic library with an efficient approach. Appl Microbiol Biotechnol 80: 823–830.
21. Sunna A, Bergquist PL (2003) A gene encoding a novel extremely thermostable 1,4-β-xylanase isolated directly from an environmental DNA sample. Extremophiles 7: 63–70.
22. Cowan D, Meyer Q, Stafford W, Muyanga S, Cameron R, et al. (2005) Metagenomic gene discovery: past, present and future. Trends Biotechnol 23: 321–329.
23. Wang G, Luo H, Meng K, Wang Y, Huang H, et al. (2011) High genetic diversity and different distributions of glycosyl hydrolase family 10 and 11 xylanases in the goat rumen. PLoS ONE 6: e16731.
24. Solomon V, Teplitsky A, Shulami S, Zolotnitsky G, Shoham Y, et al. (2007) Structure-specificity relationships of an intracellular xylanase from *Geobacillus stearothermophilus*. Acta Crystallogr Sect D 63: 845–859.
25. Li W, Godzik A (2006) Cd-hit: a fast program for clustering and comparing large sets of protein or nucleotide sequences. Bioinformatics 22: 1658–1659.
26. Bentley SD, Corton C, Brown SE, Barron A, Clark L, et al. (2008) Genome of the actinomycete plant pathogen *Clavibacter michiganensis* subsp. *sepedonicus* suggests recent niche adaptation. J Bacteriol 190: 2150–2160.
27. Flint HJ, Whitehead TR, Martin JC, Gasparic A (1997) Interrupted catalytic domain structures in xylanases from two distantly related strains of *Prevotella ruminicola*. Biochim Biophys Acta 1337: 161–165.
28. Schloss PD, Handelsman J (2005) Introducing DOTUR, a computer program for defining operational taxonomic units and estimating species richness. Appl Environ Microbiol 71: 1501–1506.
29. Gibbs MD, Reeves RA, Bergquist PL (1995) Cloning, sequencing, and expression of a xylanase gene from the extreme thermophile *Dictyoglomus thermophilum* Rt46B.1 and activity of the enzyme on fiber-bound substrate. Appl Environ Microbiol 61: 4403–4408.
30. Downes J, Sutcliffe IC, Hofstad T, Wade WG (2006) *Prevotella bergensis* sp. nov., isolated from human infections. Int J Syst Evol Microbiol 56: 609–612.
31. Lipson DA, Schmidt SK (2004) Seasonal changes in an alpine soil bacterial community in the colorado rocky mountains. Appl Environ Microbiol 70: 2867–2879.
32. Ludwig W, Strunk O, Westram R, Richter L, Meier H, et al. (2004) ARB: a software environment for sequence data. Nucl Acids Res 32: 1363–1371.
33. Dodd D, Kocherginskaya SA, Spies MA, Beery KE, Abbas CA, et al. (2009) Biochemical analysis of a β-D-xylosidase and a bifunctional xylanase-ferulic acid esterase from a xylanolytic gene cluster in *Prevotella ruminicola* 23. J Bacteriol 191: 3328–3338.
34. Millward-Sadler SJ, Davidson K, Hazlewood GP, Black GW, Gilbert HJ, et al. (1995) Novel cellulose-binding domains, NodB homologues and conserved modular architecture in xylanases from the aerobic soil bacteria *Pseudomonas fluorescens* subsp. *cellulosa* and *Cellvibrio mixtus*. Biochem J 312 (Pt 1): 39–48.
35. Ricard G, McEwan NR, Dutilh BE, Jouany JP, Macheboeuf D, et al. (2006) Horizontal gene transfer from Bacteria to rumen Ciliates indicates adaptation to their anaerobic, carbohydrates-rich environment. BMC Genomics 7: 22.
36. Polizeli ML, Rizzatti AC, Monti R, Terenzi HF, Jorge JA, et al. (2005) Xylanases from fungi: properties and industrial applications. Appl Microbiol Biotechnol 67: 577–591.
37. Lozupone C, Knight R (2005) UniFrac: a new phylogenetic method for comparing microbial communities. Appl Environ Microbiol 71: 8228–8235.
38. Lejon DP, Nowak V, Bouko S, Pascault N, Mougel C, et al. (2007) Fingerprinting and diversity of bacterial copA genes in response to soil types, soil organic status and copper contamination. FEMS Microbiol Ecol 61: 424–437.
39. Liang Y, Van Nostrand JD, Deng Y, He Z, Wu L, et al. (2011) Functional gene diversity of soil microbial communities from five oil-contaminated fields in China. ISME J 5: 403–413.
40. Saleh-Lakha S, Miller M, Campbell RG, Schneider K, Elahimanesh P, et al. (2005) Microbial gene expression in soil methods, applications and challenges. J Microbiol Methods 63: 1–19.
41. LeCleir GR, Buchan A, Hollibaugh JT (2004) Chitinase gene sequences retrieved from diverse aquatic habitats reveal environment-specific distributions. Appl Environ Microbiol 70: 6977–6983.
42. Urich T, Lanzen A, Qi J, Huson DH, Schleper C, et al. (2008) Simultaneous assessment of soil microbial community structure and function through analysis of the meta-transcriptome. PLoS ONE 3: e2527.
43. Damon C, Lehembre F, Oger-Desfeux C, Luis P, Ranger J, et al. (2012) Metatranscriptomics reveals the diversity of genes expressed by eukaryotes in forest soils. PLoS ONE 7: e28967.
44. Gilbert JA, Field D, Huang Y, Edwards R, Li W, et al. (2008) Detection of large numbers of novel sequences in the metatranscriptomes of complex marine microbial communities. PLoS ONE 3: e3042.
45. Jackson ML (1958) Soil Chemical analysis. Prentice Hall of India, New Delhi. 498.
46. Brady SF (2007) Construction of soil environmental DNA cosmid libraries and screening for clones that produce biologically active small molecules. Nat Protoc 2: 1297–1305.
47. White JR, Roberts M, Yorke JA, Pop M (2008) Figaro a novel statistical method for vector sequence removal. Bioinformatics 24: 462–467.
48. Tamura K, Dudley J, Nei M, Kumar S (2007) MEGA4: molecular evolutionary genetics analysis (MEGA) software version 4.0. Mol Biol Evol 24: 1596–1599.

9

Leaching Behavior of Heavy Metals and Transformation of Their Speciation in Polluted Soil Receiving Simulated Acid Rain

Shun-an Zheng[1,2,3]*, Xiangqun Zheng[1,2,3], Chun Chen[1,2,3]

1 Agro-Environmental Protection Institute, Ministry of Agriculture, Tianjin, People's Republic of China, **2** Key Laboratory of Production Environment and Agro-Product Safety, Ministry of Agriculture, Tianjin, People's Republic of China, **3** Tianjin Key Laboratory of Agro-Environment and Agro-Product Safety, Tianjin, People's Republic of China

Abstract

Heavy metals that leach from contaminated soils under acid rain are of increasing concern. In this study, simulated acid rain (SAR) was pumped through columns of artificially contaminated purple soil. Column leaching tests and sequential extraction were conducted for the heavy metals Cu, Pb, Cd, and Zn to determine the extent of their leaching as well as to examine the transformation of their speciation in the artificially contaminated soil columns. Results showed that the maximum leachate concentrations of Cu, Pb, Cd, and Zn were less than those specified in the Chinese Quality Standards for Groundwater (Grade IV), thereby suggesting that the heavy metals that leached from the polluted purple soil receiving acid rain may not pose as risks to water quality. Most of the Pb and Cd leachate concentrations were below their detection limits. By contrast, higher Cu and Zn leachate concentrations were found because they were released by the soil in larger amounts as compared with those of Pb and Cd. The differences in the Cu and Zn leachate concentrations between the controls (SAR at pH 5.6) and the treatments (SAR at pH 3.0 and 4.5) were significant. Similar trends were observed in the total leached amounts of Cu and Zn. The proportions of Cu, Pb, Cd, and Zn in the EXC and OX fractions were generally increased after the leaching experiment at three pH levels, whereas those of the RES, OM, and CAR fractions were slightly decreased. Acid rain favors the leaching of heavy metals from the contaminated purple soil and makes the heavy metal fractions become more labile. Moreover, a pH decrease from 5.6 to 3.0 significantly enhanced such effects.

Editor: Stephen J. Johnson, University of Kansas, United States of America

Funding: This work was financially supported by Central Public Research Institutes Basic Funds for Research and Development (Agro-Environmental Protection Institute, Ministry of Agriculture). The funders had no role in study design, data collection and analysis, decision to publish, or preparation of the manuscript.

Competing Interests: The authors have declared that no competing interests exist.

* E-mail: zhengshunan@gmail.com

Introduction

Acid rain has been a well-known environmental problem for decades and can lead to acidification of surface waters and soils. Acid deposition is formed from SO_2 and NO_x emitted to the atmosphere, largely because of fossil-fuel combustion. The most important sources are energy production, especially coal- and oil-fired power plants, and transportation sources, such as vehicles and ships. The air pollutants are transformed in the atmosphere to H_2SO_4 and HNO_3, transported across distances potentially as far as hundreds of kilometers, and deposited as precipitation (wet deposition) and as gas and particles (dry deposition) [1,2]. Acid deposition is also an environmental problem of increasing concern in China, where acid rain is mainly distributed in the areas of Yangtze River to the south, Qinghai-Tibet Plateau to the east, and in the Sichuan Basin. About 40% of the total territory of China is affected by the acid rain [3]. Sichuan Basin is one of the most severely hit-area of acid rain in southern China, where the rapid industrialization for last few decades has caused fast growth in sulfur emissions. Based on the monitoring data for 21 cities in Sichuan Province within the State-Controlled-Network of China, the number of cities with the annual average pH value of acid rain lower than 5.6 was 19. According to the environmental protection and monitoring agencies in Sichuan Province, the direct economic loss due to acid rain is estimated to be U. S. $ 3 billion for one year [4].

Sichuan basin is also known as the "Red Basin" because it is mainly covered by red or purple rock series of the Trias–Cretaceous system, from which the purple soils, one of the most important soils for agricultural production in subtropical areas of China, are developed and formed. Purple soil, classified as Eutric Regosols in FAO Taxonomy or Pup-Calric-Entisol in the Chinese soil taxonomy, is typically characterized by thin soil horizons and inherited many of the characteristics of parent materials or rocks, such as its color ranging from purple to red. But other changes in purple soil properties have taken place as a result of land use changes, agricultural practices, or eco-environment disturbances [5,6]. Due to the soil background and human activities, the soil has been severely contaminated by heavy metals (commonly including Cu, Pb, Cd and Zn) in many areas, resulting in potential risk to local human health and environment [7,8].

There is accumulating evidence that acid rain is able to enhance metal mobilization in soil ecosystems. Previous studies demonstrated that the H^+ ion in the acidic water displaces the cations from their binding sites, reduces the cation exchange capacity

(CEC), and increases the concentrations of these cations in the soil-water system. The negatively charged sulfate and nitrate ions in the acid rain can act as "counter-ions", which allow cations to be leached from the soil. Through a series of chemical reactions, cations such as K^+, Na^+, Ca^{2+} and Mg^{2+} are leached out and become unavailable to plants as nutrients [9,10]. Likewise, toxic ions, such as Cu, Pb and Cd, usually bound to the negatively charged surface of soil particles can be displaced by H^+ ion too [11].

Leaching is the process by which contaminants are transferred from a stabilized matrix to liquid medium, such as water or other solutions. However, the influence of acid rain on the leaching behavior of heavy metals and transformation of their speciation in polluted purple soil is not investigated in detail. The objectives of this study were to evaluate the leaching of heavy metals Cu, Pb, Cd and Zn in a contaminated purple soil affected by simulated acid rain (SAR) over a range of pH, and to identify how simulated acid rain influences the chemical speciation of these metals in purple soil. Column experiments were used in this study to provide information about element release, transport in soil, and chemistry of soil and leachates.

Materials and Methods

Soil Sample Collection and Analysis

The analyzed soil sample was collected from a 0 cm to 20 cm layer of agricultural purple soil in the Pengzhou Agro-ecological Station of the Chinese Academy of Agricultural Sciences in Sichuan Province, China. The study area has a subtropical humid monsoon climate with an average annual precipitation of 850 mm to 1000 mm and an average annual temperature of 15°C to 16°C. The soil samples were air-dried, ground, and passed through a 2 mm sieve. Selected soil characteristics determined by standard methods [12] were 6.27 for pH, 0.64 $g\cdot kg^{-1}$ for $CaCO_3$, 26.51 $g\cdot kg^{-1}$ for organic matter, 20.81 $g\cdot kg^{-1}$ for CEC, 265.41 $g\cdot kg^{-1}$ for clay (<0.002 mm), 51.84 $mg\cdot kg^{-1}$ for total copper, 43.92 $mg\cdot kg^{-1}$ for total lead, 0.36 $mg\cdot kg^{-1}$ for total cadmium, 121.23 $mg\cdot kg^{-1}$ for total zinc.

Contaminated Soil Preparation

Artificially contaminated soils were composed of the collected purple soil. The load quantities of Cu [$Cu(NO_3)_2\cdot 3H_2O$], Pb [$Pb(NO_3)_2$], Cd [$Cd(NO_3)_2\cdot 4H_2O$], and Zn [$Zn(NO_3)_2\cdot 6H_2O$] were as follows: Cu+Pb+Cd+Zn = 400 $mg\cdot kg^{-1}$+500 $mg\cdot kg^{-1}$+1 $mg\cdot kg^{-1}$+500 $mg\cdot kg^{-1}$. Based on the levels of polluted soil defined by the Chinese Environmental Protection Agency, the selected concentrations in this study represented moderately contaminated soils (Grade III of the National Soil Heavy Metals Standards GB15618-1995). The thoroughly mixed soil samples were stored and incubated at a 75% water holding capacity for 12 mon. The chemical speciation of aged soil was then determined by the sequential extraction procedure of Tessier et al. [13]. The chemical reagents, extraction conditions, and their corresponding fractions are defined as follows:

1. Exchangeable fraction (EXC): 2 g of the soil sample (oven-dry weight), 16 mL 1.0 $mol\cdot L^{-1}$ $MgCl_2$, pH 7, shake vigorously in a reciprocating shaker for 1 h, 20°C.
2. Carbonate-bound fraction (CAR): 16 mL of pH 5, 1.0 $mol\cdot L^{-1}$ sodium acetate, shake vigorously in the reciprocating shaker for 5 h, 20°C.
3. The Fe/Mn oxide-bound fraction (OX): 40 mL of 0.04 $mol\cdot L^{-1}$ $NH_4OH\cdot HCl$ in 25% (v/v) acetic acid at pH 3 for 5 h at 96°C with occasional agitation.
4. The OM-bound fraction: 6 mL of 0.02 $mol\cdot L^{-1}$ HNO_3 and 10 mL of 30% H_2O_2 (pH adjusted to 2 with HNO_3), water bath, 85°C for 5 h with occasional agitation. 10 mL of 3.2 $mol\cdot L^{-1}$ NH_4OAc in 20% (v/v) HNO_3, shake vigorously in the reciprocating shaker for 30 min.
5. Residual fraction (RES): Dried in a force-air oven at 40°C, 24 h. Subsamples after sieving with 0.149 mm openings were used for determining Cu, Zn, Pb and Cd contents.

Extractions were performed in 100 mL polypropylene centrifuge tubes. Between each successive extraction, the supernatant was centrifuged at 1500×*g* for 30 min and then filtered using a membrane filter (0.45-μm nominal pore size).

Simulated Acid Rain Preparation

Simulated acid rain was designed according to the main ion composition and pH of the local rain water (the pH of rainfall varied from 3.0 to 5.6) [4]. Synthetic acid rain with pH values of 3.0, 4.5, and 5.6 was prepared from a stock H_2SO_4–HNO_3 solution (4:1, v/v). The concentrations of Ca^{2+}, NH_4^+, Mg^{2+}, SO_4^{2-}, CO_3^{2+}, Cl^-, and K^+ in SAR were 1.5, 2.62, 1.00, 10.00, 2.61, 11.17, and 1.78 $mg\cdot L^{-1}$, respectively.

Leaching Experiment

The column (Fig. 1A) was oriented vertically and slowly saturated from the bottom with deionized water until it reached the field-holding capacity. The soil column was allowed to stabilize for 24 h. The feed solutions were composed of SAR at pH 3.0, 4.5, and 5.6, with the last group as the control. The feed solutions were then introduced into the system using a peristaltic pump to percolate through the packed soil columns at a flow rate of (60±5) $mL\cdot h^{-1}$, which corresponds to the field infiltration velocity of 3.0 $cm\cdot h^{-1}$ [14] (Fig. 1B). The redox potential was measured in one-third of the columns to check the aeration of columns and to avoid waterlogging conditions. Each column was flushed with 3000 mL (1530 mm) of the incoming solution, which corresponds to 1.5 yr of precipitation (rain) in the study area. Experiments were conducted in triplicate at each pH treatment.

The leachate from the soil column was filtered through a 0.45 μm membrane filter, and 200 mL of the filtered leachate (equivalent to 100 mm of added SAR) was sampled using glass collectors to determine the heavy metal concentrations, pH, and electrical conductivity. After the tests, the columns were separated and extruded. All of the soils from the entire depth range were thoroughly mixed, dried, and ground to analyze the chemical speciation of the heavy metal residues using the sequential extraction procedure [13].

Metal Determination and Quality Control

The Cu, Zn, Pb, and Cd concentrations in soil were determined by digesting 0.5 g of the soil samples (oven-dry weight) with a HNO_3–HF–$HClO_4$ mixture followed by elemental analysis. The concentrations of these metals in all the solutions were analyzed by graphite furnace atomic absorption spectrometry (AA220Z; Varian, USA). The detection limits for Cu, Zn, Pb, and Cd were 2, 2, 1, and 0.5 $\mu g\cdot L^{-1}$, respectively. All the reagents used for analysis were of analytical grade or higher. All the containers were soaked in 10% HCl, rinsed thoroughly in deionized water, and dried before use. The standard substances such as the geochemical standard reference sample soil in China (GSS-15) were used to examine the precision and accuracy of determination. The relative errors (REs) between the sum of the metal concentration in individual fractions and the measured total metal concentration in

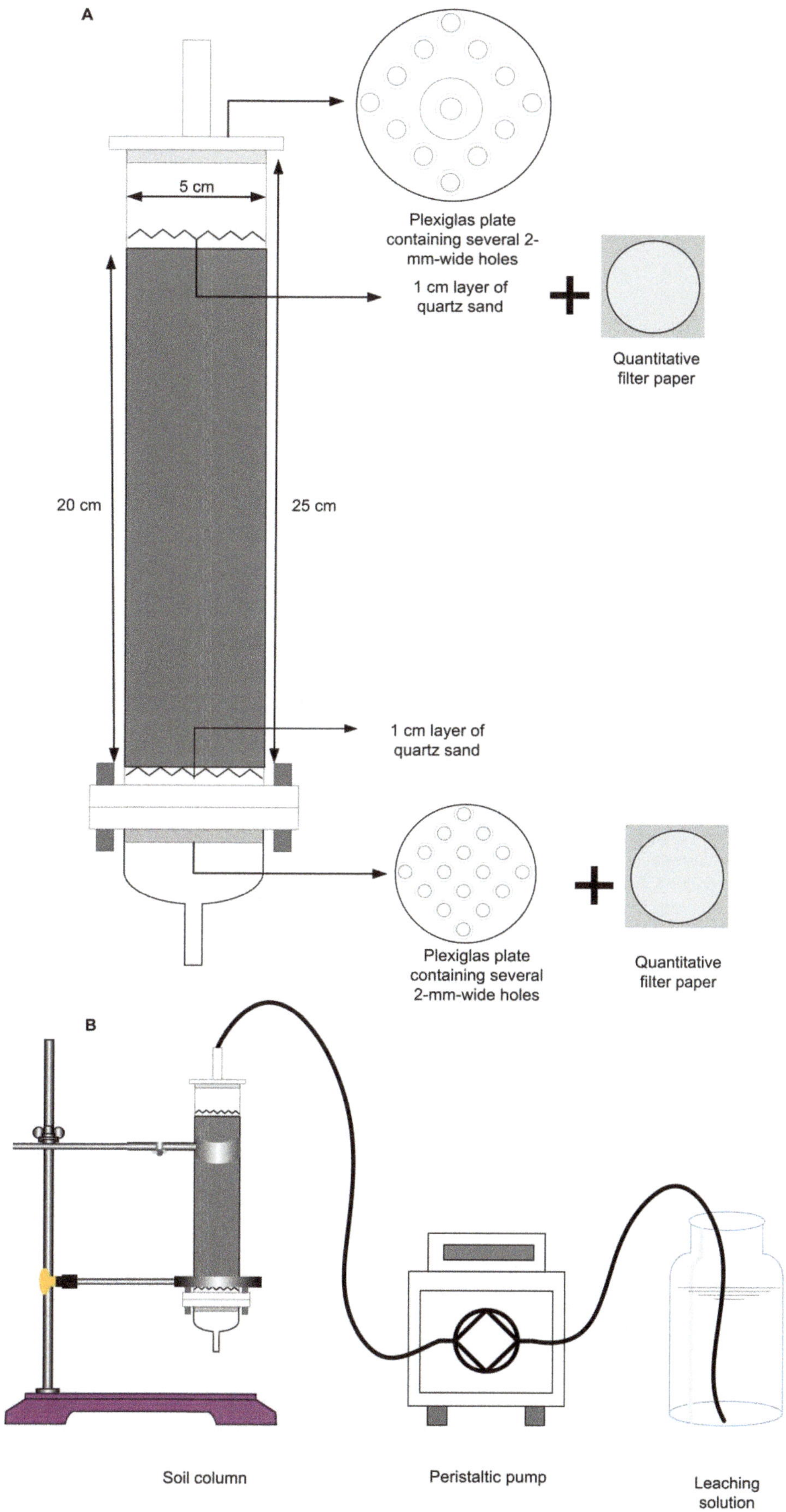
A
5 cm
Plexiglas plate containing several 2-mm-wide holes
1 cm layer of quartz sand
Quantitative filter paper
20 cm
25 cm
1 cm layer of quartz sand
Plexiglas plate containing several 2-mm-wide holes
Quantitative filter paper
B
Soil column
Peristaltic pump
Leaching solution

Figure 1. Leaching experimental design. (A) Schematic diagram of soil column. (B) Schematic analytical setup for the measurement of metal concentrations in the leaching experiment.

the soil samples, which ranged from −12.36% to 9.44%, were calculated to check the reliability of the sequential extraction procedure.

Statistical Analysis

Data were analyzed using the Origin 8.5.1 for Windows software at the 5% and 1% significance levels.

Results and Discussion

Heavy Metal Concentrations in the Leachate

The Pb and Cd concentrations in the leachate from SAR treatments at pH 3.0, 4.5, and 5.6 were generally very low throughout the leaching period, with most of them below their detection limits (data not shown). The maximum Pb and Cd concentrations in the leachate were only (17.0±2.16) and (2.8±0.42) $\mu g \cdot L^{-1}$, respectively, which are both less than the limits of the Chinese drinking water quality standards (50 and 10 $\mu g \cdot L^{-1}$, respectively). No significant differences were observed among the treatments of SAR at different pH levels. This result suggested that the polluted purple soil, which received SAR at a given pH in our study, may not cause groundwater contamination by Pb and Cd leaching. Low leachate concentrations of Pb and Cd can probably be attributed to the low mobility of Pb in soil as well as the comparatively low levels of Cd in artificially contaminated purple soil [15,16].

The Different results were observed with Cu and Zn. The leaching concentrations of Cu and Zn under SAR conditions are shown in Fig. 2. Higher Cu and Zn leachate concentrations were observed throughout the entire leaching period (ranging from 0.021 $mg \cdot L^{-1}$ to 1.49 $mg \cdot L^{-1}$ for Cu; 0.019 $mg \cdot L^{-1}$ to 2.34 $mg \cdot L^{-1}$ for Zn) as compared with those of Pb and Cd. However, the maximum Cu and Zn concentrations were still below the Chinese Quality Standards for Groundwater (Grade IV), thereby suggesting that leaching from the polluted purple soil under acid rain at the pH used in this study is unlikely to cause Cu and Zn contamination in water systems.

Fig. 2A shows the Cu concentrations in the leachates when SAR was used as the leaching reagent. Similar trends were observed for the changes in the Cu concentration of each treatment. The application of SAR produced a pulse of Cu in the leachates during the first stage of the experiment, with peaks of 1.49, 1.11, and 0.88 $mg \cdot L^{-1}$ at a pH of 3.0, 4.5, and 5.6, respectively. The concentrations decreased thereafter and reached <0.2 $mg \cdot L^{-1}$ at the end of the experiment. A small amount of Cu (<0.18 $mg \cdot L^{-1}$) was leached from the soil column in the first 200 mm of rainfall, which may be attributed to the required time for Cu to move from the upper surface layer to the bottom of the soil column and for the Cu-bound compounds to functionalize with the soil components. The SAR treatments at pH 3.0 and 4.5 yielded higher Cu concentrations than those in the control (SAR at pH 5.6), particularly before 1000 mm of SAR was added. The Cu leachate concentration increased as the pH values of SAR decreased. However, the Cu leachate concentrations for the treatments at pH 3.0 and 4.5 were approximately equal to those in the control as the leaching amounts were increased, thereby suggesting that the leachable Cu in all of the treatments was decreased. After 1200 mm of SAR was added, no significant differences were observed between the control and the treatments at pH 3.0 and 4.5.

The changes in the Zn leachate concentrations against the leaching amounts in the SAR treatments at different pH levels are shown in Fig. 2B. The variation curves of Zn with the three treatments were generally similar. Two peak values of the Zn concentration were found during the leaching experiment. The first peak appeared after approximately 300 mm to 500 mm of continuous leaching, and the second one appeared after 1200 mm of SAR was added. The peak values of the Zn concentration in the leachates were 2.34, 1.92, and 1.46 $mg \cdot L^{-1}$ for the treatments at pH 3.0, 4.5, and 5.6, respectively. These results suggested that the release of Zn from the contaminated purple soil can be observed in two stages. In the first leaching stage, leached Zn mainly existed as water soluble and exchangeable fractions. After being leached out, these mobile Zn began to adsorb and desorb in the soil column with the downward movement of leachate, and after 300–500 mm rainfall Zn concentration in the leachate reached the first peak value. During the second stage, Zn was released from different fractions such as the CAR-, OX-, and OM-bound Cu. A comparable amount of Zn was shifted from those fractions, which revealed the second peak values after 1200 mm of rainfall was added. Many studies have demonstrated that Zn in EXC and the water soluble fraction usually account for only 0.5% to 7% of the total Zn concentration [17–19], which could explain why the second peak values were higher than the first ones.

Total Leaching Amounts of Heavy Metals

An analysis of the total leaching amounts of heavy metals can directly reflect the leaching strength of the said heavy metals. The total leaching amounts of Pb and Cd were not calculated because most of their concentrations in the leachate were below the detection limits. The total leaching amounts of Cu and Zn in each treatment are shown in Table 1. Two-way ANOVA showed that the total amounts of the leached metals from the soil columns were significantly affected by the pH level and metal species as well as the interaction between these two factors (Table 2). Zn leached more easily from the soil column than Cu at each pH level (Tables 1 and 2). Similarly, Guo et al. [20] revealed that more Zn flowed out of the soil column than Cu in acidic soils under SAR treatments. The amount of leached metals increased at pH 3.0 and 4.5 as compared with that in the controls (pH 5.6), whereas the magnitude of this increase depended on the metal. Metallic elements become more soluble (leachable) under acidic conditions [21]. pH may likewise control the nature of the interactions between metals and soil surfaces [22]. Thus, higher Cu and Zn concentrations were induced by SAR and dissolved in columns at lower pH (Table 1). The H^+ ion in acid rain displaces the cations from their binding sites, which causes the increased amount of heavy metal desorption, as shown in the following reactions [Eq. (1–2)]:

$$SOH \cdots X^{2+} + H^+ \leftrightarrow SOH_2 + X^{2+} \quad (1)$$

$$SOX^{2+} + H^+ \leftrightarrow SOH + X^{2+} \quad (2)$$

The different affinities of the binding sites to metals may cause the different rates and amounts of the desorbed metal. For Zn, greater amounts were released (Table 1). H^+ ions can only replace

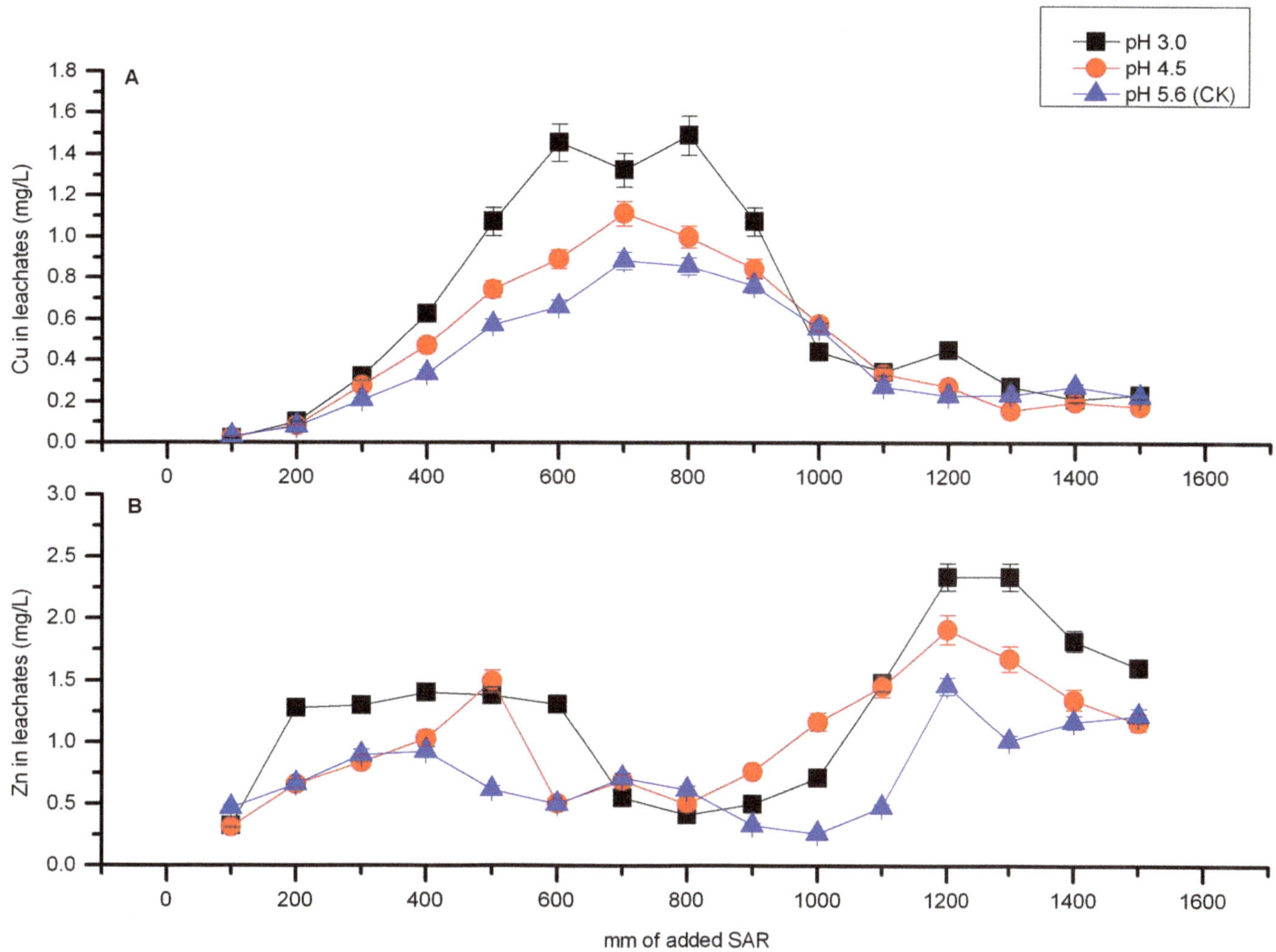

Figure 2. Metal concentrations in leachates as functions of addition of SAR. The concentrations of Cu and Zn in the leachates as functions of addition of SAR at different pHs. (A) Copper. (B) Zinc.

some metal cations, whereas the surface OH^- groups attached to the Fe or Al atoms can accept protons from the solution, thereby increasing the positive surface charges via the protonation of mineral surfaces [23,24]. Protonation is a very rapid reaction during the initial stage and depends on the amounts and properties of ferric oxides.

$$Fe\text{-}OH + H^+ \leftrightarrow Fe\text{-}OH_2{}^+ \leftrightarrow dissolution(Fe^{3+}) \qquad (3)$$

The surface protonation in soil promotes the dissolution of ferric oxides in acidic solutions [Eq. (3)], which exhibit high adsorptive capacities for heavy metals and consequently increase the

Table 1. Total amounts of the metals leached from the soil columns in mg of metal per kg of soil.

Treatment	Total leaching amounts (mg·kg^{-1})	
	Cu	Zn
pH 3.0	(3.93±0.44) a	(7.81±0.61) a
pH 4.5	(2.98±0.36) b	(6.47±0.47) b
pH 5.6	(2.27±0.39) c	(4.72±0.33) c

Note: The values are means ± standard deviation. Different lower case letters show significant differences in the same treatment (ANOVA/LSD, $P<0.05$).

Table 2. Two-way ANOVA of pH levels and metal species effects for total amounts of the metals leached from the soil columns.

Source of variance	SS	MS	F
pH	48.22	48.22	245.75**
Metal species	16.93	8.46	43.14**
pH × metal species	1.64	0.82	4.18*
Error	2.35	0.20	
Total	69.14		

Note: **significant at 99% probability level, *significant at 95% probability level.

solubilization of these metals in their oxide fractions.

Chemical Speciation of Heavy Metals in Purple Soil Affected by SAR

The chemical fractions of Cu, Pb, Cd, and Zn in soils before and after the column leaching tests are shown in Fig. 3. For the original contaminated purple soil, Cu and Cd were dominantly associated with RES (37% to 41%), followed by OM (31% to 33%). Generally, EXC- and CAR-bound Cu and Cd accounted for <9% of the total amount of the respective metals. A significant fraction of Pb in the soils was bound in RES (66.31%), and four fractions accounted for <35% of the total Pb. Some studies have shown [15,25] that Pb is mostly present in the RES of soil at the surface or profile scale and is widely considered to exhibit very low geochemical mobility. The amount of Zn in soils was mainly associated with RES (42.69%), then with the OX (25.96%), followed by the CAR (14.44%), and OM (12.23%). The percent of exchangeable Zn in the soils was relatively low (4.69%).

Some variations in the proportions of the four metals in five fractions after the tests are shown by Fig. 3. The relative concentrations of the elements in the EXC and OX fractions were generally increased after the leaching experiment was conducted for the three treatments, whereas those of RES, OM, and CAR fractions were slightly decreased. The Pb and Cd concentrations in EXC ranged from 3.28% to 4.39% and 4.47% to 6.79%, respectively. By contrast, these metals increased by more than 3% in OX. The increase in the Cu concentrations in EXC was greater than that in the OX (EXC 3% to 7% vs. OX <2%). The relative concentrations of Zn in OM and OX significantly increased, particularly in the EXC fraction (average 4.32%).

The EXC fraction was usually the first to be brought into the solution and is readily available for plant uptake. The four trace elements in EXC, with high bioavailability and mobility, were increased from ~1.2% to 5.03%, which indicated the increased direct risk of Cu, Pb, Cd, and Zn contamination in the soil/groundwater system as caused by acid rain. The amounts of the non-residual fractions represent the amounts of potentially active trace elements [25]. Generally, the high proportion of trace elements in the non-residual fractions of soils may suggest the large contributions of anthropogenic elements. The non-residual fractions of Cu, Pb, Cd, and Zn in the soil zones after the leaching tests averaged at 62.55%, 37.43%, 65.95%, and 61.61%, respectively. The non-residual fractions of Cu, Pb, Cd, and Zn in the original soil samples averaged at 59.75%, 33.69%, 62.04%, and 57.31%, respectively. The increased non-residual fractions represented the increased potential risk of Cu, Pb, Cd, and Zn contamination in the soil/groundwater system. Simple correlation

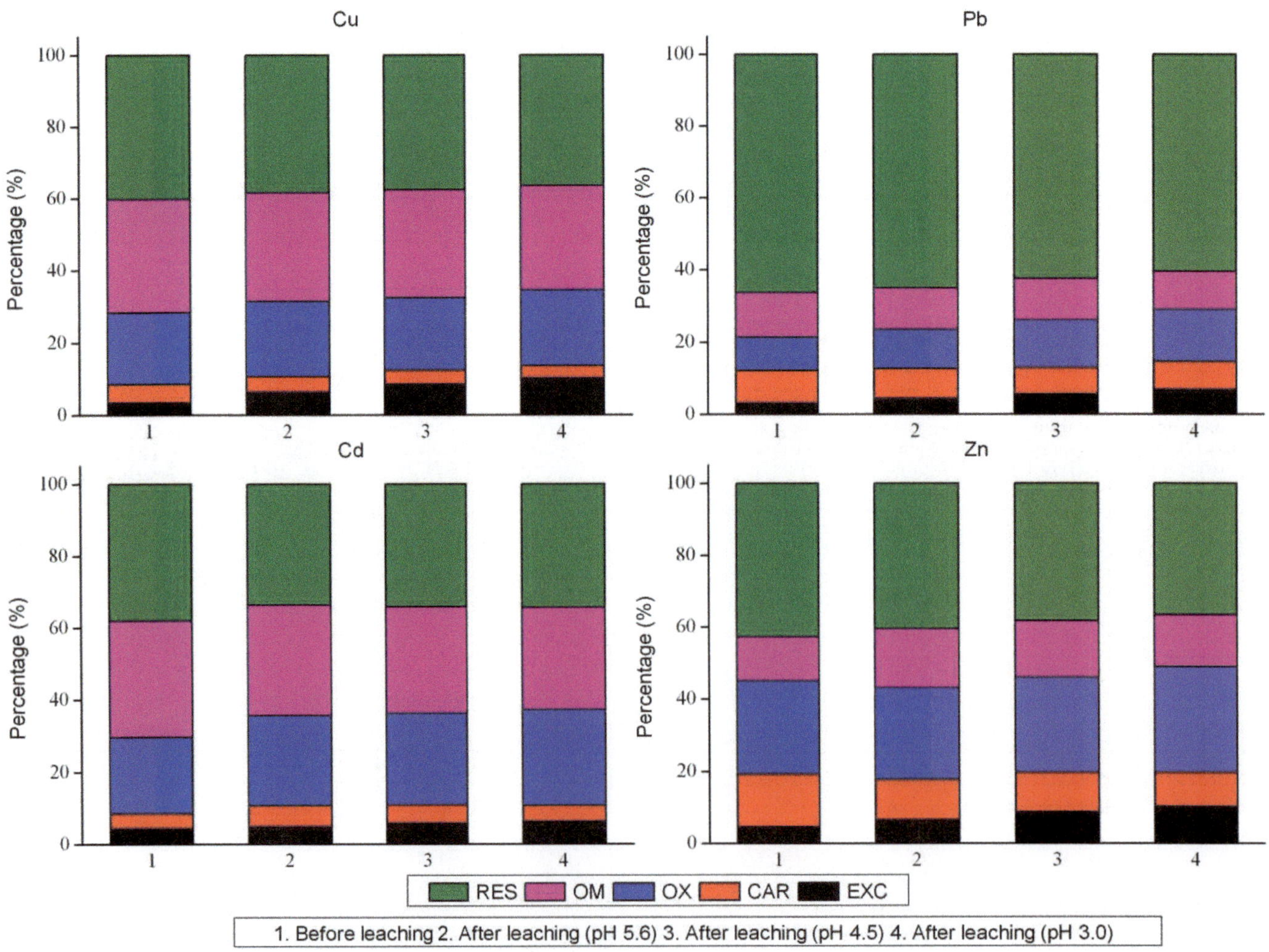

Figure 3. Variation of the percentage of metals in different fractions. Species distribution of Cu, Pb, Cd and Zn extracted with Tessier scheme before and after the column tests.

analysis indicated that the pH was significantly correlated with the exchangeable heavy metal content ($r = 0.974$, $P < 0.01$). Likewise, the pH was significantly correlated with the non-residual heavy metal content ($r = 0.968$, $P < 0.01$). These data suggested that acid rain favors the leaching of heavy metals from the contaminated purple soil. Consequently, acid rain can affect the mobilization potential of heavy metals. The decrease in the pH from 5.6 to 3.0 significantly enhanced such effects of acid rain, which are likely to cause more serious harmful effects on soil-vegetation systems.

Conclusions

The maximum leachate concentrations of Cu, Pb, Cd, and Zn from the soil that received SAR were less than those in the Chinese Quality Standard for Groundwater (Grade IV). Moreover, most of the Pb and Cd leachate concentrations were below their detection limits. By contrast, higher Cu and Zn leachate concentrations were observed because these metals were released in higher amounts as compared with Pb and Cd. The differences in the Cu and Zn leachate concentrations between the control (SAR at pH 5.6) and the treatments (SAR at pH 3.0 and 4.5) were significant. Similar trends were observed in the total leaching amounts of Cu and Zn. The proportions of Cu, Pb, Cd, and Zn in the EXC and OX fractions were generally increased after the leaching experiment at three pH levels, whereas those of the RES, OM, and CAR fractions were slightly decreased. Acid rain favors the leaching of heavy metals from the contaminated purple soil and causes the heavy metal fractions to become more labile. A decrease in the pH from 5.6 to 3.0 significantly enhanced such effects. However, the experiment is only an indoor simulation that used artificially contaminated purple soil. Thus, long-term field experiments on soils contaminated with acid rain should be conducted to study the cycling effect of heavy metal residues in the soil or those that leach into the groundwater.

Acknowledgments

The authors are especially grateful to Dr. Yirong Guo for critical review and perceptive comments on the manuscript. The comments and suggestions of the editor and two anonymous referees are appreciated.

Author Contributions

Conceived and designed the experiments: SZ. Performed the experiments: SZ CC. Analyzed the data: SZ XZ. Contributed reagents/materials/analysis tools: XZ. Wrote the paper: SZ.

References

1. Larssen T, Lydersen E, Tang D, He Y, Gao J, et al. (2006) Acid rain in China. Environ Sci Technol. 40: 418–425.
2. Singh A, Agrawal M (2008) Acid rain and its ecological consequences. J Environ. Biol. 29: 15–24.
3. Xie Z, Du Y, Zeng Y, Li Y, Yan M, et al. (2009) Effects of precipitation variation on severe acid rain in southern China. J Geogr Sci. 19: 489–501.
4. Ma L, Wang B, Yang J (2008) Spatial- Temporal Distribution of Acid Rain in Sichuan Province. Envion Sci Manage. 33: 26–29.
5. Li ZB, Li P, Han JG, Li M (2009) Sediment flow behavior in agro-watersheds of the purple soil region in China under different storm types and spatial scales. Soil Till Res. 105: 285–291.
6. Gao Y, Zhu B, Zhou P, Tang JL, Wang T, et al. (2009) Effects of vegetation cover on phosphorus loss from a hillslope cropland of purple soil under simulated rainfall: a case study in China. Nutr Cycl Agroecosys. 85: 263–273.
7. Wu F, Yang W, Zhang J, Zhou L (2010) Cadmium accumulation and growth responses of a poplar (Populus deltoids×Populus nigra) in cadmium contaminated purple soil and alluvial soil. J Hazard Mater. 177: 268–273.
8. Li Q, Wei C, Huang Y, Wang L, Wang D (2010) A study on the characteristics of heavy metals in orange ecosystem. Chin J Geochem. 29: 100–106.
9. Zhang JE, Ouyang Y, Ling DJ (2007) Impacts of simulated acid rain on cation leaching from the Latosol in south China. Chemosphere 67: 2131–2137.
10. Ling DJ, Huang QC, Ouyang Y (2010) Impacts of simulated acid rain on soil enzyme activities in a latosol. Ecotoxicol Environ Saf. 73: 1914–1918.
11. Wang M, Gu B, Ge Y, Liu Z, Jiang D, et al. (2009) Different responses of two Mosla species to potassium limitation in relation to acid rain deposition. J Zhejiang Univ-Sci B. 10: 563–571.
12. Lu RK (2000) Methods of soil and agrochemical analysis (in Chinese). Beijing: China Agricultural Science and Technology Press.
13. Tessier A, Campbell P, Bisson M (1979) Sequential extraction procedure for the speciation of particulate trace metals. Anal Chem. 51: 844–851.
14. Grolimund D, Borkovec M, Barmettler K, Sticher H (1996) Colloid-facilitated transport of strongly sorbing contaminants in natural porous media: a laboratory column study. Environ Sci Technol. 30: 3118–3123.
15. Martinez CE, Motto HL (2000) Solubility of lead, zinc and copper added to mineral soils. Environ Pollut. 107: 153–158.
16. Lee SZ, Chang L, Yang HH, Chen CM, Liu MC (1998) Adsorption characteristics of lead onto soils. J Hazard Mater. 63: 37–49.
17. Kabala C, Singh BR (2001) Fractionation and mobility of copper, lead, and zinc in soil profiles in the vicinity of a copper smelter. J Environ Qual. 30: 485–492.
18. Ma LQ, Rao GN (1997) Chemical fractionation of cadmium, copper, nickel, and zinc in contaminated soils. J Environ Qual. 26: 259–264.
19. Xiang HF, Tang HA, Ying QH (1995) Transformation and distribution of forms of zinc in acid, neutral and calcareous soils of China. Geoderma 66: 121–135.
20. Guo ZH, Liao BH, Huang CY (2005) Mobility and speciation of Cd, Cu, and Zn in two acidic soils affected by simulated acid rain. J Environ Sci. 17: 332–334.
21. Wilson MJ, Bell N (1996) Acid deposition and heavy metal mobilization. Appl Geochem. 11: 133–137.
22. Zhang H, Davison W, Tye AM, Crout NMJ, Young SD (2006) Kinetics of zinc and cadmium release in freshly contaminated soils. Environ Toxicol Chem. 25: 664–670.
23. Zhu MX, Jiang X, Ji GL (2005) Investigation of time-dependent reactions of H^+ ions with variable and constant charge soils: a comparative study. Appl Geochem. 20: 169–178.
24. Zhang FS, Zhang XN, Yu TR (1991) Reactions of hydrogen ions with variable charge soils: I. Mechanisms of reaction. Soil Sci. 151: 436–442.
25. Tack F, Verloo MG (1995) Chemical speciation and fractionation in soil and sediment heavy metal analysis: a review. Int J Environ An Ch. 59: 225–238.

10

Does *S*-Metolachlor Affect the Performance of *Pseudomonas* sp. Strain ADP as Bioaugmentation Bacterium for Atrazine-Contaminated Soils?

Cristina A. Viegas[1,2]*, Catarina Costa[2], Sandra André[4], Paula Viana[4], Rui Ribeiro[3], Matilde Moreira-Santos[3]

1 Department of Bioengineering, Instituto Superior Técnico (IST), Technical University of Lisbon (TUL), Lisboa, Portugal, **2** Institute for Biotechnology and Bioengineering (IBB), Centre for Biological and Chemical Engineering, Instituto Superior Técnico (IST), Technical University of Lisbon (TUL), Lisboa, Portugal, **3** Instituto do Mar (IMAR), Department of Life Sciences, University of Coimbra, Coimbra, Portugal, **4** Agência Portuguesa do Ambiente (APA), Amadora, Portugal

Abstract

Atrazine (ATZ) and *S*-metolachlor (*S*-MET) are two herbicides widely used, often as mixtures. The present work examined whether the presence of *S*-MET affects the ATZ-biodegradation activity of the bioaugmentation bacterium *Pseudomonas* sp. strain ADP in a crop soil. *S*-MET concentrations were selected for their relevance in worst-case scenarios of soil contamination by a commercial formulation containing both herbicides. At concentrations representative of application of high doses of the formulation (up to 50 $\mu g\ g^{-1}$ of soil, corresponding to a dose approximately 50× higher than the recommended field dose (RD)), the presence of pure *S*-MET significantly affected neither bacteria survival (~10^7 initial viable cells g^{-1} of soil) nor its ATZ-mineralization activity. Consistently, biodegradation experiments, in larger soil microcosms spiked with 20× or 50×RD of the double formulation and inoculated with the bacterium, revealed ATZ to be rapidly (in up to 5 days) and extensively (>96%) removed from the soil. During the 5 days, concentration of *S*-MET decreased moderately to about 60% of the initial, both in inoculated and non-inoculated microcosms. Concomitantly, an accumulation of the two metabolites *S*-MET ethanesulfonic acid and *S*-MET oxanilic acid was found. Despite the dissipation of almost all the ATZ from the treated soils, the respective eluates were still highly toxic to an aquatic microalgae species, being as toxic as those from the untreated soil. We suggest that this high toxicity may be due to the *S*-MET and/or its metabolites remaining in the soil.

Editor: Stephen J. Johnson, University of Kansas, United States of America

Funding: This work was supported by Fundo Europeu de Desenvolvimento Regional and Fundação para a Ciência e Tecnologia (http://www.fct.pt/), Portugal (Contracts PTDC/AAC-AMB/111317/2009, PTDC/AMB/64230/2006, PPCDT/AMB/56039/2004, and Ciência 2007-FSE and POPH funds). The funders had no role in study design, data collection and analysis, decision to publish, or preparation of the manuscript.

Competing Interests: The authors have declared that no competing interests exist.

* E-mail: cristina.viegas@ist.utl.pt

Introduction

Diverse pesticides (and their metabolites), fertilizers and organic components used in commercial formulations can be found in agricultural soils due to intensive use in crop cultivation or accidental spills. In particular, agricultural dealership sites and mix-load or disposal sites may represent potential sources of environmental contamination with high levels of diverse compounds [1,2]. For example, 205, 2272, 13, 1829 and 108 $\mu g\ g^{-1}$ of soil of the *s*-triazine herbicides atrazine (ATZ), cyanazine, simazine and the chloroacetanilide herbicides metolachlor (MET) and alachlor, respectively, were measured in soil samples from a mix-load site [1]. As a consequence, mixtures of chemicals are likely to contaminate freshwaters via leaching, drainage and runoff events [3–5], and toxicological synergies of multiple chemicals may enhance deleterious effects on ecosystems and human health [6]. Herbicidal formulations containing ATZ, either as the sole active ingredient or in combination with the racemic mixture of MET or, more recently, with the product enriched in the active *S*-enantiomer (*S*-MET), have been used worldwide for broad-spectrum weed control in several crops [7–10].

In soil, ATZ presents moderate persistency (DT_{50} = 28–150 days, field, aerobic) and mobility (soil organic carbon/water partition coefficient, K_{OC} = 100 L kg^{-1}) [8]. These properties and increasing concerns regarding potential impact of ATZ and its toxic chlorinated N-dehalkylated metabolites on human health and ecosystems [11–13], promoted active research on ATZ-degrading microorganisms and on bioremediation strategies aiming to reduce soil contamination to safe levels and to minimize dispersion into surrounding aquatic compartments [1,14–19]. *Pseudomonas* sp. ADP is the best-characterized ATZ-mineralizing bacteria and uses ATZ as sole N source by means of a catabolic pathway encoded in the plasmid pADP-1 [16,18]. It has high potential for the bioaugmentation of ATZ-contaminated soils, but, presumably due to C limitation and low survival in soil, it was found to be less effective at high ATZ concentrations that are relevant in the case of spill or careless disposal scenarios [17,18]. As part of a framework for the rational bioremediation of ATZ-contaminated land, we recently presented evidences that a clean-up strategy, combining soil bioaugmentation with this bacterial strain and biostimulation with citrate [17], was effective at a larger microcosms scale [14,15]. It led to the rapid removal of ATZ from a natural crop soil spiked with a commercial formulation containing ATZ as single active ingredient at doses mimicking worst-case scenarios (e.g. spills and/or concentrated hotspots) [14,15]. More importantly, assessment of the ecotoxicity of soil samples and eluates/leachates from the bioremediated soils proved the effective soil decontamination in less than 10 days, hence contributing to significantly diminish the toxicity impact in the aquatic compartment [14].

S-metolachlor is relatively non-persistent ($DT_{50} = 21$ days, field, aerobic) and moderately mobile ($K_{OC} = 226$ L kg^{-1}) in soil, but very persistent in water [8,9]. Both the racemic mixture and the *S*-enantiomer have the potential to cause adverse effects on microorganisms [2,20] and on higher terrestrial and aquatic organisms [8,20–22]. Freshwater macrophytes and microalgae are however the most vulnerable ones [6,8,23]. To our knowledge, the few published studies addressing the effect of MET or *S*-MET on ATZ-degradation in soil have been limited to indigenous soil microorganisms and intrinsic biodegradation. In some cases a negative effect was demonstrated [2] while in others there was apparently no impact [24].

Possible deleterious effects exerted by *S*-MET [2] could contribute to diminish the efficacy of the bioremediation tool previously shown to be effective for the clean-up of soils contaminated with high concentrations of ATZ [14,15,17]. Therefore, the present work aimed examining whether *S*-MET could negatively affect the survival and the ATZ-biodegradation performance of the bioaugmentation bacterium *Pseudomonas* sp. strain ADP in a crop soil. We used a representative crop soil from Central Portugal [14,15] spiked with mixtures of ATZ and *S*-MET (either as the commercial formulation Primextra S-Gold or as pure active ingredients). Doses mimicked worst-case scenarios of soil contamination, being thus higher than the recommended dose (RD) for weed control in corn plantations. The following issues were addressed: first, the ability of *Pseudomonas* sp. ADP, either combined or not with citrate amendment, to mineralize [ring-UL-^{14}C]ATZ mixed with increasing doses of Primextra S-Gold (up to 50×RD), in soil at small laboratory scale. Second, the effects of *S*-MET in a pure form on the bacteria survival and on its ATZ-mineralizing ability, at concentrations of the pure active ingredients representing up to 50×RD, also in small soil microcosms. Third, the rate and extent of biodegradation of ATZ from 20× or 50×RD Primextra S-Gold when applying the bioremediation tool, in larger and more realistic soil microcosms [14,15]. The efficacy of ATZ-biodegradation in soil microcosms was evaluated mainly by performing microalgae ecotoxicity tests (72-hours *Pseudokirchneriella subcapitata* growth tests) on the eluates from soil samples. At this stage, eluates from soils contaminated with the herbicidal formulation Atrazerba FL (with ATZ as single active ingredient) and previously reported to be decontaminated after a 10-days treatment with the bioremediation tool [14] were also used for comparison purposes. The extent of ATZ removal was also examined with chemical analysis of ATZ and its metabolites in soil samples. In addition, possible modifications in the concentration of *S*-MET initially present in the soil microcosms contaminated with Primextra S-Gold as well as the possible formation of its major degradation products *S*-MET oxanilic acid (OA) and *S*-MET ethanesulfonic acid (ESA) [8,25,26] were also examined during the biodegradation experiments carried out.

Results

ATZ mineralization by *Pseudomonas* sp. ADP in soil spiked with Primextra S-Gold

Following inoculation with viable cells of the bioaugmentation bacterium ($1.3\pm0.5\times10^7$ CFU g^{-1} soil dry weight) of soil spiked with mixtures of [^{14}C]ATZ plus increasing doses of non-labeled ATZ from the double formulation Primextra S-Gold (5×, 20× or 50×RD), ATZ mineralization started rapidly. The percentage of $^{14}CO_2$ produced from the ^{14}C-labeled ATZ attained maximal values within 3 days (Fig. 1A). Comparatively, ATZ mineralization was negligible (<2%) in the non-inoculated control soil during the same time-period (Fig. 1A). However, a moderate inhibition of rate and extent of ATZ mineralization occurred in the soils contaminated with doses of Primextra S-Gold increasing from 5×

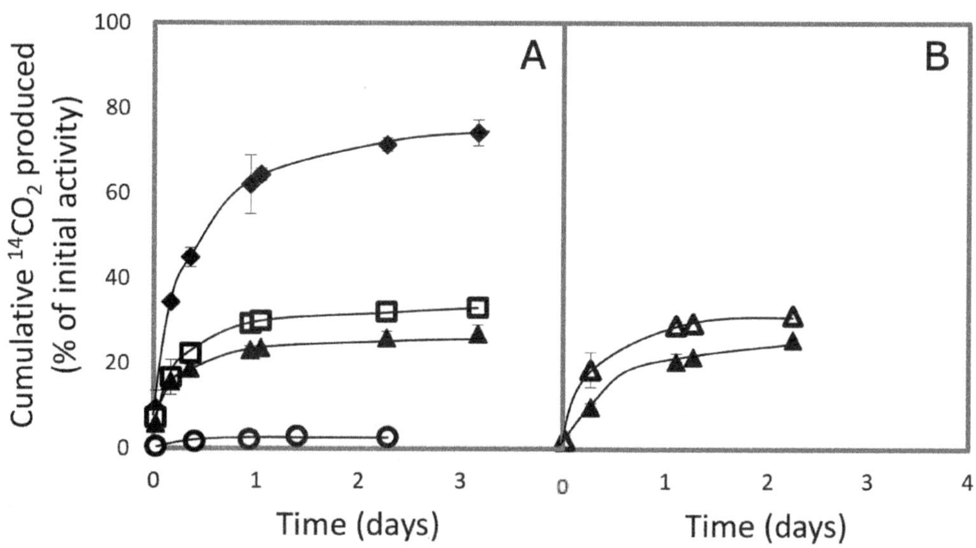

Figure 1. ATZ mineralization by *Pseudomonas* sp. ADP Rif^R in soil contaminated with Primextra S-Gold. Time-course formation of $^{14}CO_2$ from [ring-UL-^{14}C]ATZ in (**A**) soil spiked with [^{14}C]ATZ plus increasing doses of Primextra S-Gold as follows: 5× (◆), 20× (□) and 50×RD (▲), without citrate amendment; or (**B**) soil spiked with [^{14}C]ATZ plus 50×RD of Primextra S-Gold and amended with 3.4 mg g^{-1} trisodium citrate (Δ) or non-amended (▲). The average amount of $^{14}CO_2$ released from non-inoculated control soil is shown for comparison (○) in (**A**). Data are means±SD of measurements from three replicated samples from at least two independent experiments under each condition.

up to 50×RD (Fig. 1A). For example, the percentage of initial labeled ATZ evolving as $^{14}CO_2$ at day 3 were 74.3±3.0% and 27.0±2.1% in the soil microcosms with 5× and 50×RD of Primextra S-Gold, respectively (Fig. 1A). These values correspond to approximately 37 and 24%, respectively, of the total estimated amount of ATZ mixed into the soil (assuming that labeled and non-labeled ATZ may be mineralized homogeneously). The soil that had been contaminated with 50×RD of Primextra S-Gold showed the lowest percentage of [^{14}C]ATZ mineralization following bioaugmentation (Fig. 1A). In this soil, enhancement of the ratio of soluble carbon to nitrogen from atrazine (C_s:N_{atz}) from ~1 (in the crop soil used) up to ~50 [15], due to soil amendment with trisodium citrate, led to a slight but significant increase in the rate and extent of $^{14}CO_2$ formation (Fig. 1B).

Effects of *S*-MET in *Pseudomonas* sp. ADP survival and ability to mineralize ATZ in soil

Viable populations of *Pseudomonas* sp. ADP inoculated into the soil were exposed to mixtures of ATZ plus *S*-MET at concentrations representing applications of approximately 30× or 50×RD of each active ingredient (Fig. 2). For the soil contaminated with 24 μg ATZ g^{-1}, the addition of *S*-MET at 30 μg g^{-1} (representing ~30×RD of each active substance) and up to 60 μg g^{-1}, did not significantly affect the bacterial survival (Fig. 2A). Moreover, the rate and extent of [^{14}C]ATZ mineralization was essentially the same whether or not *S*-MET was added to soil previously contaminated with a total of 24 or 40 μg ATZ g^{-1} of soil (Fig. 2B), indicating that presence of *S*-MET does not significantly affect the ability of *Pseudomonas* sp. ADP to mineralize ATZ in the worst-case conditions tested herein.

Biodegradation of ATZ from Primextra S-Gold in larger soil microcosms

The performance of the bioaugmentation/biostimulation treatment for ATZ-contaminated soils, consisting on one initial inoculation with *Pseudomonas* sp. ADP (4.1±1.2×10^7 CFU g^{-1} soil dry weight) combined with soil amendment with trisodium citrate (C_s:N_{atz}~50) [15], was examined in larger soil microcosms spiked with 20× or 50×RD of Primextra S-Gold (Fig. 3). For both doses of the commercial formulation and upon soil bioaugmentation, bacterial numbers were always higher (2.5-fold, in average) in the soil amended with citrate compared with non-amended one (Figs. 3A and B). Despite that, whether or not soil was amended with citrate, most of the initial ATZ was rapidly removed from soil in up to 2 days with no lag period required (Figs. 3C and D). This high rate of ATZ biodegradation contrasted with the high levels of ATZ remaining in the untreated soils (Figs. 3C and D). Nevertheless, slight differences on the extent of ATZ biodegradation in the inoculated soils were observed depending on the initial level of soil contamination. For example, in the soil spiked with 20×RD of Primextra S-Gold, the ATZ concentration declined by ~96%, from 12.8±0.4 to less than 0.5 μg g^{-1}, in only 5 days (Fig. 3C). On the other hand, in the soil with the highest dose of Primextra S-Gold (50×RD, corresponding to an initial measured ATZ of 29.1±12.5 μg g^{-1} soil dry weight), inoculation with the bacterium without citrate amendment led to quite high levels of ATZ still remaining in the soil, namely 2.4 and 1.4±0.8 μg ATZ g^{-1} at days 5 and 8, respectively (Fig. 3D and data not shown). Combination of soil bioaugmentation with citrate amendment apparently allowed a slight but significant improvement in ATZ removal from the soil, with its concentration decreasing to 0.4±0.1 μg g^{-1} in 8 days (Fig. 3D and data not shown).

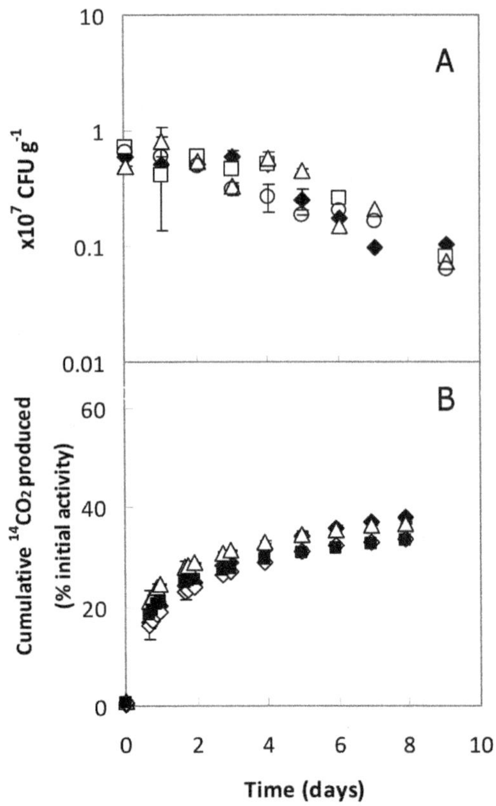

Figure 2. Effects of *S*-MET in *Pseudomonas* sp. ADP Rif^R (A) survival and (B) ATZ- mineralization in soil. In (**A**), soil previously contaminated with 24 μg ATZ g^{-1} of soil was supplemented with increasing concentrations of *S*-MET, namely 0 (◆), 15 (□), 30 (Δ) or 60 (○) μg g^{-1} of soil, prior to inoculation. In (**B**), soil was spiked with a total of 24 μg ATZ g^{-1} of soil (including [^{14}C]ATZ) plus 30 μg *S*-MET g^{-1} (Δ) or no *S*-MET (◆), or with a total of 40 μg ATZ g^{-1} plus 50 μg *S*-MET g^{-1} (■) or no *S*-MET (◇), prior to inoculation. Data are means±SD of measurements from at least duplicate determinations from two or three independent experiments under identical conditions.

Fate of *S*-MET and its major degradation products in the soil microcosms contaminated with Primextra S-Gold

Dissipation of *S*-MET in the soil microscosms was not different in untreated or inoculated soil (Figs. 3E and 3F). Indeed, in all the different conditions tested, its concentration was moderately reduced by around 40% on average during the first 5 days. For example, *S*-MET concentration declined from the initial 21±5 and 35±19 μg g^{-1} in the 20×RD and the 50×RD-contaminated soils, respectively, to 13±2 and 21±4 μg g^{-1}at day 5 (Figs. 3E and 3F), and maintained identical values after 8 days (data not shown). This decrease in the concentration of *S*-MET was accompanied by the accumulation of its derivatives *S*-MET ESA and *S*-MET OA in the soil microcosms. Namely, *S*-MET ESA concentration increased from undetectable values (<6 μg kg^{-1} soil dry weight), at time zero, to 27±9 and 46±19 μg kg^{-1} after 5 days of incubation, in the 20×RD and the 50×RD-contaminated soils, respectively. The *S*-MET OA concentration increased from <6 to 42±22 and 87±33 μg kg^{-1}, respectively, in the same period of time. At day 8, the concentrations of these compounds measured in the soil microcosms were somewhat lower, namely 22±2 and 28±7 μg *S*-MET ESA kg^{-1}, or 29±5 and 43±35 μg *S*-MET OA kg^{-1}, in the 20×RD and the 50×RD-contaminated soils, respectively (data not shown).

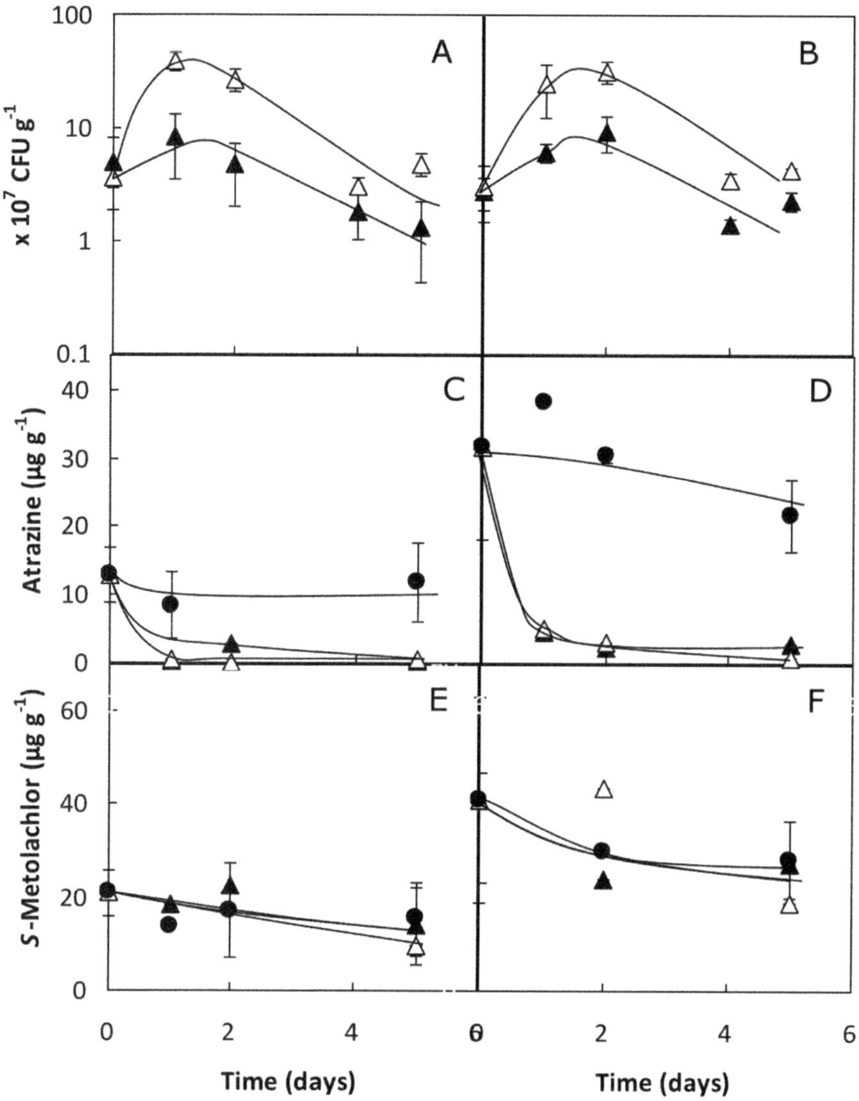

Figure 3. Biodegradation of ATZ from Primextra *S*-Gold and fate of *S*-MET in larger soil microcosms. Time-course variation of (**A, B**) the concentration of viable cells of *Pseudomonas* sp. ADP Rif[R], (**C, D**) the average concentration of ATZ and (**E, F**) the average concentration of *S*-MET, in the soil microcosms contaminated with (**A, C, E**) 20×RD or (**B, D, F**) 50×RD of Primextra S-Gold, and bioaugmented with *Pseudomonas* sp. ADP with (Δ) or without (▲) citrate amendment, during incubation at 25°C. ATZ and *S*-MET concentrations in the non-inoculated control soil (●) are also shown in (C, D) and (E, F), respectively, for comparison. Data are means±SD of measurements from at least two replicated samples from two independent experiments under each condition.

Ecotoxicity removal efficacy

To estimate potential ecotoxicological effects of addition of the two doses of Primextra S-Gold (20× and 50×RD) to soil, and to assess the efficacy of treatment of these soils with the bioremediation tool, the ecotoxicity to microalgae was assessed in eluates prepared from soil samples collected in the microcosms during the biodegradation experiments with *Pseudomonas* sp. ADP plus citrate (Fig. 3), as examined before for the case of soils contaminated with up to 20×RD of the single ATZ formulation Atrazerba FL [14]. Since there was a similarity between toxicity data obtained with the eluates from soil samples collected after 5 or 8 days upon treatment with the bioremediation tool, only the 5 days toxicity data are presented in Fig. 4.

The eluates from the soil microcosms that were spiked with Primextra S-Gold but not subjected to biaugmentation/biostimulation treatment caused serious deleterious effects on microalgae growth, compared with those from soil not contaminated with the commercial formulation (Fig. 4A). After treatment with the bioremediation tool, the toxicity of the soil eluates was still significantly higher than that of the eluates from the control soil not-contaminated with the herbicide (Fig. 4A). Indeed, microalgae growth rates were similar in the eluates from the treated or untreated soils, for both doses of Primextra S-Gold (Fig. 4A). On the contrary, treatment of Atrazerba-FL-contaminated soils (up to 50×RD) with the same bioaugmentation/biostimulation tool, led to an effective decrease on the ecotoxicity of the respective eluates, compared to the eluates from untreated soil (Fig. 4B). More importantly and corroborating previous observations [14], toxicity values were achieved which were comparable to those obtained for eluates from the soil not contaminated with the herbicide (Fig. 4B).

In addition, we examined whether the *S*-MET (and/or its degradation products) that remained in the soil during the

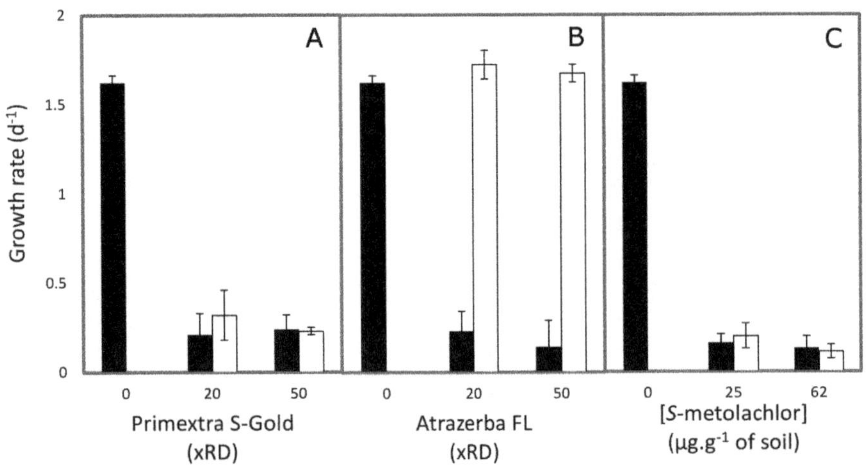

Figure 4. Ecotoxicity of eluates prepared from herbicide-contaminated soil. Mean 72-hours growth rate of *Pseudokirchneriella subcapitata* on eluates prepared with soil samples collected from microcosms spiked with the indicated doses of (**A**) Primextra S-Gold, (**B**) Atrazerba FL, or (**C**) pure *S*-MET, at day 5 after soil amendment with *Pseudomonas* sp. ADP Rif^R plus citrate (white columns) or without the bioaugmentation/biostimulation tool (black columns). Data are means±SD from toxicity tests in at least three independent eluate samples.

biodegradation experiments in the microcosms contaminated with Primextra *S*-Gold (Figs. 3E and F, and data described in previous section) could be responsible, at least partially, for the high toxicity of the respective eluates (Fig. 4A). Consistently, eluates from soils that were contaminated with *S*-MET in the pure form (25 or 62 μg g^{-1} soil dry weight) were also highly toxic to the microalgae irrespective of the soil having been subjected or not to the *Pseudomonas* sp. ADP plus citrate treatment (Fig. 4C).

Discussion

Effects of *S*-MET on *Pseudomonas* sp. ADP performance

When considering *in situ* bioremediation strategies, bioaugmentation may fail in the field as a result of the susceptibility of the specialized degrading microorganisms to high concentrations of non-target compounds; other plausible reasons may include microbial competition for a limiting nutrient and/or inhibition of degradation through catabolic repression/competitive inhibition phenomena in the soil [1,2,26,27]. The significance of pesticide interactions with intrinsic xenobiotic degraders or with bioaugmentation bacteria in soil is thus a relevant issue [2,24,26,27]. In the present work, we examined whether the presence of levels of *S*-MET representing worst-case scenarios of soil contamination with a commercial formulation containing both *S*-MET and ATZ may affect the performance of the bioaugmentation bacterium *Pseudomonas* sp. ADP for ATZ-biodegradation in soil. Soil amendment with viable cells of this bacterial strain plus an adequate provision of citrate to get a ratio C_s:N_{atz}~50, has previously been shown to be effective in the cleanup of a crop soil contaminated with high doses (up to 20×RD, mimicking similar worst-case situations) of the formulation Atrazerba FL [14]. It is worth mentioning that intrinsic ATZ-mineralization is absent or shows a very long lag-phase in the soil used ([14,15], and this work). In the present study, we demonstrated that the presence of *S*-MET in the soil at concentrations between 15 and 50 μg g^{-1} of soil, representative of applications of up to approximately 50×RD of the double herbicidal formulation Primextra S-Gold, affected neither the survival of *Pseudomonas* sp. ADP used to bioaugment the contaminated soil (~10^7 CFU g^{-1} of soil) nor the rate and extent of mineralization of ^{14}C-labeled ATZ by this bacterium. Consistently, its growth curve in liquid PADP medium (initial inoculum around 4×10^7 CFU ml^{-1}) was not affected as a result of medium supplementation with *S*-MET at concentrations up to 60 μg ml^{-1} (Costa C and Viegas CA, unpublished results). Based on the results obtained herein, we speculate that the observed moderate decrease in the bacterium ability to mineralize [^{14}C]ATZ in the presence of increasing doses of Primextra S-Gold (from 5× up to 50×RD) may mainly reflect inhibitory effects on the bacterium physiology presumably caused by high concentrations of ATZ in soil (as suggested in Silva et al. [17]) and/or by unknown substances that may be present in the commercial formulation (even though not addressed in more detail in the present work), rather than by the active ingredient *S*-MET. Other authors [24] also reported that the pattern of intrinsic degradation of ATZ (initial concentration 40 μg g^{-1} of soil) in a soil from a pesticide-contaminated site was not significantly affected by the presence of 50 μg g^{-1} of MET or of another herbicide, pendimethalin [24]. On the contrary, Moorman et al. [2] reported that the viability of indigenous ATZ-degrading microorganisms and hence ATZ-mineralization in a contaminated soil from an agricultural chemical dealership area were negatively affected by high concentrations of MET. In the latter study, however, the authors tested the effects of MET at 200 μg g^{-1} of soil [2], which is a concentration quite higher and thus presumably more toxic than the ones tested in the present work.

Consistently, based solely on soil chemical analysis, we present evidences that in soil microcosms contaminated with 20× or 50×RD Primextra S-Gold and subsequently treated with optimized quantities of *Pseudomonas* sp. ADP and citrate, ATZ removal from soil, due to its biodegradation, was extensive (>96% of the initial, to less than 0.5 μg g^{-1} of soil) and rapid (in less than 1 week) as reported before for the soils contaminated with the formulation Atrazerba FL that contains ATZ as the sole active herbicide [14,15]. Moreover, these results together with indications from the chemical data that, in the time-frame of the biodegradation experiments, deethylatrazine (DEA) and deisopropylatrazine (DIA) did not accumulate in the soils (data not shown), point to a significant potential reduction in the dispersion of the herbicide and of its highly toxic N-dealkylated metabolites into the adjacent water compartments, as reported before [14]. In agreement with previous studies, this would indicate an important environmental impact of the treatment of the contaminated soils with the bioremediation tool [14,15,18]. However, contrarily to what is reported for the case of soils contaminated with Atrazerba FL ([14], and present work), the eluates prepared from the soils spiked with Primextra S-Gold and subsequently treated with the ATZ-degrading bacterium plus citrate remained significantly toxic to the microalgae *P. subcapitata* in the time-frame of the biodegradation experiments. Based on experimental evidences, we propose that the high ecotoxicity of these eluates may be mainly associated with the presence of *S*-MET that remained in the soil and/or of its degradates, as is further discussed below.

Contribution of *S*-MET for water extracts ecotoxicity

On one hand, soil bioaugmentation with *Pseudomonas* sp. ADP apparently had no significant effect on the dissipation of *S*-MET in the soil microcosms contaminated with Primextra S-Gold, suggesting that this bacterial strain is not able to degrade the chloroacetanilide herbicide, as reported by others [19]. In spite of that, the concentration of *S*-MET in the soils decreased moderately during the 8 days of the biodegradation experiments even though keeping values always higher than 60% of the initial concentration as described above. We thus suggest that the mobilization of a significant portion of intact *S*-MET from the soil microcosms to the water extracts may have contributed at least partially for the high ecotoxicity of the eluates towards the microalgae. Even though *S*-MET levels were not measured in the eluates prepared from the soil samples, using an estimated soil/solution distribution coefficient (K_d value) approximately equal to 2.5 for *S*-MET in a sandy loam soil with 3.1% organic matter [28], it can be anticipated that ~40% of the *S*-MET present in the soil may be mobilized into the water. Since a soil:water 1:10 ratio (w/v) was used in the eluate preparation, this points for predictable *S*-MET concentrations in the water extracts considerably higher (about 50-fold) than the value 8 $\mu g\ L^{-1}$ reported as the acute 72 hours median effective concentration for *P. subcapitata* growth [8], or of similar order of magnitude of the MET concentrations reported recently as causing almost complete inhibition of the growth of this microalgae species [6]. According to published toxicological data, phytoplankton species are important targets for damage caused by *S*-MET and other chloroacetanilide herbicides, being more susceptible than aquatic organisms from higher trophic levels [6,8,21,23]. In addition to the specific mode of action of these compounds in the target plants (inhibition of fatty acids synthesis) [23], they also have potential to cause non-specific toxicity over diverse soil and aquatic non-target organisms, including microorganisms, related with the lipophilic nature of their molecules [20,21,23]. Since water extracts obtained from the soils contaminated with equivalent quantities of pure *S*-MET were as highly inhibitors of microalgae growth as those prepared from the soils contaminated with Primextra S-Gold, the high ecotoxicity of the latter may be less likely attributable to the presence of ingredients from the commercial formulation other than *S*-MET (and/or the *S*-MET derivatives herein found to accumulate in the soil microcosms, which could also be detrimental for the microalgae). Nevertheless, possible toxic effects of unknown ingredients should not be ruled out as they might have been hidden due to the high toxicity exerted by *S*-MET (and/or derivatives).

The metabolites *S*-MET ESA and *S*-MET OA did build-up in the soil microcosms concomitantly with the moderate dissipation of *S*-MET down to 60% of its initial value. MET can suffer biodegradation in soil [8,25,26] and in aquatic systems [29] to form these major metabolites, among others [26,30]. The main route of chloroacetanilides degradation in soil has been reported to be microbial [26,29,30]. For example, ESA and OA derivatives of chloroacetanilides were reported to arise primarily from aerobic dechlorination via GST-mediated reactions and further metabolism of glutathione conjugates by diverse soil bacteria such as pseudomonads and *Enterobacteriacea* [29,30]. We speculate that under the conditions used in the present work some indigenous active bacteria presumably present in the soil may have contributed for the formation of these two metabolites during the time-course of the biodegradation experiments. Their relative values measured in the soil samples (approximately in a 35:20 ratio OA:ESA) are within the range of published values for other soils [26].

ESA and OA degradates of chloroacetanilide herbicides [8] are frequently detected in ground and surface waters worldwide, often at higher concentrations than the parent compounds [29,31], and are considered of potential concern [31]. Recently, Gadabgui et al. [32] performed a toxicological risk assessment of acetochlor, alachlor and the respective ESA and OA metabolites, and concluded that the toxicity of the degradates for mammals may be lower than that of the parent compounds [32]. To our knowledge scarce information exists about the toxicological properties of ESA and OA metabolites for microorganisms and aquatic organisms. We suggest that the *S*-MET ESA and OA that accumulated in the soils examined herein (and possibly other metabolites that may have formed but were not analysed [26]) might also contribute for the toxicity of the soil water extracts to the microalgae *P. subcapitata*. Importantly, the transformation of *S*-MET into these degradates, whose mobility in soil appear to be greater than that of the parental compound [25,33], suggests that they may have potential to contaminate water compartments and hence impact water quality [31,32].

In conclusion, the results herein presented point to the complexity of bioremediating soils contaminated with mixtures of pesticides, in this particular case of ATZ and *S*-MET. Both active substances have been frequently used alone or combined in herbicidal formulations [1,8,10,26]. Due to the potential effects of these substances and their metabolites for non-target aquatic organisms, risk mitigation measures are recommended particularly when the herbicidal formulation is applied in regions with vulnerable soil [31]. To date, as far as we are aware of, effective biodegradation and detoxification of MET/*S*-MET or derivatives ESA and OA in soil have not been well succeeded, apparently because microorganisms do not easily metabolize their aromatic ring [1,19,34]. In the present work, there are evidences of the transformation of 40% initial *S*-MET in the contaminated soils, presumably performed by intrinsic microorganisms present in the soil, but this is not associated with an effective decontamination as discussed above. Further work is needed for the development of efficient bioremediation strategies for land contaminated with herbicide mixtures such as the one herein examined. It is necessary, in one hand, to focus on the isolation and optimization of performance of soil bacteria (most probably working in consortia) able to concomitantly biodegrade multiple chemicals to less toxic derivatives; and, in the other hand, to address the effects of multiple pesticides on the performance of specific degrading microorganisms [1,35]. The present work is a contribution to enhance the knowledge on the latter issue, particularly with respect to the interaction of *S*-MET with the well-known ATZ-degrading bacterium *Pseudomonas* sp. ADP [14–18]. Overall results also highlight the importance of monitoring the efficacy of the soil clean-up processes based on ecotoxicity assessments of soil aqueous extracts before and after the implementation of the bioremediation treatment, besides chemical analysis [14]. Ecotoxicological evaluation provide a more realistic glimpse over the ecological risk assessment of soil remediation, being particularly important, in aquatic ecosystems, for the evaluation of the potential impact of the mobilization of extractable and bounded fractions of ATZ, *S*-

MET and other pesticides (including their possible metabolites) via water, mainly due to leaching and runoff events.

Materials and Methods

Chemicals

Atrazine (ATZ; Pestanal, purity 99.1%), *S*-metolachlor (*S*-MET; Pestanal, purity 98.2%) and 4-morpholinepropanesulfonic acid (MOPS) were purchased from Sigma-Aldrich (Seelze, Germany), trisodium citrate from Merck (Darmstadt, Germany), and [ring-UL-^{14}C]Atrazine (purity 99%, specific activity 1.85 GBq mmol^{-1}) from American Radiolabeled Chemicals (St. Louis, MO, USA). The formulations Primextra S-Gold (320 g of ATZ L^{-1} and 400 g of *S*-MET L^{-1} as active ingredients; 3 L ha^{-1} as the recommended dose (RD) for weeds in corn plantations) and Atrazerba FL (500 g ATZ L^{-1} as single active ingredient; RD = 2 L ha^{-1}) were purchased from Syngenta Portugal (Lisbon, Portugal) and Sapec (Setúbal, Portugal), respectively. The RD of ATZ and *S*-MET (0.96 kg ha^{-1} and 1.2 kg ha^{-1}, respectively) were estimated as equivalent to approximately 0.8 and 1.0 μg g^{-1} of soil, respectively, assuming a possible field scenario of herbicide distribution through a 5 cm diameter ×10 cm height soil column and an average soil density of 1.5 g cm^{-3} [15].

Bacterial strain and culture conditions

A spontaneous rifampicin-resistant (Rif^R) mutant of *Pseudomonas* sp. ADP which can mineralize atrazine with equal efficiency than the wild-type [36] was used. The cell suspension used as inoculum was prepared from a late-exponential culture (6.5±0.4×10^8 colony forming units (CFU) ml^{-1}, corresponding to $OD_{640\ nm}$~1) grown at 30°C in liquid PADP medium (adapted from [16,17]) following a procedure reported elsewhere [15]. Briefly, the growth medium was buffered using MOPS (0.1 M; pH 6.2) and supplied with trisodium citrate (10 g L^{-1}) as C source [17], and with ATZ (300 mg L^{-1}) from Atrazerba FL as sole N source [15]. Cells were harvested by centrifugation (5 min, 9000 *g*), and the cell pellet was washed once and resuspended in sterile saline solution (0.9% w/v NaCl). The concentrated cell suspension obtained was used as inoculum in survival, mineralization and biodegradation experiments, as is described below. The inoculum density, expressed as CFU ml^{-1}, was determined by plating culture serial dilutions onto agarized Lennox Broth (LB) medium and further incubation at 30°C during 72 hours.

Soil parameters

A natural sandy loam soil (pH 6.1; organic matter 3.1%; water content 9.8±0.9%; water holding capacity (WHC) 39.4±1.6%; cation exchange capacity 0.013 cmol g^{-1} soil dry weight; soluble carbon 23.5±5.2 μg g^{-1} soil dry weight) was used [14,15]. This soil is representative of a corn production field from Central Portugal (Escola Superior Agrária de Coimbra - ESAC, Coimbra, Portugal) with no history of pesticide applications. The soil was sieved (5 mm mesh) and stored in plastic bags at −20°C. Prior to use in the survival, mineralization or biodegradation experiments, soil was defrosted for at least 4 days at 4°C.

Survival experiments

To examine the effects of *S*-MET on the survival of *Pseudomonas* sp. ADP in ATZ-contaminated soil, experiments were carried out in sterilized EPA vials (40 ml, gastight PTFE/Silicone septa, Sigma–Aldrich) each containing 5 g soil dry weight freshly spiked with ATZ to give an approximate concentration of 24 μg g^{-1} soil dry weight (equivalent to 30×RD of the active substance). Then, three different concentrations of *S*-MET (15, 30 or 60 μg g^{-1}) were added to these soils to represent doses equivalent to 15×, 30× and 60× higher than the RD of this active ingredient. Each compound was added from a stock solution in methanol (18 or 20 mg ml^{-1} for ATZ or *S*-MET, respectively), mixed with pure methanol when needed (to guarantee a similar total volume of methanol in all vials) and with sterile deionized water (to obtain an initial soil moisture of 40% soil WHC). Vials non-supplemented with *S*-MET were included as controls. Vials were vigorously stirred in a vortex to promote homogeneous distribution into the soil and were left uncapped in the laminar flow chamber for 1 hour to allow most of the methanol to evaporate. Each vial was then inoculated with an adequate amount of a suspension of bacteria viable cells prepared as described above, in order to have an initial inoculum density of approximately 10^7 CFU g^{-1} soil dry weight. Vials were stirred again, capped and incubated at 25.1±0.2°C in the dark. The total volume of liquid added to each vial (298 μl; 40% soil WHC) took in account the volumes of herbicide solutions, inoculum and deionized water. In each experiment, 16 replicates of each condition were prepared. To determine *Pseudomonas* sp. ADP Rif^R viable cells in soil, at each time interval (up to 9 days) one replicated vial was destructively sampled and processed immediately for the determination of CFU concentration as is described below (microbiological analysis). At least duplicate determinations from two or three independent experiments under identical conditions were carried out.

Mineralization experiments

ATZ mineralization assays were carried out in sterilized EPA 40-ml vials containing 5 g soil dry weight, as previously described [15] with minor adaptations. Briefly, a mixture of [ring-UL-^{14}C]ATZ (stock solution in acetonitrile: 467.7 KBq ml^{-1}) plus non-labeled ATZ was incorporated into the soil to give a total activity of 0.65 KBq g^{-1} dry weight of soil and different total concentrations of ATZ in soil. Non-labeled ATZ was supplied from Primextra S-Gold or as the pure substance, depending on the type of experiment to be carried out, as follows: first, to evaluate the effect of increasing doses of Primextra S-Gold in ATZ mineralization by *Pseudomonas* sp. ADP, soil in the vials was freshly spiked with mixtures containing the [ring-UL-^{14}C]ATZ stock solution, aqueous suspensions of Primextra S-Gold (to give approximately 5×, 20× and 50×RD corresponding to 15, 60 and 150 L ha^{-1}, respectively) and sterile deionized water (to obtain an initial soil moisture of 40% soil WHC). Second, to examine the influence of *S*-MET in ATZ-mineralization bacterial activity, soil in the vials was freshly spiked with mixtures of the [ring-UL-^{14}C]ATZ stock solution plus adequate amounts of stock solutions of herbicides in methanol to have total concentrations of 24 or 40 μg of ATZ g^{-1} soil dry weight plus, respectively, 30 or 50 μg g^{-1} of *S*-MET (equivalent to approximately 30× or 50×RD of the active ingredients, respectively) or no added *S*-MET. Pure methanol (to guarantee a similar volume of methanol in all vials) and sterile deionized water (to obtain soil moisture of 40% soil WHC) were also added. Then, soil was vigorously mixed with a vortex apparatus to incorporate the substances, followed by the inoculation with an adequate amount of the bacterial cell suspension prepared as described above (to obtain approximately 10^7 viable cells g^{-1} soil dry weight) [15]. Non-inoculated controls were included in each set of experiments to account for intrinsic mineralization activity in the soil. In addition, in experiments aiming to examine combination of soil bioaugmentation with biostimulation with citrate on ATZ mineralization, sterile concentrated solutions of trisodium citrate were added to soil to obtain a ratio of C_s:N_{atz} equal to 50, as described before [15,17]. The total volume of liquid added to each vial (298 μl; 40% soil

WHC) took in account the volumes added of herbicide solutions, inoculum, a citrate solution and deionized water. After all soil amendments, the vials were stirred once more for homogeneity. Triplicate vials were included for each treatment. A test tube containing 1 ml solution of NaOH 1 M was placed inside each vial to trap the released $^{14}CO_2$. All vials were incubated at 25.1±0.2°C in the dark for up to 8 days. At adequate time intervals, the quantity of released $^{14}CO_2$ was quantified using a Beckman LS 5000TD scintillation counter, as described before [17]. Ultima Gold (Perkin Elmer, Waltham, USA) was used as the scintillation cocktail in a 1:4 sample-to-cocktail ratio. At least triplicate determinations from two independent experiments under each condition were carried out.

Biodegradation experiments at a larger scale

Soil microcosms consisting of glass cylinders (10 cm height×4.5 cm interior diameter) containing 160 g dry weight of soil (~7 cm×4.5 cm) over a 2 cm height layer of 2-mm-diameter glass beads supported by a fine Teflon mesh were used as described elsewhere [14,15]. Briefly, the soil was freshly spiked with 5 ml of aqueous suspensions of Primextra S-Gold to obtain approximately 20× or 50×RD (60 or 150 L ha^{-1}, respectively). After homogenization with a glass rod to promote incorporation of the herbicides, soil was inoculated with an adequate quantity of the inoculum bacterial suspension (initial density: 2.5–4.1×10^{7} CFU g^{-1} soil dry weight), and either amended or not with trisodium citrate at 1.3 or 3.4 mg g^{-1} soil dry weight in the 20× or 50×RD amended soils, respectively (to give a ratio C_s:N_{atz}~50 [15]). Identical experiments with 20× or 50×RD of Atrazerba FL or with pure *S*-MET (25 or 62 μg g^{-1} soil dry weight) instead of Primextra *S*-Gold, were also carried out for comparison purposes. Microcosms non-contaminated with the herbicides or contaminated but not inoculated with the bacterium were also included in each set of experiments as controls. In all experiments, soil moisture was adjusted to 40% soil WHC as described above, taking in account the total volume of liquid (9.5 ml, comprising herbicide suspensions, inoculum, citrate solutions, and deionized water). Amended soils were again mixed, gently packed into the glass cylinders, and incubated at 25.1±0.2°C in the dark for up to 8 days. Soil microcosms were weighted every day to replace the water lost by evaporation with sterile deionized water. Soil samples were collected from the surface, at days 0, 1, 2, 5 and/or 8, and processed immediately for microbiology analysis (1.0±0.3 g dry weight), or stored at −20°C for chemical and ecotoxicological analyses (~20 g), which were performed as described below. At least two determinations with samples from two independent experiments were carried out.

Ecotoxicity analysis

Ecotoxicity tests with the model aquatic microalgae *Pseudokirchneriella subcapitata* (strain Nr.WW 15–2521; Carolina Biological Supply Company, Burlington, NC, USA) were carried out on eluates obtained from soil samples collected in the biodegradation experiments (0, 5 and 8 days) as described above. Microalgae culturing procedures were as previously outlined [37]. Soil eluates were prepared following standard methods as described elsewhere [14]. 72-hours *P. subcapitata* growth tests followed standard guidelines [38,39], on 24-well sterile microplates, at 21 to 23°C and under continuous cool-white fluorescent illumination (100 μE m^{-2} s^{-1}). Three 900 μl sub-replicate cultures per replicated eluate, and a standard control with six replicates, were set up and inoculated with 100 μl of algal inoculum, so that the cell density at the start of the tests would be 10^{4} cells ml^{-1}. Both pH and conductivity were measured at the start of each test, and measured levels were comparable across treatments and not expected to have deleterious effects on the test organisms [38,39]. At the end of the 72-hours exposure, algal growth was estimated as the mean specific growth rate (per day). Further details on testing procedures are described elsewhere [37]. For eluates from soil samples taken from non-inoculated microcosms within the same time interval (5 or 8 days), effects of the two doses (20× or 50×RD) of each herbicide application (Primextra S-Gold, Atrazerba FL, or *S*-MET) on microalgae growth were evaluated through one-way analysis of variance followed by Dunnet's test to compare each dose with the respective non-contaminated control soil. One-tailed Student *t*-tests were employed to compare microalgae growth responses between spiked soils that were treated or not with the bioremediation tool, within the same herbicide dose and time interval. The latter tests were also used to assess decontamination efficacy due to bioaugmentation/biostimulation treatments, by comparing microalgae growth in eluates from non-contaminated (controls) and treated contaminated soils, within the same pesticide dose and time interval.

Microbiological analysis of soil samples

To determine the concentration of CFU of *Pseudomonas* sp. ADP Rif^{R}, soil samples from survival and biodegradation experiments were used as a basis for 10-fold dilution series in saline solution (0.9% NaCl w/v) in triplicate. Dilutions were spread plated onto agarized selective LB medium supplemented with rifampicin (50 mg L^{-1}) and cycloheximide (100 mg L^{-1}). Petri dishes were incubated at 30°C and colonies counted after 96 hours.

Chemical analysis

Soil samples were thawed at room temperature, dried at 40°C and further processed for analysis of the herbicides and respective metabolites in soil. For analysis of ATZ, DEA and DIA, extracts preparation and analysis by GC-Electron Ionization (EI)-MS (Perkin Elmer-Clarus 500) was carried out as described elsewhere [15]. Recovery range was between 75 and 90%, and the limits of quantification were 25 ng g^{-1} soil dry weight for ATZ, DEA, and DIA. For analysis of *S*-MET and of its OA and ESA derivatives, soil samples were dried at 40°C, extracted three times with a mixture of methanol and water (1:2) using a Liarre 60 ultrasonic apparatus (15 min) and centrifuged for 15 min at 3000 *g*. Analysis of the combined extracts was performed by GC-EI-MS (Perkin Elmer-Clarus 500) for *S*-MET, and by LC-EI-MS (Agilent 1100 Series) for *S*-MET OA and *S*-MET ESA. All extracts were injected in scan mode to confirm the presence of each analyte and in SIM (single ion monitoring) for quantification purposes. Recovery ranged from 75 and 90% and the limit of quantification was 16 and 6 ng g^{-1} soil dry weight for S-MET and for *S*-MET ESA or S-MET OA, respectively.

Acknowledgments

We are greatful to S. Chelinho (IMAR, Univ Coimbra, P), J.P. Sousa (IMAR, Univ Coimbra, P), and R. Guilherme (ESAC, Coimbra, P) for providing the soil used in these studies, and to A.M. Fialho (Instituto Superior Técnico, Lisboa, P) for advice and valuable discussions.

Author Contributions

Conceived and designed the experiments: CAV MMS. Performed the experiments: CC MMS SA PV. Analyzed the data: CC MMS PV CAV. Contributed reagents/materials/analysis tools: MMS RR PV CAV. Wrote the paper: CAV MMS RR.

References

1. Chirnside AEM, Ritter WF, Radosevich M (2009) Biodegradation of aged residues of atrazine and alachlor in a mix-load site soil. Soil Biol Biochem 41: 2484–2492.
2. Moorman TB, Cowan JK, Arthur EL, Coats JR (2001) Organic amendments to enhance herbicide biodegradation in contaminated soils. Biol Fertil Soils 33: 541–545.
3. Hildebrandt A, Guillamon M, Lacorte S, Tauler R, Barcelo D (2008) Impact of pesticides used in agriculture and vineyards to surface and groundwater quality (North Spain). Water Res 42: 3315–3326.
4. Seybold CA, Mersie W (1996) Adsorption and desorption of atrazine, deethylatrazine, deisopropylatrazine, hydroxyatrazine, and metolachlor in two soils from Virginia. J Environ Qual 25: 1179–1185.
5. Southwick LM, Grigg BC, Fouss JL, Kornecki TS (2003) Atrazine and metolachlor in surface runoff under typical rainfall conditions in southern Louisiana. J Agric Food Chem 51: 5355–5361.
6. Pérez J, Domingues I, Soares AMVM, Loureiro S (2011) Growth rate of *Pseudokirchneriella subcapitata* exposed to herbicides found in surface waters in the Alqueva reservoir (Portugal): a bottom-up approach using binary mixtures. Ecotoxicology 20: 1167–1175.
7. Health Canada (2008) Proposed maximum residue limit for S-metolachlor. Available: http://www.hc-sc.gc.ca/cps-spc/pest/part/consultations/_pmrl2008-18/index-eng.php. Accessed 30 June 2010.
8. PPDB The Pesticide Properties Database (PPDB) developed by the Agriculture & Environment Research Unit (AERU), University of Hertfordshire, funded by UK national sources and the EU-funded FOOTPRINT project (FP6-SSP-022704). Available: http://sitem.herts.ac.uk/aeru/footprint/en/index.htm. Accessed 20 April 2012.
9. Shaner DL, Brunk G, Belles D, Westra P, Nissen S (2006) Soil dissipation and biological activity of metolachlor and *S*-metolachlor in five soils. Pest Manag Sci 62: 617–623.
10. Whaley CM, Armel GR, Wilson HP, Hines TE (2009) Evaluation of *S*-metolachlor and *S*-metolachlor plus atrazine mixtures with mesotrione for broadleaf weed control in corn. Weed Technol 23: 193–196.
11. DeLorenzo ME, Scott GI, Ross PE (2001) Toxicity of pesticides to aquatic microorganisms: a review. Environ Toxicol Chem 20: 84–98.
12. Oh SM, Shim SH, Chung, KH (2003) Antiestrogenic action of atrazine and its major metabolites *in vitro*. J Health Sci 49: 65–71.
13. Sanderson JT, Seinen W, Giesy JP, van den Berg M (2000) 2-chloro-*s*-triazine herbicides induce aromatase (CYP-19) activity in H295R human adrenocortical carcinoma cells: A novel mechanism for estrogenicity. Toxicol Sci 54: 121–127.
14. Chelinho S, Moreira-Santos M, Lima D, Silva C, Viana P, et al. (2010) Cleanup of atrazine-contaminated soils: ecotoxicological study on the efficacy of a bioremediation tool with *Pseudomonas* sp. ADP. J Soils Sediments 10: 568–578.
15. Lima D, Viana P, André S, Chelinho S, Costa C, et al. (2009) Evaluating a bioremediation tool for atrazine contaminated soils in open soil microcosms: The effectiveness of bioaugmentation and biostimulation approaches. Chemosphere 74: 187–192.
16. Mandelbaum RT, Allan DL, Wackett LP (1995) Isolation and characterization of a *Pseudomonas* sp. that mineralizes the *s*-triazine herbicide atrazine. Appl Environ Microbiol 61: 1451–1457.
17. Silva E, Fialho AM, Sá-Correia I, Burns RG, Shaw LG (2004) Combined bioaugmentation and biostimulation to cleanup soil contaminated with high concentrations of atrazine. Environ Sci Technol 38: 632–637.
18. Wackett LP, Sadowsky MJ, Martinez B, Shapir N (2002) Biodegradation of atrazine and related *s*-triazine compounds: from enzymes to field studies. Appl Microbiol Biotechnol 58: 39–45.
19. Zhao S, Arthur EL, Moorman TB, Coats JR (2003) Influence of microbial inoculation (*Pseudomonas* sp. strain ADP), the enzyme atrazine chlorohydrolase, and vegetation on the degradation of atrazine and metolachlor in soil. J Agric Food Chem 51: 3043–3048.
20. Papaefthimiou C, Cabral MG, Mixailidou C, Viegas CA, Sá-correia I, et al. (2004) Comparison of two screening bioassays, based on the frog sciatic nerve and yeast cells, for the assessment of herbicide toxicity. Environ Toxicol Chem 23: 1211–1218.
21. Liu H, Ye W, Zhan, Liu W (2006) A comparative study of *rac*- and *S*-metolachlor toxicity to *Daphnia magna*. Ecotoxicol Environ Saf 63: 451–455.
22. Xu D, Wen Y, Wang K (2010) Effect of chiral differences of metolachlor and its (*S*)-isomer on their toxicity to earthworms. Ecotoxicol Environ Saf 73: 1925–1931.
23. Junghans M, Backhaus T, Faust M, Scholze M, Grimme LH (2003) Predictability of combined effects of eight chloroacetanilide herbicides on algal reproduction. Pest Manag Sci 59: 1101–1110.
24. Anhalt JC, Arthur EL, Anderson TA, Coats JR (2000) Degradation of atrazine, metolachlor, and pendimethalin in pesticide-contaminated soils: effects of aged residues on soil respiration and plant survival. J Environ Sci Health – Part B 35: 417–438.
25. Aga DS, Thurman EM (2001) Formation and transport of sulfonic acid metabolites of alachlor and metolachlor in soil. Environ Sci Technol 35: 2455–2460.
26. White PM, Thomas LP, Culbreath AK (2010) Fungicide dissipation and impact on metolachlor aerobic soil degradation and soil microbial dynamics. Sci Total Environ 408: 1393–1402.
27. Qiu Y, Pang H, Zhou Z, Zhang P, Feng Y, et al. (2009) Competitive biodegradation of dichlobenil and atrazine coexisting in soil amended with a char and citrate. Environ Pollut 157: 2964–2969.
28. Weber JB, Wilkerson GG, Reinhardt CF (2004) Calculating pesticide sorption coefficients (K_d) using selected soil properties. Chemosphere 55: 157–166.
29. Graham WH, Graham DW, Denoyelles F, Jr., Smith VH, Larive CK, et al. (1999) Metolachlor and alachlor breakdown product formation patterns in aquatic field mesocosms. Environ Sci Technol 33: 4471–4476.
30. Zablotowicz RM, Hoagland RE, Locke MA, Hickey WJ (1995) Glutathione-*S*-transferase activity and metabolism of glutathione conjugates by rhizosphere bacteria. Appl Environ Microbiol 61: 1054–1060.
31. European Commission (2004) Review report for the active substance *S*-Metolachlor. Available: http://ec.europa.eu/food/plant/protection/evaluation/newactive/s_metolachlor.pdf. Accessed 30 June 2010.
32. Gadagbui B, Maier A, Dourson M, Parker A, Willis A, et al. (2010) Derived Reference Doses (RfDs) for the environmental degradates of the herbicides alachlor and acetochlor: Results of an independent expert panel deliberation. Regul Toxicol Pharmacol 57: 220–234.
33. Krutz LJ, Senseman SA, McInnes KJ, Hoffman DW, Tierney DP (2004) Adsorption and desorption of metolachlor and metolachlor metabolites in vegetated filter strip and cultivated soil. J Environ Qual 33: 939–945.
34. Saxena A, Renwu Z, Bollag J-M (1987) Microorganisms capable of metabolizing the herbicide metolachlor. Appl Environ Microbiol 53: 390–396.
35. Viegas CA, Chelinho S, Moreira-Santos M, Costa C, Gil FN, et al. (2012) Bioremediation of soils contaminated with atrazine and other *s*-triazine herbicides: current state and prospects. In: Justin A Daniels, ed. Advances in Environmental Research - Volume 6, Nova Science Publishers Inc., NY. pp 1–49.
36. Garcia-González V, Govantes F, Shaw LJ, Burns R, Santero E (2003) Nitrogen control of atrazine utilization in *Pseudomonas* sp. strain ADP. Appl Environ Microbiol 69: 6987–6993.
37. Rosa R, Moreira-Santos M, Lopes I, Silva L, Rebola J, et al. (2010) Comparison of a test battery for assessing the toxicity of a bleached-kraft pulp mill effluent before and after secondary treatment implementation. Environ Monit Assess 161: 439–451.
38. OECD (1984) Guideline for testing of chemicals 201: Alga growth inhibition test (test guideline 201, updated, adopted 7th June). Organization for Economic Co-Ordination and Development, Paris, France.
39. Environment Canada (1992) Biological test method: growth inhibition test using the freshwater alga *Selenastrum capricornutum*. EC Report EPS 1/RM/25. Environment Canada, Ottawa, Ontario, Canada.

11

Spatially Explicit Analysis of Metal Transfer to Biota: Influence of Soil Contamination and Landscape

Clémentine Fritsch[1]*, Michaël Cœurdassier[1], Patrick Giraudoux[1], Francis Raoul[1], Francis Douay[2,3], Dominique Rieffel[1], Annette de Vaufleury[1], Renaud Scheifler[1]

1 Department of Chrono-Environment, UMR UFC/CNRS 6249 USC INRA, University of Franche-Comté, Besançon, France, **2** Université Lille Nord de France, Lille, France, **3** Laboratoire Génie Civil et géoEnvironnement (LGCgE), EA 4515, Lille, France

Abstract

Concepts and developments for a new field in ecotoxicology, referred to as "landscape ecotoxicology," were proposed in the 1990s; however, to date, few studies have been developed in this emergent field. In fact, there is a strong interest in developing this area, both for renewing the concepts and tools used in ecotoxicology as well as for responding to practical issues, such as risk assessment. The aim of this study was to investigate the spatial heterogeneity of metal bioaccumulation in animals in order to identify the role of spatially explicit factors, such as landscape as well as total and extractable metal concentrations in soils. Over a smelter-impacted area, we studied the accumulation of trace metals (TMs: Cd, Pb and Zn) in invertebrates (the grove snail *Cepaea sp* and the glass snail *Oxychilus draparnaudi)* and vertebrates (the bank vole *Myodes glareolus* and the greater white-toothed shrew *Crocidura russula*). Total and $CaCl_2$-extractable concentrations of TMs were measured in soils from woody patches where the animals were captured. TM concentrations in animals exhibited a high spatial heterogeneity. They increased with soil pollution and were better explained by total rather than $CaCl_2$-extractable TM concentrations, except in *Cepaea sp*. TM levels in animals and their variations along the pollution gradient were modulated by the landscape, and this influence was species and metal specific. Median soil metal concentrations (predicted by universal kriging) were calculated in buffers of increasing size and were related to bioaccumulation. The spatial scale at which TM concentrations in animals and soils showed the strongest correlations varied between metals, species and landscapes. The potential underlying mechanisms of landscape influence (community functioning, behaviour, etc.) are discussed. Present results highlight the need for the further development of landscape ecotoxicology and multi-scale approaches, which would enhance our understanding of pollutant transfer and effects in ecosystems.

Editor: Stephen J. Johnson, University of Kansas, United States of America

Funding: The STARTT research programme was financially supported by ANR (Agence Nationale de la Recherche - http://www.agence-nationale-recherche.fr - contract n°ANR-05-ECCO-004) and ADEME (Agence de l'Environnement et de la Maîtrise de l'Energie - http://www.ademe.fr - contract n°0572C0058). CF was financially supported by a grant from ADEME and the Conseil Régional de Franche-Comté. The funders had no role in study design, data collection and analysis, decision to publish, or preparation of the manuscript.

Competing Interests: The authors have declared that no competing interests exist.

* E-mail: clementine.fritsch@univ-fcomte.fr

Introduction

Trace metals (TMs) are naturally present in the environment, however, soils can exhibit high levels of these persistent pollutants due to anthropogenic activities, and such contamination is recognised as a subject of concern for both organism and ecosystem health [1,2]. Assessment of the environmental and ecological factors that may influence the transfer of pollutants in ecosystems is a key issue in ecotoxicology. However, the understanding of this phenomenon is hampered by the frequently high spatial and temporal variability of soil and landscape factors, which dramatically affects exposure pathways of receptors [3–6]. It is recognised that the exposure of organisms to contaminants varies spatially due to heterogeneity in the level of soil contamination, the environmental availability of the pollutant, the habitat and landscape characteristics, and some "host factors" related to ecological and behavioural characteristics of the organism (habitat preferences, home range size, feeding behaviour, migratory behaviour, etc.) [6–10]. In heterogeneous landscapes, both the duration and the intensity of exposure may vary spatially because the time spent by the animal (due to foraging behaviour for instance) in the different patches constituting the landscape is likely to change according to land use [11]. Moreover, levels of TMs in soil can also vary between habitats or landscapes because interception and retention characteristics of certain habitats increase TM levels in soil. Additionally, differences in environmental TM availability between land uses may occur due to changes in soil characteristics, soil biodiversity and speciation of metals in soil [6,12–20]. Finally, landscape patterns modulate the structure and the functioning of populations and communities [21,22] and could therefore be an important factor governing the spatial heterogeneity of organism exposure.

Several authors have emphasised the necessity of studying factors that may affect pollutant bioavailability on the same spatial scale at which harmful effects on receptors occur (i.e., at the habitat or landscape level) [3,4,23–25]. Some authors have provided evidence for the influence of soil contamination, land use heterogeneity and home range size on organism exposure using modelling methods [10,26–29]. In most cases, these studies developed individual-based random walk models and calculated the cumulative exposure of animals. An effort has also been made to introduce spatially explicit data, such as home range size, contamination heterogeneity and foraging behaviour, in the risk assessment [23,30–32]. Concepts and developments for a new field in ecotoxicology referred to as "landscape ecotoxicology" were proposed in the 1990s [33], but not much research has been developed in this new field. While the influence of spatially explicit ecological variables on exposure has been explored with modelling, the effective differences in the exposure of organisms between land uses or landscapes of a specific site have rarely been assessed in the field [19]. To our knowledge, only one recent study has investigated the influence of habitat type on TM bioaccumulation, and it was found that the transfer of pollutants from soil to biota varied among habitats [19].

Land snails and small mammals have been extensively studied for use as biomonitors of TM environmental contamination due to their bioaccumulation abilities, which have shown that TM levels in tissues (referred to as "internal" levels hereafter) reflected the contamination level of sites [34–37]. The relationships between internal TM levels in animals and in soils have mainly been studied by comparing polluted sites to reference areas, which does not allow investigating the influence of the spatial heterogeneity of total or available soil TMs on bioaccumulation. By studying the accumulation of metals in small mammals from diffusely polluted flood plains, Wijnhoven et al. [38] investigated the relationship between TM concentrations in animals and soils sampled at the same locations. These authors found weak correlations between body and soil TM concentrations and suggested that dispersal of animals and changes in foraging and feeding behaviour due to periodic flooding influenced their exposure to contaminants. Conversely, in a recent study on TM accumulation in small mammals along a pollution gradient in a smelter-impacted area, we observed that internal TM levels were highly dependent upon local soil contamination [7]. While studying TM accumulation in wood mice, van den Brink et al. [39] showed intersite differences in soil-kidney Cd ratios, which they attributed to soil parameters that drive soil TM availability to plants and invertebrates. Such discrepancies highlight the potential influence of site-specific characteristics (for instance landscape, environmental particularities such as flooding or barriers, and soil properties) on TM transfer in food webs, and highlight the need for data on the influence of environmental characteristics on pollutant transfer in ecosystems, considering both the intensity and the spatial pattern of transfer.

Metal transfer from soil to biota is controlled in large part by TM bioavailability, which may be defined as the fraction of the total concentration of a metal that is available or can be made available for uptake and, as a consequence, for causing effects in organisms [8,40–42]. TM availability is often estimated using chemical extractants; among these, calcium chloride ($CaCl_2$) is considered to be relevant for the assessment of the fraction of cadmium (Cd), lead (Pb) and zinc (Zn) that is bioavailable to some plants and invertebrates [40,42–45]. Because plants and soil invertebrates may constitute the dietary items for a large variety of invertebrate and vertebrate consumers, it could be hypothesized that $CaCl_2$-extracts of TMs reflect their potential to be transferred in food webs. It is commonly admitted that organisms respond (accumulation and/or effects) to bioavailable rather than total metal concentrations in soils, and from a risk assessment perspective, it has been recommended that studies be developed based on bioavailable rather than total TM concentrations for risk estimates [45]. In the present work, we studied four sympatric species in a large polluted site surrounding a former Pb and Zn smelter. These species differed based on their phylogeny (invertebrates versus vertebrates), physiology (TM storage abilities), diet (herbivorous, omnivorous, and carnivorous), spatial mobility and habitat preferences. Small mammals have a larger spatial range of mobility than snails [10,46–50]. The grove snail *Cepaea sp* is an herbivorous species [51,52] while the glass snail *Oxychilus draparnaudi* is carnivorous [53–55]. The bank vole *Myodes* (ex-*Clethrionomys*) *glareolus* is herbivorous/granivorous and intermediary between strictly herbivorous voles and omnivorous mice [47,56,57], while the greater white-toothed shrew *Crocidura russula* is typically carnivorous [48]. We addressed three questions: (i) Are TM concentrations in animals better related to total or chemically-extractable soil TM concentrations? (ii) Does the landscape influence the relationship between TM concentrations in animals and in soils? (iii) At which spatial scale do TM concentrations in animals correlate best with soil TM concentrations? The latter question relies upon the hypothesis that the correlation will be strongest when the spatial range considered for soil contamination will approximate the surfaces exploited by the organism of concern.

Materials and Methods

Study site and sampling strategy

This study was conducted in the surrounding area of the former "Metaleurop-Nord" smelter in Northern France (Noyelles-Godault, Nord – Pas-de-Calais, 50°25′42 N, 3°00′55 E). The contamination of agricultural and urban soils around Metaleurop has been well documented [58–61], and soils of woody habitats were also recently studied [20,62], thus showing that the area is highly polluted by Cd, Pb and Zn for both levels of contamination and surfaces of concern.

We defined a 40 km^2 (8×5 km) study area, which was divided into 160 squares (500×500 m), that constituted our sampling units. In each square, soils and animals were sampled in woody habitats. Habitats classified as "woody" consisted of natural forests, tree plantations (e.g., poplar groves), woodlots or copses and hedgerows in natural or cultivated lands and urban parks. Of the 160 squares, nine were excluded from the sampling; one was excluded because it corresponds to the location of the former smelter, which was undergoing rehabilitation process and the others because they were occupied by ploughed fields only and thus lacked any woody habitats. The number of squares actually sampled was therefore 151.

Landscape analysis

A land use analysis was performed to determine the landscape composition of each square. For this purpose, the study area was extended to 9×6 km by adding a line of squares around the initial 8×5 km grid with the aim of avoiding an edge effect in further statistical analyses. Land use mapping was accomplished using ArcGIS 8 (ESRI Co., USA) on the basis of the CORINE Land Cover (CLC) database (European Commission, 2000). The resolution of CLC is 25 ha. Based on aerial photographs (BD ORTHO® database from the *Institut Géographique National*, resolution of 0.5 m) and on field reconnaissance when necessary, the limits of the different units and the units smaller than 25 ha (such as hedgerows or copses) were manually digitalised in order to obtain a resolution of four meters. The resulting vector map was converted into a raster map (1 pixel = 4×4 m), which was composed of eight categories of land use: urban and industrial, ploughed field, short grass, shrub and tall grass, hedgerow and copse, forest, river and pond, and former Metaleurop Nord smelter. For each square, the number of pixels of each category of land use was computed. Seven groups of squares were identified using correspondence analysis and clustering, and these groups corresponded to seven landscape types. These types (hereafter referred to as "landscape") were named according to their land use matrix: agricultural lands, urban areas, woodlands, shrublands, mixed urban areas and agricultural lands, and mixed woodlands and grasslands. The last type was constituted of the former Metaleurop smelter itself. The four prevailing landscape types were agricultural lands, urban areas, woodlands and shrublands. These four landscape types were used to answer the second question of this work, which addressed the influence of landscape on the relationship between TM concentrations in soils and in individuals of the different species studied.

Soil use analysis

To answer the third question of this study, which addressed the scale at which TM concentrations in soils and animals correlate best, we needed to obtain soil contamination data in buffers of increasing radius size. To achieve this aim with the best accuracy, our data on "woody" soils were gathered along with data on agricultural and urban soils that were obtained from the soil database of the *Equipe Sols et Environnement* (LGCgE, Groupe ISA, Lille, France).

The land use map was used to build a soil use map with four types of soil use: woody soils (soils of woody patches), agricultural soils (soils of ploughed fields and grasslands) and urban soils (soils of urban areas such as gardens and parks). Dredged sediment deposits, which are extremely contaminated and do not tech

Soil, small mammal and snail sampling

In each of the 151 squares, one to 10 composite soil samples were obtained from woody patches during the autumn of 2006 [20,62]; this sample number varied depending on logistical issues, such as the number of available woody patches and the accessibility of the patches. Each composite sample in a woody patch comprised 15 randomly placed elementary samplings, and these 15 sub-samples represented the same weight part in the composite sample. The first 25 cm of soil were sampled, and the litter layer (OL layer, which is a layer on the soil surface that accumulates little to no decomposed leaves and woody fragments) was removed. However, the humus layer (OF layer, which consisted of fragmented residues) was sampled with the top mineral soil material, consistent with the most frequently recommended protocol in Europe. A total of 262 soil samples were analysed over the 40 km^2 study area. Detailed data regarding soil physico-chemical parameters and contamination are published elsewhere [20,62].

The sampling of animals, which could not have been performed on all 160 squares for logistical reasons, was conducted on 30 squares. These 30 squares were chosen in order to obtain three replicates from three levels of pollution (low, medium and high) in the four prevailing landscape types defined above (agricultural lands, urban areas, woodlands and shrublands). Because the landscape type "shrublands" was present only in the vicinity of the former smelter, we obtained only one level of pollution (high) for this type of landscape. Small mammals were captured during the autumn of 2006 using break-back traps baited with a mix of water, flour and peanut butter. Sampling authorisation was obtained from the DIrection Régionale de l'ENvironnement (DIREN) of Nord – Pas-de-Calais. In each square, 10 lines of 10 traps, each spaced three meters apart, were placed throughout the woody patches where the soils were sampled. In three squares, the available surfaces of woody patches were insufficient to place 10 lines of traps; the number of lines was therefore reduced to six or seven. The 289 trap lines were checked every morning for three consecutive days and were re-set/rebaited as necessary. The sampling effort consisted of 2820, 600, 2310 and 2940 trap-nights in agricultural lands, shrublands, urban areas and woodlands, respectively (one trap set for one night corresponds to a "trap-night"). The percentage of captures was calculated as the number of individuals trapped per 100 trap-nights. Snails were collected by hand searching in the same woody patches and at the same time as small mammals. Snails and small mammals were stored at −20°C until dissection. Each capture location (trap line for small mammals, woody patch for snails) was georeferenced using a GPS (Garmin eTrex®).

Animal preparation

Animals were identified at the species level using morphometric criteria [47,54,63,64] and were dissected. Crystalline lens weighing is a standard method for estimating the relative age of small mammals and is relevant for the bank vole [65]. However, it was not used here because a significant proportion (10%) of the crystalline lenses of bank voles was not usable because they were broken by trapping and/or freezing. Moreover, to our knowledge, no validation of this age determination method has been published for the greater white-toothed shrew. Therefore, we decided to classify the small mammals (shrews as well as voles) into three classes of relative age (juveniles, non-reproductive adults and reproductive adults) on the basis of body size, body weight and reproductive status [7]. We found significant differences in crystalline weight between these three classes, suggesting that our age classification was acceptable.

Small mammals were dissected to sample the liver, which was dried in a 60°C oven to achieve a constant dry weight before acid digestion and TM analyses (see next paragraph). Snails of the *Cepaea* genus (*C. hortensis* and *C. nemoralis*) were grouped and will be referred to hereafter as *Cepaea sp* because specific determination is not possible for juveniles. Furthermore, preliminary analyses of our data showed that TM accumulation did not differ between adult *C. hortensis* and *C. nemoralis*. *Cepaea sp* snails were classified according to two classes of relative age (juveniles and adults) based on the presence of a clear lip at the mouth of their shell. The presence of this lip indicates that the snail has attained adulthood [66]. The age of *Oxychilus draparnaudi* specimens could not be determined because no published method was available. The soft bodies of the snails were separated from the shells and dried in a 60°C oven to achieve a constant dry weight before TM analysis. Three to five individuals of *Oxychilus draparnaudi* were combined to achieve a sufficient sample weight for TM analysis.

Analyses of TM concentrations in soils and animals

Woody soils were analysed for total and $CaCl_2$-extractable metal concentrations. Samples were dried, disaggregated and homogenised before being sifted using a 250 µm mesh. Cd and Pb concentrations were measured with inductively-coupled argon plasma mass spectrometry (ICP-MS), and Zn concentrations were measured with inductively-coupled argon plasma atomic emission spectrometry (ICP-AES) after acid digestion according to the NF X31-147 standard procedure [67]. Measurements were performed by the *Laboratoire d'Analyse des Sols* of the *Institut National de la Recherche Agronomique* (INRA) of Arras (France), which benefits from the COFRAC accreditation n°1–1380 for the analytical quality of its TM measurements in soils. All precautions were taken with respect to protocol application and calibration; quality control was achieved using procedural blanks, Certified Reference Materials (CRM, namely: BCR 141 and 142; GBW 07401, 07402, 07404, 07405 and 07406), samples from inter-laboratory comparisons, internal control samples and duplicates of the analysis. Selective extractions using $CaCl_2$ (0.01 M) were conducted in triplicate on

3 g sub-samples of soil. Extractable metal concentrations were quantified using atomic absorption spectrometry (AAS, AA-6800, Shimadzu) by the *Equipe Sols et Environnement* (LGCgE, Groupe ISA, Lille, France). Quality control was performed on these measurements, using procedural blanks and CRM (BCR 483). The average recoveries of the CRM varied between 90% and 110%.

Metal concentrations in snail soft bodies and small mammal livers were measured using furnace (Cd, Pb) or flame (Zn) atomic absorption spectrometry (VARIAN 220Z and 220FS, respectively) at the Chrono-Environment department (UMR 6249 UFC/CNRS, Besançon, France). Digestion of samples was performed by dissolution in nitric acid (HNO_3, 65%, Carlo Erba analytical quality) in a dry oven (65°C) for 72 h. After digestion, samples were diluted by the addition of ultra-pure water (18.2 $M\Omega/cm^2$). Blanks (acid + ultra-pure water) and CRM (TORT-2 and DOLT-3, National Research Council, Canada) were prepared and analysed using the same methods as the samples. Average recoveries of the CRM were calculated at 95%±10% ($n = 38$) for Cd, 101%±17% ($n = 42$) for Pb, 80%±2% ($n = 22$) for Zn. Detection limits of the spectrometers (median±3 SD of blanks) were 0.17, 1.2 and 2.8 $\mu g.l^{-1}$ in the acid digests for Cd, Pb and Zn, respectively. Detection limits in snail soft bodies were 0.02, 0.12 and 0.29 $\mu g.g^{-1}$ in the grove snail and 0.09, 0.65 and 1.5 $\mu g.g^{-1}$ in the glass snail for Cd, Pb and Zn, respectively. Detection limits in small mammal livers were 0.02, 0.13 and 0.30 $\mu g.g^{-1}$ in the bank vole and 0.03, 0.24 and 0.56 $\mu g.g^{-1}$ in the greater white-toothed shrew for Cd, Pb and Zn, respectively. When measured values were under the detection limits, half of the detection limit value was used for statistical analyses. For both soil and animal samples, metal concentrations were expressed as microgram per gram dry weight ($\mu g.g^{-1}$ DW). Hereafter, body and hepatic TM concentrations of snails and small mammals will be referred to as "internal TM concentrations".

Prediction of TM concentrations in soils

To determine at which spatial scale the correlation between TM concentrations in animals and in soils was the best, a circular buffer technique was used in order to calculate soil metal concentration in the area surrounding the animal capture location. Several buffer radiuses were used: 50, 75, 100, 175, 250, 350, 500, 750 and 1000 meters around the sampling point. For each snail or small mammal, at each buffer size, median predicted Cd, Pb and Zn concentrations in the soil were computed. For this purpose, metal concentrations in soils were predicted (universal kriging) throughout the whole 54 km^2 study area at each node of a regular (100 m) grid. Detailed information about this procedure is given in Supporting Information files (Text S1, Tables S1 and S2, Figure S1).

Statistical analyses of TM concentrations in animals

For each sampled individual, the following data were available: internal TM concentration (i.e., TM concentration in the liver for the mammals and in the whole soft body for the snails), total and $CaCl_2$-extractable TM concentration in the soil of the patch where the individual as was caught (referred to hereafter as TM concentration "at sampling point"), species, age (except for *Oxychilus draparnaudi*), landscape where the animal was found (i.e., landscape type of the square, as defined in the section "Landscape analysis") and median predicted total Cd, Pb and Zn concentrations in soils at several buffer sizes (as defined in the previous section). The normality of the data distribution was checked with the test of Shapiro-Wilk. Because metal concentrations in soils and animals were skewed, variables were log-transformed for statistical analyses using $\log_{10}(x+1)$. The analysis of the data was divided into three steps to address the three aims of this paper. Statistical analyses were performed using general linear models (LMs, [68]).

The purpose of the first step was to test whether animal and soil TM concentrations at the sampling point were related and whether total or $CaCl_2$-extractable soil TM concentrations best explained TM concentrations in animals. As age can influence TM accumulation in small mammals and snails [7,34,69,70], we modelled the relationship between internal and soil TM concentrations controlling for age; residuals of the internal concentration/age relationship (which will be referred to hereafter as "internal concentration normalized to age") were modelled against total or $CaCl_2$-extractable TM concentrations in soils. On the basis of the R^2 of these models, we determined which soil TM concentration (total or $CaCl_2$-extractable) better explained the internal TM concentrations. Subsequently, the best variable was kept for further analyses. We analysed the differences in accumulation between species using LMs with internal metal as the dependent variable and total soil TM concentrations and species with interactions as explanatory variables. Comparisons of accumulation level between species were therefore performed conditionally to total TM concentrations in soils (i.e., placing the variable "total soil TM concentration" first in the model).

In the second step, we considered the influence of landscape on internal TM concentration by adding the variable "landscape" with interaction into the models. Comparisons of accumulation level between landscape types were therefore performed conditionally to age and total soil TM concentrations (i.e., placing these variables first in the model).

Finally, we determined at which spatial scale the animal and soil TM concentrations were most strongly correlated. We hypothesized that this correlation will be strongest when the buffer size approximates the surface exploited by the organism of concern. For this purpose, LMs were used with internal TM concentrations normalized to age as the dependent variable and median predicted total soil TM concentrations in buffers around sampled animals as independent variables. For each species, the R^2 of the models and the values of the coefficients (slopes) were computed for each buffer size, both on the whole dataset and by landscape. Because the sample size was small for the shrew in woodlands ($n = 4$), the grove snail in urban areas ($n = 3$) and null for snails in shrublands (no specimens), these landscape types were not analysed individually for these species. For each buffer radius size, we recorded the highest and lowest of the calculated partial R^2 values for soil concentrations and computed the differences between the values, which will be referred to hereafter as "ΔR^2". If the value of ΔR^2

Table 1. Total and $CaCl_2$-extractable trace metal concentrations in soils of woody habitats from Metaleurop-impacted area.

Metal	Min.	Med.	Max.	DSD
Cd	0.10	5.0	236	2 402
Pb	16	303	7 331	41 960
Zn	44	460	7 264	38 760
Cd-$CaCl_2$	0.02	0.20	5.8	59
Pb-$CaCl_2$	0.27	0.66	14	1.7
Zn-$CaCl_2$	0.02	4.1	143	112

Min.: minimum, Med.: median, Max.: maximum. Concentrations in woody soils ($n = 261$) and dredged sediment deposit (DSD, $n = 1$). Concentrations expressed as $\mu g.g^{-1}$ DW.

was lower than 0.05, we assumed that the R^2 did not show straightforward variations with buffer size increase.

n all cases, the significance of the variables was checked via permutation test (Monte-Carlo, 1000 permutations), the partial R^2 values were calculated using an analysis of variance (ANOVA). Pairwise differences were determined using the Tukey's post-hoc multiple comparison test (Tukey's "Honest Significant Difference" method).

All statistical analyses were performed using the software R 2.7.1 [71] with the following additional packages: ade4, geoR, gstat, maptools, pgirmess, pvclust, spdep, splancs, and vegan.

Results

Total soil TM concentrations varied from background levels to very high values, particularly in the dredged material deposit (Table 1, Table S1). $CaCl_2$-extractable TM concentrations also varied widely within woody soils (Table 1). Spatial distribution and predictions of TM concentrations in soils showed a similar spatial pattern for the three TMs (Text S1, Figure S1, Table S2). Concentric circles were found around the former smelter with an enhancement of the contamination in downwind areas (Figure 1). Hot spots corresponding to dredged sediment deposits were also found within the study area (Figure 1).

A total of 131 grove snails (*Cepaea sp.*), 78 glass snails (*Oxychilus draparnaudi*), 248 bank voles (*Myodes ex-Clethrionomys glareolus*) and 163 greater white-toothed shrews (*Crocidura russula*) were captured (Table 2). Even if these four species were present over the whole area, the species were not evenly distributed among landscapes and pollution levels (Table 2). Neither of the two snail species selected were found in the shrubland landscape. Grove snails were abundant in both agricultural and woodland landscapes but rare in urban areas. The glass snail was often present in urban areas but less often in woodlands, and it was rarely found in agricultural areas. The bank vole was clearly more abundant in woodlands than in the three other landscapes, where it was roughly similar in abundance. Greater white-toothed shrews were equally abundant in agricultural and urban areas and were rare in both shrubland and woodland landscapes.

Influence of species, age and TMs in soils at the sampling point on internal TM concentrations

Considering all the species together, internal concentrations showed a large range of variation; they varied from under detection limits for Cd and Pb and 0.88 $\mu g.g^{-1}$ DW for Zn to 741, 443 and 6619 for Cd in the greater white-toothed shrew, Pb in the grove snail and Zn in the glass snail, respectively (Table 3).

Internal Cd concentrations, conditionally to soil total concentrations (i.e., placing the variable "total soil TM concentration" first in the model), ranked in the following order: greater white-toothed shrew ~ glass snail > bank vole > grove snail ($p<0.001$, Figure 2). Regarding Pb, internal concentrations, conditionally to total soil concentrations, were higher in small mammals than in snails and ranked in the following order: greater white-toothed shrew > bank vole ~ glass snail > grove snail ($p<0.001$), while for Zn the rankings were as follows: glass snail > grove snail > greater white-toothed shrew ~ bank vole ($p<0.001$).

Hepatic Cd concentrations slightly increased with age in small mammals ($p<0.001$, $R^2=0.07$ for the vole; $p=0.002$, $R^2=0.06$ for the shrew, Table 4). Similarly, adult grove snails exhibited higher levels than juveniles ($p<0.001$, $R^2=0.22$). Pb concentrations, however, did not increase with age in small mammals although they differed with age in the grove snail, with juveniles exhibiting slightly higher concentrations than adults ($p=0.045$, $R^2=0.03$). Zn concentrations weakly increased with age in the bank vole ($p<0.001$, $R^2=0.03$) but did not vary for the shrew or for the grove snail.

Internal TM concentrations increased with total soil concentrations at the sampling points in the glass snail and the small mammals, although the grove snail did not show any increase in internal concentrations along the pollution gradient, regardless of the metal considered (Figure 2). Total TM concentrations explained 7 to 37% of the variation in internal concentrations (Table 5). Internal concentrations were always better correlated with total rather than $CaCl_2$-extractable TMs (Table 5), except for the grove snail, for which internal concentrations were not related to either total or extractable TM concentrations (except for Cd-$CaCl_2$). Internal concentrations were not correlated with extractable TMs in the glass snail. Similarly, Zn in the bank vole and Pb in the greater white-toothed shrew were not significantly related to Zn-$CaCl_2$ and Pb-$CaCl_2$, respectively.

Influence of landscape on internal TM concentrations

Accumulation of metals differed between landscapes for both the levels of internal TM (internal concentration normalized to age and conditionally to total soil TM concentration at the sampling point) and the evolution of internal TM concentrations along the

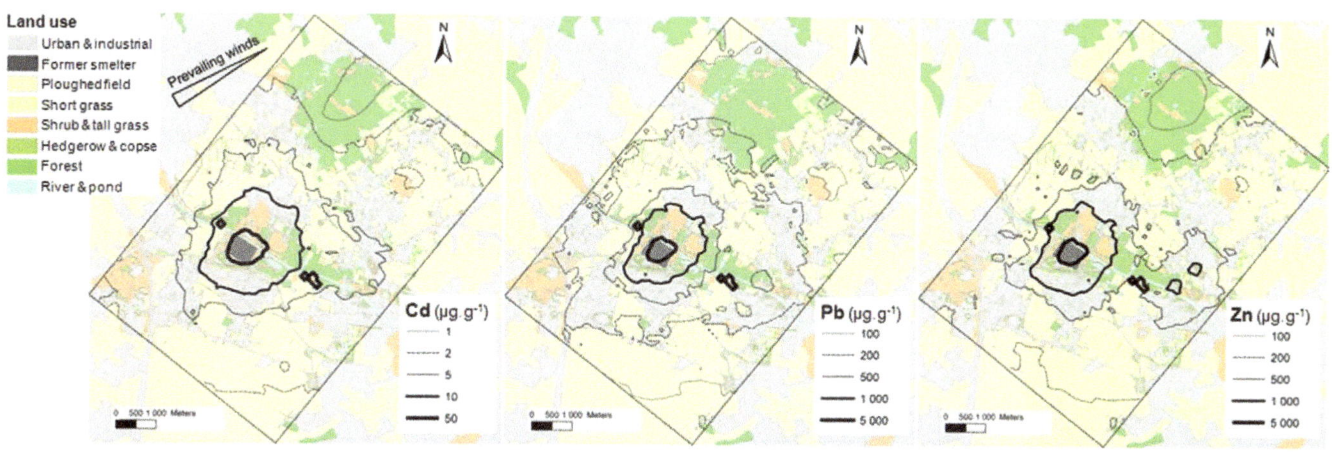

Figure 1. Iso-concentration lines of predicted total Cd, Pb and Zn concentrations in topsoils from Metaleurop-impacted area. Concentrations are expressed as $\mu g.g^{-1}$ DW.

Table 2. Distribution of specimens by species, age and landscape and percentage of capture for small mammals.

Species	Age	Landscape				Total
		Agricultural lands	Shrublands	Urban areas	Woodlands	
Cepaea sp	Juv	9	0	2	19	30
	Ad	43	0	1	57	101
	Total	52	0	3	76	131
Oxychilus draparnaudi	Total	6	0	50	22	78
Myodes glareolus	Juv	7	10	1	54	72
	NR ad	2	7	9	44	62
	Ad	8	4	4	98	114
	Total	17	21	14	196	248
	PC	0.6	3.5	0.6	6.7	2.9
Crocidura russula	Juv	10	0	13	0	23
	NR ad	50	9	56	4	119
	Ad	15	0	6	0	21
	Total	75	9	75	4	163
	PC	2.7	1.5	3.2	0.1	1.9

Juv: juveniles, NR ad: non-reproductive adults and Ad: reproductive adults, PC: number of individuals trapped per 100 trap-nights.

soil pollution gradient, i.e., the slopes of the regressions between internal TM concentrations (normalized to age) and total soil TM concentrations at the sampling points (Table 6, Figure 3). The variable "landscape" accounted for 5 to 16% of the variation in internal TM concentrations, and the interaction between soil contamination and landscape was sometimes significant, particularly for the glass snail (Table 4).

For *Cepaea* snails in urban areas, the sample size was insufficient and did not allow for reliable analysis. Internal Pb levels were higher in woodlands than in agricultural lands ($p<0.001$), while internal Cd and Zn levels did not differ among landscapes (Figure 3). Internal TM concentrations did not vary with soil TM concentrations, except in the case of Cd in woodlands where a negative correlation was revealed (Table 6, Figure 3).

For sample size reasons, the influence of landscape on TM accumulation in *O. draparnaudi* could only be studied by comparing urban areas and woodlands. Internal Cd and Zn levels did not differ between landscapes, while internal Pb levels were higher in woodlands than in urban areas ($p=0.003$, Figure 3). Internal Cd concentrations increased with soil TM concentrations in both urban areas and woodlands (without significant differences in slopes between landscapes), whereas internal Pb and Zn concentrations significantly increased along the pollution gradient in urban areas only (Table 6, Figure 3).

For the greater white-toothed shrew, too few animals were collected on woodland and shrubland landscapes and thus no interpretation could be made. However, data from agricultural lands and urban areas showed that the level of internal Cd was higher in urban than in agricultural landscapes ($p=0.014$, Figure 3), while the slopes of accumulation along the pollution gradient did not differ significantly (Table 6). There were no differences for Pb and Zn in either the level of accumulation or in the increase of internal concentrations along the gradient (Figure 3).

Concerning the bank vole, we found higher levels of internal TMs in shrublands and woodlands compared to agricultural lands and urban areas for Cd ($p<0.004$) and urban areas for Zn ($p<0.003$, Figure 3). Internal Pb levels were higher in shrublands

Table 3. Trace metal concentrations measured in snails (soft body) and small mammals (liver) from Metaleurop-impacted area.

Species	*n*	Cd			Pb			Zn		
		Min	Med	Max	Min	Med	Max	Min	Med	Max
Cepaea sp	131	UDL	42	170	1.0	23	443	0.88	447	1 927
Oxychilus draparnaudi	78	UDL	55	258	10	55	313	206	91	6 619
Myodes glareolus	243	0.13	11	116	0.08	1.8	200	19	89	173
Crocidura russula	163	3.0	72	741	UDL	8.8	159	65	139	271

Min.: minimum, Med.: median, Max.: maximum, UDL: under detection limit. Concentrations expressed as $\mu g.g^{-1}$ DW.

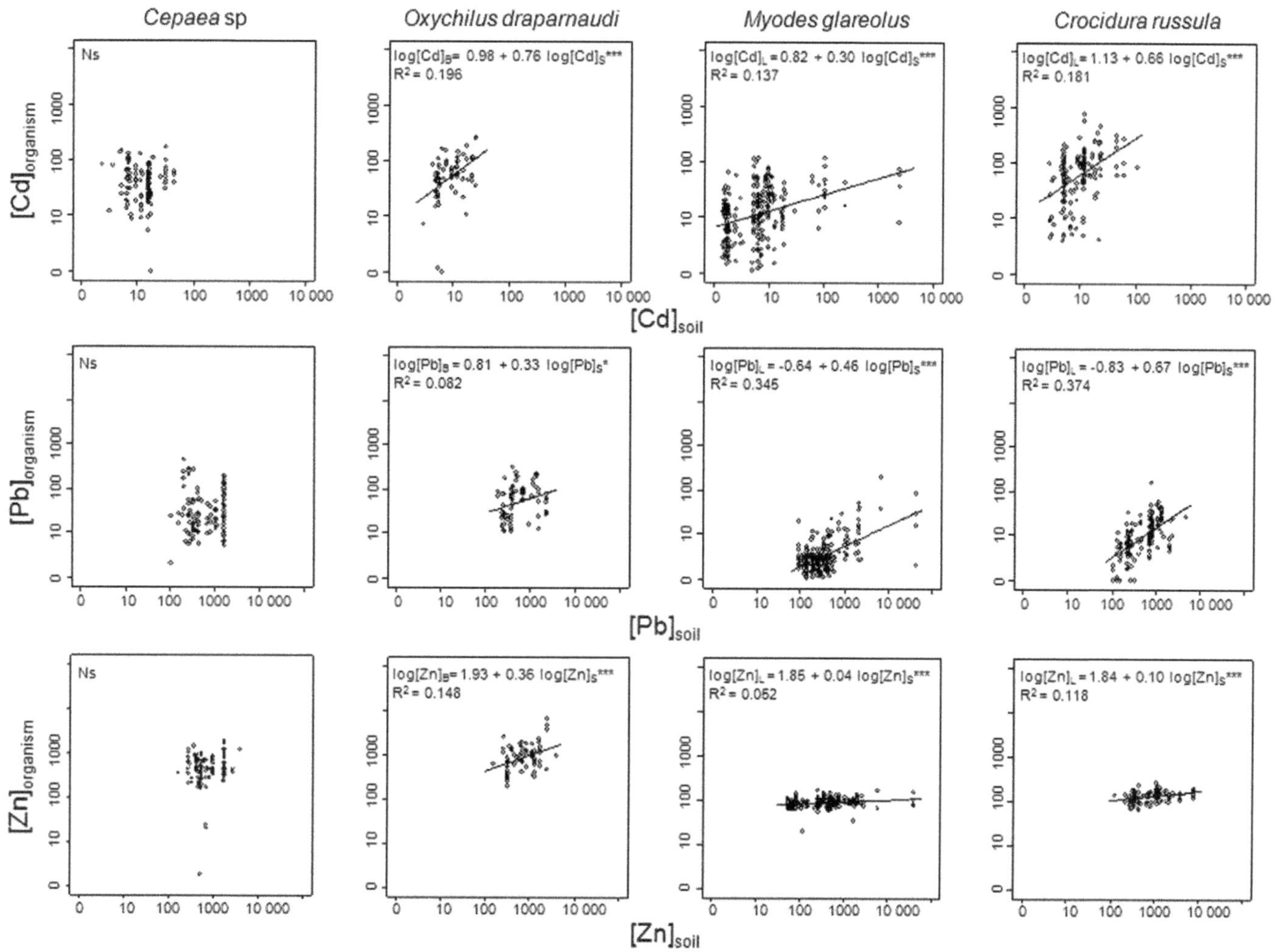

Figure 2. Relationships between Cd, Pb and Zn concentrations in organisms and in soils. TM concentrations in the soft body for snails ($[C]_B$) and in the liver for small mammals ($[C]_L$), total TM concentrations measured in soil at the sampling point ($[C]_S$) ($\mu g.g^{-1}$ DW). Statistical significance: general linear model, Ns for $p>0.05$, * for $p<0.05$, ** for $p<0.01$, *** for $p<0.001$, R^2: R-squared value of the model.

than in other landscape types ($p<0.01$). We did not detect a correlation between internal and soil Cd concentrations in shrublands, while regressions were significant in other landscapes (Table 6, Figure 3). Although the increase in internal Cd concentrations with soil contamination was highest in urban areas and lowest in agricultural lands, the coefficients did not differ significantly between landscapes (Table 6). In contrast, we found a higher increase in internal Pb along the soil pollution gradient in shrublands compared to woodlands and agricultural areas ($p<0.040$, Table 6, Figure 3). The relationships between internal and soil Zn concentrations were rarely significant and did not exhibit differences in slopes between landscapes (Figure 3).

Different patterns could be noticed between non-essential (Cd, Pb) and essential (Zn) metals in small mammals. First, most of the regressions were significant for non-essential metals, in contrast to what was observed for Zn. Second, the increase in internal concentrations along the pollution gradient was higher for non-essential metals (the lowest regression coefficient was 0.22 for Cd and Pb whereas the highest for Zn was 0.10). Finally, inter-individual variability was lower for Zn.

Several patterns linking internal and soil TM concentrations existed, showing various combinations of relatively high/low internal TM levels with sharp/slight increases in internal TMs along the pollution gradient. The relationships between internal and soil TM concentrations varied among landscapes, but these differences were metal-specific and species-specific, thus hampering any generalization (Figure 3, Tables 4 and 6).

The partial R^2 for the landscape variable was found to be somewhat less important than soil contamination but as important as age for explaining internal TM concentrations (Table 4). The models gathering all the variables (total soil TM concentration, age (except for glass snail) and landscape) explained up to 39% of the variance of the dataset for Cd, 50% for Pb and 38% for Zn (Table 4). Therefore, if the studied variables significantly accounted for internal TM concentrations in snails and small mammals, a part of the inter-individual variability remained unexplained.

Spatial range of correlation between animal and soil TM concentrations

Globally, TM concentrations in animals were related to soil TM concentrations in buffers, except for the grove snail (for which few relations were significant) and for Zn, which also showed few significant regressions. The number of significant relationships slightly increased from 27 when considering concentrations in

Table 4. Multiple and partial R-squared for significant parameters of models linking internal TMs with studied variables.

Species	$[TM]_{internal}$	Partial R^2				R^2 model
		Age	$log[TM]_{soil}$	Landscape	$[TM]_{soil}$*Landscape	
Cepaea sp	log[Cd]	0.22			0.08	0.30
	log[Pb]	0.03		0.16		0.21
	log[Zn]					ns*
Oxychilus draparnaudi	log[Cd]		0.20		0.09	0.29
	log[Pb]		0.08	0.12	0.12	0.33
	log[Zn]		0.15	0.06	0.17	0.38
Myodes glareolus	log[Cd]	0.07	0.19	0.11	0.02	0.39
	log[Pb]		0.35	0.08	0.02	0.46
	log[Zn]	0.03	0.08	0.07		0.18
Crocidura russula	log[Cd]	0.06	0.17	0.07		0.31
	log[Pb]		0.37	0.09	0.03	0.50
	log[Zn]		0.12	0.05		0.19

R^2 model: multiple R-squared, Partial R^2: partial R-squared, $[TM]_{internal}$: TM concentrations in animals, $[TM]_{soil}$: TM concentrations in soils at sampling location, $[TM]_{soil}$*Landscape: interaction between TM concentrations in soils and landscape, significance of parameters: *p-value*<0.05.
*ns: non significant *p-value*>0.05.

soil at the sampling point (Table 6) to 31 (data not shown) when considering concentrations in soil at different buffer sizes. Figures 4 and 5 show the R^2 values of the regressions between internal concentrations (normalized to age) and concentrations in soils at sampling points, as well as the evolution of the R^2 values of the models, taking into account the TM concentration in soils at different buffer sizes.

We found four major patterns in the relationships between internal TM concentrations and soil TM concentrations in buffers (Figures 4 and 5). The first pattern is represented by better correlations between internal TM and soil TM concentrations at the sampling point or in buffers of small sizes (<100 m) compared to buffers of larger sizes. This pattern is illustrated by the case of Cd and Pb in bank voles from agricultural lands. A second pattern showed better correlations at the lowest buffer size (50 m) than at the sampling point, followed by a decrease in R^2 with increasing buffer size. Notably, this pattern can be seen in urban areas, for Cd and Pb in the bank vole, the shrew and the glass snail. However, in this case, the R^2 values for Cd in bank voles and glass snails were almost equal using TMs at the sampling point and in the 50 m buffer. A third pattern exhibited a better correlation using TM concentrations in buffers rather than at the sampling point, with the R^2 value remaining stable regardless of the buffer size. This is the case for Cd in the bank vole and in the glass snail from woodlands. Finally, a fourth pattern showed the lowest R^2 at a 50 m buffer size, but exhibited roughly similar R^2 values for TMs at the sampling point and in buffers from 75 to 1000 m. This pattern can be seen for Cd and Pb in shrews from agricultural lands. For the grove snail, calculations could be performed for Cd in woodlands only, and showed that internal concentrations were better correlated using Cd concentrations in soils at the sampling point rather than in buffers, albeit with a negative correlation.

As for the landscape effect, spatial patterns and correlations between animal and soil TM concentrations were found to be metal-dependent, species-dependent and landscape-dependent.

Discussion

Relationships between internal and soil TM concentrations

The extensive, high-level TM contamination in soils surrounding the former Metaleurop smelter was reflected by the high metal levels measured in sampled organisms, which appeared among the highest reported in the literature for the various species studied [34,36,38,69,70,72–75].

In contrast to the study of Notten et al. [74] on *Cepaea* snails in the Biesbosch floodplain, we did not observe an increase of TMs in soft bodies with total soil TMs, which could be due to a seasonal effect [70]. Differences in bioaccumulation between seasons may be due to several factors, most notably to variations in snail physiology, snail behaviour and/or contamination of their diet. Our study was conducted in autumn while the study of Notten et al. [74] was performed during the summer, and indeed, *Cepaea* snails sampled in the field or exposed in microcosms in the surroundings of Metaleurop during springtime exhibited an increase of internal TM concentrations along the pollution gradient (unpublished data and [76]). Previous field and laboratory studies have also shown that internal TM concentrations in *Cepaea* snails were better related to diet than to soil TM concentrations [74,77]. The only significant relationship between TM concentrations in *Cepaea* snails and in soils was observed for $CaCl_2$-extractable Cd. This could be due to the fact that $CaCl_2$, which is usually a good extractant for the estimation of phytoavailable Cd concentrations [43,44], may have estimated the concentration of their vegetal diet relatively well. This was not the case for the glass snail, which has a carnivorous diet and had internal TM concentrations that were only related to total soil TMs. *Cepaea* snails have been shown to excrete both Pb and Zn [70,76]. Their ability to excrete these metals may explain the lack of a relationship between internal and soil concentrations. *Oxychilus* snails exhibited a stronger increase of internal Zn along the pollution gradient

Table 5. Regressions between TM concentrations in animals and in soils (total and $CaCl_2$-extractable) at sampling location.

Species	$\log[TM]_{internal} \sim \log[TM]_{soil}$		*b*	*p*	R^2
Cepaea sp	Cd	Total	−0.06	0.571	
		$CaCl_2$	0.65	0.002	0.07
	Pb	Total	0.06	0.590	
		$CaCl_2$	0.13	0.866	
	Zn	Total	0.19	0.070	
		$CaCl_2$	0.12	0.059	
Oxychilus draparnaudi	Cd	Total	0.76	<0.001	0.20
		$CaCl_2$	0.74	0.090	
	Pb	Total	0.33	0.011	0.08
		$CaCl_2$	0.48	0.275	
	Zn	Total	0.36	<0.001	0.15
		$CaCl_2$	0.05	0.430	
Myodes glareolus	Cd	Total	0.34	<0.001	0.19
		$CaCl_2$	0.37	<0.001	0.05
	Pb	Total	0.45	<0.001	0.33
		$CaCl_2$	−0.19	0.016	0.02
	Zn	Total	0.05	<0.001	0.07
		$CaCl_2$	−0.02	0.121	
Crocidura russula	Cd	Total	0.64	<0.001	0.18
		$CaCl_2$	0.85	0.015	0.04
	Pb	Total	0.67	<0.001	0.37
		$CaCl_2$	0.08	0.886	
	Zn	Total	0.10	<0.001	0.12
		$CaCl_2$	0.05	0.037	0.03

b:coefficient of the regression, *p* value: statistical significance, R^2: partial R-squared. TM concentrations measured in soft body for snails and liver for small mammals, normalized to age except for *O. draparnaudi*.

compared to *Cepaea* snails, suggesting the existence of different Zn regulation strategies among these taxa.

Recent research has been devoted to the study of the accumulation of TMs in small mammals in polluted sites, taking into account both total and extractable TM concentrations in soils In contrast to our results, in a study in Dutch floodplains, no correlation between internal Cd, Pb or Zn in and total concentrations in soils were found for both the bank vole or the greater white-toothed shrew [38], while correlations between internal Cd and soil Cd-$CaCl_2$ for the greater white-toothed shrew and between internal Zn and soil Zn-$CaCl_2$ for the bank vole were observed. In the same environmental context of the Dutch floodplains, van Gestel [41] found that soil invertebrates and small mammals showed increases in levels of Cd (and sometimes Cu and Pb) while soil pore water and $CaCl_2$ extracts of metals were low, highlighting that available soil metal concentrations were not suitable indicators of metal accumulation in the food chain. In a smelter-impacted site, Rogival et al. [78] used ammonium nitrate as chemical extractant and showed significant relationships between Cd and Pb concentrations in organs of the wood mouse and in its diet (acorns and earthworms). They found stronger relationships between the diet and soil TMs using total rather than extractable concentrations for several metals including Cd, Pb and Zn. Thus, if total concentrations in soils are better estimates of the contamination of the diet than extractable ones, this may explain why we found here better correlations between internal TMs and soil TMs using total rather than extractable concentrations. The slight increase in hepatic Zn concentrations in small mammals along the pollution gradient is consistent with previous studies, which have provided evidence for slight differences between polluted and control field sites and the ability of small mammals to regulate internal levels of this essential element [7,34,69,79,80].

The use of chemical extractants to estimate bioavailability of metals is strongly dependent on the site, the metal, and the species under consideration, rendering any generalization risky. Based on our results and the discrepancies observed in the literature, it seems, as it has been demonstrated for invertebrates and vegetation [43,81–83], that the relevance of chemical extractants for assessing the bioavailability of metals in field situations is also a great matter of concern for vertebrate wildlife. The availability of TMs in soils may not be the main factor governing TM transfer in food webs because bioavailability to herbivorous, omnivorous and carnivorous organisms cannot be predicted by a single chemical extract from a polluted soil. This phenomenon is probably due to the variety of exposure routes [6,84,85] and to the fact that factors

Table 6. Parameters of significant regressions between TM concentrations in animals and in soils (total at sampling point) by landscape.

Species	Landscape	*n*	Cd			Pb			Zn		
			b	*p*	R^2	*b*	*p*	R^2	*b*	*p*	R^2
Cepaea sp	All	131									
	A	52									
	S	0									
	U	3									
	W	76	−0.72	0.002	0.15						
Oxychilus draparnaudi	All	78	0.76	<0.001	0.20	0.33	0.010	0.08	0.36	<0.001	0.15
	A	6									
	S	0									
	U	50	0.81	<0.001	0.31	0.61	<0.001	0.34	0.63	<0.001	0.40
	W	22	1.37	0.047	0.23						
Myodes glareolus	All	243	0.34	<0.001	0.19	0.45	<0.001	0.33	0.05	<0.001	0.08
	A	16	0.36	<0.001	0.58	0.38	0.007	0.47			
	S	21				0.91	0.028	0.24			
	U	14	1.31	<0.001	0.78	0.76	0.015	0.40			
	W	192	0.58	<0.001	0.19	0.22	0.018	0.04	0.07	<0.001	0.08
Crocidura russula	All	163	0.64	<0.001	0.18	0.67	<0.001	0.37	0.10	<0.001	0.12
	A	75	0.54	0.012	0.10	0.65	<0.001	0.24			
	S	9									
	U	75	0.71	<0.001	0.17	0.79	<0.001	0.42	0.08	0.001	0.11
	W	4									

b:coefficient of the regression, *p* value: statistical significance, R^2: partial R-squared. TM concentrations measured in soft body for snails and liver for small mammals, normalized to age except for *O. draparnaudi*. Type of landscape: "A" for agricultural lands, "U" for urban areas, "S" for shrublands and "W" for woodlands.

affecting transfer to secondary consumers are not only related to the partitioning of metals in soils, but also to the parameters that modulate bioavailability along food webs, notably the levels and the chemical storage forms of TM in dietary items as well as the availability of these items [86,87].

Influence of landscape on animal TM concentrations

Our results show that landscape composition represents a significant variable that, together with the soil TM concentration and the age of the animals, contribute to explain internal TM concentrations in snails and small mammals. The density, availability and diversity of dietary items are likely to vary among landscape types and it has been proven, for both snails and small mammals, that diet varies according to the availability of dietary items [48,51,57,88]. The feeding and foraging behaviour of animals can therefore change according to the landscape type. Resulting variations in diet composition could affect the amount and the bioavailability of metals transferred along food webs, leading to variable levels of bioaccumulation in the organism considered. Indeed, it has been shown that TM bioavailability for an organism depends not only on its own digestive characteristics but also on the amount and the sequestration of metals within the food [87,89–92]. It has been emphasised that metal bioavailability can be affected by foraging and feeding behaviour of the organism [6,8,93]. The different patterns of Cd and Pb accumulation between landscapes found in the present work could thus be partially related to such food chain effects. This was also suggested by Hendrickx et al. [86], who observed site-specific TM accumulation in invertebrates (spiders and amphipods) and hypothesized that biological characteristics of sites, via alterations of trophic webs, can modify TM transfer.

Using individual-based models, Schipper et al. [10] modelled the influence of environmental heterogeneity in Dutch floodplains (soil contamination, habitat availability and suitability) on metal exposure using four species of small mammals, including the bank vole. These authors concluded that environmental heterogeneity governed only a minor part of the variation in metal exposure and that intra-species differences in exposure should mainly be due to inter-individual variations in species traits. In our study, soil contamination and landscape explained a non-negligible amount (around 30%) of the variation in internal Cd concentration. This could be due to the fact that we examined a larger pollution range and we considered not only the land use of the sampling point, but the influence of the landscape (a complex mosaic of land uses) surrounding the sampling point. The examination of exposure heterogeneity by combining these two scales of perception (i.e., heterogeneity within and between landscapes) could improve the understanding of ecological factors affecting transfer of TMs in ecosystems. Multi-scale approaches have been shown to be relevant within the context of biological contaminant (e.g. parasite) transfer in eco-epidemiology [94–96] and also appear promising for studies on chemical transfer within a landscape ecotoxicology framework.

Our data revealed a high inter-individual variability in TM concentrations, and although we considered age, soil TMs and landscape, more than half of the variance remained unexplained. Therefore, our results are also partly in accordance with those of

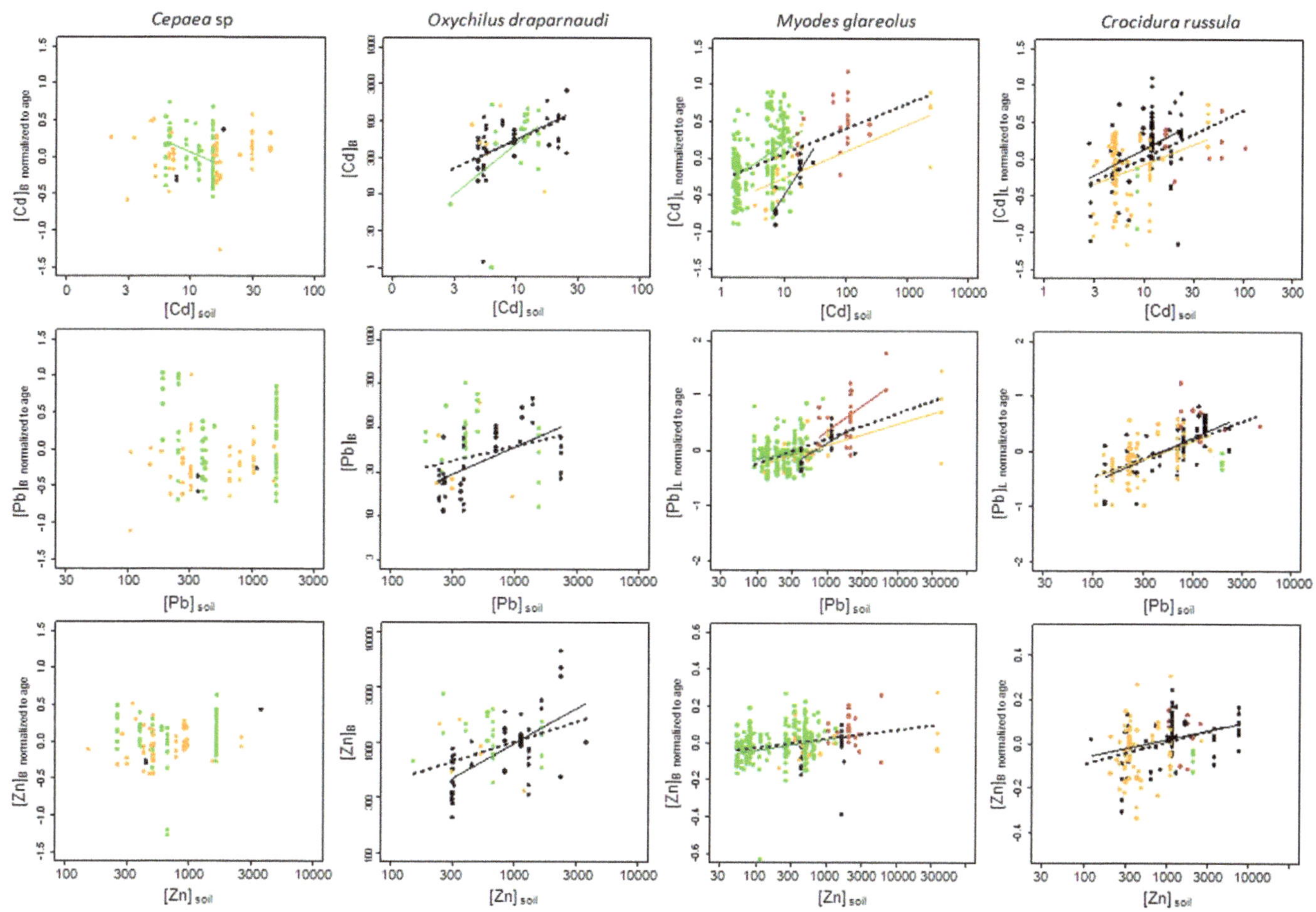

Figure 3. Relationships between TM concentrations in organisms (normalized to age) and in soils by landscape type. TM concentrations in the soft body for snails ($[C]_B$) and in the liver for small mammals ($[C]_L$) normalized to age (except for *O. draparnaudi*) and total TM concentrations measured in soil at the sampling point ($[C]_{soil}$) ($\mu g.g^{-1}$ DW). Points are coloured according to the type of landscape where animals were captured (agricultural lands: orange, urban areas: black, shrublands: red, woodlands: green). Significant regressions ($p<0.05$) are plotted with coloured lines that correspond to the landscape of concern, and black dashed lines symbolise the regression for the whole dataset including all landscape types.

Schipper et al. [10] concerning the strong influence of intra-species variability on exposure and TM accumulation. This phenomenon suggests that individual characteristics and behaviour govern a large part of the variations in internal concentrations.

In fact, accumulation abilities, inter-individual variability, spatio-temporal variations in exposure, and heterogeneity in levels of soil contamination may affect the strength of the relationship between accumulated and soil TM concentrations.

Spatial scale of correlation between animal and soil TM concentrations

Because internal and soil TM concentrations at the sampling point scale were correlated, we argue that the present study provides evidence for a spatially explicit relationship between TM concentrations in animals and in soils. Our results stress that this spatial correlation is due to both the levels of metals in soils and the landscape composition around the habitats where the animals were sampled. The improvement of the correlation between internal and external concentrations using an increasing buffer size was not straightforward, in contrast to what was expected. Marinussen and van der Zee (1996) modelled exposure to and accumulation of Cd in fictitious organisms with various sizes of home-range (10 to 400 m^2) and showed that home-range size greatly affected exposure [26]. We assume that the range of the strongest correlation between internal and soil metal concentrations depends on numerous biological and ecological parameters.

Overall interpretation of the results with the example of Pb

The movements of small mammals are known to be variable among landscapes because of the spatial heterogeneity of suitable habitats, their connectivity and the characteristics of the ecological barriers [49,97]. Presence and survival of animals in heterogeneous landscapes are related to both site-specific characteristics, which define habitat quality on a local scale, and to factors acting at the landscape and/or meta-population scale [98]. Landscape features, animal habitat preferences and resource requirements govern the spatial repartition of suitable/unsuitable patches over an area [22]. This can result in different exposure patterns among landscapes for a given species, and can lead to different inter-species exposure within the same landscape type [4,25]. Within this context, we propose a synthetic interpretation of the results we obtained from small mammals, using Pb as an example (Figure 6). We hypothesize that relationships between internal and soil TMs are modulated by both landscape and the ecological characteristics

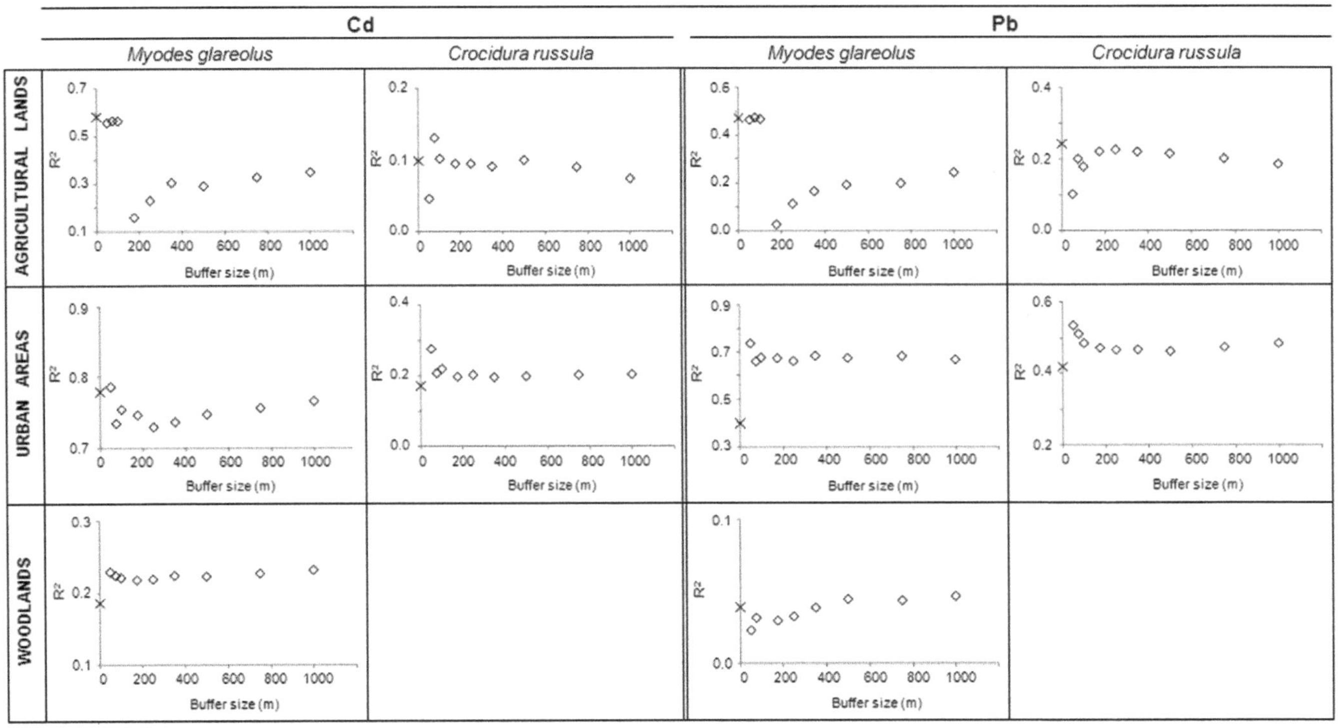

Figure 4. R^2 values of regressions between TMs in small mammals and in soils at several buffer sizes. Partial R^2 values for soil Cd and Pb concentrations in significant regressions between TM concentrations (normalized to age) in small mammals and total soil TM concentrations using measured values at the sampling point (×) and predicted values at several buffer sizes (◇) in agricultural lands, urban areas and woodlands.

of the species. In other words, we propose that the exposure of each species and the resulting bioaccumulation are dependent upon its ecological characteristics (such as spatial behaviour and diet) which are likely to be landscape-specific.

We have built hypothetical maps of suitable and unsuitable habitats on the basis of habitat preference data found in the literature. The bank vole is a forest specialist species. It can live in other woody habitats (for instance woodlots and hedgerows) in heterogeneous landscapes (agricultural, urban or others) where it is confined to woody patches and rarely moves to surrounding habitats unless it needs to cross them to reach other suitable woody habitats [49,99–101]. The greater white-toothed shrew is more eurytopic. This species is mostly present in inhabited areas and open landscapes such as agricultural fields [48,97,101,102]. Therefore, within each landscape, habitat suitability differs between voles and shrews.

We propose the existence of three different cases (Figure 6). When almost the entire area is suitable (case 1), home-ranges can overlap several habitats that differ in metal contamination and individuals can easily move between more or less contaminated patches. Therefore, internal concentrations show high inter-individual variability and are poorly related to soil contamination, as evidenced by the relatively low R^2 values and slopes of the relationships). Moreover, the strength of the relationship between internal and soil TMs does not substantially vary regardless the considered scale for soil TM concentration (the R^2 value was roughly similar between the sampling point and buffers as well as among buffers). This is the case for shrews in agricultural lands and voles in woodlands. Conversely, when animals are confined to suitable isolated patches within a hostile matrix (case 2), the absence or the presence of high local contamination in occupied patches would greatly condition animal exposure, since individuals

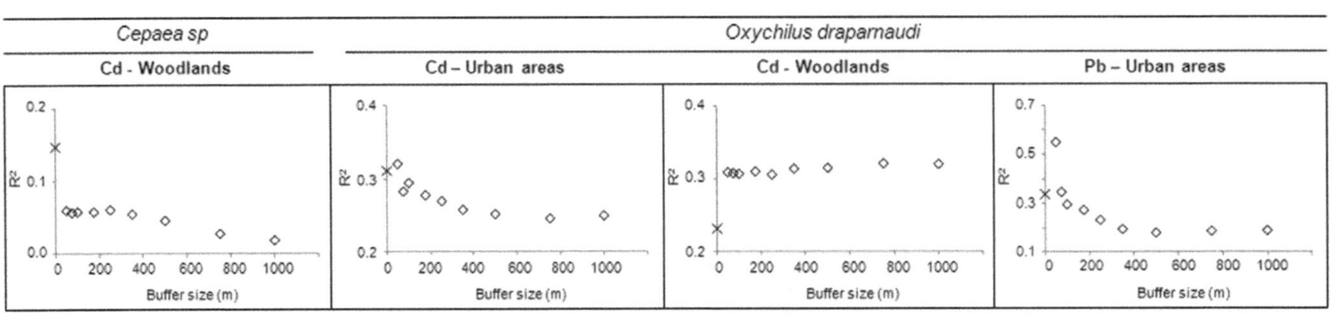

Figure 5. R^2 values of regressions between TMs in snails and in soils at several buffer sizes. Partial R^2 values for soil Cd and Pb concentrations in significant regressions between TM concentrations in snails (normalized to age for *Cepaea*) and soil TM concentrations using measured values at the sampling point (×) and predicted values at several buffer sizes (◇) in urban areas and woodlands.

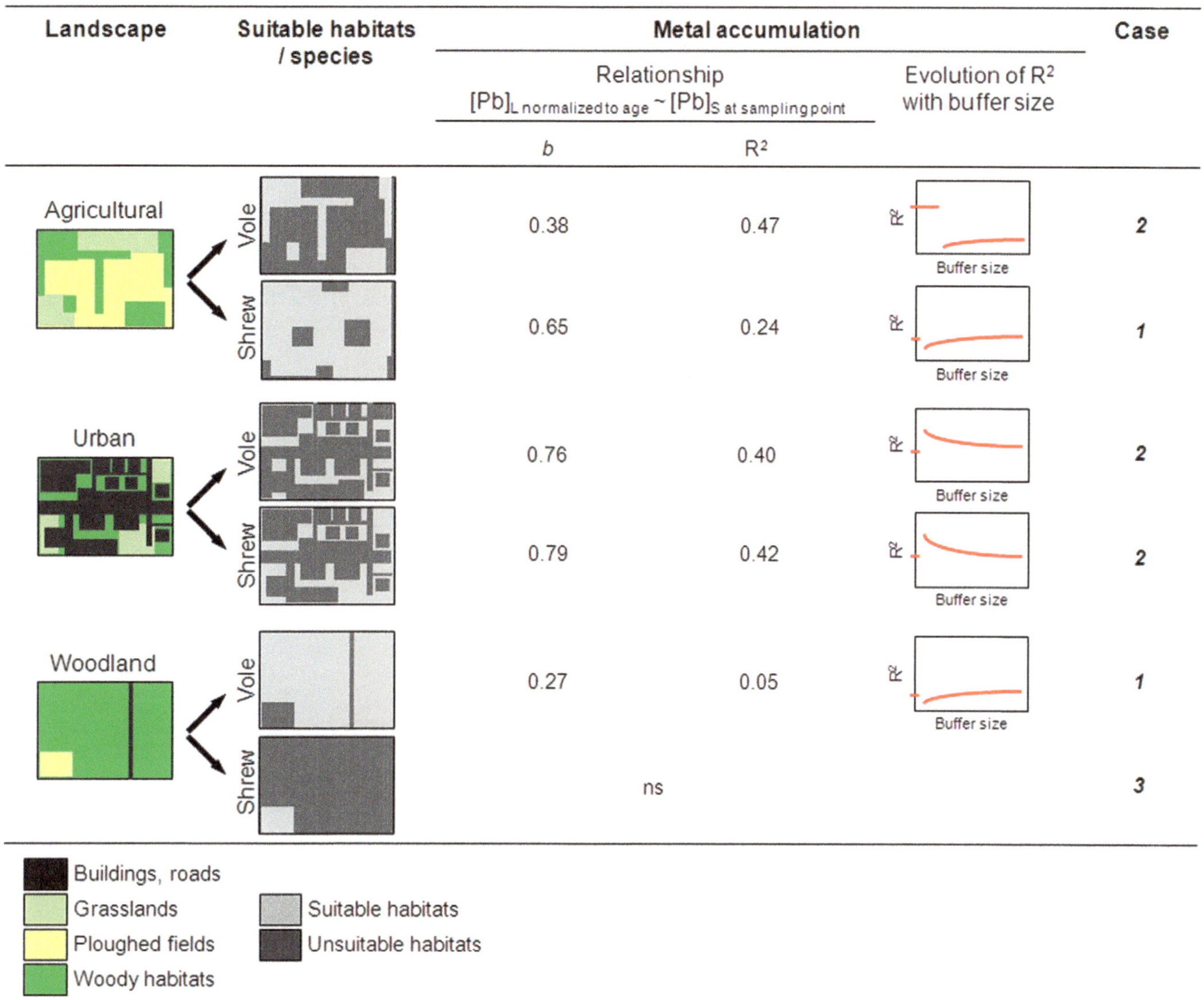

Figure 6. Summary and overall interpretation of the data with the example of Pb in small mammals. The data concerning Pb in the bank vole and in the greater white-toothed shrew is summarised, thus illustrating our overall interpretation of the data. Data concerning metal accumulation are R^2 values and slopes of the regressions between Pb concentrations in the liver (normalized to age) and total Pb concentrations in soil measured at the sampling point in addition to patterns of evolution of the R^2 values with buffer size. Maps of landscapes are based upon observed situations within the study area. Maps of suitable habitats for each species are hypothesized on the basis of habitat preferences described in the literature. Three patterns linking internal TM concentrations to soil TM concentrations have been uncovered, these patterns are named "Case 1, 2 and 3".

cannot easily move or forage in surrounding habitats. Internal concentrations therefore would more strongly depend on soil TM concentration in occupied patches, resulting in better correlations between internal and soil contamination compared to case 1. In this case, represented by voles in agricultural lands and voles and shrews in urban areas, we found globally strong relationships between TMs in animals and in soils (high R^2 values and/or slopes). Moreover, we observed better correlations on local scales (at the sampling point or at a buffer size lower than 100 m). Finally, in landscapes characterised in a large part by unsuitable habitats (case 3), animals can be found during migration and dispersal or in margins, and it may be assumed that the main duration of exposure has occurred in another place. In such cases (represented by shrews in woodlands and shrublands), we captured few individuals, all of which were non-reproductive adults, and we failed to detect relationships between TMs in animals and in soils.

We observed that landscape composition influences bioaccumulation with species-specific patterns. Thus we conclude that landscape features modulate animal exposure to pollutants, and this effect might differ according to species ecological characteristics. The spatial heterogeneity of soil contamination could more or less affect exposure, depending on the species and the landscape concerned.

Conclusions

Apart from animal age and metal concentration in soil at the sampling point, the landscape around the habitat of capture influences the internal TM concentrations. Certain landscapes are therefore more at risk than others depending on the considered organisms and metals. In some cases, transfer is high even at low levels of soil contamination and result in elevated accumulated TM levels in animals (i.e., an elevated biota to soil accumulation factor). In other cases, low internal TM levels were found at low soil TM concentrations, but internal concentrations sharply increased along the pollution gradient. Both phenomena were also found to co-occur. We propose that a landscape gathering both phenomena would the most at risk for wildlife.

The lack of relationship between internal levels in animals and $CaCl_2$-extractable TM concentrations in soil, as well as the differences in relationships between animal and soil TM concentrations between studies, provide two major insights within a risk assessment framework. First, the use of chemical extracts may not be relevant when assessing TM bioavailability to herbivorous/granivorous and carnivorous species. Second, predictions of internal TM using general accumulation models, such as the use of global regression regardless of the environmental characteristics of a site (e.g. landscape) and the intra-species variability should be used cautiously.

Exposure of wildlife to contaminants and subsequent TM accumulation is determined by several parameters acting at different biological organisation levels that are integrative variables of several processes. Based on our results, we suggest that TM accumulation in snails and small mammals is governed by ecological (diet, habitat preferences, mobility, etc) and physiological (assimilation and excretion of TM) characteristics of animals. Our results strongly suggest that availability in soil does not fully determine transfer in food webs. Species ecology and landscape are other key factors that determine organism exposure. Our findings lead us to hypothesize that ecological characteristics, such as food web structure and the way that organisms exploit their environment (home-range size, migration, feeding behaviour, habitat preferences, etc), both dependent on landscape features, mainly explain TM transfer in food webs. The present study points out the need for further investigation to develop a field of "landscape ecotoxicology" and elucidate the underlying mechanisms behind landscape effects, and additionally highlights the interest for a multi-scale approach in ecotoxicology.

Supporting Information

Figure S1 Empirical omnidirectional variograms with variographic envelopes and retained fitted models for Cd, Pb and Zn.

Table S1 Total concentrations of trace metals in soils from Metaleurop-impacted area, according to soil use type.

Table S2 Parameters of the fitted variogram models, results of cross-validation and median variance of kriged values.

Text S1 Prediction of trace metal (Cd, Pb, Zn) concentrations in topsoils of Metaleurop-impacted area.

Acknowledgments

The authors gratefully thank Cécile Grand from the *Agence De l'Environnement et de la Maîtrise de l'Energie* (ADEME) for many fruitful scientific discussions. We warmly thank Nadia Crini, Elie Dhivert, Alice Labourier, Jean-Claude Lambert, Christiane Lovy, Jonathan Paris and Pierre-Yves Peseux for their technical assistance. The authors are indebted to three anonymous reviewers for their very helpful remarks.

Author Contributions

Conceived and designed the experiments: CF RS MC PG FR AdV. Performed the experiments: CF RS MC FD DR. Analyzed the data: CF PG MC RS. Contributed reagents/materials/analysis tools: CF DR FD. Wrote the paper: CF RS MC PG FR AdV.

References

1. Smith R, Pollard SJT, Weeks JM, Nathanail CP (2006) Assessing harm to terrestrial ecosystems from contaminated land. Soil Use and Management 21: 527–540.
2. Fairbrother A, Wenstel R, Sappington K, Wood W (2007) Framework for metals risk assessment. Ecotoxicology and Environmental Safety 68: 145–227.
3. Carlsen TM, Coty JD, Kercher JR (2004) The spatial extent of contaminants and the landscape scale: An analysis of the wildlife, conservation biology, and population modeling literature. Environmental Toxicology and Chemistry 23: 798–811.
4. Ares J (2003) Time and space issues in ecotoxicology: Population models, landscape pattern analysis, and long-range environmental chemistry. Environmental Toxicology and Chemistry 22: 945–957.
5. Woodbury PB (2003) Dos and don'ts of spatially explicit ecological risk assessments. Environmental Toxicology and Chemistry 22: 977–982.
6. Smith PN, Cobb GP, Godard-Codding C, Hoff D, McMurry ST, et al. (2007) Contaminant exposure in terrestrial vertebrates. Environmental Pollution 150: 41–64.
7. Fritsch C, Cosson RP, Cœurdassier M, Raoul F, Giraudoux P, et al. (2010) Responses of wild small mammals to a pollution gradient: host factors influence metal and metallothionein levels. Environmental Pollution 158: 827–840.
8. Peakall D, Burger J (2003) Methodologies for assessing exposure to metals: speciation, bioavailability of metals, and ecological host factors. Ecotoxicology and Environmental Safety 56: 110–121.
9. Clifford PA, Barchers DE, Ludwig DF, Sielken RL, Klingensmith JS, et al. (1995) An approach to quantifying spatial components of exposure for ecological risk assessment. Environmental Toxicology and Chemistry 14: 895–906.
10. Schipper AM, Loos M, Ragas AMJ, Lopes JPC, Nolte BT, et al. (2008) Modeling the influence of environmental heterogeneity on heavy metal exposure concentrations for terrestrial vertebrates in river floodplains. Environmental Toxicology and Chemistry 27: 919–932.
11. Vermeulen F (2009) Spatially explicit exposure assessment of persistent pollutants through the food chain of the European hedgehog (*Erinaceus europaeus*). Antwerp: University of Antwerp, Belgium. 211 p.
12. Fowler D, Smith RI, Leith ID, Crossley A, Mourne RW, et al. (1998) Quantifying fine-scale variability in pollutant deposition in complex terrain using ^{210}Pb inventories in soil. Water Air and Soil Pollution 105: 459–470.
13. Hope BK (2006) An examination of ecological risk assessment and management practices. Environment International 32: 983–995.
14. Fowler D, Skiba U, Nemitz E, Choubedar F, Branford D, et al. (2004) Measuring aerosol and heavy metal deposition on urban woodland and grass using inventories of ^{210}Pb and metal concentrations in soil. Water Air and Soil Pollution 4: 483–499.
15. Rieuwerts J, Farago M (1996) Heavy metal pollution in the vicinity of a secondary lead smelter in the Czech Republic. Applied Geochemistry 11: 17–23.
16. Magiera T, Zawadzki J (2007) Using of high-resolution topsoil magnetic screening for assessment of dust deposition: Comparison of forest and arable soil datasets. Environmental Monitoring and Assessment 125: 19–28.
17. Ettler V, Vanek A, Mihaljevic M, Bezdicka P (2005) Contrasting lead speciation in forest and tilled soils heavily polluted by lead metallurgy. Chemosphere 58: 1449–1459.
18. Kapusta P, Sobczyk L, Rozen A, Weiner J (2003) Species diversity and spatial distribution of enchytraeid communities in forest soils: effects of habitat characteristics and heavy metal contamination. Applied Soil Ecology 23: 187–198.
19. Vermeulen F, Van den Brink NW, D'Havé H, Mubiana VK, Blust R, et al. (2009) Habitat type-based bioaccumulation and risk assessment of metal and As contamination in earthworms, beetles and woodlice. Environmental Pollution 157: 3098–3105.
20. Fritsch C, Giraudoux P, Cœurdassier M, Douay F, Raoul F, et al. (2010) Spatial distribution of metals in smelter-impacted soils of woody habitats: Influence of landscape and soil properties, and risk for wildlife. Chemosphere 81: 141–155.
21. Lidicker WZJ (2008) Levels of organization in biology: on the nature and nomenclature of ecology's fourth level. Biological Reviews 83: 71–78.
22. Burel F, Baudry J (2003) Landscape ecology - Concepts, methods and applications. Enfield, New Hampshire, USA: Sciences Publishers, Inc. 362 p.

23. Gaines K, Boring C, Porter D (2005) The development of a spatial explicit model to estimate radiocaesium body burdens in raccoons (*Procyon lotor*) for ecological risk assessment. Science of the Total Environment 341: 15–31.
24. Kapustka LA (2008) Limitations of the current practices used to perform ecological risk assessment. Integrated Environmental Assessment and Management 4: 290–298.
25. Barnthouse LW, Munns WRJ, Sorensen MT, eds (2008) Population-level ecological risk-assessment. Boca Raton, Florida, USA: CRC Press Taylor & Francis Group. 346 p.
26. Marinussen MPJC, van der Zee SEATM (1996) Conceptual approach to estimating the effect of home-range size on the exposure of organisms to spatially variable soil contamination. Ecological Modelling 87: 83–89.
27. Hope BK (2001) A case study comparing static and spatially explicit ecological exposure analysis methods. Risk Analysis 21: 1001–1010.
28. Hope BK (2000) Generating probabilistic spatially-explicit individual and population exposure estimates for ecological risk assessments. Risk Analysis 20: 573–589.
29. Purucker ST, Welsh CJE, Stewart RN, Starzec P (2007) Use of habitat-contamination spatial correlation to determine when to perform a spatially explicit ecological risk assessment. Ecological Modelling 204: 180–192.
30. Topping CJ, Sibly RM, Akçakaya HR, Smith GC, Crocker DR (2005) Risk Assessment of UK Skylark Populations Using Life-History and Individual-Based Landscape Models. Ecotoxicology V14: 925–936.
31. Kooistra L, Leuven RSEW, Nienhuis PH, Wehrens R, Buydens LMC (2001) A Procedure for Incorporating Spatial Variability in Ecological Risk Assessment of Dutch River Floodplains. Environmental Management 28: 359–373.
32. Kooistra L, Huijbregts MAJ, Ragas AMJ, Wehrens R, Leuven R (2005) Spatial variability and uncertainty in ecological risk assessment: A case study on the potential risk of cadmium for the little owl in a Dutch river flood plain. Environmental Science and Technology 39: 2177–2187.
33. Cairns Jr. J, Nlederlehner BR (1996) Developing a field of landscape ecotoxicology. Ecological Applications 6: 790–796.
34. Shore RF, Rattner BA (2001) Ecotoxicology of Wild Mammals. London, UK: John Wiley & Sons. 730 p.
35. Talmage S, Walton B (1991) Small mammals as monitors of environmental contaminants. Review of Environmental Contaminants and Toxicology 119: 47–145.
36. Berger B, Dallinger R (1993) Terrestrial snails as quantitative indicators of environmental metal pollution. Environmental Monitoring and Assessment 25: 65–84.
37. de Vaufleury AG, Pihan F (2000) Growing snails used as sentinels to evaluate terrestrial environment contamination by trace elements. Chemosphere 40: 275–284.
38. Wijnhoven S, Leuven R, van der Velde G, Jungheim G, Koelemij E, et al. (2007) Heavy-metal concentrations in small mammals from a diffusely polluted floodplain: importance of species- and location-specific characteristics. Archives of Environmental Contamination and Toxicology 52: 603–613.
39. van den Brink N, Lammertsma D, Dimmers W, Boerwinkel M-C, van der Hout A (2010) Effects of soil properties on food web accumulation of heavy metals to the wood mouse (*Apodemus sylvaticus*). Environmental Pollution 158: 245–251.
40. Baker S, Herrchen M, Hund-Rinke K, Klein W, Kordel W, et al. (2003) Underlying issues including approaches and information needs in risk assessment. Ecotoxicology and Environmental Safety 56: 6–19.
41. Van Gestel CAM (2008) Physico-chemical and biological parameters determine metal bioavailability in soils. Science of the Total Environment 406: 385–395.
42. Peijnenburg WJGM, Jager T (2003) Monitoring approaches to assess bioaccessibility and bioavailability of metals: Matrix issues. Ecotoxicology and Environmental Safety 56: 63–77.
43. Harmsen J (2007) Measuring bioavailability: from a scientific approach to standard methods. Journal of Environmental Quality 36: 1420–1428.
44. Meers E, Samson R, Tack FMG, Ruttens A, Vandegehuchte M, et al. (2007) Phytoavailability assessment of heavy metals in soils by single extractions and accumulation by *Phaseolus vulgaris*. Environmental and Experimental Botany 60: 385–396.
45. ISO (2008) Soil quality - Requirements and guidance for the selection and application of methods for the assessment of bioavailability of contaminants in soil and soil materials. ISO 17402:2008. Geneva, Switzerland: International Organization for Standardization.
46. Baur A, Baur B (1993) Daily movement patterns and dispersal in the land snail *Arianta arbustorum*. Malacologia 35: 89–98.
47. Le Louarn H, Quéré J, Butet A (2003) Les Rongeurs de France - Faunistique et biologie; INRA, editor. Paris, France. 256 p.
48. Lugon-Moulin N (2003) Les Musaraignes - Biologie, Ecologie, Répartition en Suisse; Porte-Plumes, editor. Ayer, Switzerland. 308 p.
49. Kozakiewicz M, Van Apeldoorn R, Bergers P, Gortat T, Kozakiewicz A (2000) Landscape approach to bank vole ecology. Polish Journal of Ecology 48: 149–161.
50. Cameron RAD, Williamson P (1977) Estimating migration and effects of disturbance in mark-recapture studies on snail *Cepaea nemoralis* L. Journal of Animal Ecology 46: 173–179.
51. Williamson P, Cameron RAD (1976) Natural diet of landsnail *Cepaea nemoralis*. Oikos 27: 493–500.
52. Wolda H, Zweep A, Schuitema KA (1971) The role of food in the dynamics of populations of the landsnail *Cepaea nemoralis*. Oecologia 7: 361–381.
53. Mahtfeld K (2000) Impact of introduced gastropods on molluscan communities, northern North Island. Wellington, New Zealand: Department of Conservation.
54. Kerney MP, Cameron RAD (2006) Guide des escargots et limaces d'Europe. Paris, France: Delachaux et Niestlé. 370 p.
55. Vonproschwitz T (1994) *Oxychilus cellarius* (Müller) and *Oxychilus draparnaudi* (Beck) as predators on egg clutches of *Arion lusitanicus mabille*. Journal of Conchology 35: 183–184.
56. Sadowska ET, Baliga-Klimczyk K, Chrzascik K, Koteja P (2008) Laboratory model of adaptive radiation: a selection experiment in the bank vole. Physiological and Biochemical Zoology 81: 627–640.
57. Abt KF, Bock WF (1998) Seasonal variations of diet composition in farmland field mice *Apodemus* spp. and bank voles *Clethrionomys glareolus*. Acta Theriologica 43: 379–389.
58. Frangi J-P, Richard D (1997) Heavy metal soil pollution cartography in northern France. Science of the Total Environment 205: 71–79.
59. Sterckeman T, Douay F, Proix N, Fourrier H, Perdrix E (2002) Assessment of the contamination of cultivated soils by eighteen trace elements around smelters in the North of France. Water Air and Soil Pollution 135: 173–194.
60. Douay F, Pruvot C, Roussel H, Ciesielski H, Fourrier H, et al. (2008) Contamination of urban soils in an area of Northern France polluted by dust emissions of two smelters. Water Air and Soil Pollution 188: 247–260.
61. Sterckeman T, Douay F, Proix N, Fourrier H (2000) Vertical distribution of Cd, Pb and Zn in soils near smelters in the North of France. Environmental Pollution 107: 377–389.
62. Douay F, Pruvot C, Waterlot C, Fritsch C, Fourrier H, et al. (2009) Contamination of woody habitat soils around a former lead smelter in the North of France. Science of the Total Environment 407: 5564–5577.
63. Charissou I (1999) Identification des restes trouvé dans les pelotes de réjection de rapaces. Supplément scientifique de la revue Epops 44: 1–31.
64. Chaline J, Baudvin H, Jammot D, Saint Girons M-C (1974) Les proies des rapaces. Paris: Petits mammifères et leur environnement. 143 p.
65. Kozakiewicz M (1976) The weight of eye lens as the proposed age indicator of the bank vole. Acta Theriologica 21: 314–316.
66. Williamson P (1979) Age determination of juvenile and adult *Cepaea*. Journal of Molluscan Studies 45: 52–60.
67. AFNOR (1996) Qualité des sols - Méthodes chimiques - sols sédiments, mise en solution totale par attaque acide - NF X31-147. Paris, France: Association Française de Normalisation.
68. Grafen A, Hails R (2002) Modern Statistics for the Life Sciences: Oxford University Press. 351 p.
69. Sánchez-Chardi A, Ribeiro CAO, Nadal J (2009) Metals in liver and kidneys and the effects of chronic exposure to pyrite mine pollution in the shrew *Crocidura russula* inhabiting the protected wetland of Doñana. Chemosphere 76: 387–394.
70. Williamson P (1980) Variables affecting body burdens of lead, zinc and cadmium in a roadside population of the snail *Cepaea hortensis* Müller. Oecologia 44: 213–220.
71. R Development Core Team (2006) R: A language and environment for statistical computing. Vienna, Austria: Foundation for Statistical Computing. http://www.R-project.org.
72. Sánchez-Chardi A, Nadal J (2007) Bioaccumulation of metals and effects of landfill pollution in small mammals. Part I. The greater white-toothed shrew, *Crocidura russula*. Chemosphere 68: 703–711.
73. Sánchez-Chardi A, Lopez-Fuster MJ, Nadal J (2007) Bioaccumulation of lead, mercury, and cadmium in the greater white-toothed shrew, *Crocidura russula*, from the Ebro Delta (NE Spain): Sex- and age-dependent variation. Environmental Pollution 145: 7–14.
74. Notten MJM, Oosthoek AJP, Rozema J, Aerts R (2005) Heavy metal concentrations in a soil-plant-snail food chain along a terrestrial soil pollution gradient. Environmental Pollution 138: 178–190.
75. Damek-Poprawa M, Sawicka-Kapusta K (2004) Histopathological changes in the liver, kidneys, and testes of bank voles environmentally exposed to heavy metal emissions from the steelworks and zinc smelter in Poland. Environmental Research 96: 72–78.
76. Fritsch C, Cœurdassier M, Gimbert F, Crini N, Scheifler R, et al. (2011) Investigations of adaptation to metallic pollution in land snails (*Cantareus aspersus* and *Cepaea nemoralis*) from a smelter-impacted area. Ecotoxicology 20: 739–759.
77. Dallinger R, Lagg B, Egg M, Schipflinger R, Chabicovsky M (2004) Cd accumulation and Cd-metallothionein as a biomarker in *Cepaea hortensis* (Helicidae, Pulmonata) from laboratory exposure and metal-polluted habitats. Ecotoxicology 13: 757–772.
78. Rogival D, Scheirs J, Blust R (2007) Transfer and accumulation of metals in a soil-diet-wood mouse food chain along a metal pollution gradient. Environmental Pollution 145: 516–528.
79. Milton A, Cooke JA, Johnson MS (2003) Accumulation of Lead, Zinc, and Cadmium in a Wild Population of (*Clethrionomys glareolus*) from an Abandoned Lead Mine. Archives of Environmental Contamination and Toxicology 44: 405–411.
80. Loos M, Ragas AMJ, Tramper JJ, Hendriks AJ (2009) Modeling zinc regulation in small mammals. Environmental Toxicology and Chemistry 28: 2378–2385.

81. Migeon A, Richaud P, Guinet F, Chalot M, Blaudez D (2009) Metal accumulation by woody species on contaminated sites in the north of France. Water Air and Soil Pollution 204: 89–101.
82. Bleeker EAJ, van Gestel CAM (2007) Effects of spatial and temporal variation in metal availability on earthworms in floodplain soils of the river Dommel, The Netherlands. Environmental Pollution 148: 824–832.
83. Hobbelen PHF, Koolhaas JE, van Gestel CAM (2006) Bioaccumulation of heavy metals in the earthworms *Lumbricus rubellus* and *Aporrectodea caliginosa* in relation to total and available metal concentrations in field soils. Environmental Pollution 144: 639–646.
84. Cœurdassier M, Gomot-de Vaufleury A, Lovy C, Badot P-M (2002) Is the cadmium uptake from soil important in bioaccumulation and toxic effects for snails? Ecotoxicology and Environmental Safety 53: 425–431.
85. Scheifler R, de Vaufleury A, Coeurdassier M, Crini N, Badot PM (2006) Transfer of Cd, Cu, Ni, Pb, and Zn in a soil-plant-invertebrate food chain: A microcosm study. Environmental Toxicology and Chemistry 25: 815–822.
86. Hendrickx F, Maelfait J-P, Bogaert N, Tojal C, Du Laing G, et al. (2004) The importance of biological factors affecting trace metal concentration as revealed from accumulation patterns in co-occurring terrestrial invertebrates. Environmental Pollution 127: 335–341.
87. Vijver MG, Van Gestel CAM, Lanno RP, Van Straalen NM, Peijnenburg WJGM (2004) Internal metal sequestration and its ecotoxicological relevance: A review. Environmental Science and Technology 38: 4705–4712.
88. Mason CF (1970) Food, feeding rates and assimilation in woodland snails. Oecologia 4: 358–373.
89. Wallace WG, Lopez GR, Levinton JS (1998) Cadmium resistance in an oligochaete and its effect on cadmium trophic transfer to an omnivorous shrimp. Marine Ecology Progress Series 172: 225–237.
90. Wallace WG, Luoma SN (2003) Subcellular compartmentalization of Cd and Zn in two bivalves. II. Significance of trophically available metal (TAM). Marine Ecology Progress Series 257: 125–137.
91. Hispard F, De Vaufleury A, Cosson RP, Devaux S, Scheifler R, et al. (2007) Comparison of transfer and effects of Cd on rats exposed in a short experimental snail–rat food chain or to $CdCl_2$ dosed food. Environment International 34: 381–389.
92. Monteiro MS, Santos C, Soares A, Mann RM (2008) Does subcellular distribution in plants dictate the trophic bioavailability of cadmium to *Porcellio dilatatus* (Crustacea, Isopoda)? Environmental Toxicology and Chemistry 27: 2548–2556.
93. Burger J, Diaz-Barriga F, Marafante E, Pounds J, Robson M (2003) Methodologies to examine the importance of host factors in bioavailability of metals. Ecotoxicology and Environmental Safety 56: 20–31.
94. Giraudoux P, Delattre P, Takahashi K, Raoul F, Quéré JP, et al. (2002) Transmission ecology of *Echinococcus multilocularis* in wildlife: what can be learned from comparative studies and multi-scale approaches? In: Craig P, Pawlowski Z, eds. Cestode zoonoses: Echinococcosis and Cystercosis An emergent and global problem. Amsterdam, The Netherlands: NATO Sciences Series, IOS press. pp 251–262.
95. Kitron U, Clennon JA, Cecere MC, Gurtler RE, King CH, et al. (2006) Upscale or downscale: applications of fine scale remotely sensed data to Chagas disease in Argentina and schistosomiasis in Kenya. Geospatial Health 1: 49–58.
96. Danson FM, Graham AJ, Pleydell DRJ, Campos-Ponce M, Giraudoux P, et al. (2003) Multi-scale spatial analysis of human alveolar echinococcosis risk in China. Parasitology 127: S133–S141.
97. Vuilleumier S, Fontanillas P (2007) Landscape structure affects dispersal in the greater white-toothed shrew: Inference between genetic and simulated ecological distances. Ecological Modelling 201: 369–376.
98. Burel F, Butet A, Delettre YR, de la Pena NM (2004) Differential response of selected taxa to landscape context and agricultural intensification. Landscape and Urban Planning 67: 195–204.
99. Pollard E, Relton J, Hedges V (1970) A study of small mammals in hedges and cultivated fields. Journal of Applied Ecology 7: 549–557.
100. Paillat G, Butet A (1996) Spatial dynamics of the bank vole (*Clethrionomys glareolus*) in a fragmented landscape. Acta Oecologica – International Journal of Ecology 17: 553–559.
101. Wijnhoven S, van der Velde G, Leuven R, Smits A (2006) Modelling recolonisation of heterogeneous river floodplains by small mammals. Hydrobiologia 565: 135–152.
102. Millān de la Peña N, Butet A, Delettre Y, Paillat G, Morant P, et al. (2003) Response of the small mammal community to changes in western French agricultural landscapes. Landscape Ecology V18: 265–278.

12

Phenotypic Plasticity of Early and Late Successional Forbs in Response to Shifts in Resources

Yingxin Huang[1]*, Xueyong Zhao[2], Daowei Zhou[1], Hongxiang Zhang[1], Wei Zheng[1]

1 Northeast Institute of Geography and Agroecology, Chinese Academy of Sciences, Changchun, China, **2** Cold and Arid Regions Environmental and Engineering Research Institute, Chinese Academy of Sciences, Lanzhou, China

Abstract

We compared the phenotypic plasticity of two early successional forbs of nutrient-poor mobile dunes (*Agriophyllum squarrosum* and *Corispermum macrocarpum*) and two later successional forbs (weeds) of stabilized, higher nutrient dunes and cropland (*Chenopodium acuminatum* and *Salsola collina*) to variations in environmental factors. A controlled (including soil nutrients, water, and population density) greenhouse experiment was conducted in Horqin sandy land, China. Late successional species had high plasticity in growth response to nutrients and water or high performance in high soil nutrients and water, reflecting their higher nutrient habitat. In contrast, the early successional species have low plasticity, reflecting their adaptation to resource-poor early successional soil. Late successional species did not always have higher reproductive effort than early successional species. Plants did not have a uniform strategy of increasing reproductive effort with any environmental stressors. Reproductive effort increased with increasing water availability and decreasing nutrient levels, while density had no effect. Patterns of plasticity traits for late successional species exhibited a complex of Master-of-some and Jack-of-all-trades. Late successional species had higher performance or higher plasticity than early successional species.

Editor: Ive De Smet, University of Nottingham, United Kingdom

Funding: This research was funded by the Knowledge Innovation Program of the Chinese Academy of Sciences (NO. KZCX2-EW-QN315), the National Science Foundation of China (31000216), and Northeast Institute of Geography and Agroecology, Chinese Academy of Sciences (KZCX3-SW-NA09-). The funders had no role in study design, data collection and analysis, decision to publish, or preparation of the manuscript.

Competing Interests: The authors have declared that no competing interests exist.

* E-mail: huangyx@neigae.ac.cn

Introduction

Phenotypic plasticity is usually defined as the ability of an individual genotype to modify its growth and development when exposed to different environmental conditions [1–3]. As an important strategy for plants to maximize or maintain fitness in variable biotic and abiotic environments [4], many ecologists have embraced the idea that under many circumstances such phenotypic plasticity can be adaptive [5,6]. Phenotypic plasticity allows individuals to exhibit advantageous phenotypes in a breadth of environmental conditions [7].

High plasticity is also commonly invoked to explain invasion success [8]. Invasive species may have increased plasticity relative to non-invasive species [9–11]. Richards et al. (2006) summarizes that the invader may benefit from three idealized scenarios: (1) a Jack-of-all-trades, in which invaders maintain fitness in unfavorable environments; (2) a Master-of-some, in which invaders increase fitness in favorable environments; or (3) a Jack-and-master, in which invaders combine some level of both above scenarios. Support for the scenarios is found in later studies [12–14].

In succession, early successional species colonize the disturbed site. As colonizers replace indigenous communities, and late successional species replace early successional species. The late successional species have some of the same advantages as invaders. Therefore, under some conditions late successional species prevail in competition with early successional species leading to a shift in community composition. There are different patterns of phenotypic plasticity between early and late successional species. Some studies point out that early successional species showed higher photosynthetic plasticity than late successional species due to selection pressures [15,16]. However it is also found that the late successional species have higher phenotypic plasticity than early successional species. For example, shade tolerant, late successional species exhibit high morphological plasticity [17,18], and late successional forbs have stronger plastic responses to nutrient patchiness compared to mid-successional forbs [19]. Additional studies comparing phenotypic plasticity between different successional species are necessary to understand patterns of response to successional environments.

Morphological traits are important to resource partitioning in many plants [20–22]. Reproductive effort, which shows the proportion of the plant's resources allocated to reproductive activities, is an important component of plant fitness [23–25]. Reproductive effort of plants varies across different seral environments by the adjusting partitioning of resources and morphological traits (in relation to reproductive fitness). Evolutionary selection should favor different patterns of reproductive effort during early or late stages of ecological succession [26]. Invasive species showed higher reproductive effort that varies across a range of resource levels [27,28]. Recent studies have suggested that invasive or late successional species might not always have a greater competitive advantage over their non-invasive native counterparts [29–32]. Abrahamson and Gadgil (1973) demonstrated that late successional species did not have higher reproductive effort compared to early successional species. In addition, Sans et al. (2004) compared three congeners including two exotic invasive and one native species and found that the reproductive effort of the native species was the highest in all treatments. High reproductive effort is usually associated with stressed environments [33–35]. Thus, plants in stressed environments should invest more resources into reproductive and not vegetative structures. However, studies have provided evidence that reproductive effort did not hold constant with changes in the different environments [36–39]. Hence, there are still inconsistencies in studies of reproductive effort.

In a previous study we compared one early and one late successional species by ANCOVA to distinguish allometric exponents in response to environmental treatments that showed higher plasticity of the late successional species [40]. However, the results are limited in extrapolation to broader patterns of succession. The present study includes two early and two late successional species, because the trends of all four species provide more information on phenotypic plastic responses of early and late successional species than previous study. We used SMATR (Standardised Major Axis Tests & Routines, [41] to evaluate allometric relationships between biomass and other plant measurements for increased statistical inference that our first study could not provide [40].

By comparing morphological traits (in relation to resource partitioning) and reproductive effort, we conducted a controlled (including nutrients, water, and density) greenhouse experiment comparing early successional species (*Agriophyllum squarrosum* and *Corispermum macrocarpum*) that colonize a resource-poor environment, with later successional species (*Chenopodium acuminatum* and *Salsola collina*) that colonize a higher- resource environment and replace the early successional communities, in Horqin sandy land. We tested three hypotheses: (1) species that exhibit different invasion scenarios are better adapted to different successional stages; (2) late successional species have higher reproductive effort; and, (3) environmental stresses do not uniformly increase reproductive effort.

Materials and Methods

Study area

This study was conducted in mobile and stabilized sand dunes of south-western Horqin Sandy Land, Inner Mongolia, China (42°55′N, 120°42′E; elevation approx. 345 m). The climate of this region is semi-arid. The average annual temperature is 6.8°C with monthly averages ranging from a minimum of −13.1°C in January to a maximum of 23.7°C in July. The average annual precipitation is 360 mm with 75% of this in the growing season of June–September [42–44].

Study species

Four annual forbs were selected from the family Chenopodiaceae: two early successional status species (*Agriophyllum squarrosum* and *Corispermum macrocarpum*) of mobile sand dunes, and two later successional species (*Chenopodium acuminatum* and *Salsola collina*) of stabilized dunes and cropland, which are commonly found in abandoned habitats [45], often in inter-dune lowland and cropland [46], which play an important role in restoration succession in degraded land in Horqin Sandy Land. *A. squarrosum* and *C. macrocarpum* are adapted to extremely degraded sandy soil, and are also the pioneer species in this habitat. Later successional dunes become stabilized by shrubs such as *Artemisia halodendrom* Turcz. ex Bess and *Salix gordejevii* Chang et Skv., and *C. acuminatum* and *S. collina* colonize with the gradual disappearance of *A. squarrosum* and *C. macrocarpum* [42,46,47].

Experimental design

All species' seeds, which were separately collected from a population for each species within the early and later successional community, were sown on May 9th 2007 in plastic plates with sandy soil obtained from the severely degraded sand dune. The seedlings were transferred into 13.8-cm-radius×26.5-cm-deep plastic pots with the same sandy soil after 4 weeks of sowing (i.e. two-leaf stage). Treatments were imposed 2 weeks after planting. Because of the increasing soil nutrient, soil water and population density during this ecological succession, each of four species was treated with different levels of soil nutrients, soil water and population density. Each factor had two levels: N+: High nutrient level, 20 g of slow-release fertilizer (Osmocote, containing N 14%,P 14% and K 14% and microelements, The Scotts Company), mixed with sandy soil at the beginning of the experiment (total nitrogen concentration was nearly 200 mg/kg soil, similar to stabilized dunes, which ranged from 200 to 300 mg/kg soil) [48]; N–: Low nutrient level, no nutrients added (the total nitrogen concentration of sandy soil ranged from 49 to 53 mg/kg soil, measured using the Kjeldahl method) [49–51]; W+: High water level, equivalent to 400 mm of rainfall during the growing season applied as 530 mL of water every 3 days; W–: Low water level, equivalent to 200 mm of rainfall during the growing season applied as 265 mL of water every 3 days (which is approximately equal to the rainfall during growth season in Horqin Sandy Land); D+: High density level, six plants in each pot arranged is an equilateral hexagonal fashion (which was equivalent to 100 plants/m^2, similar to the density in a crowded environment); D–: Low density level, three plants in each pot arranged by equilateral triangular fashion (which was equal to 50 plants/m^2, similar to the density in normal environment). A full-factorial design of the 32 possible combinations of four species, two levels of soil nutrients, water, and population density was constructed. Each treatment combination had 102 plants (34 pots in each low population density treatment and 17 pots in each high population density). All pots were placed in a naturally ventilated greenhouse in order to reduce differences between the inside and outside of the greenhouse. In order to eliminate positional effects, the pots in each combination were regarded as a group and placed together, group positions were rotated every two weeks.

Measurements

The plants were harvested from 13th to 19th September, when the reproduction of plants was completed [52], and 12 to 20 intact plants from each treatment were randomly selected for measurements. The roots were washed gently until no soil was visible. Each plant was separated into vegetative organs (including roots, stems, and leaves) and reproductive organs. These organs were oven dried at 80°C to a constant mass to determine the respective dry mass. The shattered seeds on the ground were not collected at the final harvest (seed losses were not more than 0.1% of reproductive biomass). The following traits were recorded for individual plant: absolute height, number of primary branches, number of secondary branches, vegetative biomass (including roots, stems, and leaves), reproductive biomass, total biomass, reproductive effort (reproductive biomass/total biomass). The phenotypic plasticity index (PPI, (Maximum mean-minimum mean)/maximum mean) was calculated for each trait.

Statistical analysis

A four-way factorial analysis of variance (ANOVA) was performed to test the effects of species (S), nutrients (N), water (W) and density (D) on plant traits with SPSS statistical software (version 15.0) (SPSS Inc., Chicago, Illinois). The phenotypic plasticity was quantitatively estimated by the slope of norm of reaction. Interaction between species and environments showed the difference of phenotypic plasticity between different species for each trait.

Regressions of reproductive biomass (*Y*) on vegetative biomass (*X*) were conducted to characterize the allometry of reproduction in different environmental conditions. The allometric relationship can be described by $\log y = \log b + a * \log x$, where *a* is the scaling exponent (slope) and *b* is the allometric coefficient or "scaling factor" (*y* intercept). Differences in shifts of the slope and in elevation of slopes (*y*-intercept) were assessed using standardized major axis regression (SMA, also known as reduced major axis, RMA) using the SMATR package in R software [41,53]. Standardized major axis regression (SMA) is commonly required for allometric studies [54,55]. SMA analyses are appropriate for summarizing the relationship between two variables in terms of a single slope [56]. In SMATR heterogeneity between SMA slopes is tested via a permutation test. Differences in SMA slope, elevation (intercept) and plant size are estimated [55].

Results

Phenotypic plasticity of plant morphological traits

Four way factorial ANOVA showed that morphological traits varied significantly between the four species and among treatments (all $P<0.001$, Table 1). All morphological traits increased with high soil nutrient and water availability for the four species (all $P<0.001$, Table 1). Only the number of secondary branches for individual species was significantly affected by population density, and decreased with increasing population density (Table 1; Fig. 1). Interactions between species and environmental factors (S×N, S×W and S×D) varied across morphological traits, which imply shifts in plasticity.

The *C. acuminatum* height was 65.8 cm, which was highest among the four species in the high soil nutrient treatment ($P=0.05$ level). *S. collina* height significantly increased 0.71 times (the phenotypic plasticity index was 0.42) in response to soil nutrients ($P<0.001$). The significant interactions between species and soil nutrients showed that the plasticity of *S. collina* was greater with higher nutrients than *A. squarrosum* and *C. macrocarpum* (both $P<0001$, Fig. 1). In response to soil water, *C. acuminatum* (PPI = 0.26) had the same plasticity as *A. squarrosum* (PPI = 0.26) ($P=0.925$), but had higher plasticity than *C. macrocarpum* (PPI = 0.17) ($P<0001$). In the high soil water treatment, there was no significant difference in height performance between *S. collina* and *A. squarrosum*, but the value of *S. collina* was significantly higher than *C. macrocarpum*. Although the interaction between species and population density was not significant, *C. acuminatum* (PPI = 0.09) and *S. collina* (PPI = 0.08) both had higher plasticity in height than *C. macrocarpum* (PPI = 0.02) by comparing between two species ($P=0.033$ and 0.048, respectively). The heights of *C. acuminatum* and *S. collina* were higher than *C. macrocarpum*, but not higher than *A. squarrosum* in the high population density (Fig. 1).

The numbers of primary branches in high nutrient soil for *A. squarrosum* (PPI = 0.58) and *S. collina* (PPI = 0.53) were higher than *C. macrocarpum* (PPI = 0.37) and *C. acuminatum* (PPI = 0.40), as was the plasticity in response to soil nutrients (all $P<0.001$). The increase in numbers of primary branches for *A. squarrosum* (PPI = 0.25) and *C. acuminatum* (PPI = 0.20) was significant in response to soil water ($P=0.002$ and 0.006, respectively), while the numbers of primary branches of *C. acuminatum* and *S. collina* were not higher than *A. squarrosum*, but higher than *C. macrocarpum*. However, there was no difference in plasticity of plant height among four forbs in response to population density ($P=0.480$).

S. collina had a higher number of secondary branches and higher plasticity than the other three species, ($P=0.006$, <0.001 and <0.001, respectively for plasticity). The number of secondary branches of *C. acuminatum* in high soil nutrients was not higher than *C. macrocarpum*, but higher than *A. squarrosum*. The plasticity of number of secondary branches for the two late successional species increased (by 27.5 and 25.6 branches respectively) in response to soil water compared to *A. squarrosum* ($P=0.020$ and 0.044, respectively). The decreases in number of secondary branches of all four species were significant in response to density ($P=0.004$, 0.011, 0.031 and 0.002, respectively), but the value of *S. collina* was higher than both late successional species, and value of *C. acuminatum* was higher than *A. squarrosum* (Fig. 1).

Plasticity of biomass traits and reproductive effort

The total biomass was significantly affected by all three environmental factors as well as their interactions except for W×D. Total biomass was significantly affected by S×N and S×W interactions (Table 1). The biomass of *S. collina* was 8.96 g in high soil nutrients and increased 2.63 times compared to low nutrients (Fig. 2). In response to soil nutrients, the phenotypic plasticity indices of *A. squarrosum*, *C. macrocarpum*, *C. acuminatum* and *S. collina* were 0.82, 0.35, 0.52 and 0.72, respectively. Except for similar plasticity between *C. acuminatum* and *C. macrocarpum* ($P=0.177$), the late successional species were more plastic in response to soil nutrients than both early successional species (Fig. 2). Biomass values in high soil water and plasticity of late successional species (the phenotypic plasticity indices of *A. squarrosum*, *C. macrocarpum*, *C. acuminatum* and *S. collina* were 0.60, 0.34, 0.62 and 0.50, respectively.) were higher than early successional species (Fig. 2). The plasticity of total biomass was not significantly different among four forbs in response to population density except between *S. collina* and *A. squarrosum* ($P=0.005$). In high population density, the value of *S. collina* was higher than the other three species, and the value of *C. acuminatum* was not significantly higher than early successional species ($P>0.05$ level) (Fig. 2).

Reproductive effort of late successional species was not always higher than early successional species (Fig. 2). Only the interaction between species and soil nutrients on reproductive effort was significant (Table 1). The values of *C. acuminatum* and *C. macrocarpum* were highest, the value of *S. collina* was intermediate, and the value of *A. squarrosum* was lowest across all high level treatments ($P=0.05$). Reproductive biomass was positively correlated with vegetative biomass in all groups (Table 2, Fig. 3). Individual SMA (standardized major axis) slopes varied significantly among four species. When combining the same successional status species, individual SMA slopes varied significantly between the two successional status groups ($P<0.001$) (Table 2, Fig. 3a). When comparing individual species, soil nutrient and soil water treatments had a significant affect on the individual SMA slopes or *y*-intercepts. However the population density treatment only affected the plant size (shift along on the common slope), and did not have an effect on the slopes or *y*-intercepts (Table 3). Fig. 3(b, c, d) showed the result of combining all species in a common treatment. Individual SMA slopes varied significantly between the two soil water treatments ($P=0.030$) (Table 2, Fig. 3c). Individual SMA slopes did not vary significantly between the two soil nutrient groups ($P=0.195$) (Table 2, Fig. 3b). There was a significant difference in *y*-intercepts among groups ($P<0.001$), as

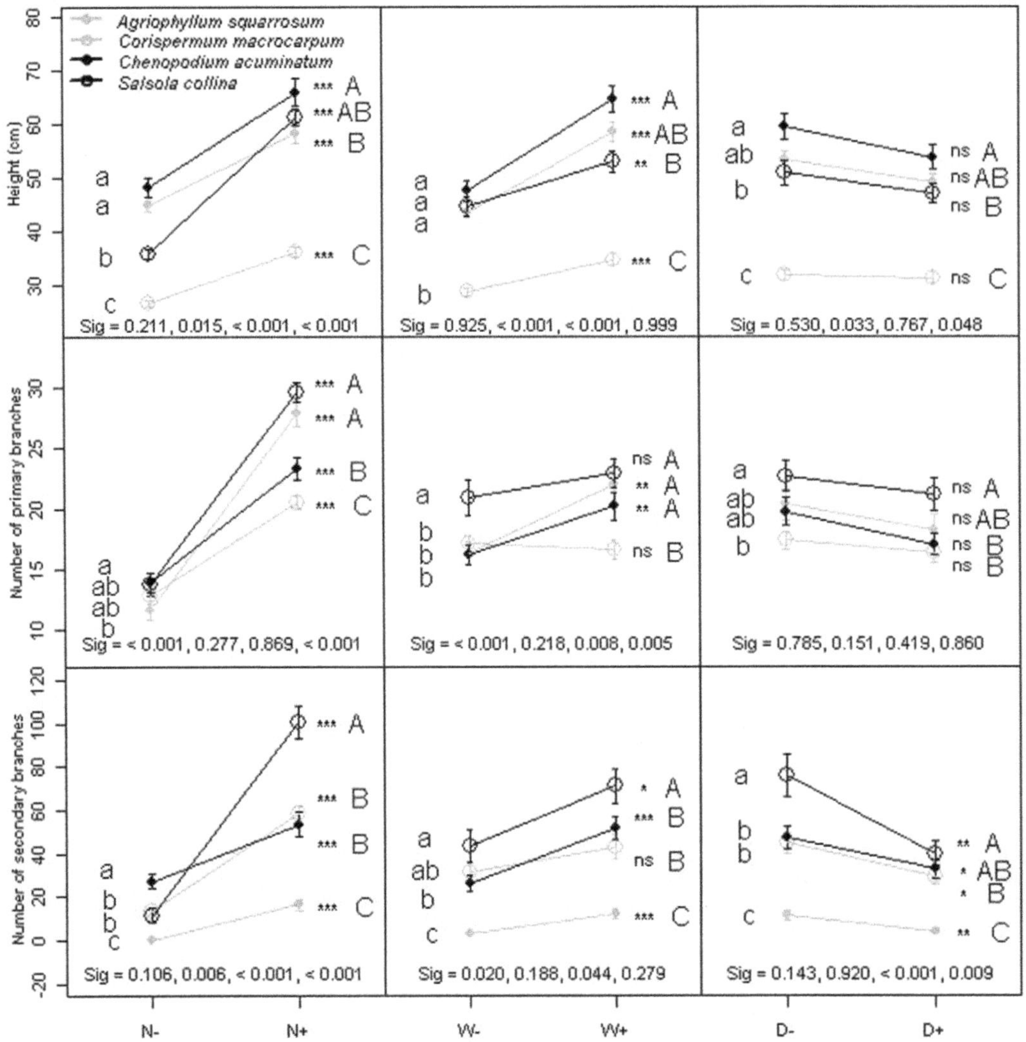

Figure 1. Norm of reaction for four species at two soil nutrients (N), water (W) and population density (D) levels on morphological traits; + and − indicate two levels of factor; The data represent means ± SD of average across treatments; ns denote not significance ($P>0.05$), asterisks denote significance at $P<0.05$ (*), $P<0.01$ (**) and $P<0.001$ (***) for each species in different level environmental treatments; the same lower case letters within columns are not significantly different at $P<0.05$ in low level environmental treatment for four species; the same capital letters within columns are not significantly different at $P<0.05$ in high level environmental treatments for four species. Sig, the *P* value of interactions between traits and environmental factors when compared between *C. acuminatum* and early succession species (*A. squarrosum* and *C. macrocarpum*) and between *S. collina* and early succession species (*A. squarrosum* and *C. macrocarpum*).

well as significant differences in group shifts along a common SMA ($P<0.001$), indicating that shifts in vegetative biomass resulted in associated shifts in reproductive biomass. High nutrient plants had higher reproductive biomass and vegetative biomass than low nutrient plants (Fig. 3b). Individual SMA slopes did not vary significantly between two population densities ($P=0.715$) (Table 2). There was no significant difference in *y*-intercepts among the SMA slopes ($P=0.390$), allowing a common SMA to be fitted to examine group shifts. Significant differences in population density shifts were evident along the common SMA ($P<0.001$), with low population density plants having higher reproductive biomass and vegetative biomass than high population density (Table 2, Fig. 3d).

Discussion

As with previous studies [57–60], the response of morphological traits to environmental variation was investigated here. Our experiment clearly demonstrated that both early and late successional species exhibited significantly taller plants, more branches and more biomass under increased nutrients and water and decreased density [52,61,62].

Late successional sand dunes have higher soil water holding capacity and higher soil moisture, and they also have higher nutrients [51,63]. The late successional species had greater growth responses to higher soil nutrients and water. The height of the two late successional species in the high resource treatments was not significantly higher than *A. squarrosum*. It may be related to the habitat of *A. squarrosum*, which is a pioneer species on the mobile dune, where the competition of intra- and interspecific competi-

Table 1. Analysis of variance for plant traits of four species under soil nutrients, water, and population densities.

	d.f.	Height	Number of primary branches	Number of secondary branches	Total biomass	Reproductive effort
Species (S)	3	160.51***	14.53***	73.40***	20.86***	470.42***
Nutrients (N)	1	366.64***	498.54***	342.69***	236.08***	350.69***
Water (W)	1	168.53***	20.52***	48.37***	124.90***	64.25***
Density (D)	1	23.82***	15.36***	66.54***	79.48***	0.01
S * N	3	14.03***	16.10***	43.50***	14.62***	3.62*
S * W	3	10.4***	5.86***	2.02	3.18*	1.27
S * D	3	2.10	0.83	6.14***	2.51	1.13
N * W	1	42.23***	8.72**	11.52***	72.69***	11.31***
N * D	1	0.26	0.52	45.71***	29.06***	0.12
W * D	1	0.80	0.25	0.34	1.85	0.96
S * N * W	3	6.47***	5.90***	3.83**	2.59	6.55***
S * N * D	3	6.75***	1.11	14.13***	8.39***	0.79
S * W * D	3	1.34	2.86*	2.76*	0.67	1.61
N * W * D	1	2.41	0.27	1.79	7.43**	0.11
S * N * W * D	3	10.30***	3.50*	3.53*	4.32**	2.88*

F-values are presented.
*refers to significance at $P<.05$;
**to significance at $P<.01$;
***to significance at $P<.001$.

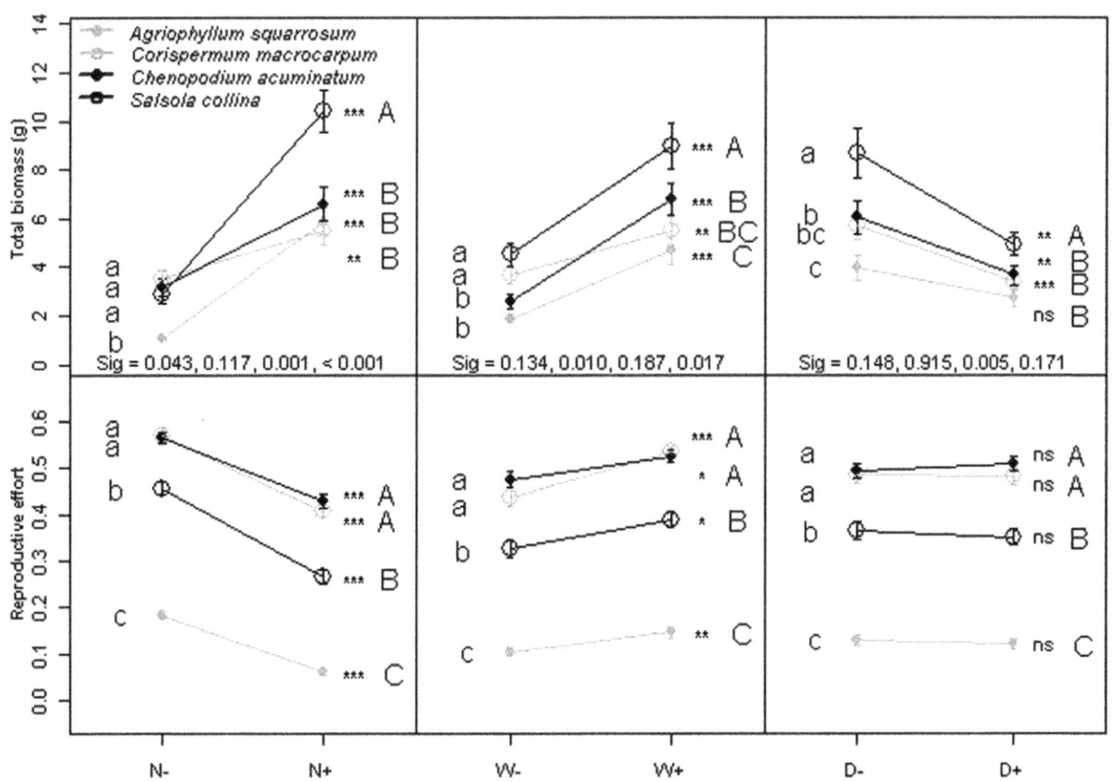

Figure 2. Norm of reaction for four species at two soil nutrients (N), water (W) and population density (D) levels on total biomass and reproductive effort of four species; + and − indicate two levels of factor. The data represent means ± SD of average across treatments; Asterisks denote significance at $P>0.05$ (ns), $P<0.05$ (*), $P<0.01$ (**) and $P<0.001$ (***) for each species in different level environmental treatment; the same lower case letters within columns are not significantly different at $P<0.05$ in low level environmental treatments for four species; Same capital letters within columns are not significantly different at $P<0.05$ in high level environmental treatment for four species. Sig, the *P* value of interactions of between total biomass and environmental factor when compared between *C. acuminatum* and early succession species (*A. squarrosum* and *C. macrocarpum*) and between *S. collina* and early succession species (*A. squarrosum* and *C. macrocarpum*).

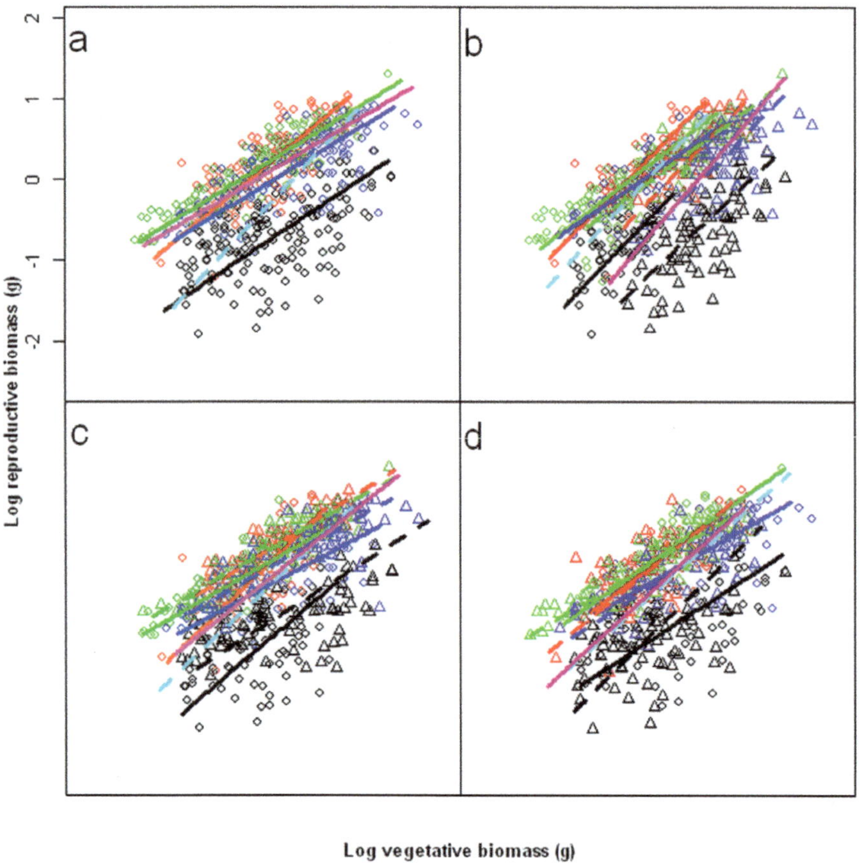

Figure 3. Log10-log10 plots showing the relationship between reproductive biomass and vegetative biomass of (a) *A. squarrosum* (black), *C. macrocarpum* (red), *C. acuminatum* (green) and *S. collina* (blue); and SMA fit line for combining all early successional species (cyan dashed line) and combining all late successional species (magenta solid line). For b (soil nutrient treatments), c (soil water treatments), and d (population density treatments), low environmental level (circle points and solid line) and high soil nutrient treatment level (triangle points and dashed line), black, red, green and blue, respectively for *A. squarrosum*, *C. macrocarpum*, *C. acuminatum* and *S. collina*, and SMA fit line for combining all low environmental levels (cyan dashed line) and combining all high environmental levels (magenta solid line).

tion is weak [61,64,65]. *C. macrocarpum* had lower values for plasticity than late successional species, except for secondary branch production. On the whole, late successional species had higher values, which indicates they exhibit the traits of a Jack-of-all-trades, showed better performance in favorable environments. They also fit another idealized scenario Master-of-some, as they are better able to increase fitness in favorable environments, than early successional species in response to environmental factors [66].

Table 2. Results of standardized major axis regression (SMA) analysis of pairwise combinations of reproductive biomass and vegetative biomass for combining all species in different treatments.

Group	Level	*n*	r^2	*p*	Slope	Intercept	Slope homogeneity (*P*)	Shift in elevation (*P*)	Shift along slope (*P*)
Succession	Early	280	0.161	**<0.001**	1.490	−0.651	**<0.001**	**<0.001**	**<0.001**
	Late	268	0.627	**<0.001**	0.819	−0.092			
Nutrient	Low	278	0.434	**<0.001**	1.480	−0.138	0.195	**<0.001**	**<0.001**
	High	270	0.182	**<0.001**	1.623	−0.884			
Water	Low	271	0.205	**<0.001**	1.308	−0.455	**0.030**	0.072	**<0.001**
	High	277	0.276	**<0.001**	1.112	−0.320			
Density	Low	266	0.297	**<0.001**	1.222	−0.423	0.715	0.390	**<0.001**
	High	282	0.236	**<0.001**	1.256	−0.388			

Significant results (P<0.05) are shown in bold.

Table 3. Results of standardized major axis regression (SMA) analysis of pairwise combinations of reproductive biomass and vegetative biomass for each species in different treatments.

Species	Nutrient			Water			Density		
	P1	*P2*	*P3*	*P1*	*P2*	*P3*	*P1*	*P2*	*P3*
A. squarrosum	0.369	**<0.001**	**<0.001**	**<0.001**	**0.002**	**<0.001**	0.744	0.552	**0.016**
C. macrocarpum	0.223	**<0.001**	**<0.001**	**<0.001**	**<0.001**	0.004	0.133	0.589	**<0.001**
C. acuminatum	0.097	**<0.001**	**<0.001**	0.225	**<0.001**	**<0.001**	0.223	0.763	**0.002**
S. collina	**<0.001**	**<0.001**	**<0.001**	0.739	**<0.001**	**<0.001**	0.232	0.430	**0.015**

P1 is the test of slope homogeneity, *P2* is the test of shift in elevation, *P3* is the test of Shift along slope. Significant results ($P<0.05$) are shown in bold.

Thus, the average performance of both late successional species was always higher than both early successional species, and the effect of interactions between succession and environmental factors also were significant. Based on our study, during the succession of degraded sandy sites, late successional species were better able to increase fitness for some traits or maintain fitness for other traits. This may be related to the higher resource levels of the stabilized dunes. The results of this study are in agreement with the first hypothesis; patterns of invasive species' plasticity can be applied to succession.

An earlier study [30] indicated that the reproductive effort of exotic species was lower than that of the native plants in all treatments, which conflicts with other reports [27,67] that invasive species exhibit higher reproductive effort. The reason for this inconsistency could be because species [67] or development time [68]. Stastny et al (2006) were compared between populations of the same species, and the same patterns may not hold true between different species, since each species will have its own intrinsic regulation of reproductive effort [30]. Because under high resource levels plants may postpone reproduction, increased resource levels can give the impression of reduced reproductive effort if plants do not have time to complete their life cycles [68]. More investigations on different species of plants, with completed reproduction, are therefore necessary in order to fully understand the mechanism of reproductive effort. However, in a successional study, the late successional species often had high reproductive effort [69]. The mid-late successional species allocated most resources to reproductive effort in order to continue the population [70]. In this study, late successional species have higher reproductive effort than *A. squarrosum.* However the ecological amplitude of *C. macrocarpum* is broad, it produces a lot of seeds to colonize in mobile, semi- mobile and semi-fixed dunes. The average performance showed that late successional species had higher reproductive effort, but at the species level, the second hypothesis was not true, late successional species did not have higher reproductive effort than early successional species. This may be related to the fact that *C. macrocarpum* can also be an early successional species.

Reproductive effort can vary with different environmental factors [71], and either increase [33] or decrease [72] in response to an increasing plant density. In the present study, density had no effect on biomass allocation and allometry of reproductive biomass. The plants exhibited similar biomass allocation ratios across all densities. Reproductive effort greatly depends on nutrients and water. The scaling exponent (slope) between vegetative and reproductive biomass was consistent, but the allometric coefficient and plant size varied with different soil nutrients. So under high soil nutrient, plants allocated less biomass to reproductive organs than to other organs. Besides plant size, soil water had significant effects on the scaling exponent, which was the allometric coefficient between vegetative and reproductive biomass in this study, so with high soil water availability, plants allocated less biomass to other organs than to reproductive organs at the same plant size. Previous reports showed that plants increase reproduction in response to environmental stress [33–35], or that plants have similar levels of reproductive output in resource-poor environments. This means that the ability of plants to maintain fecundity at low resource levels or to "make the best of a bad job" [73]. In contrast, our study supports the hypothesis that plants do not always increase reproductive effort for any environmental stressors. Subtraction of nutrients can increase reproductive effort, while addition of water can increase reproductive effort, and altering density may not affect reproductive effort or the allometry of reproductive biomass.

In conclusion, the plasticity patterns were related to successional status. For late successional species, it was a complex of Master-of-some and Jack-of-all-trades. Late successional species had higher performance in high soil nutrients and water or higher plasticity in response to soil nutrients and water than early successional species. Our results show that reproductive effort for late successional species was not always higher than for each early successional species. In contrast to the effect of water, the addition of nutrients lead to a decline in reproductive effort, while density did not have any effect on reproductive effort. Early successional species were in general less plastic in their response, perhaps a reflection of their adaptation to resource-poor early successional soil.

Acknowledgments

We are particularly grateful to Edith B. Allen, Guangdi Li, Bridget Hilbig, Justin Valliere, Matt ONeill, and Alexandria Pivovaroff for their highly relevant comments on the manuscript.

Author Contributions

Conceived and designed the experiments: YH XZ. Performed the experiments: YH. Analyzed the data: YH HZ WZ. Contributed reagents/materials/analysis tools: YH. Wrote the paper: YH DZ.

References

1. Bradshaw AD (1965) Evolutionary significance of phenotypic plasticity in plants. Advances in Genetics 13: 115–155.
2. Strand JA, Weisner SEB (2004) Phenotypic plasticity - contrasting species specific traits induced by identical environmental constraints. New Phytologist 163: 449–451.

3. Pigliucci M, Murren CJ, Schlichting CD (2006) Phenotypic plasticity and evolution by genetic assimilation. Journal of Experimental Biology 209: 2362–2367.
4. Bouvet JM, Vigneron P, Saya A (2005) Phenotypic plasticity of growth trajectory and ontogenic allometry in response to density for eucalyptus hybrid clones and families. Annals of Botany 96: 811–821.
5. Agrawal AA (2001) Ecology - Phenotypic plasticity in the interactions and evolution of species. Science 294: 321–326.
6. Dudley SA, Schmitt J (1996) Testing the adaptive plasticity hypothesis: Density-dependent selection on manipulated stem length in Impatiens capensis. American Naturalist 147: 445–465.
7. Dorken ME, Barrett SCH (2004) Phenotypic plasticity of vegetative and reproductive traits in monoecious and dioecious populations of *Sagittaria latifolia* (Alismataceae): a clonal aquatic plant. Journal of Ecology 92: 32–44.
8. Lee CE (2002) Evolutionary genetics of invasive species. Trends in Ecology & Evolution 17: 386–391.
9. McDowell SCL (2002) Photosynthetic characteristics of invasive and non-invasive species of *Rubus* (Rosaceae). American Journal of Botany 89: 1431–1438.
10. Durand LZ, Goldstein G (2001) Photosynthesis, photoinhibition, and nitrogen use efficiency in native and invasive tree ferns in Hawaii. Oecologia 126: 345–354.
11. Williams DG, Mack RN, Black RA (1995) Ecophysiology of introduced *Pennisetum setaceum* on Hawaii: the role of phenotypic plasticity. Ecology 76: 1569–1580.
12. Muth N, Pigliucci M (2007) Implementation of a novel framework for assessing species plasticity in biological invasions: responses of *Centaurea* and *Crepis* to phosphorus and water availability. Journal of Ecology 95: 1001–1013.
13. Lande R (2009) Adaptation to an extraordinary environment by evolution of phenotypic plasticity and genetic assimilation. Journal of Evolutionary Biology 22: 1435–1446.
14. Funk JL (2008) Differences in plasticity between invasive and native plants from a low resource environment. Journal of Ecology 96: 1162–1173.
15. Portes MT, Damineli DSC, Ribeiro RV, Monteiro JAF, Souza GM (2010) Evidence of higher photosynthetic plasticity in the early successional *Guazuma ulmifolia* Lam. compared to the late successional *Hymenaea courbaril* L. grown in contrasting light environments. Brazilian Journal of Biology 70: 75–83.
16. Haeberli W, Beniston M (1998) Climate change and its impacts on glaciers and permafrost in the Alps. Ambio 27: 258–265.
17. Canham CD (1988) Growth and canopy architecture of shade-tolerant trees: response to canopy gaps. Ecology 69: 786–795.
18. Williams H, Messier C, Kneeshaw DD (1999) Effects of light availability and sapling size on the growth and crown morphology of understory Douglas-fir and lodgepole pine. Canadian Journal of Forest Research 29: 222–231.
19. Johnson HA, Biondini ME (2001) Root morphological plasticity and nitrogen uptake of 59 plant species from the Great Plains grasslands, U.S.A. Basic and Applied Ecology 2: 127–143.
20. Coleman JS, McConnaughay KDM (1995) A non-functional interpretation of a classical optimal-partitioning example. Functional Ecology 9: 951–954.
21. Wang T, Zhou D, Wang P, Zhang H (2006) Size-dependent reproductive effort in *Amaranthus retroflexus*: The influence of planting density and sowing date. Canadian Journal of Botany/Revue Canadienne de Botanique 84: 485–492.
22. Kindscher K, Wells PV (1995) Prairie plant guilds: a multivariate analysis of prairie species based on ecological and morphological traits. Plant Ecology 117: 29–50.
23. Bazzaz FA, Chiariello NR, Coley PD, Pitelka LF (1987) Allocating resources to reproduction and defense. BioScience 37: 58–67.
24. Weiner J (1988) The influence of competition on plant reproduction; Lovett DJ, editor. New York: Oxford University. 228–245 p.
25. Hartnett DC (1990) Size dependent allocation to sexual and vegetative reproduction in four clonal composites. Oecologia 84: 254–259.
26. Ophélie Ronce, Isabelle Olivieri (1997) Evolution of reproductive effort in a Metapopulation with local extinctions and ecological succession. The American Naturalist 150: 220–249.
27. Baker HG (1974) The evolution of weeds. Annual Review of Ecology and Systematics 5: 1–24.
28. Bazzaz FA (1986) Life history of colonizing plants: Some demographic, genetic, and physiological features. In: Mooney HA, Drake JA, editors. Ecology of biological invasions of North America and Hawaii. New York.: Springer. pp. 96–110.
29. Jiang M, Kadono Y (2001) Growth and reproductive characteristics of an aquatic macrophyte *Ottelia alismoides* (L.) Pers. (Hydrocharitaceae). Ecological Research 16: 687–695.
30. Sans FX, Garcia - Serrano H, Afán I (2004) Life-history traits of alien and native senecio species in the Mediterranean region. Acta Oecologica 26: 167–178.
31. Throop HL (2005) Nitrogen deposition and herbivory affect biomass production and allocation in an annual plant. Oikos 111: 91–100.
32. Hawkes CV (2007) Are invaders moving targets? The generality and persistence of advantages in size, reproduction, and enemy release in invasive plant species with time since introduction. American Naturalist 170: 832–843.
33. Hickman JC (1977) Energy allocation and niche differentiation in four co-existing annual species of *Polygonum* in western North America. Journal of Ecology 65: 317–326.
34. Li B, Shibuya T, Yogo Y, Hara T, Yokozawa M (2001) Interclonal differences, plasticity and trade-offs of life history traits of *Cyperus esulentus* in relation to water availability. Plant Species Biology 16: 193–207.
35. Wu KK, Jain SK (1979) Population regulation in *Bromus rubens* and *B. mollis*: Life cycle components and competition. Oecologia 39: 337–357.
36. Sultan SE, Bazzaz FA (1993) Phenotypic plasticity in *Polygonum persicaria*. III. The evolution of ecological breadth for nutrient environment. Evolution 47: 1050–1071.
37. Sultan SE, Bazzaz FA (1993) Phenotypic plasticity in *Polygonum persicaria*. II. Norms of reaction to soil moisture, ecological breadth, and the maintenance of genetic diversity. Evolution 47: 1032–1049.
38. Voils S, Mendlinger SL, Ward D (2002) Differentiation in populations of Hordeum spontaneum Koch along a gradient of environmental productivity and predictability: plasticity in response to water and nutrient stress. Biological Journal of the Linnean Society 75: 301–312.
39. Meekins JF, Mccarthy BC (2000) Responses of the biennial forest herb *Alliaria petiolata* to variation in population density, nutrient addition and light availability. Journal of Ecology 88: 447–463.
40. Huang YX, Zhao XY, Zhang HX, Huang G, Luo YY, et al. (2009) A comparison of phenotypic plasticity between two species occupying different positions in a successional sequence. Ecological Research 24: 1335–1344.
41. Falster DS, Warton DI, Wright IJ SMATR: standardised major axis tests and routines, ver 2.0. Available: http://www.bio.mq.edu.au/ecology/SMATR/. Accessed 2006.
42. Su YZ, Zhao Hl (2003) Soil properties and plant species in an age sequence of Caragana microphylla plantations in the Horqin Sandy Land, north China. Ecological Engineering 20: 223–235.
43. Zuo X, Zhao H, Zhao X, Zhang T, Guo Y, et al. (2008) Spatial pattern and heterogeneity of soil properties in sand dunes under grazing and restoration in Horqin Sandy Land, Northern China. Soil and Tillage Research 99: 202–212.
44. Li F-R, Zhao L-Y, Zhang H, Zhang T-H, Shirato Y (2004) Wind erosion and airborne dust deposition in farmland during spring in the Horqin Sandy Land of eastern Inner Mongolia, China. Soil and Tillage Research 75: 121–130.
45. Zhao WZ, Xiao HL, Liu ZM, Li J (2005) Soil degradation and restoration as affected by land use change in the semiarid Bashang area, northern China. Catena 59: 173–186.
46. Peng Y, Jiang GM, Liu MZ, Niu SL, Yu SL, et al. (2005) Potentials for combating desertification in Hunshandak Sandland through nature reserve. Environmental Management 35: 453–460.
47. Li FR, Kang LF, Zhang H, Zhao LY, Shirato Y, et al. (2005) Changes in intensity of wind erosion at different stages of degradation development in grasslands of Inner Mongolia, China. Journal of Arid Environments 62: 567–585.
48. Su YZ, Zhao HL, Zhao WZ, Zhang TH (2004) Fractal features of soil particle size distribution and the implication for indicating desertification. Geoderma 122: 43–49.
49. Zuo X, Zhao H, Zhao X, Guo Y, Yun J, et al. (2009) Vegetation pattern variation, soil degradation and their relationship along a grassland desertification gradient in Horqin Sandy Land, northern China. Environmental Geology 58: 1227–1237.
50. Su YZ, Zhang TH, Li YL, Wang F (2005) Changes in Soil Properties After Establishment of *Artemisia halodendron* and *Caragana microphylla* on Shifting Sand Dunes in Semiarid Horqin Sandy Land, Northern China Environmental Management 36: 272–281.
51. Li F-R, Zhao W-Z, Liu J-L, Huang Z-G (2009) Degraded vegetation and wind erosion influence soil carbon, nitrogen and phosphorus accumulation in sandy grasslands. Plant and Soil 317: 79–92.
52. Huang Y, Zhao X, Zhou D, Luo Y, Mao W (2010) Allometry of *Corispermum macrocarpum* in response to soil nutrient, water, and population density. Botany-Botanique 88: 13–19.
53. Warton DI, Wright IJ, Falster DS, Westoby M (2006) Bivariate line-fitting methods for allometry. Biological Reviews 91: 259–291.
54. Niklas KJ (1994) Plant Allometry: The Scaling of Form and Process. Chicago: University of Chicago Press.
55. Wright IJ, Reich PB, Cornelissen JHC, Falster DS, Garnier E, et al. (2005) Assessing the generality of global leaf trait relationships. New Phytologist 166: 485–496.
56. Wright IJ, Reich PB, Cornelissen JHC, Falster DS, Groom PK, et al. (2005) Modulation of leaf economic traits and trait relationships by climate. Global Ecology And Biogeography 14: 411–421.
57. Fichtner K, Schulze ED (1992) The effect of nitrogen nutrition on growth and biomass partitioning of annual plants originating from habitats of different nitrogen availability. Oecologia 92: 236–241.
58. Elberse IAM, van Damme JMM, van Tienderen P (2003) Plasticity of growth characteristics in wild barley (*Hordeum spontaneum*) in response to nutrient limitation. Journal of Ecology 91: 371–382.
59. Jennifer MD, Jill AJ, Marcus TB, Cynthia W (2007) Constraints on the evolution of adaptive plasticity: costs of plasticity to density are expressed in segregating progenies. New Phytologist 176: 874–882.
60. Zhou D, Wang T, valentine I (2005) Phenotypic plasticity of life-history characters in response to different germination timing in two annual weeds. Canadian Journal of Botany 83: 28–36.

61. Huang Y, Zhao X, Zhang H, Japhet W, Zuo X, et al. (2009) Allometric effects of *Agriophyllum squarrosum* in response to soil nutrients, water, and population density in the Horqin Sandy Land of China. Journal of Plant Biology 52: 210–219.
62. Huang Y, Zhao X, Zhou D, Zhao H, Zhang H, et al. (2009) Allometry of *Salsola collina* in response to soil nutrients, water supply and population density. Nordic Journal Of Botany 27: 539–547.
63. Guo K, Dong X, Liu Z (2000) Characteristics of soil moisture content on sand dunes in Mu Us Sandy Grassland: why *artemisia ordosicadeclines* on old fixed sand dunes. Acta Phytoecologica Sinica 24: 275–279.
64. Nemoto M, Lu X (1992) Ecological characteristics of *Agriophyllum squarrosum* a pioneer annual on sand dunes in eastern Inner Mongolia, China. Ecological Research 7: 183–186.
65. Liu Z, Yan Q, Liu B, Ma J, Luo Y (2007) Persistent soil seed bank in Agriophyllum squarrosum (Chenopodiaceae) in a deep sand profile: Variation along a transect of an active sand dune. Journal of Arid Environments 71: 236–242.
66. Richards CL, Bossdorf O, Muth NZ, Gurevitch J, Pigliucci M (2006) Jack of all trades, master of some? On the role of phenotypic plasticity in plant invasions. Ecology Letters 9: 981–993.
67. Stastny M, Schaffner U, Elle E (2005) Do vigour of introduced populations and escape from specialist herbivores contribute to invasiveness? Journal of Ecology 93: 27–37.
68. Weiner J, Campbell LG, Pino J, Echarte L (2009) The allometry of reproduction within plant populations. Journal of Ecology 97: 1220–1233.
69. Abrahamson WG, Gadgil MD (1973) Growth form and reproductive effort in goldenrods (*Solidago*, Compositae). American Naturalist 107: 651–661.
70. Scheiner SM (1989) Variable selection along a successional gradient. Evolution 43: 548–562.
71. van Kleunen M, Fischer M, Schmid B (2001) Effects of intraspecific competition on size variation and reproductive allocation in a clonal plant. Oikos 94: 515–524.
72. Snell TW, Burch DG (1975) The effects of density on resource partitioning in *Chamaesyce hirta* (Euphorbiaceae). Ecology Letters 56: 742–746.
73. Sultan SE (2001) Phenotypic plasticity for fitness components in Polygonum species of contrasting ecological breadth. Ecology 82: 328–343.

13

Soil Eukaryotic Microorganism Succession as Affected by Continuous Cropping of Peanut - Pathogenic and Beneficial Fungi were Selected

Mingna Chen, Xiao Li, Qingli Yang, Xiaoyuan Chi, Lijuan Pan, Na Chen, Zhen Yang, Tong Wang, Mian Wang, Shanlin Yu*

Shandong Peanut Research Institute, Qingdao, China

Abstract

Peanut is an important oil crop worldwide and shows considerable adaptability but growth and yield are negatively affected by continuous cropping. Soil micro-organisms are efficient bio-indicators of soil quality and plant health and are critical to the sustainability of soil-based ecosystem function and to successful plant growth. In this study, 18S rRNA gene clone library analyses were employed to study the succession progress of soil eukaryotic micro-organisms under continuous peanut cultivation. Eight libraries were constructed for peanut over three continuous cropping cycles and its representative growth stages. Cluster analyses indicated that soil micro-eukaryotic assemblages obtained from the same peanut cropping cycle were similar, regardless of growth period. Six eukaryotic groups were found and fungi predominated in all libraries. The fungal populations showed significant dynamic change and overall diversity increased over time under continuous peanut cropping. The abundance and/or diversity of clones affiliated with *Eurotiales*, *Hypocreales*, *Glomerales*, *Orbiliales*, *Mucorales* and *Tremellales* showed an increasing trend with continuous cropping but clones affiliated with *Agaricales*, *Cantharellales*, *Pezizales* and *Pyxidiophorales* decreased in abundance and/or diversity over time. The current data, along with data from previous studies, demonstrated that the soil microbial community was affected by continuous cropping, in particular, the pathogenic and beneficial fungi that were positively selected over time, which is commonplace in agro-ecosystems. The trend towards an increase in fungal pathogens and simplification of the beneficial fungal community could be important factors contributing to the decline in peanut growth and yield over many years of continuous cropping.

Editor: Boris Alexander Vinatzer, Virginia Tech, United States of America

Funding: This study was supported by grants from China Agriculture Research System (CARS-14), the National Natural Science Foundation of China (31000728; 31100205), the Natural Science Fund of Shangdong Province (ZR2009DQ004; ZR2011CQ036), the Promotive Research Fund for Young and Middle-aged Scientisits of Shandong Province (BS2010NY023), Qingdao Municipal Science and Technology Plan Project (11-2-4-9-(3)-jch; 11-2-3-26-nsh). The funders had no role in study design, data collection and analysis, decision to publish, or preparation of the manuscript.

Competing Interests: The authors have declared that no competing interests exist.

* E-mail: yshanlin1956@163.com

Introduction

Changes in micro-organism populations in soils are critical to the sustainability of soil-based ecosystem function because of their involvement in such key processes as mineral nutrition cycling, organic matter turnover,soil structure formation and toxin removal or accumulation [1,2]. The productivity and sustainability of agricultural soil have a substantial affect on plant growth and final crop quality and both are significantly affected by the diversity and abundance of soil micro-organisms [3,4].

Microbial community structures in soils are affected by the treatment or management of the soil. Amendment with fertilizer [5], application of pesticides [6], different tillage regimes [7] and biological amendments, such as the introduction of genetically modified micro-organisms [8,9], have all been shown to affect soil microbial community structures. Soil type, crop selection and cropping sequence are key determining factors on soil micro-organisms [10–12]. Due to their quick response to environmental changes, micro-organisms are seen as efficient bio-indicators of soil condition and land management success [13,14].

Soil micro-organisms are also related to soil borne plant diseases and plant disease suppression. Fungi (true fungi and oomycetes) and nematodes are major causes of soil borne diseases [15] and considerable research attention has been paid to potential bio-control agents against plant fungal pathogens in recent years [16–18]. Many fungal populations and bacterial groups have been identified as antagonists of soil borne plant pathogens and play important roles in promoting plant growth [19–21]. Interestingly, it has been shown that several bacterial groups suppress soil borne plant pathogens through synergistic relationships with arbuscular mycorrhiza fungi [22,23]. The mechanisms by which these micro-organisms suppress plant disease can be divided into several categories: nutrient competition, amensalism, microbial antagonism, parasitism and systemic induced resistance [10]. In addition, it has been suggested that fungal diversity also has a beneficial effect on plant pathogen suppression and plant productivity [24]. Because of the interaction between soil micro-organisms and soil borne plant disease, the assessment and management of soil microbial community structure are central to the successful suppression of plant pathogens [25].

Due to the relationships between micro-organisms, soil quality and plant health, active management of soil microbial communities such as the manipulation of microbial populations as linked to agricultural production [26] could be a promising approach to improving soil conditions, developing natural suppression of soil borne diseases and improving crop productivity. However, relatively little is known about the specific populations, characteristics, interactions, and relationships between plants and soil micro-organisms, especially the fungal plant pathogens and their antagonists.

Peanut, one of the four most important oil crops, can adapt to a wide range of climatic conditions, and grows in many different environments between latitudes 40°N and 40°S [27]. However, its productivity drops significantly when subjected to continuous cropping. In addition, the yield of other main world crops, such as wheat, rice, corn and soybean, also decline under continuous cropping [28–30]. The negative impact of continuous cropping on soil productivity has been shown to have a significant effect on soil microbial biomass and community structure [3,31,32]. However, this decline is not yet well understood. A comprehensive knowledge of specific inhabitants, characteristics, soil microbial succession and how soil micro-organisms interact with continuous cropping practices is still lacking. This is particularly true for eukaryotic soil micro-organisms. Sudini *et al.* [11] studied bacterial communities over two peanut cropping cycles, but there have been no relevant reports on eukaryotic micro-organisms to date.

This investigation is a first attempt to study the succession of eukaryotic soil micro-organisms under continuous peanut cultivation. This was achieved by using cultivation-independent 18S rRNA gene clone library analyses. The goals of this study were to: (i) determine the phylogenetic affiliations of the most common and predominant populations of soil microbes at various peanut growth stages under a continuous cropping regime; (ii) identify the diversity and succession patterns of eukaryotic micro-organisms, with particular emphasis on the fungi.

Materials and Methods

Management Description and Soil Sampling

The study was conducted between 2008 and 2010 at Shandong Peanut Research Institute in Qingdao, China (36.10°N, 120.41°E). Peanuts were planted in May each year and harvested in October. Pot culture experiments were undertaken with forty seedlings each year. The pot diameter was 0.45 meter. The soil consisted of 28.5% clay, 41% loam, and 30.5% sand, with no known history of peanut cropping. In order to reduce interference from outside factors on the soil microbial community and truly reflect the relationship between soil eukaryotic microbial succession and continuous cropping with peanut: two rain-proof plastic buildings with two opposite opening sides were built, the plants were irrigated with sterile water and no fertilizer was applied during the growing seasons and fallow periods.

Samples were collected at four different peanut growth stages in 2008: at the seedling stage, full-flowering stage, pod bearing stage and pod-maturing stage. In 2009 and 2010 samples were only collected at the seedling stage and pod-maturing stage. The soils were sampled using a soil probe (1.5 cm diameter) from individual plant at 5–10 cm soil depths, and at distances of 3–5 cm, 8–10 cm and 13–15 cm from the main root, respectively. These distances were chosen because roots at this depth and these distances were relatively abundant and the soil microbial community around the peanut root system could be well characterized. In addition, external interference due to the sampling of the under-surface layer soil could be reduced. At each stage, five randomly selected replicates with six cores in each (two cores for each distance) were pooled together to form a homogenous sample. The samples were stored at −80°C and were used for eukaryotic micro-organism analysis.

Genomic DNA Extraction and 18S rRNA Gene Library Construction

DNA was extracted from the pooled material for each sample with Ultra-Clean soil DNA kits (MoBio, USA). The quality and quantity of the genomic DNA extracted were determined by electrophoresis on 0.8% (wt/vol) agarose gels. Primers, nu-SSU-0817 and nu-SSU-1536 [33], were used for 18S rRNA gene amplification. The reaction was performed for 25 cycles using the 'Easy-A High-Fidelity' PCR master mix from Stratagene (USA). Products from six PCRs were pooled, purified, using a Gel Extraction kit (Omega, USA), and ligated into pMD19-T simple vectors (Takara, Japan). The hybrid vectors were cloned into *Escherichia coli* TOP10. Clone libraries were constructed by random selection of approximately 110 white colonies from a single plate for each library. All clone libraries were named on the basis of the plant growth stage in the crop cycle and time of soil sampling. 08S, 08F, 08B and 08M were libraries constructed respectively for the seedling stage, full-flowering stage, pod-bearing stage and pod-maturing stage in 2008; 09S and 09M were libraries constructed respectively for the seedling stage and pod-maturing stage in 2009 and 10S and 10M were libraries constructed respectively for the seedling stage and pod-maturing stage in 2010.

18S rRNA Gene Sequencing and Data Analysis

Cloned 18S rRNA gene fragments were re-amplified by using primers RV-M (5′-GAGCGGATAACAATTTCACACAGG-3′) and M13-D (5′-AGGGTTTTCCCAGTCACGACG-3′) that flanked the insertion site of the cloning vector. All positive clones were sequenced to improve data accuracy and integrity.

Operational taxonomic units (OTUs) in each library were defined by using the DOTUR program at the 3% distance level [34]. Chimeric sequences were identified as described by Berney *et al.* [35]. The library coverage was calculated as follows: $[1-(n1/N)]\times 100$, where n1 is the number of unique OTUs (frequency, 1) detected in the library and N is the total number of clones in the same library [36]. The value obtained approximately corresponded to the species' coverage level in a given sample. Sampling efficiency and the sequence phylotype diversity of the libraries were analyzed using the rarefaction curve method and the aRarefactWin program [36].

Similarities between the soil eukaryotic assemblages at different stages of peanut growth during different crop cycles were established by performing cluster analyses on the constructed libraries using the software package SPSS 11.5 (SPSS, United States). The frequency (F) for the OTUs was calculated as follows: $F=(m/N)\times 100$, where m is the clone number of an OTU in a library and N is the total number of clones in the same library. Log-transformed frequency data were used to minimize the influence of preferential PCR amplification [37].

The nearest phylogenetic neighbor sequence for each OTU sequence was determined in the GenBank database using the BLASTN program [38]. Sequences were aligned using ClustalX v.1.81 [39]. Phylogenetic trees were constructed using the DNADIST and NEIGHBOR programs in the PHYLIP package (version 3.65c) [40] and bootstrap support was evaluated using 100 replicates.

Nucleotide Sequence Accession Numbers

The 18S rRNA gene sequences determined have been deposited in the GenBank database under accession numbers JQ320555 - JQ321347.

Results

Statistical and Comparative Analyses for the Eight Libraries

Eight 18S rRNA gene libraries were constructed over three continuous cropping cycles and representative growth stages of peanut. A total of 841 clones were screened, and 802 of them contained a proper insert. A total of 125 OTUs were identified, nine of which (10 clones) were probably chimeras, based on the chimera-check results. The library coverage ranged from 77.8 to 89.7% (Table 1). Rarefaction analysis confirmed the OTU phylotype diversity of the soil eukaryotic micro-organisms (Figure S1). The analysis results showed that the 116 valid OTUs accounted for most of the microbial eukaryote diversity present in the soil samples studied.

Phylotype diversity increased during continuous cropping with peanut. The OTU number ranged from 25 to 29 in the four libraries for the first cropping cycle; was 35 and 33, respectively, in the second cycle libraries and was 38 and 34, respectively, in the third cycle libraries (Table 1). In addition, clone libraries obtained from the same cycle but from different growth stages typically shared 10 to 22 common OTUs, accounting for 34.5 to 64.7% of the OTUs in each library (Table 1). The dynamics of the microbial eukaryotic community were further confirmed by cluster analyses. The hierarchical (Figure 1) and k-means (data not shown) clustering results were consistent and showed that libraries obtained from different growing stages, but from the same cropping cycle, were grouped together. The four clone libraries (S1, F1, B1, M1) obtained from the first cycle were grouped into a class, the two libraries (S2, M2) obtained from the second cycle were grouped into a second class, and the two libraries (S3, M3) from the third cycle were grouped into a third class. These results demonstrated that the eukaryotic soil microbial populations changed when peanut was grown continually, regardless of growth stage.

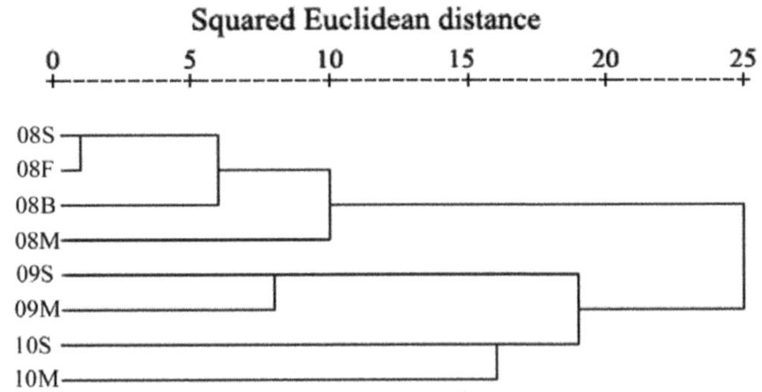

Figure 1. Clustering analysis of the 18S rRNA gene clone libraries. Dendrogram based on a hierarchical clustering analysis of the 18S rRNA gene clone libraries, constructed using the squared Euclidean distance similarity and Ward linkage procedures.

18S rRNA Gene Sequence Analyses

All the 116 valid OTUs were used for sequence analysis, most (77.6%) of which showed high levels of similarity (≥97% identity) with known sequences in the GenBank database and 7.0% were moderately related (≤94% identity) to known sequences. Phylogenetic analysis showed that 95 of the OTUs were affiliated with fungi, which predominated in all the libraries, accounting for 86.9 to 94.1% of obtained clones in each library (Table S1). The remaining 21 OTUs were related to five groups: *Alveolata*, *Metazoa*, *Rhizaria*, *Stramenopiles*, and *Viridiplantae* (Table S1). The phylotype diversity of fungi increased under continuous cropping. OTU numbers ranged from 19 to 25 for the four measurements in the first cropping cycle libraries, 29 and 28 in the second cycle libraries and increasing to 30 and 29 in the third cycle libraries. Apart from the population dynamics of the predominant fungi, the *Stramenopiles* and *Rhizaria* also showed succession changes with continuous cropping. OTUs affiliated with *Stramenopiles* only occurred in the four libraries of the first cropping cycle. Corresponding sequences had 92 to 97% similarity to known GenBank sequences and were moderately related to genera of *Nerada* and *Rhizidiomyces*. However, only OTUs related to *Rhizaria* occurred in the four libraries covering the 2009 and 2010 continuous cropping cycles. Sequences for the second and third cropping cycles had 92 to 98% similarity to known sequences and were moderately affiliated

Table 1. Analysis of the clone libraries constructed over three continuous cropping cycles and representative growing stages of peanut.

Library	No.of clones	No. of total OTUs	No. of unique OTUs	% Coverage	No. of OTUs shared by libraries from same cropping cycle	% shared OTUs
08S	99	29	12	87.9	10	34.5
08F	98	29	14	85.7		34.5
08B	100	25	13	87.0		38.5
08M	97	26	10	89.7		38.5
09S	99	35	22	77.8	18	51.4
09M	96	33	19	80.2		54.5
10S	103	38	20	80.1	22	57.9
10M	101	34	22	78.2		64.7

The coverage values were calculated as follows: $[1-(n1/N)]\times100$, where n1 is the number of unique OTUs (frequency, 1) detected in the library and N is the total number of clones in the same library.

with genera of *Cercomonas*, *Lecythium*, *Paracercomonas* and *Rhogostoma*, all of whom belonged to *Cercozoa*.

Four fungal groups were identified from the clone libraries, including the following phyla or divisions: *Ascomycota*, *Basidiomycota*, *Fungi incertae sedis* and *Glomeromycota* (Figure 2). The phylum *Ascomycota* dominated all the libraries accounting for 47.2% of the total 793 clones (excluding the 10 chimeric clones) and showed the greatest diversity, accounting for 38.8% of the 116 valid OTUs. *Basidiomycota* were also relatively abundant, accounting for 30.5% of the obtained valid clones and were also highly diverse, accounting for 21.6% of OTUs analyzed. The *Fungi incertae sedis* group and *Glomeromycota* phylum were not so abundant and diverse, accounting for 8.8% and 4.8%, respectively, of the 793 clones analyzed and each accounted for 13.8 and 7.8%, respectively, of the valid OTUs characterized. However, they both showed an increase in abundance and diversity as the number of cropping cycles increased.

1) Sequences of the *Ascomycota*

Forty five OTUs (47.2% of the total clones) were affiliated to *Ascomycota*, which predominated in all the libraries and showed the greatest diversity. Seven classes were affiliated with the *Ascomycota* phylum and these were: *Dothideomycetes*, *Eurotiomycetes*, *Laboulbeniomycetes*, *Leotiomycetes*, *Orbiliomycetes*, *Pezizomycetes* and *Sordariomycetes*. There were 13 correlative orders or subdivisions presented in the libraries (Figure S2). The *Hypocreales* sequences were most abundant and diverse, accounting for 58.8% of the *Ascomycota* clones and for 31.1% of the corresponding OTUs. The *Eurotiales* and *Pezizales* sequences were relatively diverse, accounting for 6.7% and 11.3% of the OTUs affiliated with *Ascomycota*, respectively. Comparison of the three cropping cycle libraries indicated that the OTU diversity and/or abundance of the three orders, *Eurotiales*, *Hypocreales* and *Orbiliales*, increased over the period of continuous cropping, but the order *Pezizales* and *Pyxidiophorales* decreased significantly in diversity and/or abundance (Figure 3 and Figure S2).

The *Hypocreales* sequences predominated in all the analyzed libraries, accounting for 24.1, 29.7, and 32.8%, respectively, of the the first, second and third cycle library clones (Figure 2). The sequences also showed great diversity, accounting for 12.1% of the 116 OTUs analyzed. These sequences had a high similarity level to known sequences (95–99%) and were related to genera of *Acremonium*, *Bionectria*, *Fusarium*, *Geosmithia*, *Paecilomyces*, *Trichoderma*, *Cordyceps* and an uncultured *Hypocreales* genus. The most common

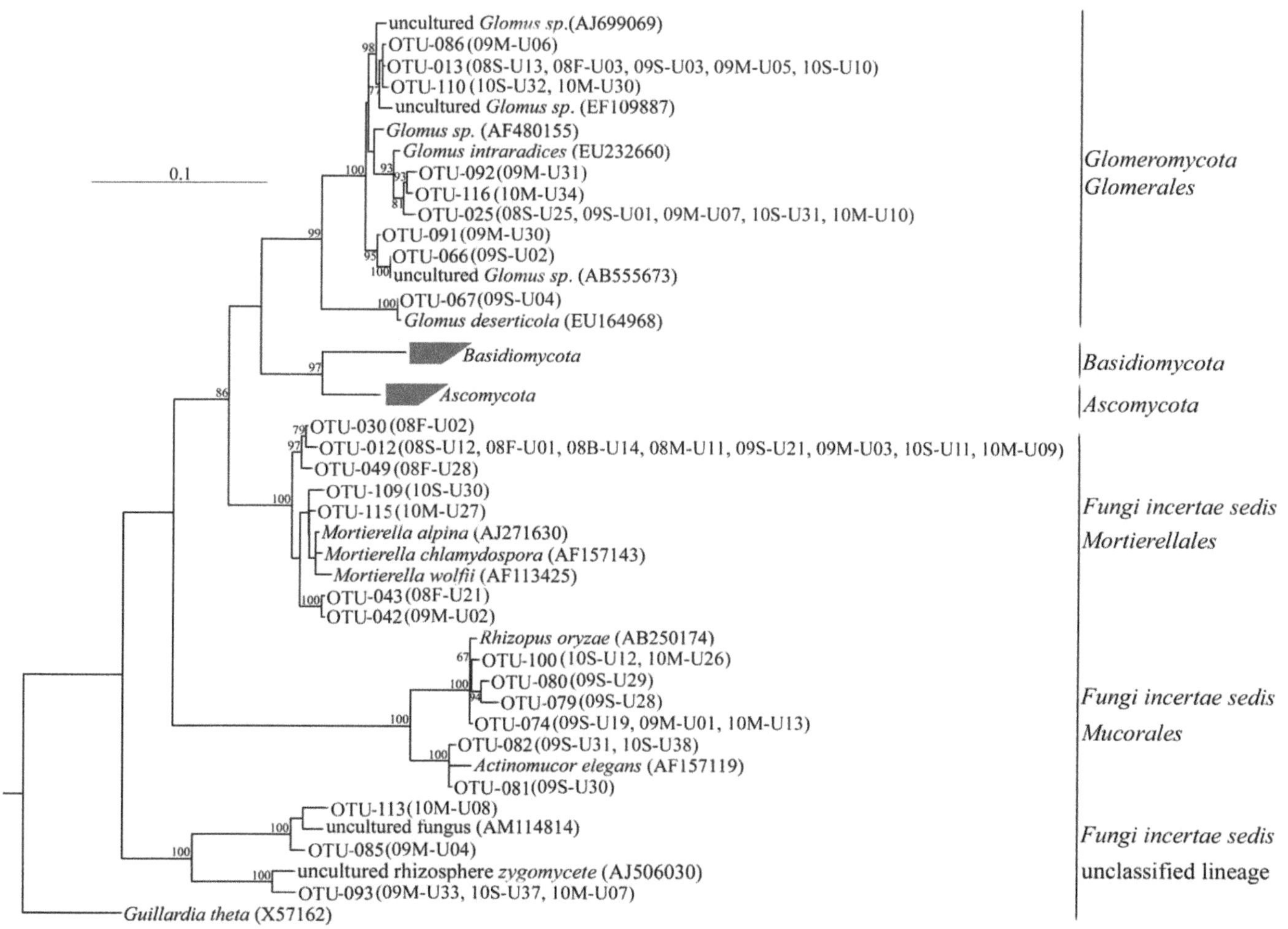

Figure 2. Phylogenetic tree of the *Fungi* sequences. Phylogenetic tree of the *Fungi* sequences recovered from the eight 18S rRNA gene clone libraries, constructed using the neighbor-joining method with the Kimura two-parameter model for nucleotide change. The *Ascomycota* and *Basidiomycota* which were just denoted the phylogenetic positions in this tree, presented in other separate phylogenetic trees. The libraries OTUs occurred were labeled. Scale bar, 0.1 substitutions per nucleotide position. Bootstrap values (100 replicates) above 60% are indicated at the nodes. The tree was rooted using the sequence related to *Guillardia theta* (X57162).

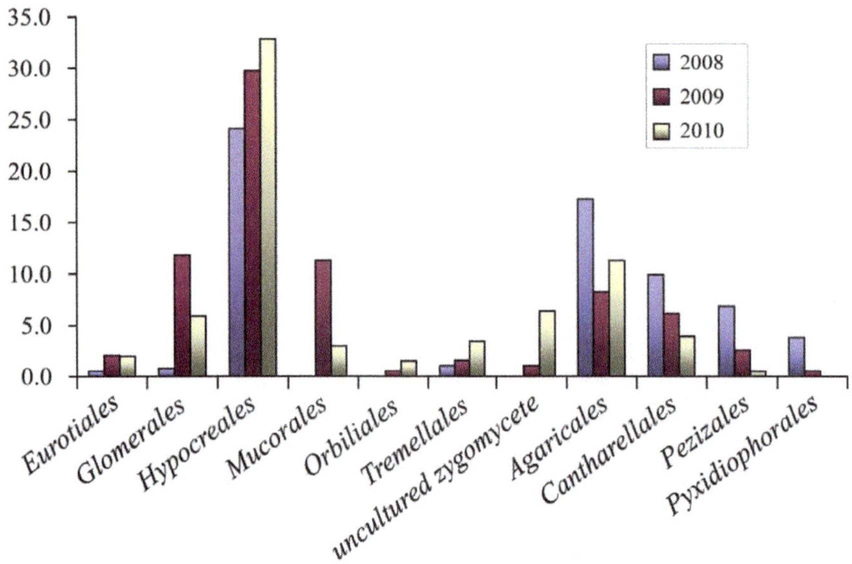

Figure 3. The clone abundance analyses of the fungal orders or phylotypes. The clone abundance analyses of the fungal orders or phylotypes which showed succession changes in populations with continuous cropping cycles. (The clone abundance was calculated as: ($m1/M \times 100$, where $m1$ is the number of related clones detected in the libraries for the same cropping cycle and M is the total number of clones in the same libraries.).

and abundant *Hypocreales* OTUs belonged to *Fusarium*, which was represented in all eight libraries and accounted for 88.2% of the 220 *Hypocreales* clones detected in the libraries. Clone abundance accounted for 22.1, 26.7 and 27.0% of the first, second and third cycle libraries, respectively. Five OTUs were related to *Bionectria*, and accounted for 0.5, 2.1 and 4.4%, respectively, of the clones obtained in the three cycle libraries. OTU-001 affiliated with *Fusarium solani* with a 98% similarity and was common in all clone libraries. The other OTUs related to *Hypocreales* only occurred in one or two of the libraries. OTUs related to *Paecilomyces* and *Acremonium* were only found in the first cycle libraries; OTUs related to the uncultured *Hypocreales* genus, were found in the second and third cycle libraries, but not in the first cycle libraries, and OTUs related to *Geosmithia* and *Trichoderma* were only found in the library S2 and M3, respectively.

Three and two OTUs, respectively, were related to the order of *Eurotiales* and *Orbiliales*. The analysis indicated that the two order phylotype diversity both increased with continuous peanut cropping, despite of low abundance. Only one OUT related to *Eurotiales* occurred in the first cycle libraries, and two OTUs occurred in the second and third cycle libraries, respectively. OUT related to *Orbiliales* only were found in the second and third cycle libraries, but not found in the first cycle libraries. The clone abundance of the two orders also showed the same tendency. The *Eurotiales* clones accounted for 0.5, 2.1, and 2.1% of the first, second and third cycle library clones, respectively. These sequences had 98 to 99% similarity to known sequences, including genera of *Aspergillus* and *Penicillium*. *Penicillium* clones (OTU-077) occurred in the libraries of the second and third continuous cropping cycles, but not in the first cycle libraries. The *Orbiliales* clones accounted for 0, 0.5 and 1.5% of the first, second and third cycle libraries. The sequences had 97 to 98% similarity to known sequences, including genera of *Orbilia* and an uncultured *Orbiliaceae* clone.

Six OTUs (33 clones) were related to the order *Pezizales*. Both the diversity and abundance of these clones decreased over time with continuous peanut cropping. Four OTUs were found in the first cycle libraries, two OTUs were found in the second cycle libraries and only one OTU was found in the third cycle libraries, which only occurred in the peanut initial growth library S3 and not in the pod-maturing stage library M3. In addition, the *Pezizales* clones accounted for 6.8, 2.6, and 0.5% of clones in the first, second and third cycle libraries, respectively. These sequences had 95–99% similarity to known Genbank sequences and were related to four genera: *Chalazion*, *Marcelleina*, *Peziza* and *Sarcosphaera*. The *Chalazion* clones were present in all the first cycle libraries, but were not found in any second and third cycle libraries. Clones related to *Peziza* occurred in five libraries, which were the four first cycle libraries and the initial growth library of the second cycle (library S2). The *Sarcosphaera* clones only occurred in the S1 library. In contrast, clones related to *Marcelleina* only occurred in the second and third cycle libraries and only in libraries for the initial growth stages S2 and S3.

Only one OTU (16 clones) belonged to the order *Pyxidiophorales* and this was 98% related to *Pyxidiophora arvernensis*. The results showed that this OTU mainly appeared in the four first cycle libraries and only one clone was found in the M2 library.

2) Sequences from the *Basidiomycota*

The *Basidiomycota* OTUs were relatively abundant and diverse in the libraries, with 25 OTUs accounting for 30.5% of the 793 valid clones that were affiliated. These OTUs were affiliated with *Agaricomycetes* and *Tremellomycetes*. The *Agaricomycetes* OTUs made up the majority of the members of the *Basidiomycota* detected in the libraries, accounting for 88.0% of the related OTUs and 94.2% of the related clones. Four orders or phylotypes were present: *Agaricales*, *Cantharellales*, *Geastrales* and *Sebacinales* (Figure S3). OTUs related to the class *Tremellomycetes* detected in the libraries were only affiliated with the order *Tremellales*. When the three cycle libraries were compared, it was found that the phylotypes of *Agaricales*, *Cantharellales* and *Tremellales* showed the great dynamic population changes (Figure3 and Figure S3).

The *Agaricales* sequences were the most dominant *Basidiomycota* phylotype in the libraries, accounting for 44.0% of the 25 OTUs and 44.2% of the 242 related clones. *Agaricales* clone abundance significantly decreased in the second cycle libraries compared to the first cycle libraries but there were no significant differences between the second and third cycle libraries. The *Agaricales* clones accounted for 17.3% of the clones in the first cycle libraries but only 8.2, and 11.3% of the clones in the second and third cycle libraries, respectively. Clone diversity also showed a similar tendency. There were 10 OTUs identified in the first cycle libraries but only three OTUs identified in the second and third cycle libraries, respectively. These sequences had high similarity to known sequences (96–99%) and were affiliated with the genera *Camarophyllopsis*, *Coprinus*, *Hemimycena*, *Panaeolus* and *Psathyrella*. OTU-016 (50 clones), affiliated with the genus *Psathyrella*, was the most common and abundant OTU, appearing in all eight libraries. OTUs from the remaining genera showed considerable dynamic population changes and occurred in between one and five libraries. OUT-028 related to *Camarophyllopsis* only occurred in the library S1; OTUs related to *Panaeolus* occurred in the first and third cycle libraries; OTUs related to *Coprinus* only occurred in library F1, B1 and S2 and OTUs related to *Hemimycena* occurred in most of the first cycle libraries, only occurred in the second library M2, but not occurred in any third library.

Only five OTUs were affiliated with the order *Cantharellales*, but were relatively abundant, accounting for 24.4% of the *Basidiomycota* clones. The clone abundance decreased significantly with continuous cropping, accounting for 9.9, 6.2 and 3.9% of the clones detected in the first, second and third cycle libraries, respectively. These sequences were associated with genus of *Ceratobasidium*, *Rhizoctonia* and *Thanatephorus* with 96 to 99% similarity. OTU-017 affiliated with *Thanatephorus* was common in all eight libraries. The other four OTUs only appeared in one or two libraries.

In contrast to the decreasing population trend shown by the *Agaricales* and *Cantharellales*, the *Tremellales* showed an increasing population trend with continuous cropping. Despite only three OTUs (14 clones) being related to *Tremellales*, clone abundance still showed a significant increase accounting for 1.0, 1.5, and 3.4% respectively, of the first, second and third cycle library clones. These OTUs were all affiliated with genus *Cryptococcus* with 99% similarity. OTU-26 was relatively common, appearing in six libraries except B1 and M3, but the other two OTUs only occurred in the third cycle libraries.

3) Sequences of the *Fungi Incertae Sedis*

Sixteen OTUs (70 clones), accounting for 13.8% of the 116 valid OTUs, were affiliated with the *Fungi incertae sedis* where there were two classes or subdivisions, *Mucoromycotina* and an unclassified lineage, present. The *Mucoromycotina* sequences were the predominant members of the *Fungi incertae sedis* detected in the libraries, accounting for 78.6% of the characterized clones. Two orders related to *Mucoromycotina* were identified, including *Mortierellales* and *Mucorales*. Three OTU (15 clones) were affiliated with an early diverging fungal lineage and were unclassified. Comparison analysis indicated that phylotypes related to *Fungi incertae sedis* changed significantly in the three libraries over the three continuous cropping cycles (Figure 1 and Figure 3).

The *Mortierellales* and *Mucorales* OTUs were relatively abundant, accounting for 43.8 and 37.5% of the *Fungi incertae sedis* OTUs and for 38.6 and 40.0%, respectively, of the 70 related clones. The *Mortierellales* OTUs were common in all eight libraries, and were all affiliated with the genus *Mortierella* with 98–99% similarity. Clone abundance showed no significant change over the three cycle libraries and only OUT-012 was common, the other six OTUs occurred in only one library. The *Mucorales* OTUs occurred in all four libraries covering the second and third cropping cycles but were not found in any first cycle library. The clone abundance rose and fell over the three year cropping period, accounting for 0, 11.3 and then 2.9% of the clones detected in the first, second and third cycle libraries, respectively. These sequences were affiliated with genera of *Actinomucor* and *Rhizopus* with 98–99% similarity. Three OTUs were affiliated with an early diverging fungal lineage and these were related to uncultured *zygomycete*. OUT-093 began to appear in the second cycle library at the pod-maturing stage (library M2). Clone abundance increased successively accounting for 1.0, 3.9 and 7.9% of the clones identified in libraries M2, S3 and M3, respectively. The other two OTUs only occurred in the library M2 and M3, respectively.

4) Glomeromycota Sequences

Nine OTUs (38 clones) were affiliated with the *Glomeromycota*, with 97–99% similarity to known GenBank sequences, and all sequences were related to the genus *Glomus* belonging to the order *Glomerales*. Analysis results indicated that clone diversity and abundance both increased significantly with continuous peanut cropping (Figure 1 and Figure 3). However, they slightly decreased in the third cycle libraries compared to the second cycle libraries. The OTU number was 2, 1, 4, 5, 3 and 3 in libraries S1, F1, S2, M2, S3 and M3, respectively, but they did not occur in the later growth stage libraries, B1 and M1. The clones accounted for 2.0, 1.0, 13.1, 10.4, 6.8 and 5.0%, respectively, of the clones detected in the corresponding libraries. This result indicated an increasing trend in related clone abundance with continuous peanut cropping but a decrease in peanut growth with each cropping cycle.

Discussion

In this study, 18S rRNA gene clone library analyses were undertaken to study the succession of soil eukaryotic micro-organisms in a soil under continuous peanut cultivation. In order to minimize the occurrence of artifacts and reduce the drawbacks of PCR-based rRNA analysis [41], homogenous samples were prepared and six PCR products were mixed for clone library construction. It is known that soil microbial communities are often difficult to fully characterize, mainly because of their considerable phenotypic and genotypic diversity as well as their heterogeneity [42]. In addition, most of them are generally unculturable also being a reason for that [43]. Based on the combination of rarefaction analysis and library coverage value calculations (ranging from 77.8 to 89.7%), these results show that most of the microbial eukaryotic phylotypes in the test soil, especially the common and dominant members, were accurately estimated using library analyses based on direct DNA extraction. Despite the disadvantages inherent in using this method [10], it does offer the possibility of assessing the total microbial diversity present by modulation.

Soil micro-organism communities are affected by a wide range of factors [44]. It has been reported that the treatment or management of soil, especially the soil type, the choice of plant species and cropping patterns, are the factors that most affect the microbial community structure in soil [10,12]. In order to truly reflect the relationship between soil eukaryotic microbial succession and continuous cropping with peanut, three measures were employed to reduce interference from other factors, as described in the material and methods. Cluster analyses indicated that soil microbial eukaryotic assemblages obtained from the same cropping cycle were similar, regardless of growth stage. Phylotype diversity was also shown to increase with continuous cropping. These findings indicated that dynamic changes in soil microbial eukaryotic communities do occur with continuous peanut cropping. Soil microbial biomass and structure has also been found to be significantly influenced by continuous cropping with cucumber, rice and pea [3,31,32]. Different rotation sequences also had significant effects on soil microbial community structures [9,32]. Therefore, it has been shown that microbial populations in soil are significantly selected by cropping patterns. However, the successional progress of specific microbial phylotypes and the interaction mechanisms between micro-organisms and plants need to be investigated further.

The data from this study showed that the fungal clones were the most common and dominant microbial eukaryotes in the soil during every growth stage in the three cropping cycles and the related phylotypes indicated significant dynamic changes in microbial populations with continuous cropping. It has also been reported that fungal biomass altered in continuously cropped rice fields [31] and arbuscular mycorrhizal fungi (AMF) declined in a continuous pea cropping regime relative to a pea–wheat rotation [3]. These results indicate that the soil fungal community was affected by continuous cropping and this may be a common phenomenon in agro-ecosystems.

Soil fungi have a high functional diversity. Some fungi improve soil quality and plant growth, such as AMF, by taking up nutrients and transferring them to the plants, thereby enhancing plant nutrition [45,46]. Some have been identified as plant pathogens and impair plant growth while some others are antagonistic against pathogens [24,47]. Consequently, it can be supposed that the changes in the fungal community with continuous cropping can disturb ecological function and balance, which may reduce growth over time.

Many fungal phylotypes identified in the libraries created by this study showed dynamic changes in populations with the continuous peanut cropping. The trends were complex and can be mainly divided into four types: trends showing a continuous increase or decrease; a trend where the population increases and then decreases or a trend where the population deceases and then increases. The complexity of microbial populations seen in this study raises many questions, such as what specific phylotypes were present? Why did populations change over time and what roles did the changing micro-organism population play during continuous cropping?

The abundance and/or diversity of clones affiliated with *Eurotiales*, *Hypocreales*, *Orbiliales* and *Tremellales* all increased with continuous cropping. The *Hypocreales* clones were common and predominant in all libraries and showed a high diversity. These diverse genera showed various trends during continuous cropping. *Fusarium* clones were abundant and ubiquitous and abundance increased in the second and third cropping cycles; Abundance of *Bionectria* clones also increased in the second and third cropping cycles. Clones related to an uncultured *Hypocreales* genus began to appear from the second cropping cycle but *Geosmithia* and *Trichoderma* clones did not appear until the third cycle. In contrast, clones related to genera *Acremonium* and *Paecilomyces* disappeared in the second cycle. Clones from the orders *Eurotiales* (genera *Aspergillus* and *Penicillium*) *Orbiliales* (genus *Orbilia*) and *Tremellales* (genus *Cryptococcus)* were not so diverse but their abundance increased with continuous cropping of peanut. Most of the related genera identified have been shown to be pathogenic to plants, including *Fusarium*, *Geosmithia*, *Acremonium*, *Aspergillus*, *Penicillium* and *Cryptococcus* [10,17,48–51]. In contrast, many species affiliated with *Trichoderma* and *Paecilomyces* have been reported to be suppressive of plant fungal pathogens [47,52,53]. The complex trends shown by these populations may imply complex interactions between them.

The relationships between pathogenic fungi and antagonists are complex. Competition, antagonism and hyperparasitism have been suggested as being possible interaction mechanisms [24,54]. Beneficial micro-organisms have a wide range of mechanisms that inhibit the growth or activity of pathogenic micro-organisms. However, pathogens also have an array of mechanisms to counteract antagonists. An example of pathogen-antagonist interaction has been described for *Fusarium* and *Trichoderma asperellum*, *Trichoderma asperellum* isolates were suppressive to *Fusarium* wilt of tomato, but the mycotoxin, deoxynivalenol, produced by *Fusarium* species, acted as a negative signal repressing the expression of the nag1 chitinase gene in *Trichoderma atroviridae* [55,56]. So the pathogen-antagonist interaction is mutual.

The dynamic population changes in *Mucorales* clones further suggests the complexity in interactions between pathogens and antagonists. Members of the order *Mucorales* have been identified as plant pathogens [57] and their related clones showed a relatively complex succession tendency. They were not found in any first cycle libraries; the abundance ratio rose to 11.3% in the second cycle libraries but then decreased to 2.9% in the third cycle libraries. This result, combined with the upper tendency analyses of pathogen related clones, all indicated that soil fungal pathogens increased under a continuous cropping regime, but the succession models were non-single.

The increase in fungal pathogens in the soil may result in more intense competition by some antagonistic fungi. *Glomerales* species are extensive root colonizers, and the genus *Glomus,* belonging to the AMF, has been shown to have the ability to protect host plants against pathogens [58,59]. The abundance and diversity of *Glomus* clones increased significantly with continuous cropping in the first and second cycles but slightly decreased in the third cycle libraries. But clone abundance showed a decreasing trend with each subsequent cropping cycle. Analyses using clones related to pathogen antagonists predict a more complex dynamic succession of beneficial micro-organisms in soil under a continuous cropping regime.

Based on these analyses, it can be supposed that soil pathogens and antagonists first existed in a balanced equilibrium but under the stress of a continuous peanut cropping regime, the original pathogenic micro-organisms became more abundant, and as the soil micro-environment changed, new pathogens appeared and some original antagonists decreased or disappeared. At the same time, the original antagonistic inhabitants increased and very few new antagonist members appeared. This meant that few pathogen populations decreased or disappeared. However, in general, populations of fungal pathogens increased under continuous cropping.

Numerous diseases and symptoms are displayed by plants infected with soil borne fungal pathogens. However, these diseases are difficult to diagnose, because most of the symptoms occur below ground [60]. Consequently, the relationships between soil borne diseases and continuous cropping have been little reported but other recent studies have confirmed the results produced by this investigation. Root rot is the primary disease caused by soil borne pathogens [54]. It has been demonstrated that under a continuous pea, regime, root abundance was reduced two-fold compared to a pea–wheat rotation and *Fusarium* root rot was more severe [3]. This was consistent with the current analyses that *Fusarium* clone abundance increased with continuous peanut cropping. In addition, continuous cropping also increased the incidence of potato diseases, such as black scurf, stem canker and common scab [61], again implying an increase in plant pathogen populations, which is also consistent with the results from this study. These results suggest that the increasing trend in fungal pathogens related to *Eurotiales*, *Hypocreales*, *Mucorales* and *Tremellales* could be an important factor limiting the growth of peanut under a continuous cropping regime.

In contrast to the increase in the population of clones related to pathogens, the abundance and diversity of clones affiliated with *Agaricales* and *Pezizales* decreased under continuous peanut cropping. The *Agaricales* and *Pezizales* identified in the libraries were diverse. Most of the related genera, including *Chalazion, Peziza, Sarcosphaera, Camarophyllopsis, Coprinus* and *Hemimycena,* mainly occurred in the libraries from the first cropping cycle. The *Panaeolus* clone succession was unusual in that they existed in the first cycle, disappeared in the second cycle, and re-appeared again in the third cycle. The orders *Agaricales* and *Pezizales* have been reported to be beneficial fungi [62,63]. Despite that, their roles in soil quality and plant growth are not well understood. Based on these fungi, it can be suggested that beneficial fungi decreased under continuous cropping and the recurrence of a few members like *Panaeolus* may be due to the recovery ability of the soil ecosystem. Simplification of the beneficial fungal community may also contribute to a decline in peanut growth under continuous cropping.

The results have highlighted the pathogenic and beneficial fungi that were significantly selected by continuous cropping and the overall trends that took place. There were dynamic population changes in the phylotypes present in the libraries. The *Cantharellales* and *Pyxidiophorales* clones identified were not so diverse, but were ubiquitous in every cropping cycle and continuously decreased over time. In addition, some clones were related to unclassified *Fungi incertae sedis* lineages. None of these clones existed until the late stage of the second cropping cycle and the clone abundance successively increased. Despite their unknown function, they must be connected with continuous cropping stress and should not be dismissed. They could be suppressed in more balanced environments but are able to thrive in environments influenced by intensive continuous cropping. Many phylotypes that decreased continuously, such as *Agaricales, Pezizales, Cantharellales* and *Pyxidiophorales,* could reflect genotype homogenization in the soil. The environmental and genetic uniformity of the agricultural ecosystem may promote the emergence of new pathogens [64], so these unclassified fungi that appeared after intensive monocropping, may be related to new pathogenic species.

The current study, along with previous studies, demonstrated that the soil microbial community was affected by continuous cropping; especially with regards to the pathogenic and beneficial fungi that were significantly selected for, and this process could be common in agro-ecosystems. The fungal populations were the most predominant and common members of the soil micro-eukaryotic populations and showed significant dynamic changes under continuous cropping, especially the pathogenic and beneficial fungi. The increase in the fungal pathogens and simplification of the beneficial fungal community could be important factors contributing to the decline in peanut growth under continuous cropping.

It is the first step toward fully understanding the microbe-soil-peanut interaction under continuous cropping stress. The goal of this kind of study is to manipulate or alter the soil microbial characteristics through various management practices that increase soil microbial activity, diversity, populations of plant-beneficial organisms, and pathogen antagonists, resulting in suppression of the negative factors that bring about the decline in plant growth under a continuous cropping regime. There are many related areas of research that need further investigation. For example, how do pathogenic and beneficial fungi interact with the soil, the plant and with one another? What roles do soil bacteria play in the microbial community? What management practices should be adopted so that the soil microbial characteristics can be beneficially altered or manipulated and which plant growth stages would most benefit from any changes?

Supporting Information

Figure S1 Rarefaction curves for the 18S rRNA gene libraries constructed from each of the soil samples.

Figure S2 Phylogenetic tree representing affiliations of the 18S rDNA sequences related to the *Ascomycota* phylum.

Figure S3 Phylogenetic tree representing affiliations of the 18S rDNA sequences related to the *Basidiomycota* phylum.

Table S1 The abundance and diversity analyses of clones affiliated with six eukaryotic groups determined in libraries.

Author Contributions

Conceived and designed the experiments: SY MC. Performed the experiments: MC XL QY XC. Analyzed the data: MC XL LP NC. Contributed reagents/materials/analysis tools: ZY TW MW. Wrote the paper: MC SY XL.

References

1. Brussaard L, De Ruiter PC, Brown GG (2007) Soil biodiversity for agricultural sustainability. Agr Ecosyst Environ 121(3): 233–244.
2. Van Elsas JD, Jansson JK, Trevors JT (2006) Modern soil microbiology, second ed. CRC Press, Boca Raton: W. B. Whitman.
3. Nayyar A, Hamel C, Lafond G, Gossen BD, Hanson K, et al (2009) Soil microbial quality associated with yield reduction in continuous-pea. Appl Soil Ecol 43(1): 115–121.
4. Pankhurst CE, Ophel-Keller K, Doube BM, Gupta VVSR (1996) Biodiversity of soil microbial communities in agricultural systems. Biodivers Conserv 5(2): 197–209.
5. Jangid K, Williams MA, Franzluebbers AJ, Sanderlin JS, Reeves JH, et al (2008) Relative impacts of land-use, management intensity and fertilization upon soil microbial community structure in agricultural systems. Soil Biol Biochem 40(11): 2843–2853.
6. Klose S, Acosta-Martinez V, Ajwa HA (2006) Microbial community composition and enzyme activities in a sandy loam soil after fumigation with methyl bromide or alternative biocides. Soil Biol Biochem 38: 1243–1254.
7. Caesar T, Lenssen AW, Caesar AJ, Sainju UM, Gaskin JF (2010) Effects of tillage on microbial populations associated to soil aggregation in dryland spring wheat system. Eur J Soil Biol 46: 119–127.
8. De Leij FAAM, Sutton SJ, Whipps JM, Fenlon JS, Lynch JM (1995) Impact of field release of genetically modified *Pseudomonas fluorescens* on indigenous microbial population of wheat. Appl Environ Microbiol 61(9): 3443–3453.
9. Larkin RP (2008) Relative effects of biological amendments and crop rotations on soil microbial communities and soilborne diseases of potato. Soil Biol Biochem 40(6): 1341–1351.
10. Garbeva P, Van Veen JA, Van Elsas JD (2004) Microbial diversity in soil: Selection of microbial populations by plant and soil type and implications for disease suppressiveness. Annu Rev Phytopathol 42: 243–270.
11. Sudini H, Liles MR, Arias CR, Bowen KL, Huettel RN (2011) Exploring soil bacterial communities in different peanut-cropping sequences using multiple molecular approaches. Phytopathology 101(7): 819–827.
12. Wieland G, Neumann R, Backhaus H (2001) Variation of microbial communities in soil, rhizosphere, and rhizoplane in response to crop species, soil type, and crop development. Appl Environ Microbiol 67(12): 5849–5854.

13. Avidano L, Gamalero E, Cossa GP, Carraro E (2005) Characterization of soil health in an Italian polluted site by using microorganisms as bioindicators. Appl Soil Ecol 30: 21–33.
14. Kandeler E, Mosier A, Morgan J, Milchunas D, King J, et al (2006) Response of soil microbial biomass and enzyme activities to the transient elevation of carbon dioxide in a semi-arid grassland. Soil Biol Biochem 38: 2448-2460.
15. Schardl CL, Craven KD (2003) Interspecific hybridization in plant-associated fungi and oomycetes: a review. Mol Ecol 12(11): 2861–2873.
16. Heydari A, Pessarakli M (2010) A review on biological control of fungal plant pathogens using microbial antagonists. J Biol Sci 10: 273–290.
17. Pliego C, Ramos C, De Vicente A, Cazorla FM (2011) Screening for candidate bacterial biocontrol agents against soilborne fungal plant pathogens. Plant Soil 340: 505–520.
18. Wehner J, Antunes PM, Powell JR, Caruso T, Rillig MC (2011) Indigenous arbuscular mycorrhizal fungal assemblages protect grassland host plants from pathogens. PLoS ONE 6(11): e27381.
19. Blakney AJ,PattenCL (2011) A plant growth-promoting *pseudomonad* is closely related to the *Pseudomonas syringae* complex of plant pathogens. FEMS Microbiol Ecol 77(3): 546–557.
20. Ownley BH, Gwinn KD, Vega FE (2010) Endophytic fungal entomopathogens with activity against plant pathogens: ecology and evolution. Biocontrol 55(1): 113–128.
21. Gao G, Yin D, Chen S, Xia F, Yang J, et al (2012) Effect of biocontrol agent *Pseudomonas fluorescens* 2p24 on soil fungal community in cucumber rhizosphere using T-RFLP and DGGE. PLoS ONE 7(2): e31806.
22. Akhtar MS, Siddiqui ZA (2008) Biocontrol of a root-rot disease complex of chickpea by *Glomus intraradices*, *Rhizobium sp.* and *Pseudomonas straita*. Crop Prot 27: 410–417.
23. Budi SW, Van Tuinen D, Martinotti G, Gianinazzi S (1999) Isolation from the *Sorghum bicolor* mycorrhizosphere of a bacterium compatible with Arbuscular Mycorrhiza development and antagonistic towards soilborne fungal pathogens. Appl Environ Microbiol 65(11): 5148–4150.
24. Wehner J, Antunes PM, Powell JR, Mazukatow J, Rillig MC (2010) Plant pathogen protection by arbuscular mycorrhizas: a role for fungal diversity? Pedobiologia 53(3): 197–201.
25. Mazzola M (2004) Assessment and management of soil microbial community structure for disease suppression. Ann Rev Phytopathol 42: 35–59.
26. Welbaum GE, Sturz AV, Dong Z, Nowak J (2004) Managing soil microorganisms to improve productivity of agro-ecosystems. Crit Rev Plant Sci 23: 175–193.
27. Sharma KK, Bhatnagar-Mathur P (2006) Peanut (Arachis hypogaea L.). Methods Mol Biol (Clifton NJ) 343: 347–358.
28. Dobermann A, Dawe D, Roetter PR, Cassman KG (2000) Reversal of rice yield decline in a long-term continuous cropping experiment. Agron J 92: 633–643
29. Johnson NC, Copeland PJ, Crookston RK, Pfleger FL (1992) Mycorrhizae: possible explanation for yield decline with continuous corn and soybean. Agron J 84: 387–390.
30. Lithourgidis AS, Damalas CA, Gagianas AA (2006) Long-term yield patterns for continuous winter wheat cropping in northern Greece. Eur J Agron 25(3): 208–214.
31. Reichardt W, Mascarina G, Padre B, Doll J (1997) Microbial communities of continuously cropped, irrigated rice fields. Appl Environ Microbiol 63(1): 233–238.
32. Yao HY, Jiao XD, Wu FZ (2006). Effects of continuous cucumber cropping and alternative rotations under protected cultivation on soil microbial community diversity. Plant Soil 284: 195–203.
33. Borneman J, Hartin RJ (2000). PCR primers that amplify fungal rRNA genes from environmental samples. Appl Environ Microbiol 66: 4356–4360.
34. Schloss PD, Handelsman J (2005) Introducing DOTUR, a computer program for defining operational taxonomic units and estimating species richness. Appl Environ Microbiol 71(3): 1501–1506.
35. Berney C, Fahrni J, Pawlowski J (2004) How many novel eukaryotic 'kingdoms'? Pitfalls and limitations of environmental DNA surveys. BMC Biol 2: 13.
36. Dang HY, Li TG, Chen MN, Huang GQ (2008) Cross-ocean distribution of *Rhodobacterales* bacteria as primary surface colonizers in temperate coastal marine waters. Appl Environ Microbiol 74(1): 52–60.
37. Dang H, Lovell CR (2000) Bacterial primary colonization and early succession on surfaces in marine waters as determined by amplified rRNA gene restriction analysis and sequence analysis of 16S rRNA genes. Appl Environ Microbiol 66(2): 467–475.
38. Altschul SF, Madden TL, Schäffer AA, Zhang J, Zhang Z, et al (1997) Gapped BLAST and PSI-BLAST: a new generation of protein database search programs. Nucleic Acids Res 25(17): 3389–3402.
39. Thompson JD, Gibson TJ, Plewniak F, Jeanmougin F, Higgins DG (1997) The CLUSTAL_X windows interface: flexible strategies for multiple sequence alignment aided by quality analysis tools. Nucleic Acids Res 25(24): 4876–4882.
40. Felsenstein J (1989) PHYLIP–phylogeny inference package (version 3.2). Cladistics 5: 164–166.
41. Von Wintzingerode F, Göbel UB, Stackebrandt E (1997) Determination of microbial diversity in environmental samples: pitfalls of PCR-based rRNA analysis. FEMS Microbiol Rev 21(3): 213–229.
42. Nannipieri P, Ascher J, Ceccherini MT, Landi L, Pietramellara G, et al (2003) Microbial diversity and soil functions. Eur J Soil Sci 54: 655–670.
43. Inceoglu O, Hoogwout EF, Hill P, Van Elsas JD (2010) Effect of DNA extraction method on the apparent microbial diversity of soil. Appl Environ Microbiol 76: 3378–3382.
44. Van Veen JA, Van Overbeek LS, Van Elsas JD (1997) Fate and activity of microorganisms introduced into soil. Microbiol Mol Biol Rev 61(2): 121–135.
45. Leigh J, Hodge A, Fitter AH (2009) Arbuscular mycorrhizal fungi can transfer substantial amounts of nitrogen to their host plant from organic material. New Phytol 181(1): 199–207.
46. Bonfante P, Anca I-A (2009) Plants, mycorrhizal fungi, and bacteria: a network of interactions. Annu Rev Microbiol 63: 363–383.
47. Verma M, Brar SK, Tyagi RD, Surampalli RY, Val o JR (2007) Antagonistic fungi, *Trichoderma spp.*: Panoply of biological control. Biochem Eng J 37: 1–20.
48. Anaissie EJ, Stratton SL, Dignani MC, Lee Ck, Summerbell RC, et al (2002) Pathogenic *Aspergillus* species recovered from a hospital water system: a 3-year prospective study. Clin Infect Dis 34(6): 780–789.
49. Benhamou N (2004) Potential of the mycoparasite, *Verticillium lecanii*, to protect citrus fruit against *Penicillium digitatum*, the causal agent of green mold: a comparison with the effect of chitosan. Phytopathology 94(7): 693–705.
50. Čížková D, Šrůtka P, Kolařík M, Kubátová A, Pažoutová S (2005) Assessing the pathogenic effect of *Fusarium*, *Geosmithia* and *Ophiostoma* fungi from broad-leaved trees. Folia Microbiol 50: 59–62.
51. D'Souza CA, Alspaugh JA, Yue C, Harashima T, Cox GM, et al (2001) Cyclic AMP-dependent protein kinase controls virulence of the fungal pathogen *Cryptococcus neoformans*. Mol Cell Biol 21: 3179–3191.
52. Kubicek CP, Mach RL, Peterbauer CK, Lorito M (2001) *Trichoderma*: from genes to biocontrol. J Plant Pathol 83: 11–23.
53. Ying SH, Feng MG (2004) Relationship between thermotolerance and hydrophobin-like proteins in aerial conidia of *Beauveria bassiana* and *Paecilomyces fumosoroseus* as fungal biocontrol agents. J Appl Microbiol 97(2): 323–331.
54. Raaijmakers JM, Paulitz TC, Steinberg C, Alabouvette C, Moënne-Loccoz Y (2009) The rhizosphere: a playground and battlefield for soilborne pathogens and beneficial microorganisms. Plant Soil 321: 341–361.
55. Cotxarrera L, Trillas-Gay MI, Steinberg C, Alabouvette C (2002) Use of sewage sludge compost and *Trichoderma asperellum* isolates to suppress *Fusarium wilt* of tomato. Soil Biol Biochem 34(4): 467–476.
56. Lutz MP, Feichtinger G, Défago G, Duffy B (2003) Mycotoxigenic *Fusarium* and deoxynivalenol production repress chitinase gene expression in the biocontrol agent *Trichoderma atroviridae*. Appl Environ Microbiol 69(6): 3077–3084.
57. Morace G, Borghi E (2012) Invasive mold infections: virulence and pathogenesis of *Mucorales*. Int J Microbiol DOI: 10.1155/2012/349278.
58. Powell JR, Parrent JL, Hart MM, Klironomos JN, Rillig MC, et al (2009) Phylogenetic trait conservatism and the evolution of functional tradeoffs in arbuscular mycorrhizal fungi. Proc Biol Sci 276: 4237–4245.
59. Sikes BA, Cottenie K, Klironomos JN (2009) Plant and fungal identity determines pathogen protection of plant roots by arbuscular mycorrhizas. J Ecol 97: 1274–1280.
60. Bruehl GW (1987) Soilborne plant pathogens. Macmillan Pub, NY.
61. Larkin RP, Honeycutt CW, Griffin TS, Olanya OM, Halloran JM, et al (2011) Effects of different potato cropping system approaches and water management on soilborne diseases and soil microbial communities. Phytopathology 101(1): 58–67.
62. Curl EA (1988) The role of soil microfauna in plant-disease suppression. CRC Crit Rev Plant Sci 7(3): 175–196.
63. Tedersoo L, Hansen K, Perry BA, Kjøller R (2006) Molecular and morphological diversity of *Pezizalean ectomycorrhiza*. New Phytol 170: 581–596.
64. Stukenbrock EH, McDonald BA (2008) The origins of plant pathogens in agro-ecosystems. Annu Rev Phytopathol 46: 75–100.

14

Desert Farming Benefits from Microbial Potential in Arid Soils and Promotes Diversity and Plant Health

Martina Köberl[1], Henry Müller[1], Elshahat M. Ramadan[2], Gabriele Berg[1]*

1 Institute for Environmental Biotechnology, Graz University of Technology, Graz, Austria, **2** Faculty of Agriculture, SEKEM and Heliopolis University, Cairo, Egypt

Abstract

Background: To convert deserts into arable, green landscapes is a global vision, and desert farming is a strong growing area of agriculture world-wide. However, its effect on diversity of soil microbial communities, which are responsible for important ecosystem services like plant health, is still not known.

Methodology/Principal Findings: We studied the impact of long-term agriculture on desert soil in one of the most prominent examples for organic desert farming in Sekem (Egypt). Using a polyphasic methodological approach to analyse microbial communities in soil as well as associated with cultivated plants, drastic effects caused by 30 years of agriculture were detected. Analysing bacterial fingerprints, we found statistically significant differences between agricultural and native desert soil of about 60%. A pyrosequencing-based analysis of the 16S rRNA gene regions showed higher diversity in agricultural than in desert soil (Shannon diversity indices: 11.21/7.90), and displayed structural differences. The proportion of Firmicutes in field soil was significantly higher (37%) than in the desert (11%). *Bacillus* and *Paenibacillus* play the key role: they represented 96% of the antagonists towards phytopathogens, and identical 16S rRNA sequences in the amplicon library and for isolates were detected. The proportion of antagonistic strains was doubled in field in comparison to desert soil (21.6%/12.4%); disease-suppressive bacteria were especially enriched in plant roots. On the opposite, several extremophilic bacterial groups, e.g., *Acidimicrobium, Rubellimicrobium* and *Deinococcus-Thermus,* disappeared from soil after agricultural use. The N-fixing *Herbaspirillum* group only occurred in desert soil. Soil bacterial communities were strongly driven by the a-biotic factors water supply and pH.

Conclusions/Significance: After long-term farming, a drastic shift in the bacterial communities in desert soil was observed. Bacterial communities in agricultural soil showed a higher diversity and a better ecosystem function for plant health but a loss of extremophilic bacteria. Interestingly, we detected that indigenous desert microorganisms promoted plant health in desert agro-ecosystems.

Editor: Jack Anthony Gilbert, Argonne National Laboratory, United States of America

Funding: This project was funded by the EU-Egypt Innovation Fund (C2/S1/190). The funders had no role in study design, data collection and analysis, decision to publish, or preparation of the manuscript.

Competing Interests: The authors have declared that no competing interests exist.

* E-mail: gabriele.berg@TUGraz.at

Introduction

In contrast to desertification, which is recognised as a major threat to biodiversity, to convert deserts into arable, green landscapes is a global vision as well as competent answer to world hunger and climate change [1,2]. Desert farming, which generally relies on irrigation, is one way to this vision. Agriculture systems were already developed in arid landscapes by ancient cultures, yet nowadays, there is a dramatically increasing need for large-scale desert farming to feed the population. For example, in Egypt, desert farmland is expected to grow about 40% till 2017, but this needs about five billion m^3 of water a year [3]. These enormous amounts of water and the expected impact on the climate conditions are the major disadvantages of agriculture in the desert. While these problems are well-investigated, the effect on the bio-resource soil was yet not assessed. Moreover, there is still a gap of knowledge about the effect of management and land uses on the bacterial diversity of soils, which new molecular tools like metagenomics can help to close [4,5].

Deserts represent extreme environments for microorganisms [6]. Although the conditions varied strongly in the different regions of the world, all of them are characterised by a combination of extreme temperatures and desiccation, high soil salinity, low nutrient levels, high summer UV radiation levels, and physical instability caused by strong winds: all factors contribute to the visual appearance of a sterile environment. While early studies supported this "sterility" by very low levels of viable/cultivable microorganisms, applications of new methods in microbial ecology led to interesting new findings and showed a contrasting picture [6,7]. For example, in their global-scale study, Fierer & Jackson [8] found that the acidic soils of tropical forests harbour fewer bacterial taxa than the neutral pH soils of deserts. In McMurdo Dry Valleys, a hyperarid polar desert, microbial soil communities were relatively depauperate but harboured a broad range of previously unreported bacteria and fungi from polar regions [9]. In different sites in the Negev Desert, archaeal and bacterial diversity analysed by fingerprints using T-RFLP of the 16S rRNA genes was not constrained by precipitation, although the taxonomic composition differed [10]. In soil of the Atacama Desert, a high diversity of microorganisms known for life in hypersaline environments was found by analysis of DGGE profiles [11]. Most of the desert microbial communities seem to be structured solely by a-biotic processes [6,7]. But, if adapted desert plants occurred, e.g. *Panicum* and *Stipagrostis* in Sinai or *Reaumuria negevensis* in Negev, they strongly shaped soil microbial diversity [12,13]. However, all these investigations showed a unique and extraordinary microbial diversity in desert soils. An understanding of diversity in such microbial communities can be used to assess potential effects of desert farming on soil ecosystem services like plant health [14]. Emerging problems with soil-borne pathogens limited the plant yield after several years often drastically. Due to their specific ecology soil-borne pathogens are difficult to suppress; disease-suppressive bacteria, which are able to antagonise and biologically control them, provide a promising and sustainable solution [15].

The objective of this study was to analyse the effect of desert farming on soil microbial diversity and on disease-suppressive bacteria. We studied microbial diversity in native Egyptian desert soil in comparison to the agricultural soil, which was used more than 30 years for organic agriculture in Sekem farms (www.sekem.com; Egypt). To study the role of plant-associated bacteria in the agricultural soil, we analysed microbial communities in the rhizosphere and endorhiza of cultivated medical plants. Sekem is not only one of the most prominent examples of organic farming in the desert; they were assigned for social entrepreneurship [16,17]. For this study, we used a broad set of methods including i) bacterial fingerprints using 16S rRNA PCR-SSCP (Single Strand Conformational Polymorphism) analysis to compare the communities at statistical level and identify the dominant bacterial taxa, ii) pyrosequencing-based 16S rRNA profiling to get a deeper insight into the soil communities, iii) a cultivation approach to assess the impact on disease-suppressive bacteria, and iv) a multivariate statistical analysis to identify environmental factors driving microbial communities. We demonstrate that long-term organic agriculture had a strong impact on microbial community structure and function, and identified highly specialised communities in all microenvironments.

Results

Molecular fingerprinting of microbial communities

To get a first overview about the structure of the bacterial communities, fingerprints were performed by SSCP analysis of 16S rRNA genes amplified from DNA obtained from desert and agricultural soil. In addition, we analysed bacterial communities from rhizosphere and endorhiza of the dominant plants German chamomile (*Matricaria chamomilla* L.), pot marigold (*Calendula officinalis* L.) and *Solanum distichum* Schumach. & Thonn. cultivated on farms. In comparison to the desert, in field soil an impressive diversity of bacteria was found (Fig. S1). According to cluster analyses, the composition of the bacterial community of agricultural soil differed significantly from the desert soil by approximately 60% of the bacterial strains (Fig. 1). In the bacterial community of desert soil two dominant bands could be detected, which were also abundant in all samples from the rhizosphere and endorhiza of all three investigated medical plants (Fig. S1). The two dominant bands were identified by partial 16S rRNA gene sequence analysis as *Ochrobactrum* sp. (closest database match *O. grignonense*, 99% similarity to NR_028901) and *Rhodococcus* sp.

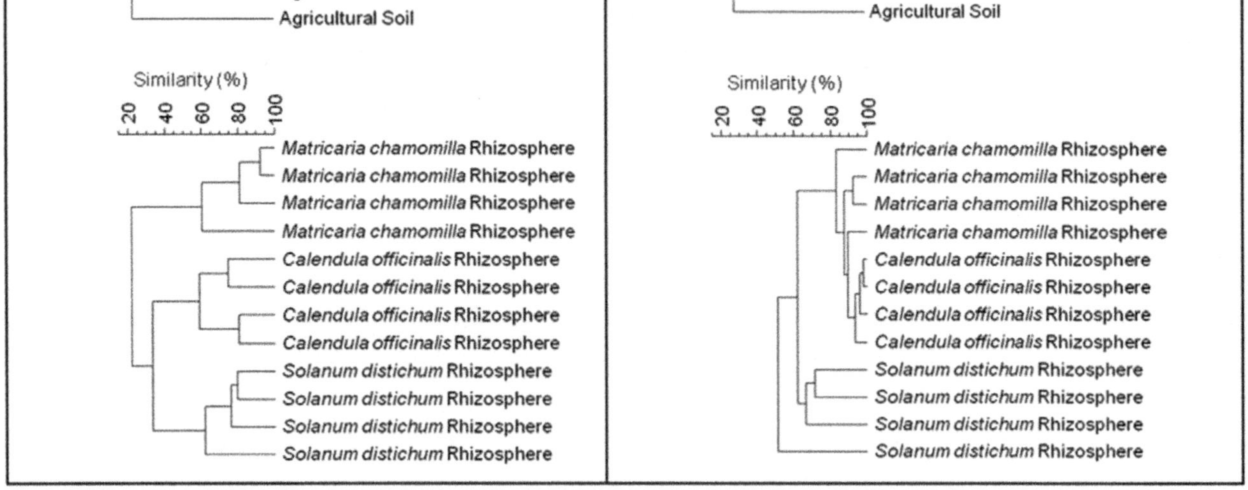

Figure 1. UPGMA dendrograms of total bacterial (A) and *Pseudomonas* (B) communities in soil and rhizosphere of the medical plants. The dendrograms were generated from the SSCP community profiles with GelCompar II. Following settings were used: dendrogram type: unweighted pair group method with arithmetic mean (UPGMA); similarity coefficient: curve based: Pearson correlation; position tolerances: optimisation: 4%, position tolerance: 1%.

(closest database match *R. erythropolis*, 99% similarity to NR_037024). Further, *Bacillus* sp. was found nearly in all samples (closest database match *B. subtilis* subsp. *subtilis*, 99% similarity to NR_027552). For the rhizosphere as well as for the endorhiza of the medical plants a clear plant-specific effect of the bacterial communities was found (Fig. 1). They shared only 20% of the bacterial community, whereas the majority was determined by plant-specific bacteria.

Pyrosequencing-based 16S rRNA profiling of the bacterial community in soil

To deeply survey the diversity and the composition of the bacterial communities present in untreated desert soil and after 30 years of organic agriculture, a pyrosequencing-based analysis of partial 16S rRNA gene sequences (V4-V5 region) has been employed. In desert soil, we recovered 19,244 and in agricultural soil 33,384 quality sequences with a read length of $\geq$ 150 bp. Of all quality sequences 83.0% could be classified below the domain level; this proportion is in accordance with other pyrosequencing-based studies [18-20]. To determine rarefaction curves, operational taxonomic units (OTUs) were identified at sequence divergences of 3% (species level), 5% (genus level) and 20% (phylum level). The rarefaction analysis of the bacterial community in the desert soil in comparison to the agricultural soil is shown in Fig. 2. At a dissimilarity level of 20%, both curves show a clear saturation. Thus the surveying effort covered almost the full extent of taxonomic diversity at this level of genetic distance. Additionally, a comparison of rarefaction analyses with the number of OTUs estimated by the Chao1 richness estimator [20] revealed that in both soils over 90% of the estimated taxonomic richness was covered by the sequencing effort (Table 1). At the genus level (5% dissimilarity) the full extent of taxonomic diversity was not surveyed (42% and 31%). The computed Shannon indices of diversity (H') were much higher for the agricultural soil than for the desert soil, this indicates a higher bacterial diversity in soil due the agricultural use of the desert.

The 43,673 classifiable sequences obtained from both soil types together were affiliated with 18 different phyla. Proteobacteria (30.2%), Firmicutes (27.3%) and Actinobacteria (10.5%) were the dominant phyla (Fig. 3, Table S1). These dominant phyla were present in both soils. In detail, Firmicutes are highly enriched in agricultural soil (from 11.3% in desert soil to 36.6% in agricultural soil), Proteobacteria (46.0% in desert soil and 21.0% in agricultural soil) and Actinobacteria (20.7% in desert soil and 4.6% in agricultural soil) occurred in farmland in lower concentrations than in the surrounding desert. Further, in both soils Bacteroidetes (4.6% and 5.3%) and Gemmatimonadetes (1.4% and 1.9%) were present. Considering only phyla covering more than 1% of quality sequences, Acidobacteria (7.9%) and Planctomycetes (1.1%) were only found in the agricultural soil, and *Deinococcus-Thermus* (1.1%) was only detectable in the desert sand. These abundances of the phyla coincided with results from previously reported meta-analysis of bacterial community composition in soils and, despite the special soil type of the desert, the composition covers rather well with studies of completely different soils [18,20-22] with exception of Firmicutes. Most of the Firmicutes sequences were classified as belonging to the genus *Bacillus*; in the agricultural soil also *Paenibacillus* was found (5% of classified Firmicutes). In desert soil *Ochrobactrum* was the most abundant genus within the (Alpha-)Proteobacteria (79% of classified Proteobacteria) and *Rhodococcus* among the Actinobacteria (90% of classified Actinobacteria). The Acidobacteria in the agricultural soil are affiliated only with subdivision 6.

Quantitative analysis of bacterial abundances

A cultivation-independent approach (total 16S rRNA gene copy numbers using quantitative PCR) was combined with cultivation, which was also the basis for isolate screening, to assess bacterial abundances in the different microhabitats. Copy numbers in agricultural soil were 9.4±0.5 $\log_{10}$ g^{-1} compared to 8.1±1.1 $\log_{10}$ g^{-1} in desert soil (Fig. S2), and were not statistically significantly different. In contrast, abundances of culturable bacteria determined on R2A resulted in statistically significant higher abundances in agricultural soil (7.7±0.4 $\log_{10}$ CFU g^{-1}) in comparison to desert soil (4.6±0.6 $\log_{10}$ CFU g^{-1}). The rhizosphere of all three investigated medical plants was highly colonised by culturable bacteria: $\log_{10}$ CFU ranged from 7.8±0.3 to 8.0±0.2 g^{-1} fw. In contrast, in the endorhiza significantly lower CFUs were detected with $\log_{10}$ 2.0±0.2 to 3.7±0.8 g^{-1} fw.

Antagonistic potential of the bacterial community towards pathogenic fungi

A cultivation approach was used to analyse a functional aspect of the bacterial communities. To assess the indigenous anti-phytopathogenic potential, the antagonistic activity against three major soil-borne phytopathogenic fungi *Verticillium dahliae*, *Rhizoc-*

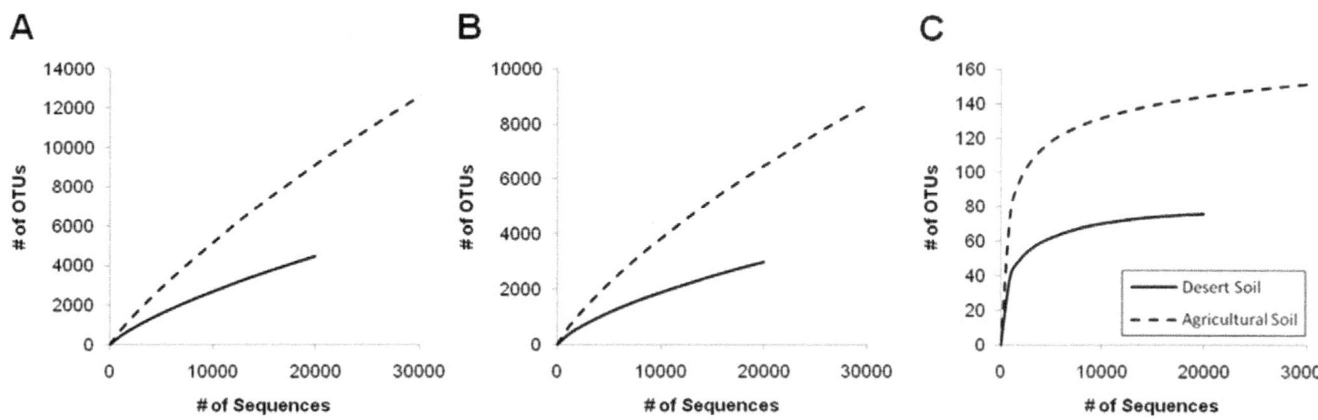

Figure 2. Rarefaction analyses of the two soil types. Rarefaction curves indicate that the diversity of bacterial phylotypes is higher in the agricultural soil compared to the surrounding desert soil. OTUs are shown at genetic distance levels of 3% (A), 5% (B) and 20% (C). Data were calculated by employing tools of the RDP pyrosequencing pipeline (http://pyro.cme.msu.edu).

Table 1. Species richness estimates obtained at 3%, 5% and 20% genetic dissimilarity from pyrosequencing of 16S rRNA from metagenomic DNA extracted from desert soil and agricultural soil.

	Shannon index[a] (H')			Rarefaction[b] (no. of OTUs)			Chao1[c] (no. of OTUs)			Coverage (%)		
	3%	5%	20%	3%	5%	20%	3%	5%	20%	3%	5%	20%
Desert Soil	7.90	7.04	3.02	4,465	2,967	76	13,278	7,012	77	33.6	42.3	98.8
Agricultural Soil	11.21	9.94	3.91	9,112	6,474	144	38,985	20,838	161	23.4	31.1	90.0

[a] a higher number indicates more diversity;
[b] the results from the rarefaction analyses are also depicted in Fig. 2;
[c] nonparametric richness estimator based on the distribution of singletons and doubletons.

tonia solani and *Fusarium culmorum* was determined. From each microenvironment up to 200 isolates were randomly selected and assessed regarding their anti-phytopathogenic capacity *in vitro*. All isolated soil bacteria (199 isolates from desert soil and 155 isolates from agricultural soil) were screened by dual testing regarding their antagonistic activity towards *V. dahliae*, *R. solani* and *F. culmorum* (Table 2). In general, bacterial isolates obtained from the soil of the farm exhibited a higher *in vitro* antagonistic potential towards soil-borne phytopathogenic fungi in comparison to the bacteria isolated from the desert soil (agricultural soil 21.6±0.8%; desert soil 12.4±0.7%). From the agricultural soil, 17.4% (27 isolates) demonstrated *in vitro* broad-spectrum antagonism towards all three pathogens, from the desert soil 10.6% (21 isolates) were able to suppress the growth of all fungi tested. No enrichment of antagonists in the rhizosphere and endorhiza of the investigated medical plants was detected. In general, *M. chamomilla* and *S. distichum* showed a higher antagonistic potential than *C. officinalis*. Especially the endorhiza from *M. chamomilla* harboured a high proportion of antagonists. Whereas in the soil and in the rhizosphere could be found most antagonistic bacteria towards *F. culmorum*, in the endorhiza of the medical plants most antagonists showed antagonism towards *V. dahliae*.

To assess the diversity of bacterial antagonists, isolates with an activity towards at least two of the soil-borne pathogenic fungi (162 isolates) were characterised genotypically and identified by partial 16S rRNA gene sequencing (Fig. 4). Using restriction fragment length polymorphism, of the 16S rRNA (= amplified ribosomal RNA gene restriction analysis [ARDRA]), the antagonistic isolates could be clustered into six groups: (1) *Bacillus subtilis*, (2) *Bacillus cereus*, (3) *Bacillus endophyticus*, (4) *Paenibacillus/Brevibacillus*, (5) *Streptomyces*, and (6) *Lysobacter*. With the exception of the *Lysobacter* strain (only one isolate from the rhizosphere of *M. chamomilla*, closest database match *L. enzymogenes*), only Gram-positive antagonists were found. All antagonistic populations were dominated by Firmicutes; *Bacillus* and *Paenibacillus* could be isolated from all habitats. Interestingly, antagonistic isolates of the genus *Streptomyces* were found exclusively in desert soil.

To analyse the genotypic diversity within the taxonomic groups at population level, BOX PCR patterns of the whole bacterial genome were used. Especially within the large *Bacillus subtilis* cluster (123 isolates), a high genotypic diversity was found. At a cut off level of 80%, they could be divided into 37 genotypic groups. By partial 16S rRNA gene sequencing isolates were identified as *B. subtilis* subsp. *subtilis* and *spizizenii*, *B. vallismortis*, *B. mojavensis* and *B.*

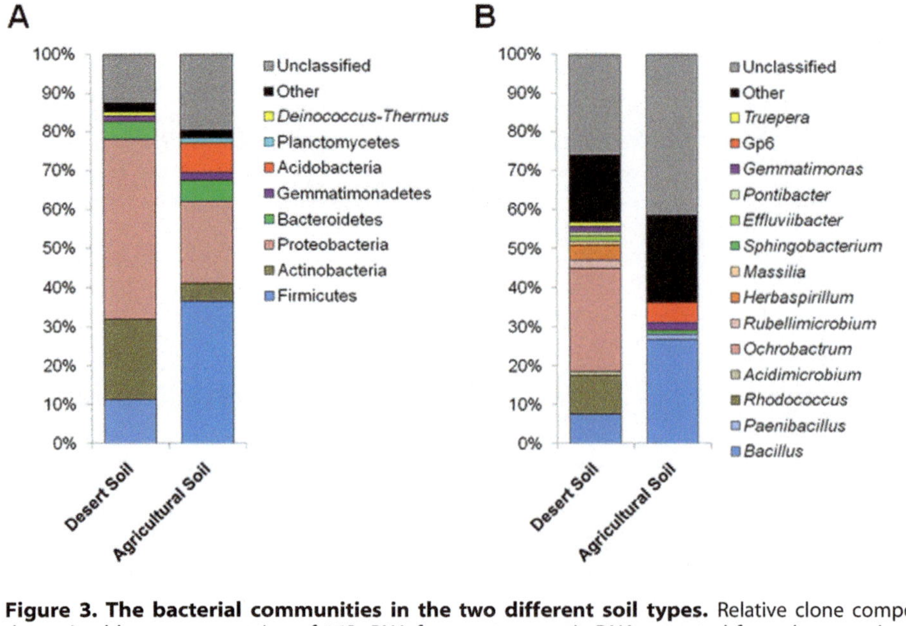

Figure 3. The bacterial communities in the two different soil types. Relative clone composition of major phyla (A) and genera (B) was determined by pyrosequencing of 16S rRNA from metagenomic DNA extracted from desert and agricultural soil. The identification of the closest strain based on 16S rRNA sequence similarity was achieved using the web server SnoWMAn 1.7 (http://snowman.genome.tugraz.at). Phylogenetic groups accounting for ≤1% of all quality sequences are summarised in the artificial group others.

Table 2. Proportions of bacterial isolates antagonistic towards the soil-borne fungal pathogens *Verticillium dahliae*, *Rhizoctonia solani* and *Fusarium culmorum*.

		Proportion of antagonists (%)[a]		
Microhabitat	**Origin**	***V. dahliae***	***R. solani***	***F. culmorum***
Soil	Desert Soil	11.1±1.8	12.8±0.6	13.4±0.1
	Agricultural Soil	20.0±1.6	21.9±2.2	22.6±1.4
Rhizosphere	*Matricaria chamomilla*	12.5±2.9	8.3±0.7	13.0±1.8
	Calendula officinalis	9.0±0.5	7.1±0.1	10.1±3.1
	Solanum distichum	13.7±2.3	13.8±3.8	15.7±0.0
Endorhiza	*Matricaria chamomilla*	19.9±1.8	16.4±2.3	18.8±2.6
	Calendula officinalis	4.2±2.9	0.0±0.0	1.4±1.0
	Solanum distichum	13.5±5.1	10.4±5.8	12.5±5.8

[a]Data are averages of 1st and 2nd sampling ± confidences.

atrophaeus. The *Paenibacillus*/*Brevibacillus* isolates could be divided into eight BOX clusters and *Bacillus endophyticus* into two. *Streptomyces* was subdivided in three genotypes, the closest database matches were *S. peucetius*, *S. scabiei* and *S. subrutilus*. Surprisingly, among the *Bacillus subtilis* group, isolates with identical BOX patterns could be detected in desert soil as well as in the agricultural soil, and also in rhizosphere and endorhiza of the medical plants (Fig. 5). Based on unique genotypic patterns and antagonistic potential, 45 promising biocontrol strains were selected of which 89% belonged to the Bacillales (Table S2).

Ecological factors driving the bacterial communities in soil

Indirect correspondence analysis (CA) based on the OTUs of soil species obtained by microbial fingerprinting showed the

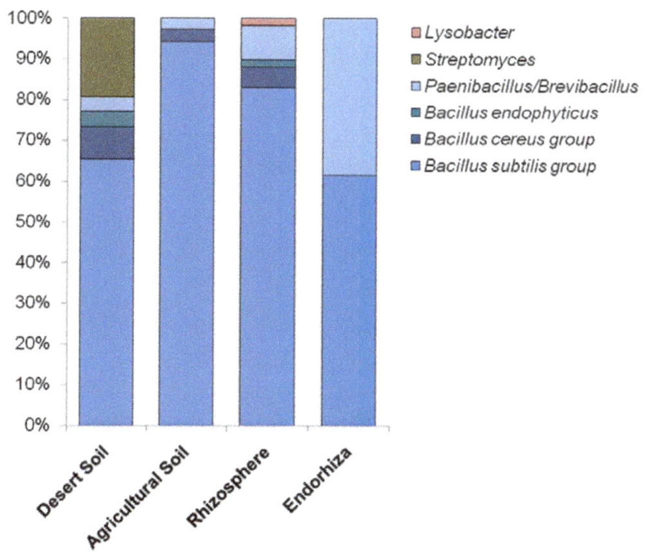

Figure 4. Diversity of bacterial antagonists with an activity towards pathogenic fungi. Isolates with activity against two pathogens were identified by partial 16S rRNA gene sequencing. Samples from rhizosphere and endorhiza include isolates from the medical plants *Matricaria chamomilla*, *Calendula officinalis* and *Solanum distichum*.

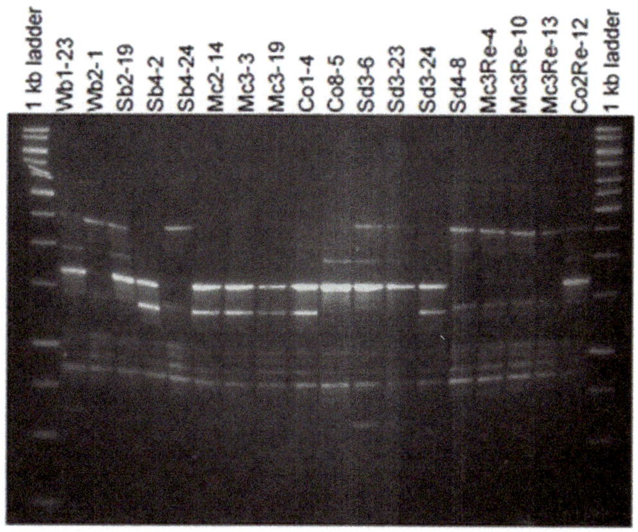

Figure 5. BOX PCR fingerprints of genetically very similar antagonists of the *Bacillus subtilis* group isolated from different microenvironments. The similarity in the dendrogram between them was more than 80%. Isolates were encoded by abbreviations: (1) soil type or plant species (Wb = desert soil; Sb = Sekem soil, Mc = *Matricaria chamomilla*, Co = *Calendula officinalis*, Sd = *Solanum distichum*), (2) replicate (1–4), (3) microenvironment (Re = endorhiza, rhizosphere and soil have no further designation), and (4) consecutive number of the isolate per replicate.

coherence and similarity of the different samples indicated by crowding points at a CA biplot (data not shown). Furthermore, the influence of the environmental factors on the bacterial soil communities was examined using the multivariate statistical analysis. A significant effect was proved for water supply (precipitation + irrigation) (regression coefficient: 0.3760), pH (0.3719) and, to a lower extent for organic carbon (0.1600) and soil quality (0.1011).

Discussion

Agriculture in deserts open new ways to solve diverse problems: produce enough food for poor regions e.g. in Africa, produce renewable crops for industrial applications and to capture and restore CO_2 in soil. However, agricultural use induces a drastic shift for the whole ecosystem, and risk assessments to evaluate the function are necessary. Here we analysed differences of microbial communities in undisturbed desert soil in comparison to desert soil, which was cultivated under organic (biodynamic) conditions for 30 years. Altogether, a strong impact of long-term agriculture on microbial community structure and function was identified, which will be discussed and assessed in detail.

The composition of the bacterial communities in desert and agricultural soil differed strongly. In microbial fingerprints, both communities showed about 60% differences. Using a pyrosequencing-based approach of the 16S rRNA gene region, reasons for these differences could be identified. The relative abundance of Firmicutes in agricultural soil was significantly enhanced from 11 to 37%. Especially the proportion of 37% is remarkable because Janssen [22] reported them to contribute only a mean of 2% (range 0-8%) in the total bacterial soil community. *Bacillus* and *Paenibacillus* play the key role to explain this difference; they were dominant in the 16S rRNA gene amplicon library and represented 96% of the antagonists towards phytopathogens identified in the culturable fraction. In addition, by microbial fingerprints we showed that this Gram-positive group was enriched in the rhizosphere as well as endorhiza of medical plants cultivated in Sekem. Interestingly, we found *Bacillus* isolates with the same BOX pattern in desert and field soil as well as in the rhizosphere and endorhiza of medical plants, which was confirmed also by our bacterial fingerprint analysis. Furthermore, identical 16S rRNA gene sequences were found for isolates as well as in the amplicon library. This is further evidence for enrichment of plant rhizosphere-specific bacteria from the soil. Moreover, it indicates that the antagonistic bacteria were enriched from desert soil and not from the compost treatment. Both, *Bacillus* and *Paenibacillus* are well-characterised plant-associated genera with antagonistic properties towards fungal plant pathogens [23]. While the proportion of Firmicutes was enhanced in field soil, several extremophilic bacterial groups, e.g. *Acidimicrobium*, *Rubellimicrobium* and *Deinococcus-Thermus* disappeared. Bacteria from all of these genera/phyla are either impossible or else extremely difficult to cultivate and only found in extreme environments by molecular analysis. For example, bacteria from the phylum *Deinococcus-Thermus* possess important adaptations such as resistance to environmental hazards, e.g., desiccation, ultraviolet radiation, high salinity, and high temperatures [24]. In general, the proportion of cultivable bacteria was lower in desert soil than in field soil, which was shown in the comparison between results obtained by cultivation and qPCR analysis. Based on our pyrosequencing data, bacterial communities in agricultural soil were characterised by a higher diversity than in desert soil (Shannon diversity indices: agricultural soil 11.21; desert 7.90). The high bacterial diversity found in the organically managed soil was shown for agriculture in the desert for the first time but was already reported for another organically managed system [14].

Additionally, for the function of the bacterial communities in desert and agricultural soil we found strong differences. The proportion of strains with antagonistic *in vitro* activity against soil-borne phytopathogens was statistically significantly enhanced in agricultural soil in comparison to desert soil. Other current studies showed also that organic farming methods can mitigate ecological damages caused by pests and pathogens by promoting natural enemies, analysed for example in the pathosystem potato – potato beetle [25] or grape – *Botrytis cineria* [26]. Although the proportion of antagonistic strains was higher in agricultural soil, their diversity was much lower. All of the isolated antagonists belong to the *Bacillus/Paenibacillus* group. In contrast, in desert soil, diverse antagonistic *Streptomyces* species were identified, including *Streptomyces peucetius*, a species known to produce anthracycline antibiotics [27]. Another interesting fact is that members of the *Herbaspirillum* group, most of them known as N-fixing species, only occurred in desert soil. In most of the deserts, plants have a very short period to develop. This fact is well-known, when shortly after a rainfall millions of seedlings occur and colour the whole desert in green. Therefore, plants need plant growth promoting rhizobacteria, and *Herbaspirillum* strains belong to this group. Owing to compost treatment agricultural soils are saturated with nitrogen. These facts could explain that this important functional group had a lower abundance in field soil. Another genus with an interesting occurrence is *Orchrobactrum*. In desert soil *Ochrobactrum* was the most abundant genus within Proteobacteria and also in microbial fingerprints this genus was found in high abundance in soil but also in the rhizosphere/endorhiza of medical plants. Bacteria of this genus are known for its ambivalent interaction with eukaryotes, while they show plant growth promotion effects on plants, they can cause opportunistic infections in humans [28].

What are the reasons for the changes in structure and function of the bacterial community? The main factor, which explains the differences, is the continuous irrigation of farm land. This factor was identified by an indirect correspondence analysis. Precipitation in this arid region is general low (21-52 mm). The agriculture is completely dependent on irrigation water coming from the Nile or from local ground water. Irrigation systems were used to supply about 2,500-2,600 l m^{-3} per year. The aridity level was also one of the main factors that shaped the microbial community structure in patchy desert landscapes of Negev [29]. By the multivariate statistic, the pH of soil was identified as the second impact factor. This factor was often reported as main driver, e.g. in global studies of microbial communities in soil [8,18]. Another factor, which contributed to the shift in the bacterial community in a lower extend, is compost treatment. This was already shown for other examples in organic agriculture: due to the use of compost, studies have found that biodynamic farms have a significantly better soil quality than conventionally farmed soils but comparable to the soil quality achieved by other organic methods [14,17]. The compost treatment is responsible for nutrient and organic matter supply. On the other side, compost is known for an extremely high but also specific bacterial diversity. No evidence was found for an impact of these specifically adapted bacteria on soil communities. One factor, which could be not included in the statistical analyses, is plant-specific enrichment of bacterial communities. The extent of plant specificity was shown in a study of *Verticillium* host plants published by Smalla et al. [30], and later described for many other plant species [31]. In our study, we found a highly pronounced effect for each of the medical plants investigated. All three medical plants, which belong to the dominant herbs in Sekem and were included in the study, are known for their production of secondary metabolites. For example, German chamomile, for which we found the strongest effect, is used medicinally to treat sore stomach and irritable bowel syndrome. Chamomile plants produce the terpene bisabolol, and other active ingredients like farnesene, chamazulene, flavonoids and coumarin [32]. Some of them are known for their anti-microbial properties, and others, such as flavonoids often serve as signals in plant-microbe interactions [33].

In a final assessment, bacterial communities in agricultural soil showed a higher diversity and a better ecosystem function for plant health, which was measured as proportion of disease-suppressive bacteria. On the other side, there is a loss of extremophilic bacteria, which are typical inhabitants of desert soil. However, due to the fact that all farms are still surrounded by desert, we can conclude that also this specific diversity is maintained. The most interesting fact detected in our study was that indigenous desert microorganisms fulfil important functions in desert agro-ecosystems: *Bacillus* and *Paenibacillus* strains were enriched via plant roots from desert soil. This was shown at the population level using genotypic fingerprinting by BOX pattern, at community level by microbial fingerprints as well as in the metagenome.

Materials and Methods

The experimental design comprise samples from agricultural soil, rhizosphere and endorhiza samples from main medical crops cultivated in Sekem farms as well as samples from the surrounding desert soil from two different sampling times. All sampling sites are private property of the Sekem companies. The sampling was done in cooperation with Angela Hoffmann and Elshahat M. Ramadan (Sekem) with permission of Ibrahim Abouleish, the owner of Sekem, for a joint project. Therefore, no other permit was required. Samples from agricultural used soil were taken at the Sekem farm Adleya, located in the North-eastern desert region of

Egypt near Bilbeis (30°22′88"N; 31°39′41"E). The agriculture was completely dependent on irrigation water (2,607 l m^{-3} on average per year) coming from the Nile or from local ground water drillings; sprinkler and drip irrigation systems were used. The farmland soil was fertilised with compost that was produced on their own composting facility, where rice straw, water hyacinth, wood chips, organic waste, clay, chicken and cow manure was used as input materials. The compost was applied twice a year (May and September), during the preparation of the fields for the cropping season. The soil texture at the Sekem farm was classified by Luske & van der Kamp [17] as loamy sand (pH 8.4) with an organic carbon content of 0.8% and a clay content of 4%. Desert soil was collected from two sites in the surrounding desert uninfluenced by human activities (30°35'01"N; 32°25'49"E; 29°52′26″N, 31°13′1″E) and was classified as sand (pH 7.7) with an organic carbon content of <0.2% and a clay content of 1.5%. Desert soil was characterised by a low moisture level; plants were very scarce [17]. At each site, four composite samples of soil in a horizon of 10–30 cm depth were collected. Furthermore, from three different species of medical plants (German chamomile [*Matricaria chamomilla* L.], pot marigold [*Calendula officinalis* L.] and *Solanum distichum* Schumach. & Thonn.) planted on the Adleya farm (30°22′88"N; 31°39′41"E), roots with adhering soil were obtained. From each plant four independent composite samples consisted of 5–10 plants were taken. At the first sampling time (October 2009), *Matricaria chamomilla* and *Calendula officinalis* have been in the seedling stage, whereas the samples from the perennial *Solanum distichum* were taken from lignified plants. At the second sampling time (April 2010), all medical plants were in the flowering stage.

To isolate total community DNA from soil and from rhizosphere for all cultivation independent analyses, 5 g of soil/roots with adhering soil and 45 ml of 0.85% NaCl were mixed for 5 min on the vortex. For the isolation from the endorhiza, 5 g material of roots were surface-sterilised with 4% NaOCl for 5 min, then the roots were washed three times with sterile Aqua dest. After 10 ml sterile 0.85% NaCl were added the roots were homogenised using mortar and pestle. For isolation of total DNA from soil, rhizosphere and endorhiza 4 mL of the liquid parts were centrifuged at high speed (16,000×g, 4°C) for 20 min and resulting microbial pellets were stored at −70°C. In the desert soil, a lower concentration of DNA was expected. Therefore, for the isolation of total DNA the pellets of 10 ml supernatant were used. Total community DNA was extracted using the FastDNA® SPIN Kit for Soil (MP Biomedicals, Solon, USA) according to the manufacturer's protocol and used for fingerprints and the deep-sequencing approach.

Fingerprinting of microbial communities by Single Strand Conformational Polymorphism Analysis (SSCP) was carried out as described by Schwieger & Tebbe [34]. Bacterial 16S rRNA gene sequences were amplified by PCR using the eubacterial primer pair Unibac-II-515f (5′-GTG CCA GCA GCC GC-3′) and Unibac-II-927r^{P} (5′-CCC GTC AAT TYM TTT GAG TT-3′) [35]. The PCR was performed by using a total volume of 60 µl containing 1 × Taq&Go (MP Biomedicals, Eschwege, Germany), 1.5 mM $MgCl_2$, 0.2 µM of each primer and 1 µl of template DNA (95°C, 5 min; 32 cycles of 95°C, 20 s; 54°C, 15 s; 72°C, 30 s; and elongation at 72°C, 10 min). For the analysis of the *Pseudomonas* community a nested PCR was performed. In a first PCR *Pseudomonas* were selectively amplified with primers F311Ps (5′-CTG GTC TGA GAG GAT GAT CAG T-3′) and 1459rPs^{P} (5′-AAT CAC TCC GTG GTA AAC GT-3′) [36] followed by a second PCR with the primer pair Unibac-II-515f/Unibac-II-927r^{P}. The reaction mixture for the first PCR (20 µl) was composed of 1 × Taq&Go, 2.25 mM $MgCl_2$, 0.5 mg/ml BSA, 1.5% DMSO, 0.2 µM of each primer and 1 µl of template DNA (94°C, 7 min; 30 cycles of 94°C, 45 s; 56°C, 2 min; 72°C, 2 min; and elongation at 72°C, 10 min). Samples served as templates for the second PCR. The obtained amplicons were separated using the TGGE Maxi system (Biometra, Göttingen, Germany) at 400 V and 26°C. Silver staining was used for the routine detection of DNA bands in SSCP gels [37]. Dominant bands were excised from SSCP gels as described by Schwieger and Tebbe [34]. Extracted DNA fragments were re-amplified by PCR and sequenced. For phylogenetic analysis and identification of related sequences, the obtained sequences were aligned with reference gene sequences from GenBank using BLAST algorithm.

Computer-assisted comparisons of SSCP generated community profiles were performed by using the software GelCompar II (Applied Maths, Kortrijk, Belgium). The cluster analysis was performed using following settings: dendrogram type: unweighted pair group method with arithmetic mean (UPGMA); similarity coefficient: curve based: Pearson correlation; position tolerances: optimisation: 4%, position tolerance: 1% [38]. Furthermore, correspondence analysis was used to answer the question whether a correlation exists (1) between the independently sampled microbial communities of the different sampling points and (2) between soil communities and environmental factors. The following environmental data were used: i) soil quality (sand, loam, semi-loam), ii) soil pH, iii) content of organic carbon and iv) water supply (sum of local precipitation per year [21, 52 mm] and irrigation). According to the distance of the bands, the SSCP gels were theoretically divided into operational taxonomic units (OTUs). The presence or absence of individual amplified product DNA bands in each group was scored. The obtained matrix was used to compare data statistically using the indirect correspondence analysis for unimodal data of the software package Canoco 4.5 [39].

To analyse the taxonomic composition of the soil bacterial community by a deep-sequencing approach, the hypervariable V4-V5 region of the 16S rRNA gene (*Escherichia coli* positions 515 to 927) was amplified in a nested PCR approach for pyrosequencing. In a first PCR the primer pair 27f/1492r [40] was used and in the second PCR V4-V5 region was amplified with the following primer set, containing the 454 pyrosequencing adaptors and sample specific tags (underlined): Unibac-II-515f_MID13 (5′-CGT ATC GCC TCC CTC GCG CCA TCA GCA TAG TAG TG GTG CCA GCA GCC GC-3′) respectively Unibac-II-515f_MID14 (5′-CGT ATC GCC TCC CTC GCG CCA TCA GCG AGA GAT AC GTG CCA GCA GCC GC-3′) and Unibac-II-927r_MID13-14 (5′-CTA TGC GCC TTG CCA GCC CGC TCA G CCC GTC AAT TYM TTT GAG TT-3′). The reaction mixture for the first PCR (20 µl) contained 1 × Taq&Go, 0.25 µM of each primer and 1 µl of template DNA (95°C, 5 min; 30 cycles of 95°C, 30 s; 57°C, 30 s; 72°C, 90 s; and elongation at 72°C, 5 min). The second PCR was performed by using 1 × Taq&Go, 1.5 mM $MgCl_2$, 0.4 µM of each primer and 2 µl of template DNA (95°C, 5 min; 32 cycles of 95°C, 20 s; 54°C, 15 s; 72°C, 30 s; and elongation at 72°C, 10 min). PCR products of four independent soil samples of the same habitat were pooled in equal volumes and purified by employing the Wizard® SV Gel and PCR Clean-Up System (Promega, Madison, USA). A total of 130 ng of amplified 16S rRNA gene product from each soil was required to construct the libraries for 454 pyrosequencing. For taxonomy-based analysis, the web server SnoWMAn 1.7 (http://snowman.genome.tugraz.at) [41] was employed. Sequences that were shorter than 150 bp in length or of low quality were removed from the pyrosequencing-derived data sets and following settings were used: analysis type: BLAT pipeline; reference database: greengenes_24-Mar-2010; rarefaction method: MOTHUR; taxonomy: RDP; confidence threshold: 80%; include taxa covering more than: 1%. For rarefaction analysis and ascertainment of diversity indices, the data were normalised considering the same number of sequences to all samples using default settings in the open source software package QIIME (http://qiime.sourceforge.net), which allows analysis of high-throughput community sequencing data [42]. Rarefaction curves were calculated by using the tools aligner, complete linkage clustering and rarefaction of the ribosomal database project (RDP) pyrosequencing pipeline (http://pyro.cme.msu.edu) [43]. Shannon [44] and Chao1 [45] indices were calculated based on the complete linkage clustering data.

The same region of the 16S rRNA gene was amplified by quantitative PCR to determine the total bacterial abundances in desert and agricultural soil. Reactions were conducted in a total volume of 10 µl containing 1 × KAPA™ SYBR® FAST qPCR MasterMix Universal (PEQLAB, Polling, Austria), 0.25 µM of each primer (Unibac-II-515f and Unibac-II-927r [35]) and 1 µl template DNA (95°C, 5 min; 35 cycles of 95°C, 20 s; 54°C, 15 s; 72°C, 30 s; and melt from 72 to 95°C). Rotor-Gene™ 6000 real-time rotary analyser (Corbett Research, Sydney, Australia) was used for quantification of fluorescence. For absolute quantification the PCR amplified 16S rRNA gene fragment was cloned into a pGEM®-T Easy Vector (Promega, Mannheim, Germany). Serial dilutions of PCR fragments generated with the primers usp (5'-GTAAAACGACGGCCAGT-3') and rsp (5'-CAGGAAACAGCTATGACC-3'), which specifically bind to sides flanking the multi cloning side of the Vector, were used as standard for calculation of copy number. Concentrations determined by absolute quantification were calculated to copy number per g soil. Each replicate was analysed three times in two independent runs. Significances in the difference between desert and agricultural soil were calculated using the independent samples *t* test with PASW Statistics 18 (SPSS Inc., Chicago, USA).

Same cell suspensions as used for the isolation of total community DNA were used for isolation of bacteria from soil, rhizosphere and endorhiza: They were used for dilution and plating on R2A (Roth, Karlsruhe, Germany) in duplicates. Plates were incubated for 4 days at room temperature (RT) and colony forming units were counted to calculate the means of colonies ($\log_{10}$ CFU) based on fresh weight (fw). If possible, for each replicate 24 bacterial isolates were selected and subcultured on nutrient agar (NA). The isolates were purified and then stored at −70°C in nutrient broth (NB) (Sifin, Berlin, Germany) containing 15% glycerol. Isolates were encoded using a combination of letters and numbers indicating: (1) soil type or plant species (Wb = desert soil; Sb = Sekem soil, Mc = *Matricaria chamomilla*, Co = *Calendula officinalis*, Sd = *Solanum distichum*), (2) replicate (1–4), (3) microenvironment (Re = endorhiza, rhizosphere and soil have no further designation), and (4) consecutive number of the isolate per replicate.

Altogether, 1,212 selected bacterial isolates were screened in dual-culture *in vitro* assays on Waksman agar (WA) [46] for their antagonistic potential towards *Verticillium dahliae* Kleb. V25, *Rhizoctonia solani* Kühn AG4, and *Fusarium culmorum* (Wm. G. Sm.) Sacc. E1. For *R. solani* and *F. culmorum* agar disks of 5 mm diameter with mycelia were directly cut out from PDA plates (Roth, Karlsruhe, Germany) and placed between the streaks of four bacterial isolates. *V. dahliae* was grown in liquid culture in Czapek Dox broth (Duchefa, Haarlem, Netherlands) at 20°C. 200 µl of the suspension containing hyphal fragments were plated onto the agar and after surface drying the isolates were placed on the same plate. Inhibition zones were measured after 4–7 days of incubation at RT. Each isolate was tested twice independently. From antagonistic isolates, DNA was prepared following the protocol of Berg et al. [46]. Amplified ribosomal RNA gene restriction analysis (ARDRA) using the restriction endonucleases *Hha*I (MP Biomedicals, Eschwege, Germany) and *Pst*I (New England Biolabs, Ipswich, UK) was used to group isolates at genus level. Isolates displaying similar ARDRA patterns were further analysed using BOX-PCR genomic fingerprinting. BOX-PCR fingerprints were performed using the BOX_A1R primer (5'-CTA CGG CAA GGC GAC GCT GAC G-3') as described by Rademaker and de Bruijn [47]. PCR conditions were used as specified by Berg et al. [28] and PCR products were separated by gel electrophoresis on 1.5% agarose gels. Antagonists with either individual ARDRA patterns or different BOX patterns (cut-off level 80%) were identified by partial 16S rRNA gene sequence analysis according to Berg et al. [46]. PCR product was sequenced with the Applied Biosystems 3130l Genetic Analyser sequencer, Data Collection v3.0, Sequencing Analysis v5.2 (Foster City, USA) at the sequencing core facility ZMF, Medical University of Graz, Austria. Obtained sequences were aligned with reference gene sequences from GenBank using BLAST algorithm. Sequences obtained were submitted to EMBL Nucleotide Sequence Database under accession numbers FR854236-FR854290.

Supporting Information

Figure S1 16S rRNA PCR-SSCP profiles of the bacterial communities in soil and endorhiza of the medical plants. Std.: 1 kb DNA ladder. The following bands were identified as: 1. *Ochrobactrum grignonense,* 99% similarity to NR_028901 and 2. *Rhodococcus erythropolis* 99% similarity to NR_037024.

Figure S2 Abundances of (A) total and (B) culturable bacteria in desert and agricultural soil. Data for total bacteria were ascertained by qPCR of the 16S rRNA genes and data for culturable bacteria by isolation on R2A. Averages of 16S rRNA gene copy numbers and viable counts per gram soil as $\log_{10}$ and confidences are shown.

Table S1 Relative composition of bacterial phyla, classes, orders, families and genera in desert and agricultural soil.

Table S2 Identification of selected bacterial antagonists isolated from different habitats.

Acknowledgments

This publication is dedicated to Ibrahim Abouleish, founder of SEKEM and winner of Right Livelihood Award. We would like to thank Angela Hofmann (Cairo), Kornelia Smalla (Braunschweig) and Birgit Birnstingl (Graz), for inspiring discussions and Rudolf Bauer (Graz) for advice regarding the medical plants. Furthermore, we want to thank Christin Zachow (Graz) for her helpful support in all practical questions and Christian Berg (Graz) for statistical support.

Author Contributions

Conceived and designed the experiments: GB HM EMR MK. Performed the experiments: MK. Analyzed the data: MK GB. Contributed reagents/materials/analysis tools: GB. Wrote the paper: MK GB.

References

1. Clery D (2011) Environmental technology. Greenhouse-power plant hybrid set to make Jordan's desert bloom. Science 331: 136.
2. Pereira AC, Gama VF (2010) Anthropization on the Cerrado biome in the Brazilian Uruçuí-Una Ecological Station estimated from orbital images. Braz J Biol 70: 969–976.
3. Reuters (2007) Egypt plan to green Sahara desert stirs controversy. Reuters website, Available: http://www.reuters.com/article/2007/10/09/us-desert-egypt-idUSL2651867020071009. Accessed 2011 Aug 16.
4. Barot S, Blouin M, Fontaine S, Jouquet P, Lata JC, et al. (2007) A tale of four stories: soil ecology, theory, evolution and the publication system. PLoS One 28: e1248.
5. Acosta-Martínez V, Dowd S, Sun Y, Allen V (2008) Tag-encoded pyrosequencing analysis of bacterial diversity in a single soil type as affected by management and land use. Soil Biol Biochem 40: 2762–2770.
6. Cary SC, McDonald IR, Barrett JE, Cowan DA (2010) On the rocks: the microbiology of Antarctic Dry Valley soils. Nat Rev Microbiol 8: 129–138.
7. Caruso T, Chan Y, Lacap DC, Lau MC, McKay CP, et al. (2011) Stochastic and deterministic processes interact in the assembly of desert microbial communities on a global scale. ISME J: In press.
8. Fierer N, Jackson RB (2006) The diversity and biogeography of soil bacterial communities. Proc Natl Acad Sci USA 103: 626–631.
9. Pointing SB, Chan Y, Lacap DC, Lau MC, Jurgens JA, et al. (2009) Highly specialised microbial diversity in hyper-arid polar desert. Proc Natl Acad Sci USA 106: 19964–19969.
10. Angel R, Soares MI, Ungar ED, Gillor O (2010) Biogeography of soil archaea and bacteria along a steep precipitation gradient. ISME J 4: 553–563.
11. de Los Ríos A, Valea S, Ascaso C, Davila A, Kastovsky J, et al. (2010) Comparative analysis of the microbial communities inhabiting halite evaporites of the Atacama Desert. Int Microbiol 13: 79–89.
12. Othman AA, Amer WM, Fayez M, Hegazi NA (2004) Rhizosheath of Sinai desert plants is a potential repository for associative diazotrophs. Microbiol Res 159: 285–293.
13. Saul-Tcherkas V, Steinberger Y (2011) Soil microbial diversity in the vicinity of a Negev Desert shrub – *Reaumuria negevensis*. Microb Ecol 61: 64–81.
14. Mäder P, Fliessbach A, Dubois D, Gunst L, Fried P, et al. (2002) Soil fertility and biodiversity in organic farming. Science 296: 1694–1697.
15. Mendes R, Kruijt M, de Bruijn I, Dekkers E, van der Voort M, et al. (2011) Deciphering the rhizosphere microbiome for disease-suppressive bacteria. Science 27: 1097–1100.
16. Schieffer A, Lessem R (2009) Beyond social and private enterprise: towards the integrated enterprise. Transit Stud Rev 15: 713–725.
17. Luske B, van der Kamp J (2009) Carbon sequestration potential of reclaimed desert soils in Egypt. Louis Bolk Instituut & Soil and More International. Soil and More International website, Available: http://www.soilandmore.nl/userfiles/Carbon_Sequestration_Potential_Report_Final.pdf. Accessed 2011 Mar 16.
18. Lauber CL, Hamady M, Knight R, Fierer N (2009) Pyrosequencing-based assessment of soil pH as a predictor of soil bacterial community structure at the continental scale. Appl Environ Microbiol 75: 5111–5120.
19. Lazarevic V, Whiteson K, Huse S, Hernandez D, Farinelli L, et al. (2009) Metagenomic study of the oral microbiota by Illumina high-throughput sequencing. J Microbiol Methods 79: 266–271.
20. Will C, Thürmer A, Wollherr A, Nacke H, Herold N, et al. (2010) Horizon-specific bacterial community composition of German grassland soils, as revealed by pyrosequencing-based analysis of 16S rRNA genes. Appl Environ Micobiol 76: 6751–6759.
21. Hansel CM, Fendorf S, Jardine PM, Francis CA (2008) Changes in bacterial and archaeal community structure and functional diversity along a geochemically variable soil profile. Appl Environ Microbiol 74: 1620–1633.
22. Janssen PH (2006) Identifying the dominant soil bacterial taxa in libraries of 16S rRNA and 16S rRNA genes. Appl Environ Microbiol 72: 1719–1728.
23. Emmert EAB, Handelsman J (1999) Biocontrol of plant diseases: a (Gram-) positive perspective. FEMS Microbiol Lett 171: 1–9.
24. Battistuzzi FU, Hedges SB (2008) A major clade of prokaryotes with ancient adaptations to life on land. Mol Biol Evol 26: 335–343.
25. Crowder DW, Northfield TD, Strand MR, Snyder WE (2010) Organic agriculture promotes evenness and natural pest control. Nature 466: 109–112.
26. Schmid F, Moser G, Müller H, Berg G (2011) Functional and structural microbial diversity in organic and conventional viticulture: organic farming benefits natural biocontrol agents. Appl Environ Micobiol 77: 2188–2191.
27. Niraula NP, Kim SH, Sohng JK, Kim ES (2010) Biotechnological doxorubicin production: pathway and regulation engineering of strains for enhanced production. Appl Microbiol Biotechnol 87: 1187–1194.
28. Berg G, Eberl L, Hartmann A (2005) The rhizosphere as a reservoir for opportunistic human pathogenic bacteria. Environ Microbiol 7: 1673–1685.
29. Ben-David EA, Zaady E, Sher Y, Nejidat A (2011) Assessment of the spatial distribution of soil microbial communities in patchy arid and semi-arid landscapes of the Negev Desert using combined PLFA and DGGE analyses. FEMS Microbiol Ecol 76: 492–503.
30. Smalla K, Wieland G, Buchner A, Zock A, Parzy J, et al. (2001) Bulk and rhizosphere soil bacterial communities studied by denaturing gradient gel electrophoresis: plant-dependent enrichment and seasonal shifts revealed. Appl Environ Microbiol 67: 4742–4751.
31. Berg G, Smalla K (2009) Plant species and soil type cooperatively shape the structure and function of microbial communities in the rhizosphere. FEMS Microbiol Ecol 68: 1–13.
32. McKay DL, Blumberg JB (2006) A review of the bioactivity and potential health benefits of chamomile tea (*Matricaria recutita* L.). Phytother Res 20: 519–530.
33. Shaw LJ, Morris P, Hooker JE (2006) Perception and modification of plant flavonoid signals by rhizosphere microorganisms. Environ Microbiol 8: 1867–1880.
34. Schwieger F, Tebbe CC (1998) A new approach to utilize PCR-single-strand-conformation polymorphism for 16S rRNA gene-based microbial community analysis. Appl Environ Microbiol 64: 4870–4876.
35. Zachow C, Tilcher R, Berg G (2008) Sugar beet-associated bacterial and fungal communities show a high indigenous antagonistic potential against plant pathogens. Microb Ecol 55: 119–129.
36. Milling A, Smalla K, Maidl FX, Schloter M, Munch JC (2004) Effects of transgenic potatoes with an altered starch composition on the diversity of soil and rhizosphere bacteria and fungi. Plant Soil 266: 23–39.
37. Bassam BJ, Caetano-Anolles G, Gresshoff PM (1991) Fast and sensitive silver staining of DNA in polyacrylamid gels. Anal Biochem 196: 80–83.
38. Kropf S, Heuer H, Grüning M, Smalla K (2004) Significance test for comparing complex microbial community fingerprints using pairwise similarity measures. J Microbiol Methods 57: 187–195.
39. Leps J, Smilauer P (2003) Multivariate analysis of ecological data using Canoco. Cambridge: Cambridge University Press. 282 p.
40. Lane DJ (1991) 16S/23S rRNA sequencing. In: Stackebrandt E, Goodfellow M, eds. Nucleic acid techniques in bacterial systematics. Chichester: John Wiley & Sons. pp 115–175.
41. Stocker S, Snajder R, Rainer J, Trajanoski S, Gorkiewicz G, et al. (2011) SnoWMAn: High-throughput phylotyping, analysis and comparison of microbial communities. Under revision.
42. Caporaso JG, Kuczynski J, Stombaugh J, Bittinger K, Bushman FD, et al. (2010) QIIME allows analysis of high-throughput community sequencing data. Nat Methods 7: 335–336.
43. Cole JR, Wang Q, Cardenas E, Fish J, Chai B, et al. (2009) The ribosomal database project: improved alignments and new tools for rRNA analysis. Nucleic Acids Res 37: D141–D145.
44. Shannon CE (1997) The mathematical theory of communication. 1963. MD Comput 14: 306–317.
45. Chao A, Bunge J (2002) Estimating the number of species in a stochastic abundance model. Biometrics 58: 531–539.
46. Berg G, Roskot N, Steidle A, Eberl L, Zock A, et al. (2002) Plant-dependent genotypic and phenotypic diversity of antagonistic rhizobacteria isolated from different *Verticillium* host plants. Appl Environ Microbiol 68: 3328–3338.
47. Rademaker JLW, de Bruijn FJ (1997) Characterization and classification of microbes by rep-PCR genomic fingerprinting and computer-assisted pattern analysis. In: Caetano-Anollés G, Gresshoff PM, eds. DNA markers: protocols, applications and overviews. New York: John Wiley & Sons. pp 151–171.

Comparisons of Three Methods for Organic and Inorganic Carbon in Calcareous Soils of Northwestern China

Xiujun Wang[1,2]*, Jiaping Wang[1,3], Juan Zhang[1,3]

1 State Key Laboratory of Desert and Oasis Ecology, Xinjiang Institute of Ecology and Geography, Chinese Academy of Sciences, Urumqi, Xinjiang, China, **2** Earth System Science Interdisciplinary Center, University of Maryland, College Park, Maryland, United States of America, **3** Graduate University of Chinese Academy of Sciences, Beijing, China

Abstract

With increasing interest in the carbon cycle on arid land, there is an urgent need to quantify both soil organic carbon (SOC) and inorganic carbon (SIC) thus to assess various methods. Here, we present a study employing three methods for determinations of SOC and SIC in the Yanqi Basin of northwest China. We use an elemental analyzer for both SOC and SIC, the Walkley-Black method for SOC, a modified pressure calcimeter method for SIC, and a simple loss-on-ignition (LOI) procedure for determinations of SOC and SIC. Our analyses show that all three approaches produce consistently low values for SOC (1–14 g kg^{-1}) and high values for SIC (8–53 g kg^{-1}). The Walkley-Black method provides an accurate estimate of SOC with 100% recovery for most soil samples. The pressure calcimeter method is as accurate as the elemental analysis for measuring SIC. In addition, SOC and SIC can be accurately estimated using a two-step LOI approach, i.e., (1) combustion at 375°C for 17 hours to estimate SOC, and (2) subsequent combustion at 800°C for 12 hours to estimate SIC. There are strong linear relationships for both SOC and SIC between the elemental analysis and LOI method, which demonstrates the capability of the two-step LOI technique for estimating SOC and SIC in this arid region.

Editor: David L. Kirchman, University of Delaware, United States of America

Funding: This study is financially supported by the Program of 100 Talented Young Scientists with the Chinese Academy of Sciences (O972021001)(http://www.cas.cn/ggzy/rcpy/brjh/). The funders had no role in study design, data collection and analysis, decision to publish, or preparation of the manuscript.

Competing Interests: The authors have declared that no competing interests exist.

* E-mail: wwang@essic.umd.edu

Introduction

Soil carbon storage, as the third largest carbon pool in the Earth System, plays an important role in the global carbon cycle and climate change [1]. The majority of carbon, in most soils, is held as soil organic carbon (SOC) whereas in soils of the arid and semiarid regions, the most common form is inorganic carbon, primarily carbonate [2]. Quantifying both SOC and soil inorganic carbon (SIC) is essential to our understanding of the carbon cycle at regional to global scales.

Soil organic carbon is commonly measured by dry combustion with automated analyzers, or a wet chemical oxidation method, i.e., the Walkley and Black method [3]. The automated technique is simple and accurate, but the cost is high and may not be feasible. In addition, this approach often involves pretreatment with acid to remove carbonate for calcareous soils, which may erode the instrument and destroy organic matter in soil samples [4]. On the other hand, the Walkley and Black method has been used as one of the standard methods to determine SOC [5], particularly in China [6]. This technique is less expensive than dry combustion. However, this procedure may lead to variable recovery of SOC [7,8], and hazard Cr due to the use of dichromate.

Soil inorganic carbon can be accurately measured by several methods [9]. Commonly, SIC is determined by measuring CO_2 production after adding HCl acid into soil [10–13]. Sherrod et al. [12] demonstrated that a modified pressure calcimeter method, with addition of $FeCl_2$ to minimize oxidation of organic matter, provided reliable measurement of SIC. Alternatively, dry combustion with automated analyzers can also be utilized to determine SIC. A direct measurement involves dry combustion of soils that are pre-combusted to remove organic matter in an O_2 stream [14]. An indirect measurement is to determine total carbon (without pretreatment with acid to remove carbonate) and SOC separately, then to calculate the difference. Although the automated approaches would provide accurate estimates for SIC, they might not be used widely because of the high costs associated with purchase and maintenance of automated analyzers.

Apart from these traditional methods, loss-on-ignition (LOI) provides an alternative approach, which involves heating soil samples at high temperature to combust soil organic matter (SOM) or carbonate and measuring weight losses [15,16]. The LOI techniques are simple, less expensive and less labor intensive relative to the automated techniques and chemical methods. Thus, the LOI approaches have been used widely to estimate SOM or SOC in agricultural and forest soils and sediments [16–21], and to determine carbonate content in sediments [13,15,22,23]. However, this technique is rarely applied to low fertility soils for SOM or SOC measurement, or used to measure SIC despite some studies showing that the LOI method can give accurate estimates of carbonate for sediments [13,24]. Moreover, there is little utilization of the LOI methods for estimating SOC and SIC in Chinese soils.

The arid and semiarid lands cover approximately 35% of the Earth's land surface, and may play an important role in the global carbon cycle and climate mitigation [1,25,26]. With recent interest in studying the carbon cycle on arid land [27], there is a need to quantify both SOC and SIC of arid soils thus to assess various methods. Here, we select three techniques to determine SOC and SIC contents in the calcareous soils at the Yanqi Basin of northwest China. We employ an elemental analyzer for both SOC and SIC measurements, a wet oxidation for SOC, a modified pressure calcimeter method for SIC, and a simple LOI procedure for determinations of SOC and SIC. The objective of this study is to examine the relationships between these methods, and to test the hypothesis that the LOI technique has capacity of accurately determining SOC and SIC in arid soils.

Materials and Methods

Site Description

There were no specific permits required for the described field studies. We confirm that the location was not privately-owned or protected in any way, and the field studies did not involve endangered or protected species. Our studying area is along the Kaidu River in the Yanqi Basin (Figure 1), which is located on the southeastern flank of the Tianshan Mountain. Brown desert soil is the main soil type, which was developed from limestone parent material and classified as a Haplic Calcisol [28]. The sampling area (approximately 2,200 km) spans both sides of the Kaidu River, which has various land uses, including desert land, shrub land, cropland and grassland. The area has an annual precipitation of less than100 mm and annual evaporation of approximately 2000 mm. Agricultural lands have irrigation systems that extract water from the Kaidu River, whereas the other lands rely on rainfall and underground waters. Annual mean temperature ranges from 7°C to 10°C. On average, the annual maximum temperature is 38°C, and the minimum temperature −35°C.

Soil Sampling and Analyses

We collected 70 soil samples from 15 profiles in November, 2010. While these profiles vary in texture and color because of land use history, we sampled five layers at most profiles, i.e., 0–5 cm, 5–15 cm, 15–30 cm, 30–50 cm, and 50–100 cm, and determined bulk density (BD). For each layer, soils were air-dried, thoroughly mixed, and sieved to pass a 2-mm screen for pH and electric conductivity (EC) measurements. Basic soil properties are given in Table 1. In brief, soils in this region are characterized by high pH (from 8 to 9.4) and high sand/silt contents. Bulk density is slightly lower in the surface soils (0–15 cm: 1.38–1.41 g cm^{-3}) than in below (1.46–1.54 g cm^{-3}). Electric conductivity shows a decreasing trend over depth, from 1.25–1.32 ms cm^{-1} in the 0–15 cm to 0.7 ms cm^{-1} in the 50–100 cm. Representative subsamples were crushed to pass a 0.25 mm screen for SOC and SIC measurements by the following procedures.

Modified Walkley-Black Method for SOC

This method is modified from the traditional Walkley-Black method [3]. A soil sample (~0.4 g) is treated with 5 ml concentrated H_2SO_4 for 4 hours, then mixed with 5 ml 0.5 M $K_2Cr_2O_7$. The mixture is heated at 150–160°C for 5 minutes, and then cooled at room temperature. The solution is transferred into a triangular flask with 100 ml deionized water. Unreacted $K_2Cr_2O_7$ is determined by titrating with 0.25 M $FeSO_4$. Soil organic carbon content is calculated, without a recovery factor, from the difference in $FeSO_4$ used between a blank and a soil solution.

Pressure Calcimeter Method for SIC

We use a procedure similar to that of Sherrod [12]. A subsample of 1.0 g soil and a tiny bottle with 2 ml 6 M HCl containing 3% by weight of $FeCl_2 \cdot 4H_2O$ are placed at the bottom of a 100-ml reaction vessel. After sealing the vessel with a butyl rubber stopper that has an aluminum tearoff seal cap, we mix the acid with soil by turning the vessel sideways. Two hours later, we remove the aluminum tearoff seal cap, and then insert an 18-gauge hypodermic needle that is attached to the pressure transducer and voltage meter. We record the voltage reading to two decimal places after 3 to 5 seconds, and then calculate CO_2 concentration using a calibration curve that is developed by mixing reagent grade $CaCO_3$ with oven-dried laboratory sand. Standards are made based on a final weight of 1.0000 g mixture of sand and $CaCO_3$ to obtain inorganic carbon concentrations of 1.2, 3.6, 6.0, 12.0, 18.0, 24.0, 30.0, 36.0, 42.0, 48.0, 54.0 and 60.0 g C kg^{-1}, respectively.

Elemental Analyzer for SOC and SIC

We used a CNHS-O analyzer (EuroEA3000) for SOC and SIC estimates. For SOC measurement, 20 mg soil is pretreated with 10 drips of H_3PO_4 for 12 hours to remove carbonate. The sample is combusted at 1020°C with constant helium flow carrying pure oxygen to ensure complete oxidation of organic materials. CO_2 production is determined by a thermal conductivity detector. Total soil carbon is measured, using the same procedure without pretreatment with H_3PO_4. Soil inorganic carbon is calculated as the difference between total soil carbon and SOC.

Loss-on-ignition (LOI) Procedure for SOM and SIC

Following Wang et al. [16] and Wang et al. [15], we first place 5.0000 g soil in a quartz glass and heated at 105°C for 12 hours to remove soil moisture. We then combust soil in a programmable muffle furnace (S1849, KOYO LINDBEERE LTD) at 375°C for 17 hours and 800°C for 12 hours. Soil organic matter is calculated as the weight loss between 105°C and 375°C, and SIC as the weight loss between 375°C and 800°C:

$$SOM_{LOI}\,(g\,kg^{-1}) = \left(\frac{Weight_{105^\circ C} - Weight_{375^\circ C}}{Weight_{105^\circ C}}\right) \times 1000 \quad (1)$$

$$SIC_{LOI}\,(g\,kg^{-1}) = \left(\frac{Weight_{375^\circ C} - Weight_{800^\circ C}}{Weight_{105^\circ C}}\right) \times 0.273 \times 1000. \quad (2)$$

The conversion constant of 0.273 in equation (2) is applied to convert mass of CO_2 to mass of carbon.

Results

Soil Organic Carbon

We first compare measured SOC between the automated CNS analysis and the modified Walkley-Black method, and find a strong correlation ($r^2 = 0.93$, $P<0.001$) between the two methods (Table 2). As shown in Figure 2, SOC content is below 15 g kg^{-1} in this region. It appears that the modified Walkley-Black method over-estimates SOC when soil contains extremely low SOC (<5 g

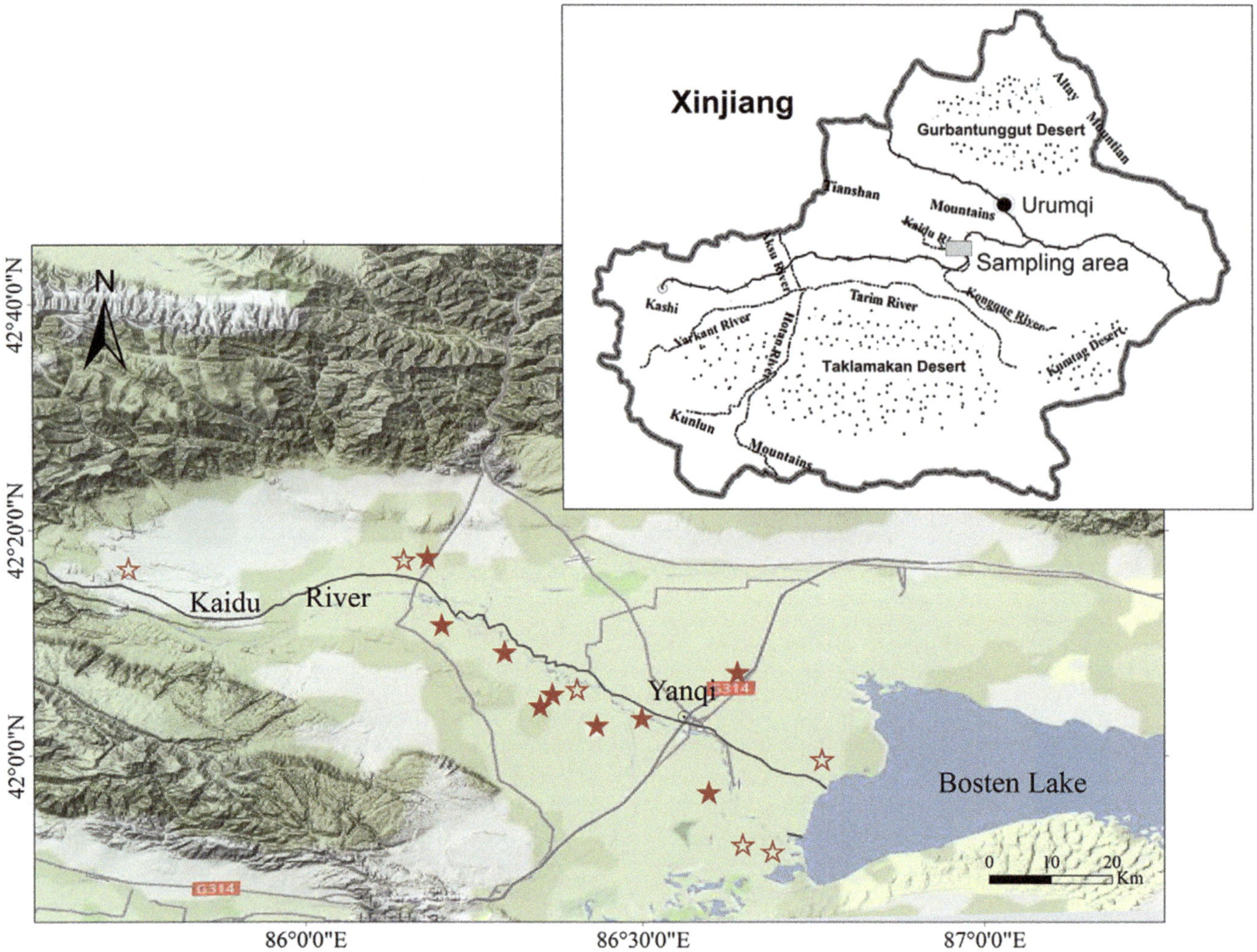

Figure 1. Sampling area with locations of 15 soil profiles. Solid stars denote the sites where topsoil texture was determined.

kg^{-1}). Excluding these data yields an averaged recovery of 0.993. These results indicate that almost 100% SOC in the brown desert soil can be oxidized, which is much higher than previous published recovery rates (less than 82%) for forest soils across Flanders region [7], but in agreement with those (~100%) in most Tasmanian soils [16]. The high recovery of SOC and large correlation coefficient in our study suggest that the modified Walkley-Black method is as accurate as the automated CNS technique except for extremely low SOC soils.

We then compare measured SOM by the LOI method with SOC by the automated CNS technique (Figure 3). As expected, SOM values (0–27 g kg^{-1}) are higher than SOC values

Table 1. Means and standard deviations (in brackets) of soil pH, bulk density (BD, g kg^{-1}) and electrical conductivity (EC, ms cm^{-1}) for each layer, and clay, silt and sand contents (%) of topsoil (0–30 cm) from 8 randomly selected soil profiles.

	pH	BD	EC	Clay	Silt	Sand
0–5	8.92 (0.3)	1.41 (0.1)	1.25 (2.4)	5.6 (2.9)	57.3 (23.0)	37.1 (25.1)
5–15	8.77 (0.3)	1.38 (0.1)	1.32 (1.5)			
15–30	8.78 (0.4)	1.50 (0.2)	1.07 (1.2)			
30–50	8.88 (0.4)	1.54 (0.1)	0.98 (1.2)	-	-	-
50–100	9.00 (0.2)	1.46 (0.1)	0.70 (1.2)	-	-	-

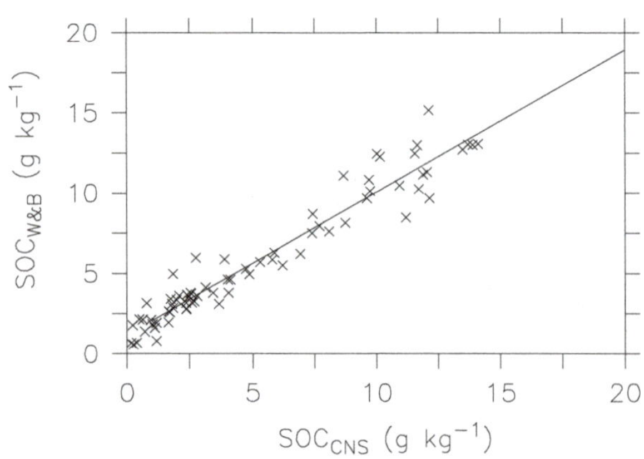

Figure 2. Linear regression of measured SOC by the CNS analyzer and Walkley-Black method.

Table 2. Relationships of soil organic carbon (SOC) and inorganic carbon (SIC) among different methods.

	n	Slope	Intercept	r^2	P *value*
$SOC_{W\&B}$ vs. SOC_{CNS}	70	0.886	1.199	0.92	<0.001
SOM_{LOI} vs. SOC_{CNS}	70	1.792	4.189	0.90	<0.001
SOC_{LOI} vs. $SOC_{W\&B}$	31	0.949	0.443	0.77	<0.001
SIC_{PC} vs. SIC_{CNS}	70	1.025	−1.718	0.99	<0.001
SIC_{LOI} vs. SIC_{CNS}	70	1.043	2.582	0.96	<0.001
SIC_{LOI} vs. SIC_{PC}	70	1.021	4.241	0.98	<0.001

(<15 g kg^{-1}). The large intercept value of 4.189 indicates that there are some non-SOM weight losses (e.g., from structural water and carbonate) during ignition. The conversion factor of 1.792 is slightly larger than the traditional values of 1.724. We find that the conversion factor is 2.71 for those samples with low SOC content (<5 g kg^{-1}) that are mainly subsoils. Based on all 70 soil samples tested, SOC may be calculated from LOI at 375°C for 17 hours, using the following equation:

$$SOC_{LOI} = (SOM_{LOI} - 4.189)/1.792 \quad (3)$$

Figure 4 shows comparison of measured SOC by the Walkley-Black method and calculated SOC by the LOI method using the equation (3). To avoid negative values in calculated SOC, we exclude samples that contain less than 4.189 g kg^{-1} SOM_{LOI} in this comparison. The slope value of 0.949 and significantly high correlation coefficient ($r^2 = 0.77$, $p<0.001$) indicate that the LOI method has potential for estimating SOC in this arid region.

Soil Inorganic Carbon

We assess measured SIC by the pressure calcimeter method and the automated CNS analysis (Figure 5). Clearly, SIC content is much higher than SOC (<15 g kg^{-1}) in this region. The pressure calcimeter method produces slightly lower SIC values (6–51 g kg^{-1}) than the automated CNS technique (9–53 g kg^{-1}). However, there is a high correlation ($r^2 = 0.99$, $p<0.001$) between the two methods. The slope value of 0.996 and large correlation coefficient indicate that the pressure calcimeter method is as accurate as the automated CNS technique for the calcareous soils.

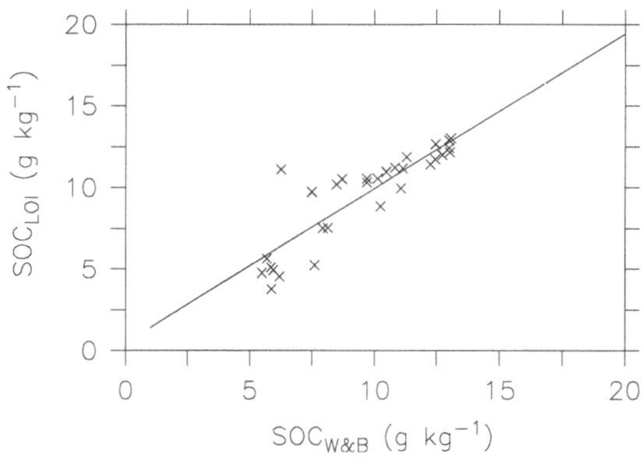

Figure 4. Relationship between SOC calculated by the LOI method and SOC measured by the Walkley-Black method.

Figure 6 presents comparison of measured SIC between the LOI method and the automated CNS analysis. Both methods give a very similar range, and form a linear relationship:

$$SICLOI = 1.043SICCNS + 2.582 \; (r^2 = 0.96, P<0.001) \quad (4)$$

The intercept value of 2.582 is significantly different from 0, suggesting that there may be some non-SIC weight losses (e.g., dehydration from other compounds) during combustion [22,23,29]. Nevertheless, the factors of slope value close to 1 and significant correlation indicate that the LOI technique can produce accurate estimates for SIC.

As shown in Figure 7, there is also a linear relationship for SIC between the pressure calcimeter method and the LOI technique:

$$SIC_{LOI} = 1.021SIC_{PC} + 4.241 \quad (5)$$

which shows a similar slope to that in the regression between the LOI method and the CNS analysis (i.e., equation 4). The relatively larger intercept in equation (5) reflects the slightly lower SIC values by the pressure calcimeter method, relative to those by the automated CNS analysis (see Table 2).

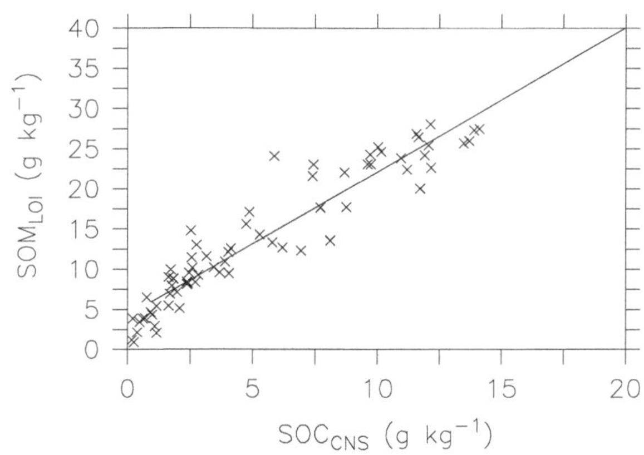

Figure 3. Relationship between SOC measured by the CNS analyzer and SOM determined by the LOI method.

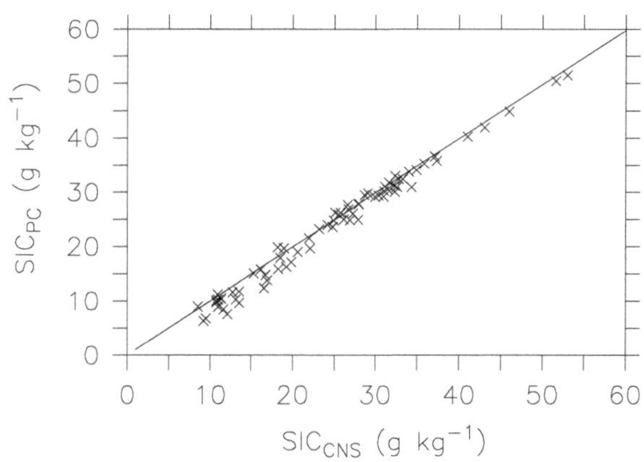

Figure 5. Liner relationship for SIC by the CNS analyzer and pressure calcimeter method.

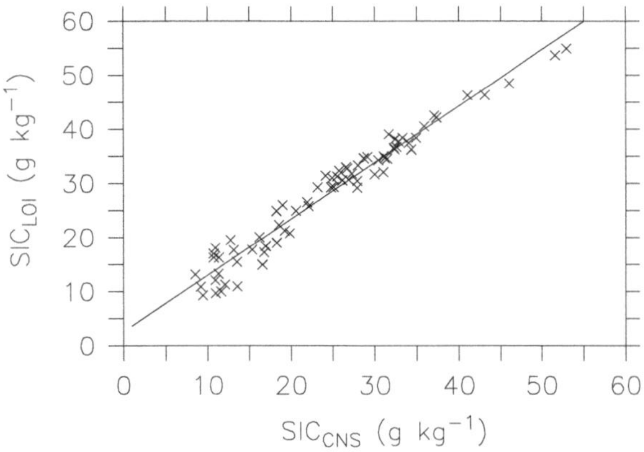

Figure 6. Liner relationship for SIC between the CNS analyzer and LOI method.

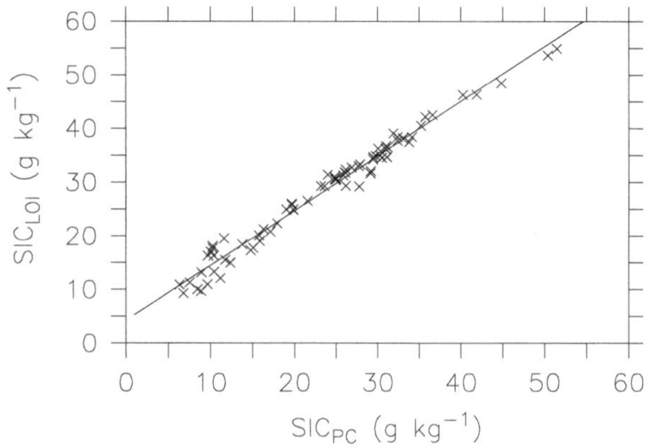

Figure 7. Liner relationship for SIC by the pressure calcimeter and LOI methods.

Discussion

Estimate of SOC

Measurement of SOC has been a common procedure in soil science. Many methods are available, each with advantages and disadvantages in terms of accuracy, expense and convenience [5,30]. The Walkley-Black procedure is considered to be cheap and easy to perform, but may exhibit variable recovery thus a correction factor is often applied to determine the total SOC content for a soil sample. For example, Sleutel et al. [31] applied a correction factor of 1.33 for estimated SOC by the Walkley-Black without external heating. However, an early study [16] showed that without external heating, the Walkley-Black method produced approximately 100% recovery for SOC in most Tasmanian soils. Many modified dichromate oxidation techniques that involve external heating do not require a correction factor [32]. Our study demonstrates that compared with the automated CNS technique, the modified Walkley-Black method with external heating yields almost 100% recovery for SOC in the calcareous soils we tested. However, the Walkley-Black method may over-estimate SOC in soils containing extremely low SOC [8], which is probably attributed to reactions of $K_2Cr_2O_7$ with inorganic soil constituents [33].

While the LOI method has been widely used to estimate SOM and SOC, there is no standard procedure in terms of heating duration and temperature. There have been various combinations with temperature ranging from 150°C to 900°C and duration from 2 hours to 17 hours in the literature, thus a regression between LOI and SOC (measured by an automated analyzer or the Walkley-Black method) varies widely [16,34,35]. In general, with increasing temperature, LOI increases whereas correlation coefficient tends to decrease [35].

One criticism of the LOI technique is that it may result in weight losses from structural water and carbonate during heating [30,34]. These non-SOM weight losses can significantly affect estimates of SOC, particularly in soils containing low SOM and high clay content. As the amount of SOM decreases, dehydroxylation of clays becomes more likely [36]. However, clay content is low in our soils (see Table 1). Thus, weight loss due to dehydroxylation of clays may be small. On the other hand, temperature is the dominant factor affecting LOI, and its relationship with SOC [35]. Therefore, it is critical to choose a temperature that is high enough to completely remove SOM but low enough to prevent dehydroxylation of clay minerals and oxidation of carbonate.

Several studies have shown that a combination of 375°C and 16–17 hours results in a good LOI-SOC relationship, thus it can provide accurate estimates of SOC in many soil groups [16,37,38]. Our study demonstrates that there is a strong relationship between SOC and LOI at 375°C for 17 hours in the calcareous soils tested. While the non-zero intercept may reflect some non-SOM weight losses, the strong relationship indicates the capacity of using the LOI method for SOC estimation.

Estimate of SIC

Soil carbonate is usually quantified by acid dissolution in the reaction below:

$$CaCO_3 + 2H^+ \rightarrow Ca^{2+} + CO_2 + H_2O \qquad (6)$$

with the determination of either H^+ consumption or CO_2 production. The methods of H^+ consumption that involve reaction with a strong acid, such as HCl addition and back titration of the unreacted acid, may be problematic owing to H^+ consumption by other soil components [9]. Indeed, our own analyses using the H^+ consumption approach lead to significant underestimates of SIC (by 20%) for the soils we tested (data not shown).

There have been many methods involving determination of CO_2 production following acid dissolution [9]. While these methods are much preferred, precautions must be taken to ensure that there is no interference from organic matter oxidation. Thus, an oxidation inhibitor (e.g., $FeCl_2$) is often used to limit the oxidation of organic matter and the subsequent evolution of CO_2 from this source. Our study shows a strong correlation between the pressure calcimeter method and the automated CNS analysis. Given that measured SIC contents by the pressure calcimeter method are slightly lower than, but not significantly different from those by the automated CNS analysis, we conclude that the pressure calcimeter approach can provide an accurate measurement of SIC for the calcareous soils of this arid region.

In contrast to wide utilization of the LOI method for SOC estimates, there is little in the literature regarding application of the LOI method for SIC estimates. However, the LOI method has been used to measure carbonate content in sediments, following combustion of organic carbon at lower temperature [13,15,22,23]. The simple two-step LOI procedure for estimates of organic carbon and carbonate in sediments [15] is similar to that for determination of organic carbon and carbonate in calcareous soils by dry combustion [14]. Thus, it could be a convenient practice for estimating SOC and SIC.

To the best of our knowledge, this study is the first attempt to evaluate the LOI method for SIC measurement. Our study shows a significantly (r^2>0.96, P<0.001) linear relationship between estimated SIC by the LOI method and determined SIC by direct measurements, i.e., the automated element analyzer and the pressure calcimeter method. Despite the non-zero intercepts and slightly higher values in SIC by the LOI method, the slope values are close to one, which are consistent with those reported for sediment analyses [15]. Our results demonstrate that the LOI method has potential for easily and accurately estimating SIC in calcareous soils of arid regions.

Conclusions

We have compared three methods for determinations of SOC and SIC concentrations in calcareous soils of the Yanqi Basin. In spite of wide ranges in soil properties, SOC is generally low (1–14 g kg^{-1}) whereas SIC is considerably high (9–53 g kg^{-1}). Our study shows that the Walkley-Black method with external heating can provide accurate estimates of SOC (with almost 100% recovery) for most arid soils. The pressure calcimeter method is as accurate as the automated elemental technique for measuring SIC. In addition to these traditional methods, we have also evaluated a two-step LOI approach, i.e., (1) combustion at 375°C for 17 hours to estimate SOC, and (2) subsequent combustion at 800°C for 12 hours to estimate SIC. Our study shows that SOC and SIC can be estimated from LOI for calcareous soils except those with extremely low SOC or SIC contents. This study demonstrates that the two-step LOI technique has potential to be used as a sufficiently accurate technique for estimating SOC and SIC in arid soils. Further testing on a wider range of arid and semi-arid soils is warranted.

Acknowledgments

We are grateful for the reviewer's constructive comments.

Author Contributions

Conceived and designed the experiments: XW JW. Performed the experiments: JW JZ. Analyzed the data: XW. Contributed reagents/materials/analysis tools: XW. Wrote the paper: XW.

References

1. Lal R, Kimble JM (2000) Pedogenic carbonate and the global carbon cycle. In: Lal R, Kimble JM, Eswaran H, Stewart BA, editors. Golbal Climate Change and Pedogenic. Boca Raton, FL, USA.: CRC press. 1–14.
2. Eswaran H, Reich PF, Kimble JM, Beinroth FH, Padmanabhan E, et al. (2000) Global carbon stocks. In: R L, Kimble JM, Eswaran H, Stewart BA, editors. Global climate change and pedogenic carbonates. Boca Raton, FL: Lewis Publishers. 15–27.
3. Walkley A, Black IA (1934) An examination of the Degtjareff method for determining soil organic matter and a proposed modification of the chromic acid titration method. Soil Sci 37: 29–38.
4. Byers SC, Mills EL, Stewart PL (1978) A comparison of methods of determining organic carbon in marine sediments, with suggestions for a standard method. Hydrobiologia 58: 43–47.
5. Nelson DW, Sommers LE, Sparks D, Page A, Helmke P, et al. (1996) Total carbon, organic carbon, and organic matter. Methods of soil analysis Part 3-chemical methods: 961–1010.
6. Zhang WJ, Wang XJ, Xu MG, Huang SM, Liu H, et al. (2010) Soil organic carbon dynamics under long-term fertilizations in arable land of northern China. Biogeosciences 7: 409–425.
7. De Vos B, Lettens S, Muys B, Deckers JA (2007) Walkley–Black analysis of forest soil organic carbon: recovery, limitations and uncertainty. Soil Use and Management 23: 221–229.
8. Conyers MK, Poile GJ, Oates AA, Waters D, Chan KY (2011) Comparison of three carbon determination methods on naturally occurring substrates and the implication for the quantification of 'soil carbon'. Soil Research 49: 27–33.
9. Loeppert R, Suarez D, Sparks D, Page A, Helmke P, et al. (1996) Carbonate and gypsum. Methods of soil analysis Part 3-chemical methods: 437–474.
10. Presley B (1975) A simple method for determining calcium carbonate in sediment samples. J Sediment Petrol 45: 745–C746.
11. Dreimanis A (1962) Quantitative gasometric determination of calcite and dolomite by using Chittick apparatus. Journal of Sedimentary Research 32: 520.
12. Sherrod LA, Dunn G, Peterson GA, Kolberg RL (2002) Inorganic Carbon Analysis by Modified Pressure-Calcimeter Method. Soil Sci Soc Am J 66: 299–305.
13. Heiri O, Lotter AF, Lemcke G (2001) Loss on ignition as a method for estimating organic and carbonate content in sediments: reproducibility and comparability of results. Journal of Paleolimnology 25: 101–110.
14. Rabenhorst MC (1988) Determination of organic and carbonate carbon in calcareous soils using dry combustion. Soil Science Society of America Journal 52: 965–969.
15. Wang QR, Li YC, Wang Y (2011) Optimizing the weight loss-on-ignition methodology to quantify organic and carbonate carbon of sediments from diverse sources. Environmental Monitoring and Assessment 174: 241–257.
16. Wang XJ, Smethurst PJ, Herbert AM (1996) Relationships between three measures of organic matter or carbon in soils of eucalypt plantations in Tasmania. Australian Journal of Soil Research 34: 545–553.
17. David MB (1988) Use of loss-on-ignition to assess soil organic carbon in forest soils Communications in Soil Science and Plant Analysis 19: 1593–1599.
18. De Vos B, Vandecasteele B, Deckers J, Muys B (2005) Capability of loss-on-ignition as a predictor of total organic carbon in non-calcareous forest soils. Communications in soil science and plant analysis 36: 2899–2921.
19. Wright AL, Wang Y, Reddy KR (2008) Loss-on-Ignition Method to Assess Soil Organic Carbon in Calcareous Everglades Wetlands. Communications in soil science and plant analysis 39: 3074–3083.
20. Konen ME, Jacobs PM, Burras CL, Talaga BJ, Mason JA (2002) Equations for predicting soil organic carbon using loss-on-ignition for north central US soils. Soil Science Society of America Journal 66: 1878–1881.
21. Ball DF (1964) Loss-on-ignition as an estimate of organic matter and organic carbon in non-calcareous soils. Journal of Soil Science 15: 84–92.
22. Dean WE (1974) Determination of carbonate and organic matter in calcareous sediments and sedimentary rocks by loss on ignition: comparison with other methods. Journal of Sedimentary Petrology 44: 242–248.
23. Santisteban JI, Mediavilla R, López-Pamo E, Dabrio CJ, Zapata MBR, et al. (2004) Loss on ignition: a qualitative or quantitative method for organic matter and carbonate mineral content in sediments? Journal of Paleolimnology 32: 287–299.
24. Wang Q, Li Y, Wang Y (2011) Optimizing the weight loss-on-ignition methodology to quantify organic and carbonate carbon of sediments from diverse sources. Environmental Monitoring and Assessment 174: 241–257.
25. Scharpenseel HW, Mtimet A, Freytag J (2000) Soil inorganic carbon and global change. In: Lal R, Kimble JM, Eswaran H, Stewart BA, editors. Global Climate Change and Pedogenic Carbonates. Boca Raton, FL: CRC Press. 27–42.
26. Xie JX, Li Y, Zhai CX, Li CH, Lan ZD (2009) CO_2 absorption by alkaline soils and its implication to the global carbon cycle. Environmental Geology 56: 953–961.
27. Stone R (2008) Have desert researchers discovered a hidden loop in the carbon cycle? Science 320: 1409–1410.
28. FAO-UNESCO (1988) soil map of the world: revised legend. World Soil Resources Report No 60 FAO: Rome.
29. Allison L, Black C (1965) Methods of soil analysis. Part 2. Chemical and microbiological properties. Agronomy monograph 9: 1367–1378.
30. Chatterjee A, Lal R, Wielopolski L, Martin MZ, Ebinger MH (2009) Evaluation of Different Soil Carbon Determination Methods. Critical Reviews in Plant Sciences 28: 164–178.
31. Sleutel S, De Neve S, Singier B, Hofman G (2007) Quantification of organic carbon in soils: A comparison of methodologies and assessment of the carbon content of organic matter. Communications in Soil Science and Plant Analysis 38: 2647–2657.
32. Lettens S, De Vos B, Quataert P, van Wesemael B, Muys B, et al. (2007) Variable carbon recovery of Walkley-Black analysis and implications for national soil organic carbon accounting. European Journal of Soil Science 58: 1244–1253.
33. Walkley A (1947) A critical examination of a rapid method for determining organic carbon in soils- effect of variations in digestion conditions and inorganic soil constituents. Soil Science 63: 251–264.
34. Szava-Kovats R (2009) Re-analysis of the Relationship between Organic Carbon and Loss-on-Ignition in Soil. Communications in Soil Science and Plant Analysis 40: 2712–2724.
35. Abella S, Zimmer B (2007) Estimating organic carbon from loss-on-ignition in northern Arizona forest soils. Soil Science Society of America Journal 71: 545–550.
36. Pribyl DW (2010) A critical review of the conventional SOC to SOM conversion factor. Geoderma 156: 75–83.
37. McKeague J (1976) Manual of soil sampling and methods of analysis. Land Resource Research Inst., Res. Br. Agriculture Canada, Ottawa, Canada.
38. Beaudoin A (2003) A comparison of two methods for estimating the organic content of sediments. Journal of Paleolimnology 29: 387–390.

In Situ Enzyme Activity in the Dissolved and Particulate Fraction of the Fluid from Four Pitcher Plant Species of the Genus *Nepenthes*

Yayoi Takeuchi[1,2]*, Michaela M. Salcher[3], Masayuki Ushio[4], Rie Shimizu-Inatsugi[1], Masaki J. Kobayashi[1], Bibian Diway[5], Christian von Mering[6], Jakob Pernthaler[3], Kentaro K. Shimizu[1]

1 Institute of Plant Biology, University of Zurich, Zurich, Switzerland, **2** Graduate University for Advanced Studies, School of Advanced Sciences, Hayama, Kanagawa, Japan, **3** Department of Limnology, Institute of Plant Biology, University of Zurich, Kilchberg, Switzerland, **4** Center for Ecological Research, Kyoto University, Otsu, Shiga, Japan, **5** Botanical Research Centre, Sarawak Forestry Corporation, Kuching, Malaysia, **6** Institute of Molecular Life Sciences, University of Zurich, Zurich, Switzerland

Abstract

The genus *Nepenthes*, a carnivorous plant, has a pitcher to trap insects and digest them in the contained fluid to gain nutrient. A distinctive character of the pitcher fluid is the digestive enzyme activity that may be derived from plants and dwelling microbes. However, little is known about *in situ* digestive enzymes in the fluid. Here we examined the pitcher fluid from four species of *Nepenthes*. High bacterial density was observed within the fluids, ranging from 7×10^6 to 2.2×10^8 cells ml^{-1}. We measured the activity of three common enzymes in the fluid: acid phosphatases, β-D-glucosidases, and β-D-glucosaminidases. All the tested enzymes detected in the liquid of all the pitcher species showed activity that considerably exceeded that observed in aquatic environments such as freshwater, seawater, and sediment. Our results indicate that high enzyme activity within a pitcher could assist in the rapid decomposition of prey to maximize efficient nutrient use. In addition, we filtered the fluid to distinguish between dissolved enzyme activity and particle-bound activity. As a result, filtration treatment significantly decreased the activity in all enzymes, while pH value and *Nepenthes* species did not affect the enzyme activity. It suggested that enzymes bound to bacteria and other organic particles also would significantly contribute to the total enzyme activity of the fluid. Since organic particles are themselves usually colonized by attached and highly active bacteria, it is possible that microbe-derived enzymes also play an important role in nutrient recycling within the fluid and affect the metabolism of the *Nepenthes* pitcher plant.

Editor: Wendy A. Peer, Purdue University, United States of America

Funding: This study was funded by the University Research Priority Program in Systems Biology/Functional Genomics of the University of Zurich (http://www.sysbio.uzh.ch/index.html) and a Japan Science Promotion Society Research Fellowship for YT (http://www.jsps.go.jp/english/). These funders had no role in the study design, data collection and analysis, decision to publish, or preparation of the manuscript. BD performed the experiments and is an employee of Sarawak Forestry Corporation (http://www.sarawakforestry.com/), which permitted and supported our field research in Lambir Hills National Park in Sarawak, Malaysia.

Competing Interests: BD is an employee of Sarawak Forestry Corporation, which permitted and supported the authors' field research in Lambir Hills National Park in Sarawak, Malaysia.

* E-mail: takeuchi_yayoi@soken.ac.jp

Introduction

As already described by Darwin [1], carnivorous plants possess special organs to consume insects and other small invertebrates. This carnivory reflects an adaptive trait to grow in low-nutrient habitats where there is a particular lack of nitrogen, phosphorous, and potassium [2–4]. Such plants compensate for the lack of these nutrients in the soil by catching prey, and then digesting and absorbing nutrients from it.

The pitcher plants develop jug-shaped leaves or leaf extensions to trap animals. Among the seven genera of pitcher plants, *Nepenthes* is the largest genus in the tropics and is mostly distributed in the Malesia region, with its highest diversity occurring in Borneo [5]. The pitchers have nectar that is attractive to ants and other arthropods [6]. Because the inner wall of the upper pitcher is waxy and slippery, insects lose their footing while foraging for nectar and slip down the steep walls, thus becoming trapped in the fluid of the lower pitcher. It is reported that more than 60% of the nitrogen of the plant body is derived from prey in *Nepenthes mirabilis* (Lour.) Druce [7], which suggests a relatively high degree of reliance on animal-derived nutrient in this genera. The efficiency of prey capture and digestion is thus crucial for plant growth and survival.

The proximate agents in digesting or decomposing the organic matter are extracellular enzymes within the pitcher fluid. These enzymes in part originate from the plant itself, where they are secreted by digestive cellular glands that are located in the surface of the lower zone within the pitcher [8,9]. The characteristics of the enzymes derived from the *Nepenthes* species have been investigated in previous studies, and the enzymes include at least proteases [10–12], chitinases [12,13], phosphatases [8,14], and glucosidases [14].

On the other hand, digestive enzymes might also be produced by microbes such as fungi and bacteria that, for example, colonize insect carcasses. So far, such putatively microbial-derived enzymes have not been examined in the pitcher fluid of *Nepenthes*, even though it is known that bacteria may form in high abundance in this habitat [3,15,16]. Microbial-derived enzymes such as phosphatases, glucosidases, and aminopeptidases typically play a critical role in the processing of carbon, nitrogen, and phosphorus in a variety of ecosystems (e.g., fresh water [17–24], sea water [25,26], and soil [27–31]), and microbes substantially contribute to ecosystem functioning, especially in terms of nutrient cycling. In addition, microbes that dwell on organic solids such as dead zooplankton, animal feces, and algal aggregates are often metabolically more active than their free-living counterparts, and the degradation activities of enzymes that are attached to bacterial cells and aggregated organic matter are often substantially higher than the respective activities of the dissolved enzymes in natural waters [32]. Likewise, it is conceivable that enzymes associated with the particulate fraction of the pitcher fluid of *Nepenthes*, which is composed of free-living microbes and complexes of bacteria and organic solids, might also substantially contribute to organic matter decomposition.

In this study, we aimed to profile the *in situ* enzyme activity and the abundances of free-living microbes within the pitcher fluid of four *Nepenthes* species. We also assessed enzyme activity before and after filtration of the fluid to establish whether the digestive enzymes were mainly dissolved, or if a substantial fraction was also associated with particles such as bacterial cells and other organic solids. We addressed the following specific questions.

1. What is the density of free-living microbes in the pitcher fluid of *Nepenthes*?
2. How prevalent is the digestive enzyme activity in the pitcher fluid?
3. What is the proportion of particle-bound enzyme activity in the fluid as compared with dissolved enzyme activity?

Materials and Methods

Study site and plant species

We used four *Nepenthes* species, *Nepenthes albomarginata* T. Lobb ex Lindl, *Nepenthes ampullaria* Jack, *Nepenthes bicalcarata* Hook. F., and *Nepenthes gracilis* Korth, which usually occur in lowland heath forest, in peat swamp forest, or at forest margins in Borneo [5]. The lid of the pitcher in *N. ampullaria* is held away from the pitcher mouth, thus allowing rainwater and fallen leaves to enter the pitcher; for the other three species, lids cover the pitcher mouth. The samples were collected within or around the Lambir Hills National Park (LHNP), Sarawak, Malaysia (4°2′N, 113°50′E; 150 m a.s.l.). *N. albomarginata* was collected near the summit of the Bukit Pantu hill (approximately 300 m a.s.l.). *N. ampullaria* was collected near a pond or at the edge of the LHNP in a relatively open and bushy area. *N. bicalcarata* and *N. gracilis* were also collected at the edge of the LHNP.

We also used the fluid of cultivated *N. ampullaria*, *N. bicalcarata*, and *N. gracilis* (a possible hybrid) from the greenhouse in the Botanical Garden of the University of Zurich, Zurich, Switzerland (47°21′N, 8°33′E; 440 m a.s.l.).

Collection of pitcher fluid, pH measurement, and filtration procedure

Collection of field samples in Borneo took place four times, in Feb., Jun., Aug., and Nov. 2009, and cultivated plants in Zurich were sampled once, during Nov. 2009. The sample size and species varied each time because of the availability of plants on each date/site; in Borneo, 6 pitchers of *N. albomarginata* in Jun.; 2, 6, 5, 6 pitchers of *N. ampullaria* in Feb., Jun., Aug., Nov., respectively; 3, 5 pitchers of *N. bicalcarata* in Jun., Aug., respectively; 3, 6, 6 pitchers of *N. gracilis* in Feb., Jun., Nov., 2009, respectively; in Zurich, 6, 5, 4 pitchers of *N. ampullaria*, *N. bicalcarata*, and *N. gracilis*, respectively.

First, we carefully transferred the pitcher fluid to Falcon tubes using a 5–10 ml pipette in the field. All the pitchers of the Borneo plants contained prey, such as ants, termites, and flies, but also living mosquito larvae, and debris. The pitchers from cultivated plants also contained prey such as flies, but no living mosquito larvae. Because of logistic restraints in the field, the samples were stored in the dark at 4°C for up to three days prior to the enzyme assay.

Each sample was split into two subsamples, one of which was left unmanipulated (subsequently termed 'untreated') while the other one was filtered through a membrane syringe filter (pore size, 0.2 μm; subsequently termed 'filtration'). Because such a filtration treatment should remove both large particles and free-living microbes from the fluid, we assumed that the filtered fluid contained only dissolved enzymes. By contrast, the untreated fluid contained both dissolved and particle-bound enzymes (subsequently termed 'total enzymes'). Thus, any difference in enzyme activity between the untreated and filtered samples should be caused by particle-bound enzymes that, at least in part, also originate from microbes. In addition, in contrast to sterile fluid of unopened pitcher in *Sarracenia*, fluid in the unopened *Nepenthes* pitcher contains bacteria [15]. Thus, unopened pitchers were not treated as a negative control. We also initially attempted fluid sterilization by adding antibiotics such as kanamycin, ampicillin, and spectinomycin, in order to distinguish between the enzyme activity in the fluid with and without microbes. However, the pH of the fluid changed significantly after addition of these antibiotics, possibly because they induced microbial cell burst. We therefore used the filtration treatment exclusively to obtain sterile fluid.

The pH values of untreated and filtered fluid were measured using a pH meter (Huber & Co. AG, Switzerland), while we used pH paper (PANPEHA, pH 0–14, Schleicher & Schuell, USA, and pH 2.8–4.4, Toyo Roshi Co. Ltd., Japan) when the amount of the sample fluid was too small for the pH meter ($N = 2$ for *N. bicalcarata* in Zurich, and $N = 4$ for each of *N. gracilis* in Borneo (Nov. 2009) and Zurich). No significant difference between the two methods was found in the pH values (data not shown).

Enzyme assay

We used three chromogenic substrates, 4-nitrophenyl phosphate, 4-nitrophenyl β-D-glucoside, and 4-nitrophenyl N-acetyl-β-D-glucosaminide, to estimate the activity of acid phosphatases (AP), β-D-glucosidases (BG), and β-D-glucosaminidases (NAG), respectively. We followed up by using a common colorimetric method for 4-nitrophenol linked substrates [33] to determine the enzyme activity in the pitcher fluid samples. All substrates were prepared in 50 mM (pH 5.0) acetate buffer, and the concentration of each was 5, 5, and 2 mM, respectively.

In Feb., Jun., and Aug. 2009, the assays were conducted for both untreated and filtered fluid in the laboratory of the LHNP field station as follows. Portions of 100, 150, and 100 μl of fluid from *N. albomarginata*, 200, 400, and 400 μl from *N. ampullaria*, 100, 400, and 400 μl from *N. bicalcarata*, and 100, 200, and 200 μl from *N. gracilis* were used for assessing the activity of AP, BG, and NAG, respectively. Each sample was added to a 1.5 ml tube, diluted, if necessary, to 400 μl with 50 mM acetate buffer, and mixed with 400 μl of each substrate. As controls, 400 μl of each substrate solution was mixed with 400 μl of 50 mM acetate buffer. Mixed samples were incubated at 25°C for two hours. Thereafter, 160 μl of 1.0 N NaOH was added to terminate the reaction. Absorbance was measured at 410 nm for AP, BG, and NAG with a Shimadzu UV-1200 spectrophotometer (Shimadzu, Kyoto, Japan). In Nov.

2009, the same treatment was carried out with fluid from cultivated plants in the laboratory at the University of Zurich. On that occasion, we reduced the total volume of measured sample to one-tenth of the volume that was required in the LHNP field station because of the availability of superior instrumentation. Absorbance was then measured using a Beckman DU800 microplate spectrometer (Beckman, Fullerton, California, USA), which has the same accuracy in terms of wavelength and photometrics as the UV-1200 according to the manufacturer's specifications.

Standard curves were determined for each set of measurements using 4-nitrophenol for AP, BG, and NAG. The activities were recalculated into mmol l^{-1} h^{-1}. If the estimate of particle-bound activity as derived from the difference of untreated and filtered samples yielded a negative result, it was set to zero (i.e., no difference). If the activity measurement in a particular sample was already below the detection limit before filtration, it was excluded from the subsequent statistical analysis.

Total bacterial abundance

Samples of pitcher fluid were fixed with formaldehyde (2% final concentration) and stored at 4°C until further processing. Five to 300 µl of formaldehyde-fixed samples were stained with DAPI (4′, 6-diamidino-2-phenylindole, 1 µg ml^{-1} sample [34], and filtered onto black polycarbonate filters (0.22 µm pore size, Osmonics) placed on cellulose nitrate support filters (0.45 µm pore size, Sartorius). Images of preparations were automatically recorded with a CCD camera (Zeiss Axiocam) mounted on an epifluorescence microscope (Zeiss AxioImager.Z1) equipped with a motorized stage, and analyzed by a custom-made automated counting routine programmed in Visual Basic for Applications [35].

Statistical analyses

We examined the differences in mean pH values and enzyme activity among dates/sites within species and among species within dates/sites by applying Kruskal–Wallis tests. When the difference was significant, we also performed multiple comparisons with a *post hoc* Scheffé-type test (nonparametric). The differences in pH values and enzyme activities between untreated and filtered fluid were examined by a Wilcoxon signed-rank test for paired observations. The ratio of enzyme activity in untreated and filtered fluid was calculated and examined by a Kruskal–Wallis test. In all cases of multiple comparisons, we adjusted the p-values by a false discovery rate control method [BH method; 36]). These statistical analyses were performed using R 2.13.0 [37]. Furthermore, to explore factors that affect enzyme activity in the Borneo samples, we conducted a generalized linear mixed model (GLMM) for each enzyme using the lmer function in R (the restricted maximum likelihood (REML) approach, package lme4. Enzyme activities were log-transformed and were included as the response variable ($N = 86$ for AP, $N = 68$ for BG, $N = 52$ for NAG). If the activity of the filtered sample was zero, we added a small positive value, 10^{-2}, because of log-transformed values ($N = 7$ for BG, $N = 5$ for NAG, i.e., closed symbols in Figures 1 and 2). We used the filtration treatment (untreated and filtration), pitcher plant species and pH (log-transformed) as explanatory variables. We included random intercepts for the pitcher ID and date that allowed intercepts to vary by them, as well as resulting in a random slope with respect to the treatment that allowed the effect to vary by pitcher ID. To evaluate the significance of parameter estimates in the fitted GLMM, we obtained a 95% confidence limit (CL) using the Markov Chain Monte Carlo (MCMC) method using the mcmcsamp function in R (package lme4). When the CL of the parameter does not include zero, this indicates that the factor differs significantly from zero at the 5% level. This multilevel modeling and MCMC procedure work reasonably well when the data contain groups with a small sample size [38].

Estimated numbers of bacterial cells per ml were log-transformed. An ANCOVA was performed to examine whether the pH and species/site were associated with the density. Posterior comparison of means after the ANCOVA used the Tukey–Kramer HSD test. ANCOVAs were implemented using the REML formula, JMP 5.0 (SAS Institute). We also tested the correlation between bacterial densities and the mean ratio of dissolved enzyme activity to total enzyme activity (filtration/untreated) by using a generalized linear model (GLM) for each enzyme.

Results

The characteristics of the *Nepenthes* fluid

The pH values of *Nepenthes* fluid were acidic in all the samples (Figure 3), and did not differ for the dates of collection within the

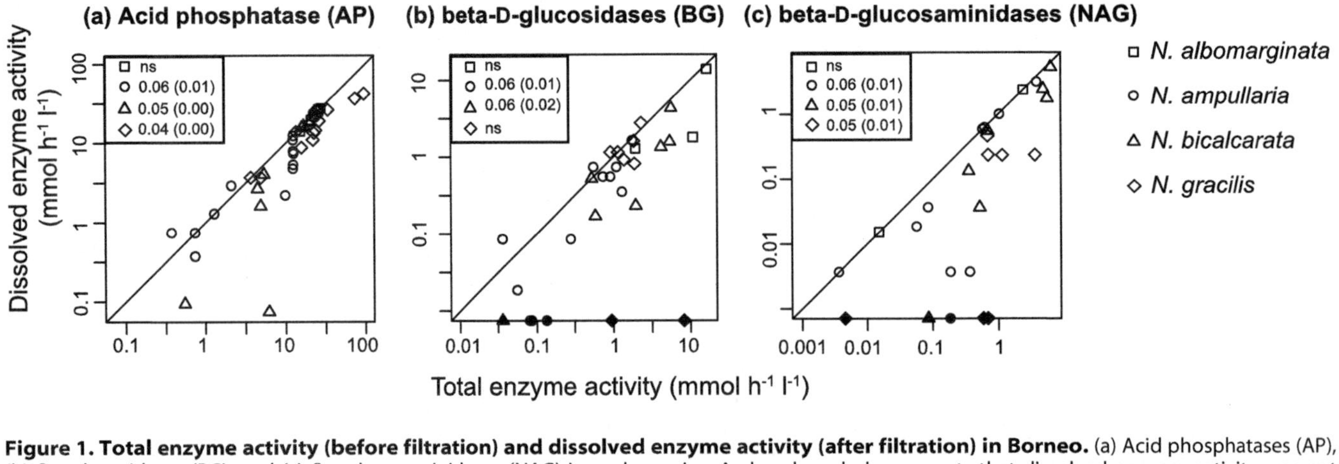

Figure 1. Total enzyme activity (before filtration) and dissolved enzyme activity (after filtration) in Borneo. (a) Acid phosphatases (AP), (b) β-D-glucosidases (BG), and (c) β-D-glucosaminidases (NAG) in each species. A closed symbol represents that dissolved enzyme activity was not detectable and was treated as zero. The differences in mean values between untreated and filtered fluid were examined by Wilcoxon signed-rank test, and p-values were adjusted by the BH method for multiple comparison and were shown in the upper box for each species (unadjusted p-values are shown in brackets). ns: not significant.

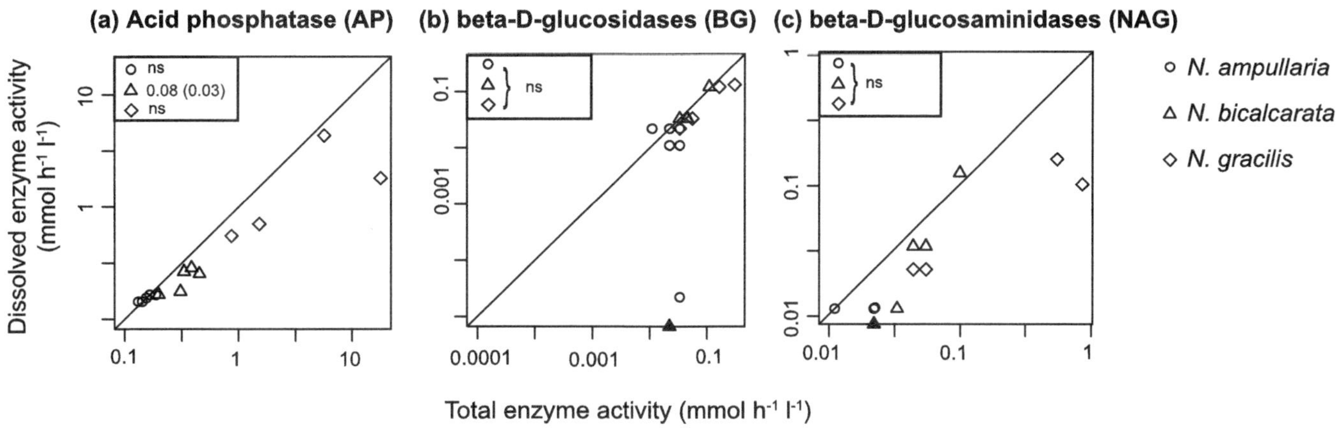

Figure 2. Total enzyme activity (before filtration) and dissolved enzyme activity (after filtration) in Zurich. (a) Acid phosphatases (AP), (b) β-D-glucosidases (BG), and (c) β-D-glucosaminidases (NAG) in each species. A closed symbol represents that dissolved enzyme activity was not detectable and was treated as zero. The differences in mean values between untreated and filtered fluid were examined by the Wilcoxon signed rank test, and p-values were adjusted by the BH method for multiple comparison and were shown in the upper box for each species (unadjusted p-values are shown in brackets). ns: not significant.

species sampled in Borneo. *N. gracilis* tended to have a lower pH compared with other species but this was not always significant through the date sequence (Figure 3). The pH did not change after the filtration treatment (Wilcoxon signed-rank test, $p = 0.97$).

In situ enzyme activity

The averaged total enzyme activity (untreated) and dissolved enzyme activity (filtration) was calculated for the four *Nepenthes* species (Table 1, Figures 1 and 2). AP activity was always detectable in all the pitchers of all species in both Borneo and Zurich (Table 1). Likewise, BG and NAG activity was also found in all species, although there was no measurable activity of this enzyme in some pitchers of all the examined species (Table 1). Altogether, total enzyme activities did not differ significantly between the dates for either enzyme in either species (Kruskal–Wallis test, $p<0.05$); therefore, they were pooled for further statistical analysis.

To assess the contribution of particle-bound enzyme activity, the mean ratio of dissolved enzyme activity to total enzyme activity was calculated. The values varied from 0.2 to 1.0 in the Borneo samples and from 0.3 to 1.0 in the ones from Zurich (Table 1). Total enzyme activity was significantly lower in the Zurich samples than in the Borneo samples for all the enzymes (Mann–Whitney test, $p<0.05$). Similarly, the Zurich samples showed a lower

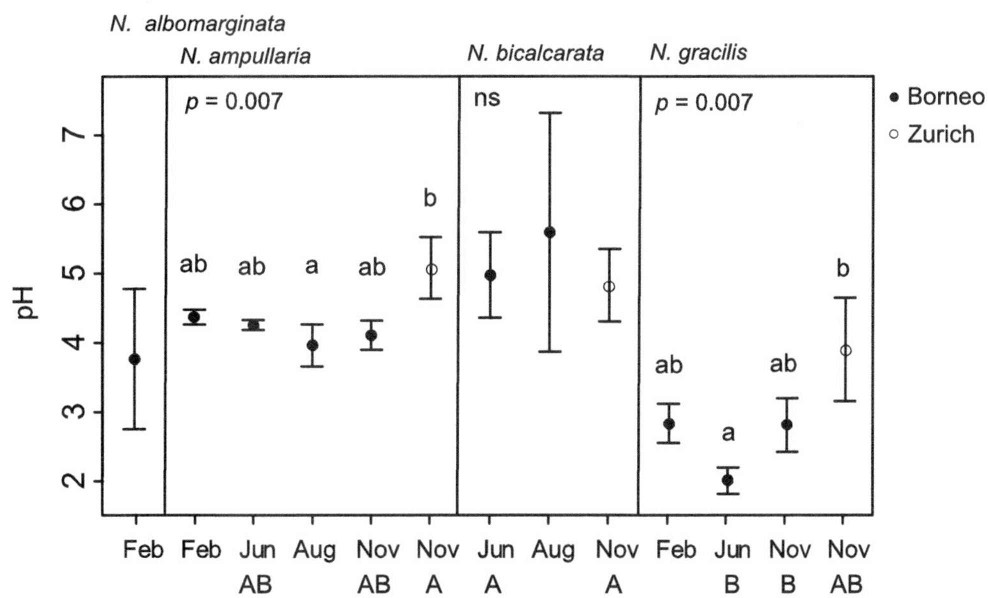

Figure 3. Average (SD) of pH values in each species per date/site. The differences in mean value of the pH between the dates/sites within a species were examined by the Kruskal-Wallis test (p-values adjusted by the BH method, shown above each species). Differences of a species/site between dates were also examined by the Kruskal-Wallis test (X-axis legend; Feb. and Aug.; not significant, Jun. and Nov.; $p = 0.005$). Results that share the same letters are not significantly different (lower case for species examined at the same date, upper case for dates between species, Scheffé-type test, $p<0.05$).

Table 1. Summary of activities of the three enzymes.

					Total (Untreated)			Dissolved (Filtration)					
Site	Enzyme	Species	No. of sampled pitchers	No. of pitchers that had positive enzyme activity	Mean enzyme activity (mmol l^{-1} h^{-1})	±	SD	Mean enzyme activity (mmol l^{-1} h^{-1})	±	SD	Mean ratio of dissolved enzyme activity to total	±	SD
Borneo	Acid phosphatase	*N. albomarginata*	6	6	23.09	±	4.61	22.60	±	5.17	0.97	±	0.04
	(AP)	*N. ampullaria*	14	14	6.52	±	5.82	4.16	±	4.10	0.82	±	0.46
		N. bicalcarata	8	8	9.28	±	7.17	7.32	±	7.89	0.59	±	0.38
		N. gracilis	15	15	28.88	±	24.47	19.24	±	11.10	0.76	±	0.22
	beta-D-glucosidases	*N. albomarginata*	6	3	9.60	±	7.14	5.68	±	7.18	0.57	±	0.36
	(BG)	*N. ampullaria*	19	15	0.59	±	0.55	0.42	±	0.46	0.64	±	0.65
		N. bicalcarata	8	7	2.61	±	2.41	1.17	±	1.51	0.40	±	0.36
		N. gracilis	15	9	2.27	±	2.28	1.11	±	0.85	0.70	±	0.47
	beta-D-glucosaminidases	*N. albomarginata*	6	2	1.21	±	1.69	1.20	±	1.68	1.00	±	0.01
	(NAG)	*N. ampullaria*	19	10	0.70	±	1.16	0.55	±	0.99	0.56	±	0.45
		N. bicalcarata	8	7	2.64	±	2.81	1.47	±	1.95	0.41	±	0.32
		N. gracilis	15	7	1.07	±	1.18	0.17	±	0.18	0.18	±	0.25
Zurich	Acid phosphatase	*N. ampullaria*	6	6	0.16	±	0.02	0.16	±	0.01	0.99	±	0.06
	(AP)	*N. bicalcarata*	5	5	0.33	±	0.09	0.23	±	0.05	0.70	±	0.13
		N. gracilis	4	4	6.48	±	7.89	1.85	±	1.75	0.49	±	0.29
	beta-D-glucosidases	*N. ampullaria*	6	6	0.03	±	0.01	0.01	±	0.01	0.75	±	0.70
	(BG)	*N. bicalcarata*	5	5	0.05	±	0.03	0.04	±	0.05	0.72	±	0.43
		N. gracilis	4	4	0.14	±	0.13	0.08	±	0.06	0.61	±	0.13
	beta-D-glucosaminidases	*N. ampullaria*	6	5	0.02	±	0.00	0.01	±	0.00	0.60	±	0.22
	(NAG)	*N. bicalcarata*	5	5	0.05	±	0.03	0.04	±	0.05	0.58	±	0.46
		N. gracilis	4	4	0.38	±	0.40	0.07	±	0.06	0.32	±	0.17

dissolved enzyme activity than the Borneo samples (Mann–Whitney test, $p<0.05$).

Dissolved vs. particle-bound enzyme activity

First, we tested whether the filtration reduced the enzyme activity significantly. In the Borneo samples, AP activity significantly decreased after filtration for *N. gracilis*, whereas this decrease was only marginally significant for *N. ampullaria* and *N. bicalcarata* (Wilcoxon signed-rank test, Figure 1a). BG activity decreased after filtration in *N. ampullaria* and *N. bicalcarata* (Figure 1b). For NAG, we found a marginally significant decrease in activity after filtration in *N. ampullaria*, *N. bicalcarata*, and *N. gracilis* (Figure 1c). These data support the contribution of particle-bound enzyme activities in Borneo samples. In the Zurich samples, we did not find any significantly decreased enzyme activity after filtration (Figure 2), with the possible marginal exception of AP activity in *N. bicalcarata* (Figure 2a), which might be partly because of the lower enzyme activity in the Zurich samples (also see Discussion).

Second, we tested whether the mean ratio of enzyme activity in untreated and filtered fluid differed among species. The ratios did not differ for any enzyme among the species from the Borneo samples (Kruskal–Wallis test, $p>0.05$ after p-value adjustment by the BH method), nor among the collection dates for the pooled species (Kruskal–Wallis test, $p>0.05$ after p-value adjustment by the BH method). Similarly, there were no differences among the species sampled in Zurich, except for the ratio of AP activity (Kruskal–Wallis test, $p<0.02$ after p-value adjustment by the BH method), which was in the order *N. gracilis*≤*N. bicalcarata*≤*N. ampullaria* (Scheffé test, $p<0.05$). The total enzyme activity differed among sites within the same species sampled on the same date (Nov. 2009). The activities in the Borneo samples were higher in *N. ampullaria* for AP, and in *N. gracilis* and *N. ampullaria* for BG (Kruskal–Wallis test, $p<0.05$ after p-value adjustment by the BH method).

To investigate the contribution of various factors including the filtration treatment, pH and species on enzyme activity, the GLMM was conducted and then confidence intervals were estimated using MCMC. The parameter estimates of the filtration treatment were negative in all three enzyme activities (-0.77 to -0.06 for AP, -7.48 to -2.13 for BG, -7.87 to -1.84 for NAG), which further supports the observation that filtration treatment decreased the activity in all enzymes (Table 2). The pH was not correlated with the enzyme activity in the three enzymes, and furthermore, the *Nepenthes* species did not have any effect to enzyme activity mostly (Table 2).

Bacterial abundance in the fluid

The mean bacterial density in pitchers was 7.1×10^{7} cells ml^{-1}. The density in *N. ampullaria* in the Zurich samples was the lowest (7×10^{6} cells ml^{-1}, Table 3), whereas that of *N. gracilis* in the Zurich

Table 2. Summary of the GLMM for enzyme activity.

Enzyme[a]	AIC	Random effect	Variance	Explanatory variables	Estimate	SE	*t*-value	LCL[b]	UCL[c]
Acid phosphatase	232.2	Intercept by Pitcher ID	0.50	(Intercept)	1.73	1.32	1.31	−0.37	4.00
(AP)		Slope (Filtration treatment) by Pitcher ID	0.03	Filtration treatment	−0.42	0.12	−3.50	−0.77	−0.06
		Intercept by date	0.83	Log(pH)	0.69	0.84	0.82	−0.70	2.01
				Species *N. ampullaria*	−1.41	0.44	−3.19	−2.11	−0.72
				Species *N. bicalcarata*	−0.69	0.61	−1.14	−1.73	0.21
				Species *N. gracilis*	0.83	0.61	1.35	−0.18	1.75
beta-D-glucosidases	425.7	Intercept by Pitcher ID	0.00	(Intercept)	4.58	9.46	0.49	−6.91	28.68
(BG)		Slope (Filtration treatment) by Pitcher ID	12.79	Filtration treatment	−4.82	1.11	−4.35	−7.48	−2.13
		Intercept by date	0.00	Log(pH)	1.79	6.82	0.26	−10.65	12.66
				Species *N. ampullaria*	−3.47	3.37	−1.03	−15.32	−2.24
				Species *N. bicalcarata*	−2.56	4.37	−0.59	−13.70	2.22
				Species *N. gracilis*	−1.07	3.94	−0.27	−17.59	0.06
beta-D-glucosaminidases	319.1	Intercept by Pitcher ID	0.00	(Intercept)	5.30	12.24	0.43	−16.31	30.78
(NAG)		Slope (Filtration treatment) by Pitcher ID	12.74	Filtration treatment	−4.88	1.22	−4.02	−7.87	−1.84
		Intercept by date	0.00	Log(pH)	1.08	7.84	0.14	−12.12	17.08
				Species *N. ampullaria*	−3.43	3.95	−0.87	−14.62	1.58
				Species *N. bicalcarata*	−2.09	4.25	−0.49	−14.72	2.81
				Species *N. gracilis*	−2.37	5.25	−0.45	−18.93	3.08

The standard error (SE) and *t* values of the parameters are shown.
[a]Enzyme activities were log-transformed.
[b]Lower and
[c]upper 95% confidence limits for each estimates (10,000 iterations).

samples was the highest (2.2×10^8 cells ml^{-1}, Table 3). There were no significant correlations between bacterial densities and pH (ANCOVA, $F = 0.11$, $df = 1$, $p = 0.74$). Bacterial densities were positively correlated with the mean ratio of dissolved enzyme activity to total enzyme activity in BG (GLM, $t = 2.09$, $p < 0.05$) and NAG (GLM, $t = 2.349$, $p < 0.05$), but were not correlated with that in AP (GLM, $t = -1.911$, $p > 0.05$).

Discussion

High *in situ* enzyme activity in the pitcher plant fluid

In this study, we profiled the *in situ* enzyme activity of the fluid in four *Nepenthes* species growing in Borneo (wild) and Zurich (cultivated). We detected activity for all of the enzymes examined in this study, AP, BG, and NAG, in the liquid of the pitchers of all four *Nepenthes* species. Compared with other environments such as freshwater, seawater, and soil (Table 4), the enzyme activity was surprisingly high in the liquid of *Nepenthes* species, up to five orders of magnitude greater than in the others. Moreover, the activity of all the studied enzymes in *Nepenthes* was much higher than that in another carnivorous plant species, *Utricularia* Table 4, [39]. Prey in the pitchers could be a most important nutrient source for plant growth and fitness in *Nepenthes*, for which it is reported that 60% of nitrogen is derived from prey [7]. A high enzyme activity in the fluid would help in the rapid decomposition of prey to maximize the efficient nutrient use within the pitcher.

Table 3. Average concentrations of bacterial cells ($\times 10^6$ ml^{-1}).

Site	Species	No. of pitchers	Average	±	SD	
Borneo	*N. ampullaria*	6	80	±	152	B
	N. gracilis	6	40	±	30	AB
Zurich	*N. ampullaria*	6	7	±	4	B
	N. bicalcarata	5	61	±	60	AB
	N. gracilis	4	218	±	106	A

In the final colum, values sharing the same letters are not significantly different among species×sites ($p < 0.05$, Tukey-Kramer HSD test).

Enzyme activity was generally higher in Borneo, which might be related to selective pressure for high enzyme production in the wild plant but also to higher activity by the indigenous microbial consortia. Moreover, the level of enzyme activity varied substantially between individual pitchers of a single species at each site and for single sampling dates. This implies that the enzyme secretion by plants and/or microbes may depend on the local condition within each pitcher; e.g., the pitcher age, the presence of plant debris, and the amount of prey could all affect the metabolic secretion of digestive enzymes. Currently, the mechanisms of accumulation or consumption of enzymes within the pitcher remain uncertain.

All investigated species exhibited high AP enzyme activity. APs are common plant enzymes of low substrate specificity that appear to be important in the production, transport, and recycling of phosphorus [40]. It has been reported that phosphorus may be the limiting nutrient in *Nepenthes* as well as in other carnivorous plants [2]. In view of such findings and our own observation of a possibly

Table 4. Comparison of enzyme activities in wide range of environments.

Enzyme		Environment	Substrate[a]	Microbial cell density ($\times 10^5$ ml^{-1})	Microbial biomass in soil (μg g-soil^{-1}) and fresh/ seawater (μg l^{-1})	Activity in soils (μmol g-soil^{-1} h^{-1}) and fresh/ seawater (μmol h^{-1} l^{-1})	Condition	Reference
Phosphatase	Acid	Soil, Hawaii	p-NP-P	-	3620–4500	30–60	forest	28
	Acid	Soil, Kenya	p-NP-P	-	112–3151	38–97	forest, agricultural area	29
	Acid	Soil, Tahiti	p-NP-P	-	193–855	302–711	forest, agricultural area, bare	30
	Acid	Soil, USA	MUF-P	-	-	1–16	forest, agricultural area, bare	27
	Acid	Wetland soil, Great Lake	MUF-P	-	-	0.1–0.3	wetland	31
	Acid	Freshwater, Lake Kinnet	p-NP-P	-	-	3.6–10.3	particle-bound	17
	Acid	Freshwater, Lake Kinnet	p-NP-P	-	-	12.9–21.4	dissolved	17
	Acid, Alkaline	Freshwater, Great Lake	MUF-P	-	-	1.5–2.8	surface, subsurface	20
	Alkaline	Freshwater, Lake Kinnet	p-NP-P	-	-	1.8–5.8	particle-bound	17
	Alkaline	Freshwater, Lake Kinnet	p-NP-P	-	-	1.1–1.6	dissolved	17
	Alkaline	Freshwater, Lake Kuc	MUF-P	ave. 22		0.63	oligo/mesotrophic	19
	Alkaline	Freshwater, Lake Szymon	MUF-P	ave.158		0.08	hypereutrophic	19
	Alkaline	Seawater, Red Sea	p-NP-P	-	-	0.04–0.15	oligotrophic	26
	Acid	*Utricularia spp.*	MUF-P	-	-	7.6–99.9	dissolved (after filtration)	39
	Acid	*Utricularia spp.*	MUF-P	470–10200		1.8–16.0	dissolved (after filtration)	45
	Acid	*Nepenthes spp.*	p-NP-P	40–390	-	132–95957	particle-bound (untreated)	this study
	Acid	*Nepenthes spp.*	p-NP-P	-	-	74–42178	dissolved (after filtration)	this study
beta-D-glucosidases		Soil, Kenya	p-NP-BG	-	112–3151	3–20	forest, agricultural area	29
		Soil, Tahiti	p-NP-BG	-	193–855	66–194	forest, agricultural area, bare	30
		Soil, USA	MUF-BG	-	-	0.1–0.9	forest, agricultural area, bare	27
		Wetland, Great Lake	MUF-BG	-	-	0.5–1.2	wetland	31
		Fresh water, Ohio	MUF-BG	-	-	0	river	21
		Sea water, Mediterranean Sea	MUF-BG	3.1–14.7		0–0.0012	sea	25
		Fresh water, Lake Gardno	MUF-BG	-	-	0.15–0.25	surface, subsurface	20
		Fresh water, Lakes	MUF-BG	3–570	-	0.007–0.368	lakes	23
		Utricularia spp.	MUF-BG	-	-	1.35–2.95	dissolved (after filtration)	39
		Nepenthes spp.	p-NP-BG	40–390	-	11–16050	particle-bound (untreated)	this study
		Nepenthes spp.	p-NP-BG	-	-	0–13958	dissolved (after filtration)	this study
beta-D-glucosaminidases		Soil, Hawaii	p-NP-NAG	-	3.62–4.5	2–9	forest	28
		Soil, Kenya	p-NP-NAG	-	112–3151	2–20	forest, agricultural area	29
		Soil, USA	MUF-NAG	-	-	0.04–0.5	forest, agricultural area, bare	27
		Soil, Great Lake	MUF-NAG	-	-	0.1–0.5	wetland	31
		Soil, USA	MUF-NAG	-	-	0.01–0.76	forest, agricultural area, bare	27
		Freshwater, Lake Gardno	MUF-NAG	-	-	0.06–0.09	surface, subsurface	20
		Freshwater, Ohio	MUF-NAG	-	-	0	river	21
		Freshwater, Piburger Lake	MUF-NAG	-	1180–9260	0–0.0085	particle-bound	24
		Freshwater, Piburger Lake	MUF-NAG	-	1180–9260	0–0.008	dissolved	24
		Utricularia spp.	MUF-NAG	-	-	0.096–0.356	dissolved (after filtration)	39
		Nepenthes spp.	p-NP-NAG	40–390	-	4–23879	particle-bound (untreated)	this study
		Nepenthes spp.	p-NP-NAG	-	-	0–23291	dissolved (after filtration)	this study

constitutive phosphatase activity in all pitchers, phosphorus may be an essential requirement in *Nepenthes*.

By contrast, we found that BG activity was not always detectable in each pitcher, especially in *N. albomarginata* and *N. gracilis*. BG is required in the degradation process of cellulose, which is a structural component of plant cell walls. Plant material such as fallen leaves could drop into the pitcher, especially with regard to *N. ampullaria* [41], which is an open-lid species, or *N. bicalcarata*, which has a large pitcher with a lid relatively distant from the pitcher mouth. For such species that are likely to receive plant material into their pitcher, BG may be essential for cellulose decomposition.

NAG is linked to the cycling of nitrogen, which is an important nutrient for *Nepenthes* [2]. Interestingly, not all examined pitchers scored positively for NAG. These enzymes are centrally involved in the breakdown of chitin, which is the hard amino sugar component of insect carapaces and microbial cell walls. The degradation of this material might also increase the ability of other enzymes to gain access to the soft body tissue within [3]. NAG should therefore be a key enzyme to obtain nitrogen and other nutrients for the plants, and it is, therefore, unexpected that some pitchers had no measurable NAG activity. Moreover, it should be noted that some of the observed NAG activity might even be related to the molting of specialized arthropods [24], as some *Nepenthes* species, for example, are known to host live mosquito larvae that complete their life cycle inside the pitchers [42].

Bacterial abundance in the pitcher plant fluid

A high bacterial density was confirmed here even in the pitcher fluid of *N. gracilis* at an extremely low pH (Table 3), in agreement with previous findings in another *Nepenthes* species, *Nepenthes alata* Blanco [15]. The cell densities of free-living microbes were as high as, or higher than, those in productive natural aquatic environments [43]. Moreover, because we did not specifically count cells attached to larger particles, it is conceivable that the total bacterial densities in the pitchers were even higher.

Importance of particle-bound enzyme activity

We found that the activity of all three studied enzymes decreased after filtration in various samples collected in Borneo (Tables 1 and 2, Figure 1). In addition, the ratio of dissolved to particulate BG and NAG activity increased with higher numbers of bacterial cells in the fluid. While this observation does not prove that the enzyme activity in the particulate fraction of the fluid was exclusively of microbial origin, it nevertheless suggests that both free-living and particle-bound microbes might have contributed substantially to the total enzyme activity. We also found that not all three types of enzyme activity decreased following filtration in *N. ampullaria* and *N. gracilis* (Table 1, Figures 1 and 2). Moreover, the ratio of enzyme activity between the untreated and filtered samples was highly variable throughout all of the pitchers, species, and enzymes. These data suggest that other factors related to the plant species or individual pitchers might also affect the contribution of particle-bound enzyme activity.

It has been suggested in other carnivorous plants that the microbial community would play a significant role in digestion or in the attraction of prey [3,44]. For example, *Utricularia* species harbored plant-associated algae and bacteria in the traps; the bacterial density in the trap was also at a similar level as *Nepenthes*, e.g. 10^7 cells ml^{-1} (Table 4, [45]). They produced significant amount of phosphatases, especially in old pitchers [45]. This might indicate that these microbes would control the nutrient flow to other components. The algae community occurring in traps of *Genlisea* species [46] also produced enzymes, but might compete for organic phosphorous with the host plant. Another example is pitchers of *Sarracenia*, which also harbored algae and bacteria producing enzymes [47]. Considered with other evidence from a nutrient tracing experiment in *Sarracenia* species [48], these microbial communities might significantly affect the nitrogen availability within the pitcher. The taxonomic composition of the bacterial community in the fluid of *Sarracenia* species has partly been investigated recently [49,50], and it has been suggested that this is similar to an animal gut-like bacterial community. Thus, the role of the community might be related to digestion of prey, although their functions are still uncertain.

Our study also suggests, albeit indirectly, that there may be activity of enzymes derived from microbes in the investigated *Nepenthes* species. Higashi et al. [14] isolated gram positive and gram negative bacteria in the fluid of *N. hybrida*. They also suggested that these microbes would also produce digestive enzymes. Moreover, microbe-derived enzymes would have different roles on degradation of prey from plant-derived enzymes because these two kinds of enzymes had different optimum pH-ranges of enzymes. These results could indicate that, to some extent, the microbial community might play a role in the nutrient cycle and metabolism of the *Nepenthes* pitcher and act cooperatively to the host plant. To substantiate this hypothesis, further evidence is required, such as that from a proteomics study or protein sequencing, to assess the origin and relative amounts of enzymes derived from microbes. Presently, we still do not know the diversity and abundance of different specialized populations within the microbial communities in the pitcher fluids. While the high concentrations of nutrients and organic material, as well as the prey variety might foster high microbial community diversity, the plant-produced antibacterial enzymes [12] and low pH conditions might, in turn, inhibit the growth of some microbial species.

In this study, we did not focus on the effect of pitcher age and prey species on enzyme activity. In fact, it is known that fluid in the unopened *Nepenthes* pitcher contains bacteria [15]. Consistently, fluid in unopened pitchers had the same level of phosphatase activity as that in opened and matured pitchers (Takeuchi et al., data not shown, also see [14]). This might indicate that the effect of pitcher age on enzyme level would not be strong during the younger to matured stage. Because it is suggested that enzyme secretion decreased with plant age in *Utricularia* species [45], the possibility that older pitchers could also decrease enzyme secretion more or less in *Nepenthes* might not be excluded. The prey species composition and the structure of the food web in the pitcher would change in space and time as well as pitcher age [6,15,51]. An experimental study on these effects would be helpful for understanding plant-microbe interaction within a pitcher. In addition, future research should aim at elucidating the causes of the observed variability in particle-bound enzyme activity to provide additional evidence for the specific contribution of microbes to this activity. Eventually, a study is required of the community composition and function of the different microbial species in the pitcher fluid.

Acknowledgments

The authors thank Ms. Lucy Chong (Forest Research Center, Sarawak) and Mr. Mohd. Shahabding Sabki (Sarawak Forest Department) for their permission for, and assistance with, this work in Sarawak, Malaysia. We are also grateful to Mr. B. Jingan for his help in the fieldwork.

Author Contributions

Conceived and designed the experiments: YT MJK KKS. Performed the experiments: YT MMS MU RSI MJK BD. Analyzed the data: YT MMS CvM JP. Contributed reagents/materials/analysis tools: YT MU BD CvM JP KKS. Wrote the paper: YT MMS MU JP KKS.

References

1. Darwin C (1875) Insectivorous plants. London: John Murray.
2. Ellison AM (2006) Nutrient limitation and stoichiometry of carnivorous plants. Plant Biology 8: 740–747.
3. Juniper BE, Robbins RJ, Joel DM (1989) The Carnivorous Plants. London: Academic Press.
4. Osunkoya OO, Daud SD, Di-Giusto B, Wimmer FL, Holige TM (2007) Construction costs and physico-chemical properties of the assimilatory organs of Nepenthes species in Northern Borneo. Annals of Botany 99: 895–906.
5. Clarke C (2006) Nepenthes of Borneo. Kota Kinabaru: Natural Histroy Publications.

6. Kato M, Hotta M, Tamin R, Itino T (1993) Inter- and intra-specific variation in prey assemblages and inhabitant communities in *Nepenthes* pitchers in Sumatra. Tropical Zoology 6: 11–25.
7. Schulze W, Schulze ED, Pate JS, Gillison AN (1997) The nitrogen supply from soils and insects during growth of the pitcher plants *Nepenthes mirabilis, Cephalotus follicularis* and *Darlingtonia californica*. Oecologia 112: 464–471.
8. Plachno BJ, Adamec L, Lichtscheidl IK, Peroutka M, Adlassnig W, et al. (2006) Fluorescence labelling of phosphatase activity in digestive glands of carnivorous plants. Plant Biology 8: 813–820.
9. Thornhill AH, Harper IS, Hallam ND (2008) The development of the digestive glands and enzymes in the pitchers of three nepenthes species: *N. alata, N. tobaica,* and *N. ventricosa* (Nepenthaceae). International Journal of Plant Sciences 169: 615–624.
10. Amagase S (1972) Digestive Enzymes in Insectivorous Plants .3. Acid Proteases in Genus *Nepenthes* and *Drosera peltata*. Journal of Biochemistry 72: 73–81.
11. Amagase S, Mori M, Nakayama S (1972) Digestive Enzymes in Insectivorous Plants .4. Enzymatic Digestion of Insects by *Nepenthes* Secretion and *Drosera peltata* Extract - Proteolytic and Chitinolytic Activities. Journal of Biochemistry 72: 765–767.
12. Hatano N, Hamada T (2008) Proteome analysis of pitcher fluid of the carnivorous plant *Nepenthes alata*. Journal of Proteome Research 7: 809–816.
13. Eilenberg H, Pnini-Cohen S, Schuster S, Movtchan A, Zilberstein A (2006) Isolation and characterization of chitinase genes from pitchers of the carnivorous plant *Nepenthes khasiana*. Journal of Experimental Botany 57: 2775–2784.
14. Higashi S, Nakashima A, Ozaki H, Abe M, Uchiumi T (1993) Analysis of feeding mechanism in a pitcher of *Nepenthes hybrida*. J Plant Res 106: 47–54.
15. Sota T, Mogi M, Kato K (1998) Local and Regional-Scale Food Web Structure in *Nepenthes alata* Pitchers. Biotropica 30: 82–91.
16. Okahara K (1933) Physiological studies on Drosera. IV. On the function of micro-organisms in the digestion of insect bodies by insectivorous plants. Scientific reports of Tohoku Imperial University Forth Series Biology. pp 151–168.
17. Berman T, Wynne D, Kaplan B (1990) Phosphatases Revisited - Analysis of Particle-Associated Enzyme-Activities in Aquatic Systems. Hydrobiologia 207: 287–294.
18. Chrost RJ, Rai H (1993) Ectoenzyme Activity and Bacterial Secondary Production in Nutrient-Impoverished and Nutrient-Enriched Fresh-Water Mesocosms. Microbial Ecology 25: 131–150.
19. Chrost RJ, Siuda W (2006) Microbial production, utilization, and enzymatic degradation of organic matter in the upper trophogenic layer in the pelagial zone of lakes along a eutrophication gradient. Limnol Oceanogr 51: 749–762.
20. Mudryk ZJ, Skorczewski P (2004) Extracellular enzyme activity at the air-water interface of an estuarine lake. Estuarine Coastal and Shelf Science 59: 59–67.
21. Sinsabaugh RL, Foreman CM (2001) Activity profiles of bacterioplankton in a eutrophic river. Freshwater Biology 46: 1239–1249.
22. Sinsabaugh RL, Findlay S, Franchini P, Fisher D (1997) Enzymatic analysis of riverine bacterioplankton production. Limnology and Oceanography 42: 29–38.
23. Vrba J, Callier C, Bittl T, Simek K, Bertoni R, et al. (2004) Are bacteria the major producers of extracellular glycolytic enzymes in aquatic environments? International Review of Hydrobiology 89: 102–117.
24. Vrba J, KofronovaBobkova J, Pernthaler J, Simek K, Macek M, et al. (1997) Extracellular, low-affinity beta-N-acetylglucosaminidases linked to the dynamics of diatoms and crustaceans in freshwater systems of different trophic degree. Internationale Revue Der Gesamten Hydrobiologie 82: 277–286.
25. Karner M, Rassoulzadegan F (1995) Extracellular Enzyme-Activity - Indications for High Short-Term Variability in a Coastal Marine Ecosystem. Microbial Ecology 30: 143–156.
26. Li H, Veldhuis MJW, Post AF (1998) Alkaline phosphatase activities among planktonic communities in the northern Red Sea. Marine Ecology-Progress Series 173: 107–115.
27. Sinsabaugh RL, Lauber CL, Weintraub MN, Ahmed B, Allison SD, et al. (2008) Stoichiometry of soil enzyme activity at global scale. Ecology Letters 11: 1252–1264.
28. Allison SD, Nielsen C, Hughes RF (2006) Elevated enzyme activities in soils under the invasive nitrogen-fixing tree Falcataria moluccana. Soil Biology & Biochemistry 38: 1537–1544.
29. Bossio DA, Girvan MS, Verchot L, Bullimore J, Borelli T, et al. (2005) Soil microbial community response to land use change in an agricultural landscape of Western Kenya. Microbial Ecology 49: 50–62.
30. Waldrop MP, Balser TC, Firestone MK (2000) Linking microbial community composition to function in a tropical soil. Soil Biology & Biochemistry 32: 1837–1846.
31. Hill BH, Elonen CM, Jicha TM, Cotter AM, Trebitz AS, et al. (2006) Sediment microbial enzyme activity as an indicator of nutrient limitation in Great Lakes coastal wetlands. Freshwater Biology 51: 1670–1683.
32. Simon M, Grossart HP, Schweitzer B, Ploug H (2002) Microbial ecology of organic aggregates in aquatic ecosystems. Aquatic Microbial Ecology 28: 175–211.
33. Tabatabai MA, Bremner JM (1969) Use of p-nitrophenyl phosphate for assay of soil phosphatase activity. Soil Biology and Biochemistry 1: 301–307.
34. Porter K, Feig Y (1980) The use of DAPI for identifying and counting aquatic microflora. Limnol Oceanogr 25: 943–948.
35. Salcher MM, Pernthaler J, Zeder M, Psenner R, Posch T (2008) Spatio-temporal niche separation of planktonic Betaproteobacteria in an oligo-mesotrophic lake. Environmental Microbiology 10: 2074–2086.
36. Benjamini Y, Hochberg Y (1995) Controlling the False Discovery Rate: A Practical and Powerful Approach to Multiple Testing. Journal of the Royal Statistical Society Series B (Methodological) 57: 289–300.
37. R Development Core Team (2009) R: A language and environment for statistical computing. R Foundation for Statistical Computing. Vienna, Austria.
38. Gelman A, Hill J (2007) Data Analysis Using Regression and Multilevel/ Hierarchical Models. New York: Cambridge University Press.
39. Sirova D, Adamec L, Vrba J (2003) Enzymatic activities in traps of four aquatic species of the carnivorous genus *Utricularia*. New Phytologist 159: 669–675.
40. Duff SMG, Sarath G, Plaxton WC (1994) The Role of Acid-Phosphatases in Plant Phosphorus-Metabolism. Physiologia Plantarum 90: 791–800.
41. Moran JA, Clarke CM, Hawkins BJ (2003) From Carnivore to Detritivore? Isotopic Evidence for Leaf Litter Utilization by the Tropical Pitcher Plant *Nepenthes ampullaria*. International Journal of Plant Sciences 164: 635–639.
42. Ratsirarson J, Silander JA (1996) Structure and dynamics in *Nepenthes madagascariensis* pitcher plant micro-communities. Biotropica 28: 218–227.
43. Gasol JM, Duarte CM (2000) Comparative analyses in aquatic microbial ecology: how far do they go? Fems Microbiology Ecology 31: 99–106.
44. Adlassnig W, Peroutka M, Lendi T (2010) Traps of carnivorous pitcher plants as a habitat: composition of the fluid, biodiversity and mutualistic activities. Annals of Botany 107: 181–194.
45. Sirova D, Borovec J, Cerna B, Rejmankova E, Adamec L, et al. (2009) Microbial community development in the traps of aquatic *Utricularia* species. Aquatic Botany 90: 129–136.
46. Plachno BJ, Wolowski K (2008) Algae commensal community in *Genlisea* traps. Acta Societatis Botanicorum Poloniae 77: 77–86.
47. Gebühr C, Pohlon E, Schmidt AR, Küsel K (2006) Development of Microalgae Communities in the Phytotelmata of Allochthonous Populations of *Sarracenia purpurea* (Sarraceniaceae). Plant Biology 8: 849–860.
48. Butler JL, Gotelli NJ, Ellison AM (2008) Linking the brown and green: Nutrient transformation and fate in the Sarracenia microecosystem. Ecology 89: 898–904.
49. Koopman MM, Fuselier DM, Hird S, Carstens BC (2010) The Carnivorous Pale Pitcher Plant Harbors Diverse, Distinct, and Time-Dependent Bacterial Communities. Appl Environ Microbiol 76: 1851–1860.
50. Koopman M, Carstens B (2011) The Microbial Phyllogeography of the Carnivorous Plant *Sarracenia alata*. Microbial Ecology 61: 750–758.
51. Mogi M, Yong HS (1992) Aquatic arthropod communities in Nepenthes pitchers: the role of niche differentiation, aggregation, predation and competition in community organization. Oecologia 90: 172–184.

Alfalfa Carbon and Nitrogen Sequestration Patterns and Effects of Temperature and Precipitation in Three Agro-Pastoral Ecotones of Northern China

Shujuan Chang[1,2¶], Nan Liu[1¶], Xiaoya Wang[1], Yingjun Zhang[1*], Yue Xie[1]

1 Institute of Grassland Science, China Agricultural University, Beijing, People's Republic of China, **2** Inner Mongolia Rangeland Survey & Design Institute, Hohhot, Inner Mongolia, People's Republic of China

Abstract

Alfalfa (*Medicago sativa* L.) is a primary and widely cultivated forage crop in China. As a perennial leguminous grass, continuous planted alfalfa may influence carbon and nitrogen sequestration in soils. We evaluated the effect of alfalfa, planted for different lengths of time, and temperature and precipitation on soil organic carbon (SOC) and total nitrogen (TN) contents, and estimated soil SOC and TN inventories from 0–60 cm in three agro-pastoral ecotones of northern China. Alfalfa SOC and TN storage patterns were significantly different with increasing soil depths between the three regions of northern China. Continuous alfalfa grassland planted had a positive effect on accumulation of both SOC and TN in the Northwest region, whereas SOC storage peaked 6–7 years after planting in the Northeast and North region. Moreover, relatively higher TN storage appeared 7 years after planting in the Northeast and North regions. This study controlled as many factors as possible, but we caution that such temporal inferences could be artifacts of site selection. The regression analysis indicated that SOC and TN accumulation was mainly dependent on temperature (≥10°C of effective total accumulated temperature) in the North region. Precipitation in the growing season was the main limiting factor for SOC storage in the Northwest region and TN accumulation in the North regions. Therefore, the different climate factors affecting SOC and TN sequestration in alfalfa occurred at a regional scale.

Editor: Ben Bond-Lamberty, DOE Pacific Northwest National Laboratory, United States of America

Funding: This research was supported by the China Forage and Grass Research System (CARS-35) and the Beijing Key Laboratory of Grassland Science. The funders had no role in study design, data collection and analysis, decision to publish, or preparation of the manuscript.

Competing Interests: The authors have declared that no competing interests exist.

* E-mail: zhangyj@cau.edu.cn

¶ These authors contributed equally to this work.

Introduction

Soil is a vital component of the global carbon cycle [1]. Soil organic carbon (SOC) is crucially significant in sustaining soil quality, crop production and environmental quality [2]. A sustainable soil C pool will mitigate increasing CO_2 levels in the atmosphere, and consequently have a substantial influence on the C content of the atmosphere [3–5]. How to enhance the potential of the soil C pool to sequester soil carbon, and thus alleviate the rate of increase of CO_2 concentrations in the atmosphere, has become an important research area globally.

Carbon sequestration is a biotic process. Carbon fixed by photosynthesis is the primary source of SOC, with the organic litter being decomposed and converted into soil organic matter [6]. In grassland ecosystems, more than 90% of the organic carbon is present in roots and soils [7]. Many studies have shown that changes in land-use are inevitably accompanied by variations in soil carbon storage [8–10]. A change of land-use from native vegetation to agriculture resulted in sharp declines in soil organic matter [11], and the conversion of annually cultivated land to forage grasses has the potential to increase C and N sequestration [12,13]. However soil carbon inventories and turnover rates are also influenced by climate, vegetation, parent material, topography, time, and the fundamental state factors of soil formation [14,15], and so soils developed under different vegetation types and climates may have different modes of soil organic matter stabilization [16]. Climate change and their interaction with vegetation have been acknowledged as one of the main driving forces of SOC variations [17]. Many authors, studying very different environments, have found that the SOC content is directly proportional to the mean annual precipitation and inversely proportional to the mean annual temperature [17–23]. The selection and operation of appropriate land management measures for mitigating atmospheric carbon concentrations will ultimately depend on knowledge of the carbon sequestration potential under various types of land-use in different ecosystems and biomes. At the same time, we need to consider potential compromises that may arise from changes in land-use with respect to food and fiber production that support global socio-economic activities [10]. Knowledge of soil carbon sequestration by particular plants in different spatial arrangements and temporal changes in that sequestration will enable better use of the land and achievement of optimum ecological and economic benefits.

Alfalfa (*Medicago sativa* L.) is a perennial leguminous grass that is widely planted through the world. In China alfalfa has been grown for more than 2000 years, primarily as a cultivated forage crop, and is widely distributed in northern regions. In 2009, it was estimated that the alfalfa-planted area in China reached 2.80 million hectares [24]. As a principal crop in legume-cereal rotations, alfalfa plays an important ecological and economic role in the agro-pastoral ecotones of China. Alfalfa is also a primary species used to restore vegetation in the National Western Development Program [25]. In China, many studies on alfalfa grassland converted by cropland showed that this land use change in favor for soil carbon sequestration, but most of those study sites were form the semiarid Loess Plateau, Northwest of China[13,25–26]. Investigating soil carbon dynamics under continuously planted alfalfa will also contribute to the taking of suitable measures to alleviate the rate of increase of CO_2 levels in the atmosphere.

The objective of the present study was to investigate at large regional scale spatial and temporal changes in the soil carbon pool under long-term plantings of alfalfa converted from annually cultivated land. To explain the differences in dynamics of the soil C and TN storage of alfalfa grassland from three main cultivation regions, temperature and precipitation data (China Meteorological Date sharing Service System: http://cdc.cma.gov.cn/) were used in stepwise multiple regression analysis to establish which factors may be responsible for the observed differences. Based on the above content, three questions are investigated. First, what are the differences among the three regions in carbon and nitrogen sequestration in soil? Second, which are the primary climatic factors in this process? And third, whether there is a threshold in the soil C and TN storage under alfalfa grassland?

Materials and Methods

Study site

The study area is shown in Fig.1. It covers the main cultivation regions of alfalfa grassland from east to west in the agro-pastoral ecotone of the north of China. Three sites were sampled along this gradient, with each region having different climate and soil types, but a similar cultivation history. One site (the Northeast region) was located near city of Qiqihar in the northeast of China, on a sandy black loam. The second site, located near city of Hohhot, in the Inner Mongolia Autonomous Region of China (the North region), was established on a light chestnut soil. The third site was located near city of Guyuan, which is on the Loess Plateau, in Ningxia province of northwest China (the Northwest region), on a light black loessial soil and a cultivated loessial soil.

The climate of the Northeast region is typified by a windy dry spring and cold winter with an average annual temperature of 4.5°C and mean annual precipitation of 437 mm according to the statistical data from 1985 to 2010. The climate of the North region has an average annual temperature of 7.5°C, and mean annual precipitation of 392 mm. The Northwest region has a moderate semiarid climate, that with an average annual temperature of 7.09°C, and mean annual precipitation of 421 mm (Fig.2).

Samples were selected from 17 fields in 2010, comprising six alfalfa grasslands from the Northeast region, six from the North region and five from the Northwest region with the samples from each region being from fields where alfalfa had been growing for different numbers of years (Table 1). In northeast China, the six alfalfa grasslands in Qiqihar were planted in different years in an area belonging to the Institute of Animal Science in Heilongjiang province. The six alfalfa grasslands from Hohhot of the North region, all planted in different years, belong to the Shaerqin forage resource field scientific observation station of the Grassland Research, Institute of the Chinese Academy of Agricultural Sciences. The five alfalfa grasslands from Guyuan in the Northwest region, all planted in different years, are mowed grassland that was planted by local farmers. All of the grasslands are cut two or three times per year, and followed maize crops.

Soil Sampling and Preparation

Soil samples were taken at the end of the grassland growth period on September 5, 2010 in the Northeast region, September 9th in the North and the 14th in the Northwest, respectively. Five sub-samples were randomly collected using a coring tube (diameter 5 cm) and mixed as one composite sample. The soil sampling was replicated five times for each field. Soil samples were collected from depths of 0–5 cm, 5–10 cm, 10–20 cm, 20–30 cm, 30–45 cm, and 45–60 cm. After air-drying, the soils were placed in self-sealing bags and returned to the laboratory for soil organic C (SOC) and total N (TN) analyses. Another three soil samples were collected from the same depth increments in the same fields using a cutting ring (volume of 100 cm^3) for measurement of the soil bulk density.

Measurements and Soil Analyses

Soil samples from the 17 fields were analyzed for soil bulk density, SOC, and TN. The soil samples for SOC and TN analyses were sieved through a 1 mm sieve. SOC concentrations were determined by the Walkley-Black method [27]. TN concentrations were determined using the Kjeldahl wet digestion procedure [28], using a 2300 Kjeltec Analyzer Unit (FOSS, Sweden). SOC and TN concentrations were multiplied by the soil bulk density to give the SOC and TN mass per unit area per 1 cm depth for the six sampled soil layers.

Statistical Methods

Regression analyses were used to investigate the relationship between SOC or TN storage and growing years in the three regions. Multiple stepwise regression was used to investigate the correlation of SOC and TN sequestration with ≥10°C of the effective total accumulated temperature, T_e, and total precipitation (mm) in the growing reason (April to October), P_g. The temperature and precipitation data were from China Meteorological Date sharing Service System: http://cdc.cma.gov.cn/. The analysis was carried out using SPSS (SPSS, 2000).

Results

Soil Bulk Density, SOC and TN Storage Changes

There were no changes in bulk density among soil layers in the Northwest region. Moreover, there was a trend of increasing bulk density with time since planting, except for the 17-year-old field. In the Northeast region, an increase in bulk density with the increase of soil depths appeared in the top 30 cm soil layer, and then declined. The soil bulk density was in the order of North>Northeast>Northwest (Fig. 3).

In the North region, SOC and TN storage was far higher in surface soil than in sub-layer soil, and decreased with increasing of soil depth. Compared with the sharp decrease of SOC and TN with the increase of soil depth in the North, SOC and TN storage had a relative slight decrease in the Northwest. In the top 30 cm, SOC and TN storage was increased with increasing of soil depth, but decrease from 30 to 60 cm soil layer in the Northeast. The SOC and TN storage were in the order of Northeast>North>Northwest (Fig. 4).The regression model (Fig. 5) showed the relationship between the growing years and SOC and TN storage on 0–60 cm layers in alfalfa grassland. In the North region, SOC storage was increased with increasing of growing years, and reached the maximal storage after growing six

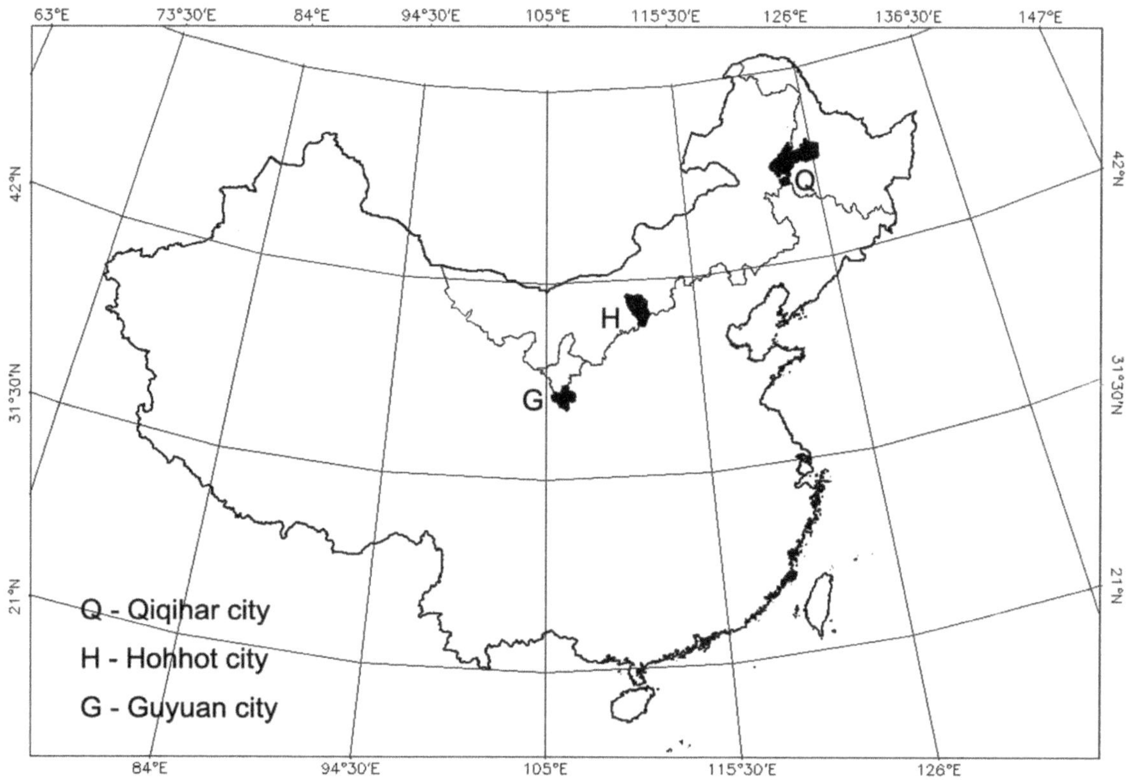

Figure 1. Location of the study areas in China.

years (41.11 Mg/ha), then begin to decline (n = 30, $P<0.05$; $R^2 = 0.495$). Similarly with TN storage, the maximal storage was appeared in 7th years of 5.99 Mg/ha in the North region (n = 30, $P<0.05$; $R^2 = 0.397$). In the Northeast region, the peak of SOC and TN storage was appeared in 7th years of 123.32 and 14.12 Mg/ha (n = 30, $P<0.05$; $R^2 = 0.299$ in SOC and 0.596 in TN). Soil carbon and nitrogen storage in the Northwest region had a significantly positive linear correlation relationship with the growing years (n = 25, $P<0.05$; $R^2 = 0.813$ in SOC and 0.775 in TN).

The Regression Model of SOC and TN Storage in Relation to Temperature and Precipitation

In the Northeast region, the stepwise regression analysis (Fig. 6) showed that SOC and TN storage cannot be quantitatively described by a combination of T_e or P_g. In the Northwest region, SOC storage increased with increasing T_e (n = 25, $R^2 = 0.819$, $P<0.001$), indicating that approximately 81.9% of the variability in SOC can be explained by variation in T_e. But TN storage can be quantitatively described by a linear combination of both T_e and P_g (n = 25, $R^2 = 0.856$, $P<0.001$). Similarly, in the North region, SOC storage was correlated with T_e in the 0–60 cm layer (n = 30, $R^2 = 0.484$, $P<0.05$), but TN storage was weakly correlated with P_g in the 0–60 cm layer (n = 30, $R^2 = 0.450$, $P = 0.05$). However, the correlation coefficients are all significant. Meanwhile, the linear regression analysis indicates that soil C in the 0–60 cm layer is more sensitive to changes in temperature than TN storage in the North region soils. But the TN (0–60 cm depths) sequestration is more sensitive to changes in precipitation than effective accumulated temperature in the Northwest region.

Discussion

Dynamics of SOC and TN

SOC and TN storage patterns among the three regions were very different in alfalfa grassland, with the Northeast having the highest storage and the Northwest the lowest. Different background conditions together with differences in soil type and climate may cause the diversity of settings on SOC and TN storage in three regions. Soil carbon and nitrogen sequestration are long-term biotic processes that are affected by climate, plant species abundance and productivity, litter decomposition, soil structure, and agricultural practices [29–34]. Changes in SOC often lead to differences in soil total N, because organic forms generally account for more than 95% of the soil N [35].

With respect to differences in the length of time that alfalfa has been planted, the amount of carbon increased with an increase in the planting time, representing a considerable accumulation of SOC in these soils following conversion of annual crops to a continuous alfalfa system. But the increase trends with the increase in planting time had a threshold in the Northeast and North regions. The Northwest region had less carbon sequestration than the other regions. Meanwhile, in the Northwest region, soil bulk density was lower than the other region, with no significant difference between the six depths. The relative lower soil bulk density led to the lower capability on fertilizer and water conservation in the soil arable layer. This means that the conversion of annual cropland to the continuous alfalfa grassland in this poor soil may have greater potential on the accumulation of soil organic carbon and nitrogen. In the Northeast region, soil bulk density on 20–30 cm layers was significant higher than other layers, that caused by the plow pan profile on this layer. Long-term cultivation of a single crop in company with surface tillage will lead

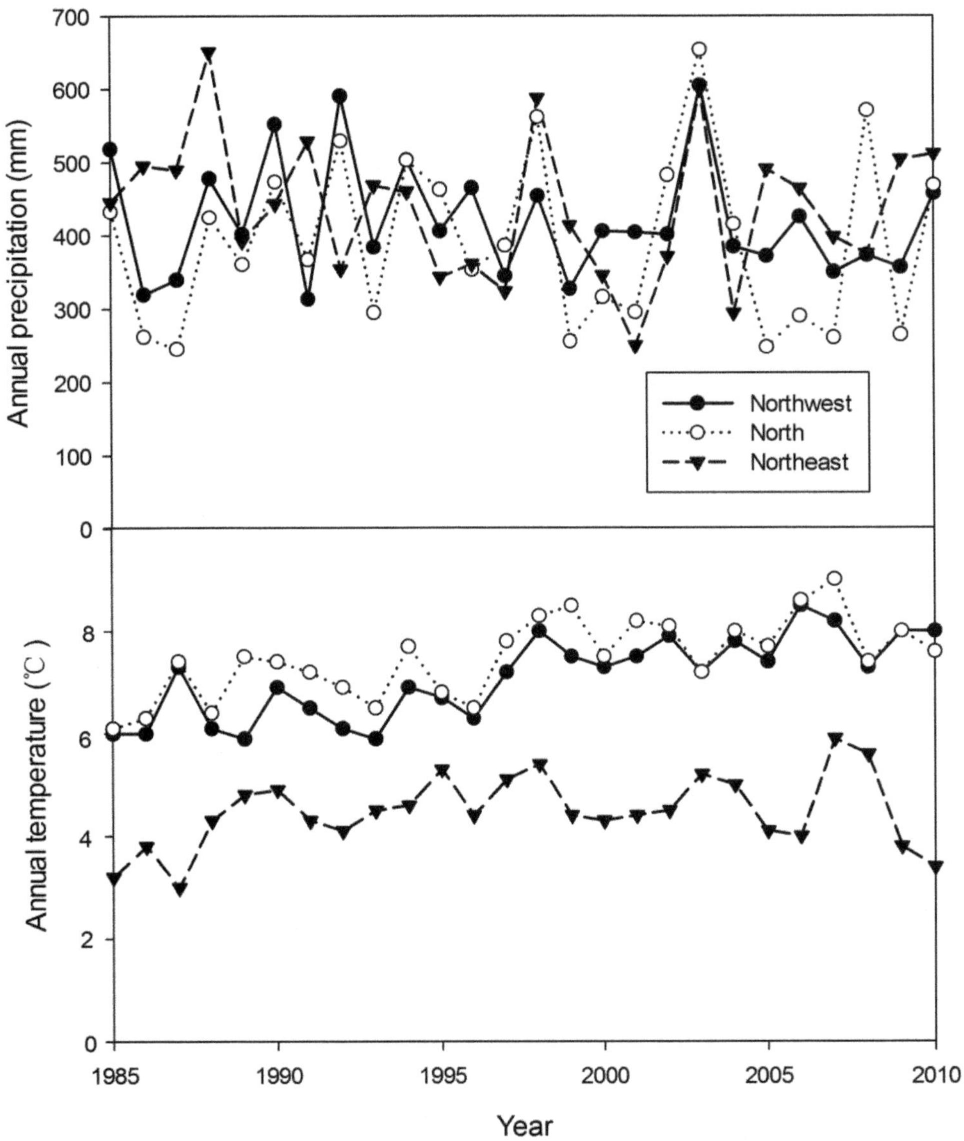

Figure 2. Annual precipitation (mm) and annual temperature (°C) in the three regions from 1985 to 2010. The data from China Meteorological Date sharing Service System: http://cdc.cma.gov.cn/.

Table 1. Details of soil samples from alfalfa grasslands growing for different numbers of years in the three regions.

(Northeast Region) Qiqihar		(North Region) Hohhot		(Northwest Region) Guyuan	
Ab.	**No. of growth years**	**Ab.**	**No. of growth years**	**Ab.**	**No. of growth years**
QM10	10-yr alfalfa grassland	HM9	9-yr alfalfa grassland	GM17	17-yr alfalfa grassland
QM8	8-yr alfalfa grassland	HM7	7-yr alfalfa grassland	GM10	10-yr alfalfa grassland
QM6	6-yr alfalfa grassland	HM5	5-yr alfalfa grassland	GM7	7-yr alfalfa grassland
QM4	4-yr alfalfa grassland	HM4	4-yr alfalfa grassland	GM4	4-yr alfalfa grassland
QM3	3-yr alfalfa grassland	HM2	2-yr alfalfa grassland	GM1	1-yr alfalfa grassland
QM2	2-yr alfalfa grassland	HM1	1-yr alfalfa grassland		
Latitude	47°15′43.70″		40°36′11.05″		36°11′41.32″
longitude	123°40′32.30″		111°45′24.45″		106°24′35.64″

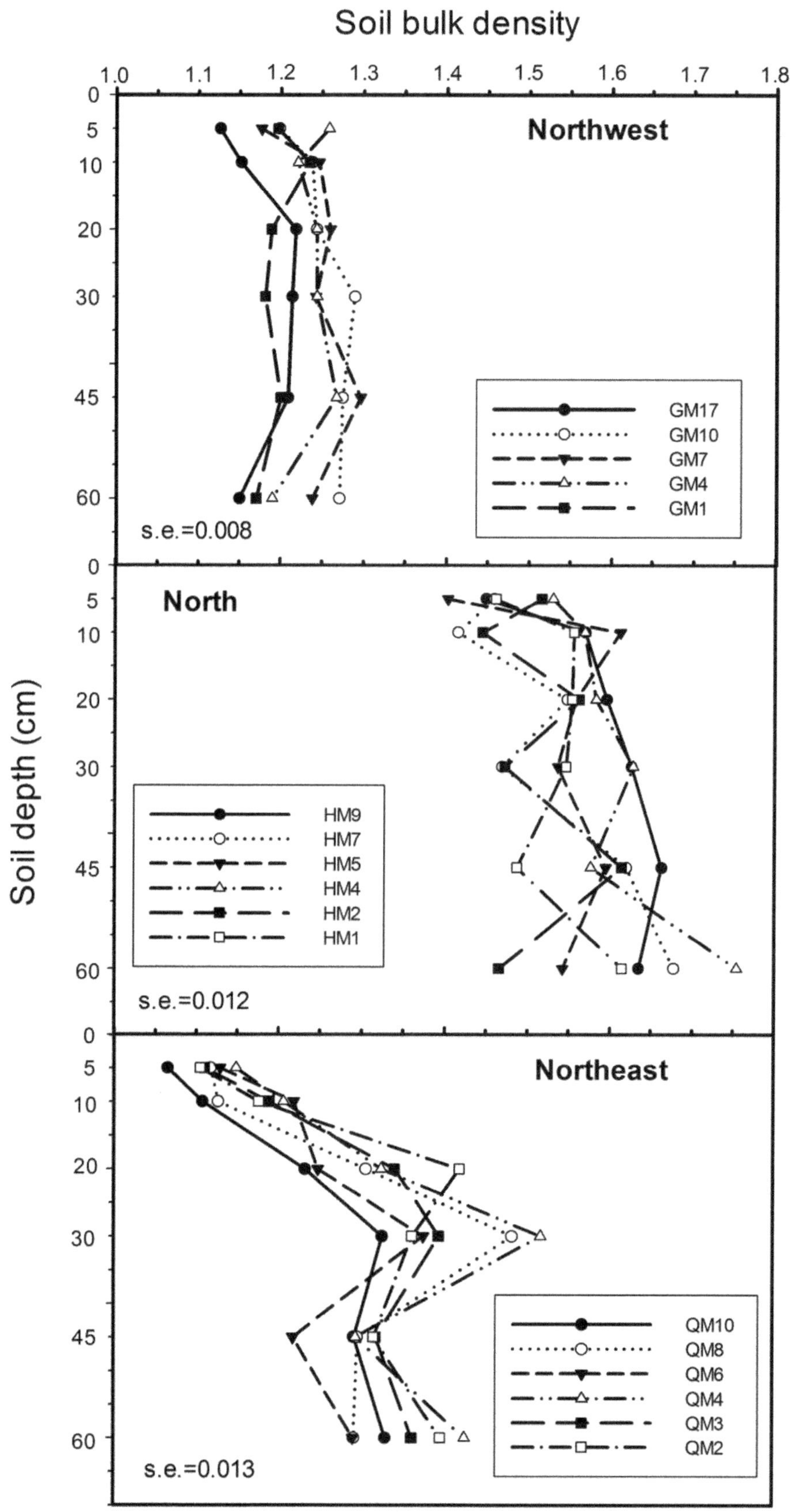

Figure 3. Mean soil bulk density in different soil layers from the three regions. The value of standard error (s.e.) for each region was stated in the graph (n = 90 for Northwest, n = 108 for North and Northeast region).

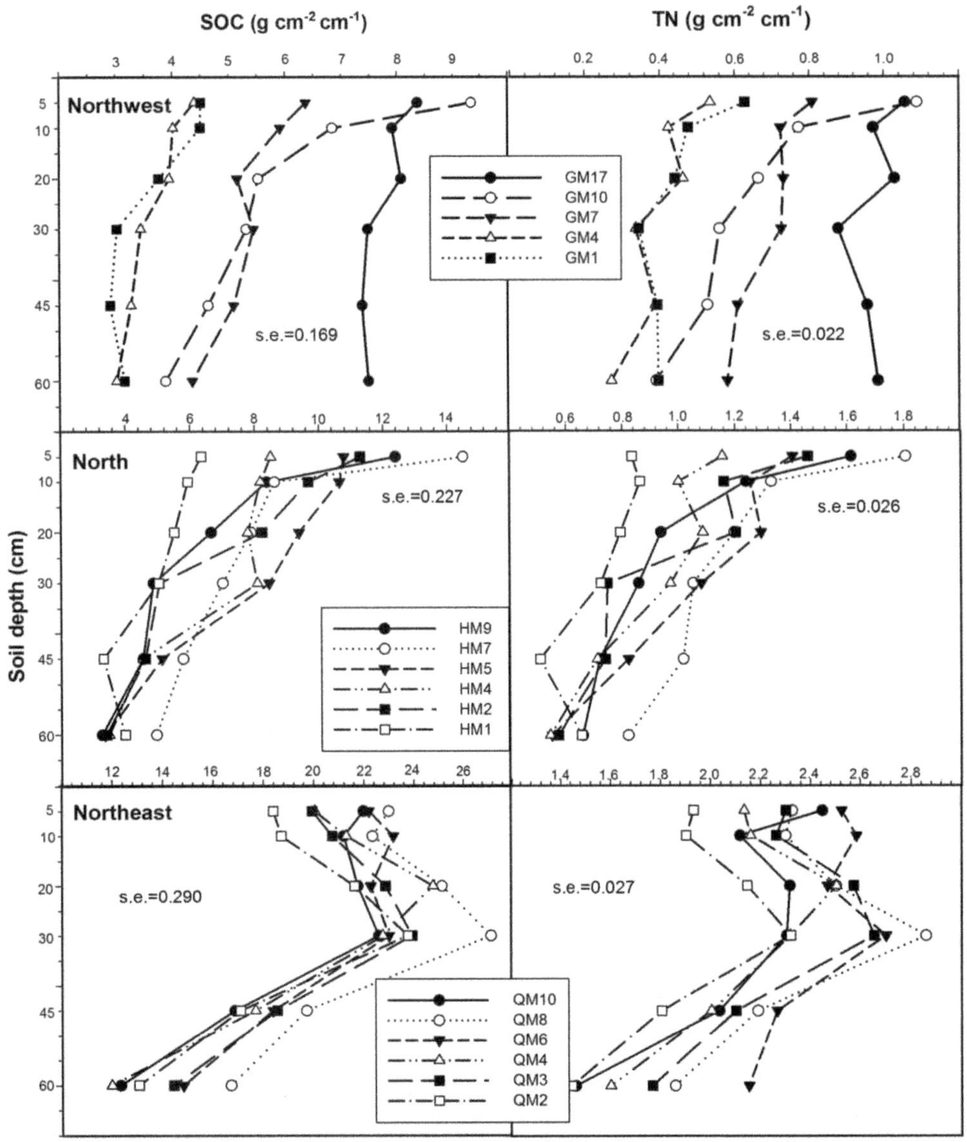

Figure 4. Mean SOC and TN storage in different soil layers from the three regions. The value of standard error (s.e.) for each region was stated in the graph (n = 150 for Northwest, n = 180 for North and Northeast region).

to thickening of the plow pan profile, soil compaction and bulk density increased [36]. Fluctuations of the SOC content in the Northeast region are gentler than in the other regions, indicated by smaller differences with increasing soil depth. In the North region, the SOC and TN contents in the 30–60 cm layer are significantly lower than in the 0–30 cm layer. This is possibly because the soil in the North region is a chestnut soil that has a calcic horizon at 30–60 cm [37]. The tight structure of calcic horizons influences root growth.

In recent years, there has been considerable research into changes in soil properties when annual cropland is converted to perennial forage. Such conversions have been shown to increase soil carbon and nitrogen contents [12–13,25]. However, Jiang et al. [26] considered that removal of large quantities of forage from alfalfa grassland systems through mowing, which is the continuous alfalfa system used in Northwest China, can exhaust soil water leading to severe soil dryness, and deplete SOC and soil nutrients in the 0–100 cm soil layer.

How to identify changes in organic matter is crucial in defining sustainable agricultural practices [38]. Changes in SOC storage in the three regions under the continuous alfalfa system are not consistent. In the North and Northeast region, SOC and TN storage changes did not show a consistent trend with the increase in the number of growing years. However, SOC storage in the Northwest region showed significant increases with increasing years of growing alfalfa. This is consistent with other research findings [13,39–40], which found considerable accumulation of SOC in soils where the land use had changed from annual crops to perennial grasses.

We used a chronosequence approach (space-for-time) to study changes of SOC and TN storage in alfalfa with different growing time. The increasing trends of increasing planting time had a threshold in the Northeast and North regions, so temporal results (e.g., strange decline in SOC after 6 years in Figure 5) are potentially confounded with site selection problems, such as spatial heterogeneity. Although we controlled for many factors, we

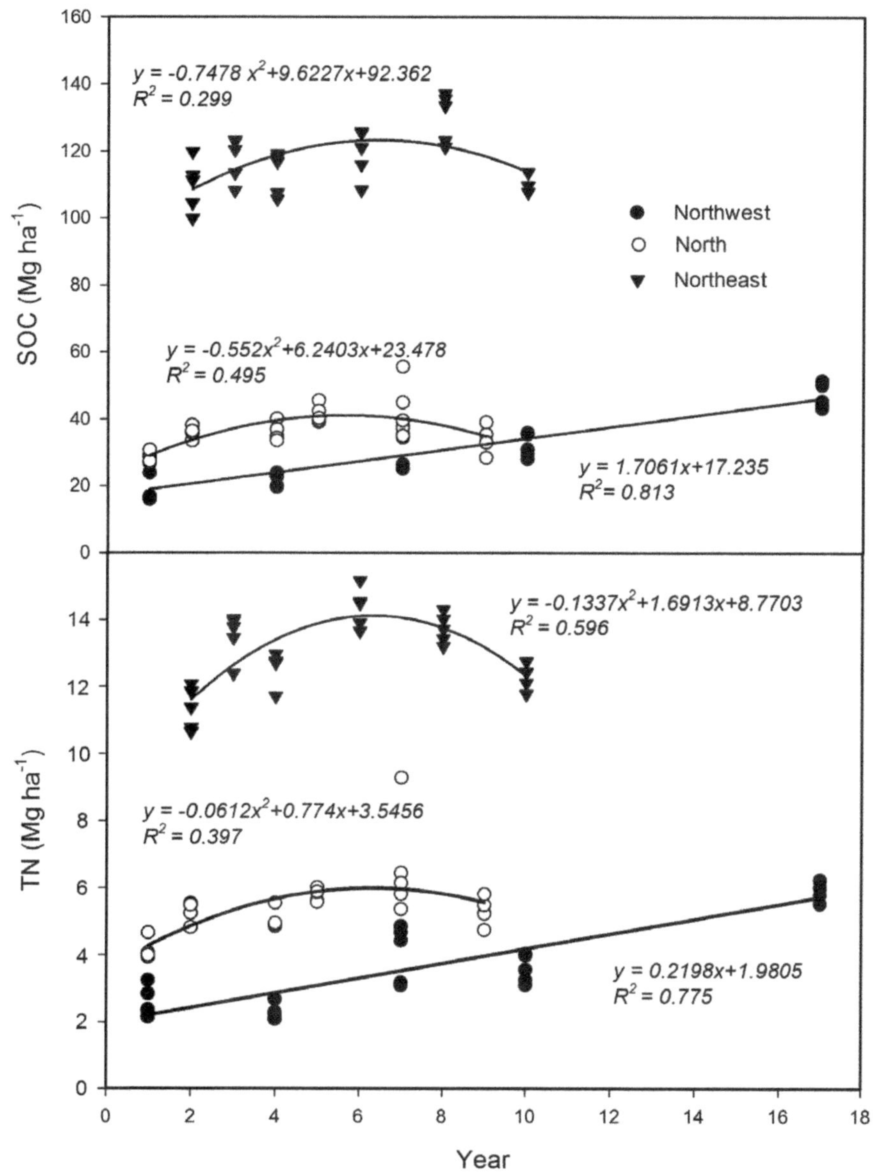

Figure 5. The relationship of soil organic carbon (SOC) and total nitrogen (TN) storage with years of growing alfalfa in North, Northeast and Northwest region of China.

caution that such temporal inferences could be artifacts of site selection. Walker et al. [41] suggested that if the date of the initial disturbance and subsequent history of the site are known, chronosequences provide the opportunity to study ecological processes over time periods that are longer than direct observation would permit. With careful site selection the chronosequence approach is thus an effective research method for evaluating.

The Relationship between Soil C and Temperature and Precipitation

Changes in temperature and precipitation may affect the soil and vegetation, with the soil affecting vegetation through its influence on water availability, elemental cycling and the soil temperature regime at an ecosystem level [42]. How to balance the effects between the temperature and the precipitation on SOC content determines what climatic factors in a specific area will be limited or decision of the soil carbon sequestration mode. In Canada, Fantappiè et al. [23] indicated that the most remarkable climate effect was that high air temperatures corresponded to low SOC values, probably because of SOC mineralization. High precipitation instead, when considered alone, had a positive effect on SOC content, but their interaction with temperature seemed to have more effect on SOC variations. In the Northeast region the accumulation of SOC and TN in the 0–60 cm layer does not agree with a regression on the regional temperature or precipitation. Precipitation is the main limiting factor in the Northwest region and the North region on TN sequestration. However, in the North region, the accumulation of SOC in the 0–60 cm layer is mainly dependent on the regional temperature. Temperature and precipitation together determine the changes in TN storage in the Northwest region. This suggests that other factors, like cultivation history, the type of cultivation, and soil properties, have more contribution to the changes of soil C and N in Northeast region. Nitrogen sequestration in soil is more sensitive

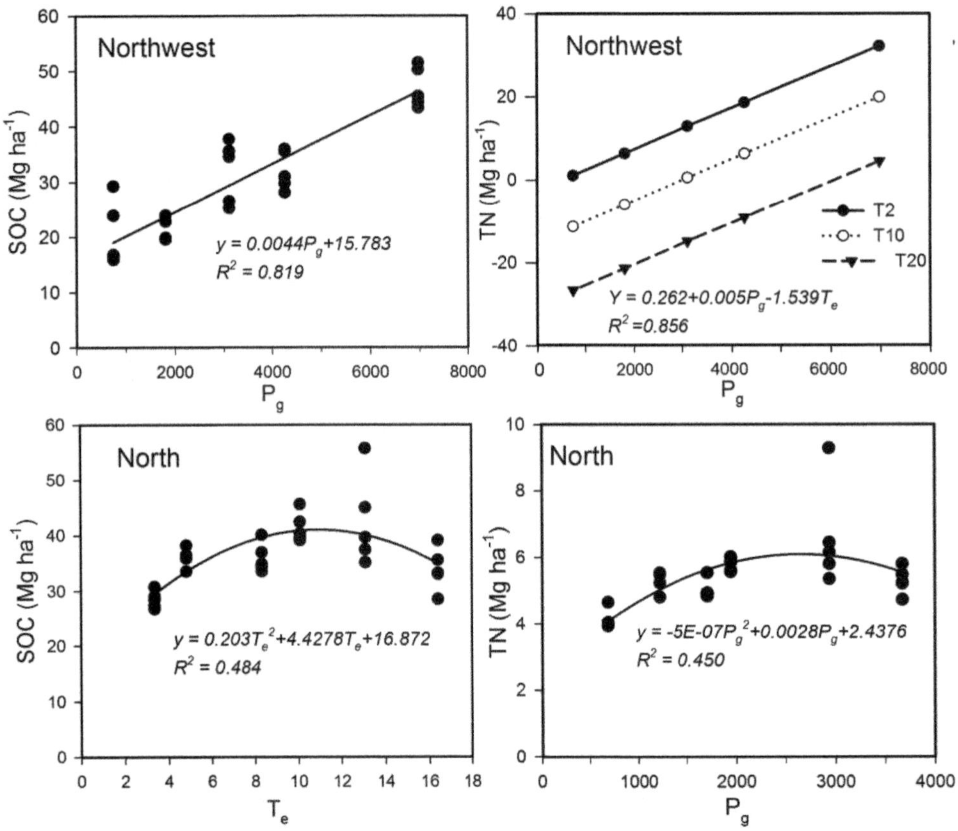

Figure 6. Relationships between SOC or TN (0-60cm) and ≥10°C of the effective total accumulated temperature (T_e) or total precipitation (mm) in the growing reason (P_g) in the North and Northwest region.

to changes in precipitation. Jenny [15] indicated a strong interdependence between climate and soil quality. Water is the key factor limiting plant growth in arid and semi-arid regions [43], and it also determines vegetation composition and distribution patterns [44,45].

Conclusion

Alfalfa SOC and TN storage patterns with increasing soil depths were significantly different between the three regions of northern China. Continuous alfalfa grassland planted had effective accumulation of both SOC and TN in the Northwest region. The peak SOC and TN storage were appeared in 6 and 7 years after alfalfa established in the North region, and both 7 years in the Northeast region. Temperature had effect on SOC storage in the North regions, whereas precipitation had more contribution to the SOC storage in the Northwest region and TN storage in the North region. Temperature and precipitation together determine the changes in TN reserves in the Northwest region. In short, SOC and TN storage has close relationship with the changes of temperature and precipitation, especially in the Northwest region.

Acknowledgments

We are grateful to the Institute of Animal Science in Heilongjiang province (Li Hong, E-mail: ; hljlihong@163.com), and the Shaerqin forage resource field scientific observation station of the Grassland Research Institute, Chinese Academy of Agricultural Sciences (Liu Lei; E-mail: ; liu4311755@163.com) for permission to use the study sites and research facilities. We are grateful to Huang YX, the owner of sites form city of Guyuan (Tel: 086-013995449140). All of study sites were got the permission. No specific permits were required for the described field studies. The location is not privately-owned or protected in any way. The field studies did not involve endangered or protected species. We wish to thank Zhou HS, Zhang XQ, Kan HM, Lin JH, Xiang M and Xu RX for their help with field sampling and data collection, and Jiang C for technical advice on the statistical analysis.

Author Contributions

Conceived and designed the experiments: YZ SC NL. Performed the experiments: SC NL. Analyzed the data: SC NL. Contributed reagents/materials/analysis tools: SC XW NL YX. Wrote the paper: SC NL.

References

1. Shi XZ, Wang HJ, Yu DS, David CW, Cheng XF, et al. (2009) Potential for soil carbon sequestration of eroded areas in subtropical China. Soil & Tillage Research 105: 322–327.
2. Doran JW, Parkin TB (1994) Defining and assessing soil quality. In: Defining Soil Quality for a Sustainable Environment, Soil Sci. Soc. Am, Madison, WI, USA. 3–21.
3. Lal R (2004) Soil carbon sequestration to mitigate climate change. Geoderma 123:1–22.
4. Bruce JP, Frome M, Haite E, Janzen H, Lal R, et al. (1999) Carbon sequestration in soils. Journal of Soil and Water Conservation 54: 382–389.
5. Upadhyay TP, Sankhayan PL, Solberg B (2005) A review of carbon sequestration dynamics in the Himalayan region as a function of land-use change and forest/soil degradation with special reference to Nepal. Agriculture Ecosystems and Environment 105: 449–465.
6. Aber J, Melillo J, Mcclaugherty C (1990) Predicting long-term patterns of mass loss, nitrogen dynamics, and soil organic matter formation from initial fine litter

chemistry in temperate forest ecosystems. Canadian Journal of Botany 68: 2201–2208.

7. Burke IC, Laurenroth WK, Milchunas DG (1997) Biogeochemistry of managed grasslands in central North America. In: Paul EA, Elliott ET, Paustian K, Cole CV, editors. Soil organic matter in temperature agroecosystems: long-term experiments in North America. CRC Press, Boca Rato, FL, 85–102.
8. Guo LB, Gifford RM (2002) Soil carbon stocks and land use change: a meta analysis. Global Change Biology 8: 345–360.
9. Grünzweig JM, Sparrow S, Yakir D, Chapin FS III (2004) Impact of agricultural land-use change on carbon storage in Boreal Alaska. Global Change Biology 10: 452–472.
10. Zhou ZY, Sun OJ, Huang JH, Li LH, Liu P, et al. (2007) Soil carbon and nitrogen stores and storage potential as affected by land-use in an agro-pastoral ecotone of northern China. Biogeochemistry 82: 127–138.
11. Wilson AT (1978) Pioneer agriculture explosion and CO2 levels in the atmosphere. Nature 273: 40–41.
12. Lal R, Follett RF, Kimble J, Cole CV (1999) Managing U.S. cropland to sequester carbon in soil. Journal of Soil and Water Conservation 54: 374–381.
13. Su YZ (2007) Soil carbon and nitrogen sequestration following the conversion of cropland to alfalfa forage land in northwest China. Soil & Tillage Research 92: 181–189.
14. Jenny H (1941) Factors of Soil Formation. McGraw–Hill, New York. 281.
15. Jenny H (1980) The Soil Resource: Origin and Behavior. Springer, New York.
16. Elliott ET (1986) Aggregate structure and carbon, nitrogen, and phosphorus in native and cultivated soils. Soil Science Society of American Journal 50: 627–633.
17. Jenny H (1980) The Soil Resource: Origin and Behavior, Ecological Studies 37. Springer-Verlag, New York. 377.
18. Kirshbaum MUF (1995) The temperature dependence of soil organic matter decomposition, and the effect of global warming on soil organic storage. Soil Biology and Biochemistry 27: 753–760.
19. Garten CT, Post WM, Hanson PJ, Cooper LW (1999) Forest soil carbon inventories and dynamics along an elevation gradient in the southern appalachian mountains. : Biogeochemistry, 45. Kluwer Academic Publishers, 115–145.
20. Jobbágy EG, Jackson RB (2000) The vertical distribution of soil organic carbon and its relation to climate and vegetation. Ecological Applications 10: 423–436.
21. Percival HJ, Parfitt RL, Scott NA (2000) Factors controlling soil carbon levels in New Zealand grasslands: is clay content important. Soil Science Society of America Journal 64: 1623–1630.
22. Lemenih M, Itanna F (2004) Soil carbon stocks and turnovers in various vegetation type and arable lands along an elevation gradient in southern Ethiopia. Geoderma 123: 177–188.
23. Fantappiè M, L'Abate G, Costantini EAC (2011) The influence of climate change on the soil organic carbon content in Italy from 1961to 2008. Geomorphology 135: 343–352.
24. Kang JM, Yang QC, GuoWS, Zhang TJ, Sun Y (2010) Yield evaluation of ten introduced alfalfa cultivars in beijing area of China. Chinese Journal of Grassland 32: 5–10.
25. Su YZ, Liu WJ, Yang R, Chang XX (2009) Changes in Soil Aggregate, Carbon, and Nitrogen Storages Following the Conversion of Cropland to Alfalfa Forage Land in the Marginal Oasis of Northwest China. Environmental Management 43: 1061–1070.
26. Jiang HM, Jin PJ, Yu J, Feng ML, Jin ZX (2006) Soil carbon pool and effects of soil fertility in seeded alfalfa fields on the semi-arid Loess Plateau in China. Soil Biology & Biochemistry 38: 2350–2358.
27. Nelson DW, Sommers LE (1982) Total carbon, organic carbon and organic matter. In: Page AL et al, editors. Methods of Soil Analysis, Part 2, 2nd edn. Agron. Monogr. ASA and SSSA Press, Madison, WI, USA, 102–129.
28. Gallaher RN, Weldon CO, Boswell FC (1976) A semi-automated procedure for total nitrogen in plant and soil samples. Soil Science Society of American Journal 40: 887–889.
29. Fu XL, Shao MA, Wei XR, Horton R (2010) Soil organic carbon and total nitrogen as affected by vegetation types in Northern Loess Plateau of China. Geoderma 155: 31–35.
30. Jimenez J, Lal R (2006) Mechanisms of C sequestration in soils of Latin America. Critical Reviews in Plant Sciences 25: 337–365.
31. Mcclaugherty CA, Pastor J, Aber JD, Melillo JM (1985) Forest litter decomposition in relation to soil–nitrogen dynamics and litter quality. Ecology 66: 266–275.
32. Posada JM, Schuur EAG (2011) Relationships among precipitation regime, nutrient availability, and carbon turnover in tropical rain forests. Oecologia 165: 783–795.
33. Post WM, Pastor J, Zinke PJ, Stangenberger AG (1985) Global patterns of soil–nitrogen storage. Nature 317: 613–616.
34. Schmitt A, Glaser B (2011) Organic matter dynamics in a temperate forest soil following enhanced drying. Soil Biology & Biochemistry 43: 478–489.
35. Stevenson FJ (1994) Humus Chemistry: Genesis, Composition, Reactions, 2nd ed. John Wiley, New York.
36. Wang CC, Chen CQ, Huang S, Deng AX (2010) Change characteristics of climatic and soil resources in Northeast China. Journal of Nanjing Agricultural University 33: 19–24.
37. Yan PJ, Wu T, Wang T (2011) Effects of calcic horizon on tree root distribution and growth in Wulanchabu city. Inner Mongolia Science Technology and Economy 235: 63–64.
38. Li XG, Li FM, Rengel Z, Bhupinderpal S, Wang ZF (2006) Cultivation effects on temporal changes of organic carbon and aggregate stability in desert soils of Hexi Corridor region in China. Soil & Tillage Research 91: 22–29.
39. Paustian K, Collins HP, Paul EA (1997) Management controls on soil carbon. In: Paul EA, Paustian KH, Elliott ET, Cole CV, editors. Soil Organic Matter in Temperate Agroecosystems: Long-Term Experiments in North America. CRC Press, Boca Raton, FL. 15–50.
40. Conant RT, Paustian K, Elliott ET (2001) Grassland management and conversion into grassland: effects on soil carbon. Journal of Applied Ecology 11: 343–355.
41. Walker LR, Wardle DA, Bardgett RD, Clarkson BD (2010) The use of chronosequences in studies of ecological succession and soil development. Journal of Ecology 98: 725–736.
42. Cheddadi R, Guiot J, Jolly D (2001) The Mediterranean vegetation: what if the atmospheric CO_2 increased? Landscape Ecology 16: 667–675.
43. Breshears DD, Myers OB, Johnson SR, Meyer CW, Martens SN (1997) Differential use of spatially heterogeneous soil moisture by two semiarid woody species: Pinusedulis and Juniperus monosperma. Ecology 85: 289–299.
44. Rutherford MC (1980) Annual plant production-precipitation relations in arid and semi-arid regions. South African Journal of Science 76: 53–56.
45. Skarpe C (1992) Dynamics of savanna ecosystems. Journal of Vegetation Science. 3: 293–400.

18

The Terrestrial Silica Pump

Joanna C. Carey[1]*, Robinson W. Fulweiler[1,2]

1 Department of Earth and Environment, Boston University, Boston, Massachusetts, United States of America, 2 Department of Biology, Boston University, Boston, Massachusetts, United States of America

Abstract

Silicon (Si) cycling controls atmospheric CO_2 concentrations and thus, the global climate, through three well-recognized means: chemical weathering of mineral silicates, occlusion of carbon (C) to soil phytoliths, and the oceanic biological Si pump. In the latter, oceanic diatoms directly sequester 25.8 Gton C yr^{-1}, accounting for 43% of the total oceanic net primary production (NPP). However, another important link between C and Si cycling remains largely ignored, specifically the role of Si in terrestrial NPP. Here we show that 55% of terrestrial NPP (33 Gton C yr^{-1}) is due to active Si-accumulating vegetation, on par with the amount of C sequestered annually via marine diatoms. Our results suggest that similar to oceanic diatoms, the biological Si cycle of land plants also controls atmospheric CO_2 levels. In addition, we provide the first estimates of Si fixed in terrestrial vegetation by major global biome type, highlighting the ecosystems of most dynamic Si fixation. Projected global land use change will convert forests to agricultural lands, increasing the fixation of Si by land plants, and the magnitude of the terrestrial Si pump.

Editor: Subho Mozumdar, University of Delhi, India

Funding: This research was partially developed under STAR Fellowship Assistance Agreement no. FP917238 awarded by the U.S. Environmental Protection Agency (EPA) to J.C. Carey. It has not been formally reviewed by EPA. The views expressed in this paper are solely those of J.C. Carey and R.W. Fulweiler, and EPA does not endorse any products or commercial services mentioned in this publication. In addition, this research was conducted under an award from the Estuarine Reserves Division, Office of Ocean and Coastal Resource Management, National Ocean Service, National Oceanic and Atmospheric Administration to J.C. Carey. The authors would also like to thank the Department of Earth Sciences at Boston University for partial funding support of J.C Carey. The funders had no role in study design, data collection and analysis, decision to publish, or preparation of the manuscript.

Competing Interests: The authors have declared that no competing interests exist.

* E-mail: jocarey@bu.edu

Introduction

Global silicon (Si) cycling regulates atmospheric carbon dioxide (CO_2) concentrations via several well-known mechanisms, particularly chemical weathering of mineral silicates [1], occlusion of carbon (C) to soil phytoliths [2], and the oceanic biological Si pump [3]. The vast majority of research on Si cycling has focused on the oceans, where Si-replete diatoms sequester 240 Tmol Si yr^{-1} [4]. Diatoms are also a critical component of the global C cycle, accounting for 35–75% of marine net primary production (NPP) [4] and serving as efficient exporters of C to the benthos [5].

However, similar to diatoms, terrestrial vegetation can also sequester large amounts of Si. Si is considered a 'quasi-essential' nutrient for plants [6]. While most plants can grow in Si-deplete soils, plant fitness is markedly improved with Si amendments [7]. In fact, Si is the only element that has been shown to never be toxic to plants even in high doses [7]. Silicon protects plants from a variety of abiotic and biotic stresses, including desiccation, predation, fungal attack, and heavy metal toxicity [6–8]. Ultimately, Si plays a critical role in plant defense [6], which we argue better facilitates plants to perform essential services, particularly C sequestration.

Si concentrations in land plants range over two orders of magnitude (<0.1 to over 10% by dry weight (by wt.)), the largest range of any element [6]. All photosynthetic plants contain some Si within their tissue, often in concentrations equal to or greater than other macronutrients, such as nitrogen (N), phosphorus (P), and potassium (K) [8]. Plants are typically broken into three modes of Si uptake: rejective, passive accumulators, or active accumulators [7]. Plants whose Si uptake is greater than that which would passively be taken up through the transpiration stream are defined as active accumulators and typically contain >0.46% Si (or 1% SiO_2) by wt. [7,9,10].

Plants take up dissolved silica (DSi) (H_4SiO_4) from soil solution via their roots. DSi is transported via the xylem for eventual deposition in transpiration termini. Upon incorporation into organisms, biogenic silica (BSi) is formed, creating siliceous bodies in cell walls known as phytoliths. While the ultimate source of Si in the biosphere is chemical weathering of mineral silicates, BSi is 7 to 20 times more soluble than mineral silicates [11], resulting in BSi being an important source of DSi on biological time scales [12]. The dynamic cycling of Si on its path from land to sea has only recently been documented [11,13], demonstrating that the biological component of the global Si cycle is driven not only by diatoms, but also by terrestrial organisms [14].

To date, the amount of Si fixed by land plants has only been estimated at the global scale [13,15], never by major terrestrial ecosystem type. Because certain plants, such as grasses, fix more Si than others, fixation of Si on land is far from uniform. Thus, we distinguish the hot spots of Si fixation, by calculating the amount of Si fixed by biome type (Table 1). Using this information, we calculate the percentage of terrestrial NPP done via active-Si accumulating organisms. To do this, we categorized Si shoot/leaf concentrations of 32 orders of angiosperms, gymnosperms, and mosses into 11 biome categories. Taxonomic order was chosen to categorize plants, as it accounts for the majority (67%) of the variability of Si concentrations within plant tissue [16].

Table 1. Amount of Si fixed by major biome type. Bold indicates on average, active Si accumulation, indicated by concentrations >0.46% Si by wt. [7].

Biome Type	NPP (C)×10^3 (Tmol yr^{-1})	Avg %Si by wt.	Si:C	Si (Tmol yr^{-1})
Tropical wet and moist forest	0.69	0.30	0.006	4.48
Tropical dry forest	0.40	0.28	0.006	2.35
Temperate forest	0.50	0.21	0.004	2.22
Boreal forest	0.53	0.23	0.005	2.65
Tropical woodland and savanna	0.93	**1.13**	0.024	22.19
Temperate steppe	0.41	**1.53**	0.032	13.26
Desert	0.12	0.26	0.006	0.65
Tundra	0.12	**1.07**	0.023	2.66
Wetland	0.32	**0.62**	0.013	4.16
Cultivated land	1.01	**1.37**	0.029	29.41
Rock and ice	0	0.00	0.000	0
Average		0.64	0.014	
Total	5.02			**84.03**
Std Dev		0.32	0.006	28.86

Si concentrations from Hodson et al. [16]. Biome categories from Houghton and Skole [17].

Materials and Methods

Leaf tissue Si concentrations were taken from supplemental data assembled in the meta-analysis completed by Hodson et al. (2005). From this dataset, 32 orders of angiosperms, gymnosperms, and mosses (representing 528 species) were assigned to a specific biome type. Fifteen orders fell into two different biome types (e.g. Pinales – temperate and boreal forest biomes). The order Poales (i.e. grasses, sedges, and rice) was the only order that fell into more than two biome types.

Given that average C concentrations in biomass consistently range from 45 to 50% [17], we assumed a wood C concentration of 47%. Dividing the calculated biome-specific average Si concentrations by 47% allowed us to calculate Si:C ratios for each biome (Table 1). In turn, our biome-specific Si:C ratios were multiplied by the known amount of C in each biome [17], creating a constrained estimate of the Si fixed by ecosystem type (Table 1).

Error estimates for the four forest biomes and the wetland biome were calculated by taking the standard deviation of the Si concentrations in the leaf/shoot tissue. Average Si concentration of the 'tropical woodland and savanna' biome was weighted for vegetation type (grasses accounted for 70%, herbaceous shrubs and trees accounted for 30% [18]). Error estimates associated with this biome include a sensitivity analysis of ±20% of each plant type. Likewise, average Si concentration of cultivated crops was weighted to account for the mass of the top four crops produced in 2010 [19] (Table 2). Average Si concentration of the tundra biome used an non-weighted average of the Poales (sedges), Sphagnales (mosses), and Ericales (herbaceous shrubs) orders. Average Si concentration of the temperate steppe biome included only Poales (grasses) as the predominant vegetation type.

Si fixed through land use change via increased agriculture was calculated by creating a correction factor that accounted for changes in land area of biome categories [20]. Correction factors were multiplied by the amount of Si currently fixed in each biome category in order to estimate the amount of Si fixed by 2050.

Table 2. Average Si concentration of cultivated crop biome, weighted by the mass of the four most produced crops [19].

Order	Genus	% Si by wt	Mass of Crop (Million Tons)	Si (Million tons)
Poales	Saccharum	0.54	1685	910
Poales	Triticum	2.42	650	1573
Poales	Zea	0.79	844	667
Poales	Oryza	3.17	672	2130
Total			3851	5280
Average Si concentration				**1.37**

Results

The amount of Si fixed annually by the 11 major biome types found on Earth was calculated (Table 1). Unsurprisingly, the least amount of Si was found in the rock and ice, and desert biomes (<1 Tmol yr^{-1}). Cultivated lands contain more Si than any other biome (29 Tmol yr^{-1}), due to both high average Si content of the most common cultivated crops (Table 2) and the large amount of biomass present on cultivated lands (Tables 1, 2).

While Si typically is deposited in the foliage, or leaves of trees, wood can also contain appreciable amounts of Si. Tropical forests contain relatively high levels of Si within their wood (Avg. 0.31% by wt.) [21,22], roughly four times more than the wood of temperate and boreal forests (Avg. 0.08% by wt.) [23]. In order to account for the different allocation of Si within various components of forest ecosystems, forest biome average Si concentrations were weighted to account for 30% of forest NPP being attributed to foliage, with the remainder allocated to woody biomass [24].

Active Si accumulation, indicated by average Si concentrations >0.46% by wt., was observed in five biomes: tropical woodland and savanna, temperate steppe, tundra, wetland, and cultivated lands (Figure 1). The elevated concentrations in these biomes were driven mostly by the predominance of the order Poales (e.g. grasses, sedges, and rice). These five biomes, which exhibit active Si accumulation, account for 55% of terrestrial NPP (Table 1). Thus, we conclude that just over half of terrestrial NPP is done by active-Si accumulating organisms, rivaling that of oceanic phytoplankton [17]. Combining our results with the estimate that 43% of ocean NPP is via Si-replete diatoms [4], we find that 59 Gton C yr^{-1}, or 49%, of current global NPP is conducted by active Si-accumulating organisms.

In order to provide a check of our results, we compared our total amount of Si fixed by land plants with those from previous studies. Summing our values of Si fixed by biome type, we find that 55–113 Tmol Si yr^{-1} is fixed globally in the terrestrial biosphere (Avg. 84±29 Tmol Si yr^{-1}). Our value is within the bounds of that provided by Conley (2002) (60–200 Tmol yr^{-1}) and almost identical to the global estimate provide by Laruelle et al. (1999) (80 Tmol yr^{-1}), providing confidence in our estimates.

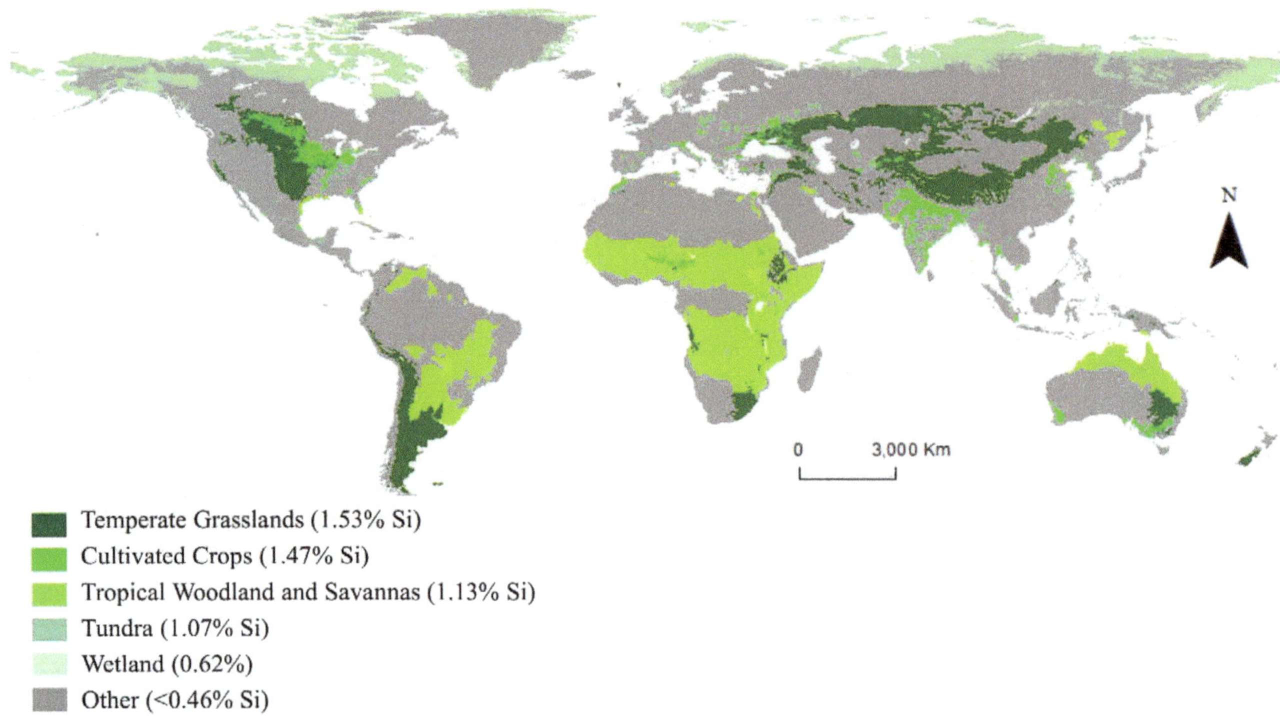

Figure 1. Global land cover by Si accumulating crops. Shades of green are directly proportional to degree of active Si accumulation (i.e., darker green indicates more Si accumulation). Gray area indicates land cover by biomes of non-active Si accumulators (<0.46% Si by dry wt.) (Table 1). Data from combination of FAO GeoNetwork Global Land Cover Distribution 2007 raster data layer with a resolution of 5-arc minutes, and ESRI, which modified the layer originally created by Olson et al. (2001). Map created using ArcMap 10.0 from ESRI.

Discussion

Human activities have already altered the amount of Si fixed in aquatic ecosystems. At the same time that the fluxes of N and P to coastal systems have markedly increased [25], river damming has led to significant declines in the amount of Si exported to coastal waters [26]. Combined, these two factors have increased the concentrations of N and P relative to Si, altering the ratio of N:P:Si in marine waters and inducing Si-limitation in several systems [27–29]. Such Si-limiting conditions can shift the phytoplankton species composition from diatom to non-diatom species, altering the base of the marine tropic structure [30,31]. Importantly, because diatoms are more efficient C exporters compared to other types of phytoplankton [5], shifting phytoplankton community structure away from diatoms reduces the amount of C exported to the benthos via the ocean biological pump.

Similarly, we hypothesize that land use change will also alter the magnitude of Si fixed in the terrestrial biosphere, although in the opposite direction. Because both cultivated crops and pasture land (included in grasslands) are the biomes of most Si sequestration, increases in agricultural production will increase the amount of BSi fixed by land plants. Our analysis shows that cultivated crop lands currently account for 35% of the BSi fixed on land each year. Area of crop and pasture land is projected to increase by 8.9×10^6 km^2, or 18%, by 2050 [20]. Assuming conversion of tropical forests to agricultural land, we calculate that an additional 7.5 Tmol Si yr^{-1} will be fixed by 2050, an increase of 9% from current levels. Thus, increased food production to feed a larger global population will heighten the magnitude of the terrestrial Si pump, as more BSi will be produced on land.

Because the Si fixed via crop cultivation is transported around the globe in large quantities, agriculture has recently been highlighted as an overlooked component of the global biological Si cycle [32]. Ultimately, much of the BSi stored in crops will be transferred to human population centers, or urban areas, which have been shown to export more Si than their forested counterparts [33]. The heightened production of the more-soluble biogenic Si within the biosphere may increase the capacity of urban systems to serve as sources of Si to coastal waters [34]. In turn, it is possible that Si-limiting conditions downstream of urban centers are buffered by the terrestrial Si pump and not as high as they would be in its absence. We hypothesize that additional Si fluxes from urban centers have the potential to improve the balance to N:P:Si ratios, at least in part, mitigating Si-limitation of nutrient-rich receiving waters.

However, the fate of the cultivated and harvest BSi in crops deserves further research attention, as environmental management practices, such as wastewater treatment, sludge disposal, and farming techniques will ultimately dictate the fate of the additional BSi produced by land plants [32]. In addition, the storage of sediment BSi and subsequent release as DSi are not directly connected processes, as they can occur over variable periods of time [35]. Soil pools contain at least two orders of magnitude more BSi than vegetation pools [13], highlighting that much of the BSi produced on land will be stored in the soils, rather than exported directly via rivers. While agricultural production will increase the amount of Si fixed by land plants, centuries of continual crop harvests have been shown to deplete soil pools of BSi [32,35,36]. Thus, conversion of more land for agricultural production could deplete soil BSi reserves over time, which could ultimately impact the magnitude of the terrestrial Si pump and long-term fluxes of Si to coastal systems.

Author Contributions

Conceived and designed the experiments: JCC RWF. Analyzed the data: JCC. Wrote the paper: JCC RWF.

References

1. Berner RA, Lasaga AC, Garrels RM (1983) The carbonate-silicate geochemical cycle and its effect on atmospheric carbon dioxide over the past 100 million years. American Journal of Science 283: 641–683.
2. Parr JF, Sullivan LA (2005) Soil carbon sequestration in phytoliths. Soil Biology and Biochemistry 37: 117–124.
3. Dugdale RC, Wilkerson FP, Minas HJ (1995) The role of a silicate pump in driving new production. Deep Sea Research Part I: Oceanographic Research Papers 42: 697–719.
4. Tréguer P, Nelson DM, Van Bennekom AJ, DeMaster DJ, Leynaert A, et al. (1995) The silica balance in the world ocean: A re-estimate. Science 268: 375–379.
5. Ittekkot V, Humborg C, SchÄFer P (2000) Hydrological Alterations and Marine Biogeochemistry: A Silicate Issue? BioScience 50: 776–782.
6. Epstein E (2009) Silicon: its manifold roles in plants. Annals of Applied Biology 155: 155–160.
7. Ma JF, Miyake Y, Takahashi E (2001) Chapter 2: Silicon as a beneficial element for crop plants. In: Datnoff LE, Snyder GH, Korndörfer GH, editors. Studies in Plant Science: Elsevier. 17–39.
8. Epstein E (1994) The anomaly of silicon in plant biology. Proceedings of the National Academy of Sciences 91: 11–17.
9. Hou L, Liu M, Yang Y, Ou D, Lin X, et al. (2010) Biogenic silica in intertidal marsh plants and associated sediments of the Yangtze Estuary. Journal of Environmental Sciences 22: 374–380.
10. Street-Perrott FA, Barker PA (2008) Biogenic silica: a neglected component of the coupled global continental biogeochemical cycles of carbon and silicon. Earth Surface Processes and Landforms 33: 1436–1457.
11. Cornelis J-T, Delvaux B, Georg RB, Lucas Y, Ranger J, et al. (2010) Tracing the origin of dissolved silicon transferred from various soil-plant systems towards rivers: a review. Biogeosciences Discuss 7: 5873–5930.
12. Derry LA, Kurtz AC, Ziegler K, Chadwick OA (2005) Biological control of terrestrial silica cycling and export fluxes to watersheds. Nature 433: 728–731.
13. Conley DJ (2002) Terrestrial ecosystems and the global biogeochemical silica cycle. Global Biogeochem Cycles 16: 1121DOI: 10.1029/2002gb001894.
14. Carey JC, Fulweiler RW (2012) Watershed land use alters riverine silica cycling. Biogeochemistry. DOI: 10.1007/s10533-012-9784-2.
15. Laruelle GG, Roubeix V, Sferratore A, Brodherr B, Ciuffa D, et al. (2009) Anthropogenic perturbations of the silicon cycle at the global scale: Key role of the land-ocean transition. Global Biogeochem Cycles 23: GB4031.
16. Hodson MJ, White PJ, Mead A, Broadley MR (2005) Phylogenetic Variation in the Silicon Composition of Plants. Annals of Botany 96: 1027–1046.
17. Houghton RA, Skole DL (1990) Carbon. In: Turner BLI, editor. The Earth As Transformed by Human Action: Global and Regional Changes in the Biosphere over the past 300 Years New York: Cambridge University Press.
18. Knoop WT, Walker BH (1985) Interactions of Woody and Herbaceous Vegetation in a Southern African Savanna. Journal of Ecology 73: 235–253.
19. FAO (2010) Food and Agriculture Organization of the United Nations (FAO). faostat.fao.org.
20. Tilman D, Fargione J, Wolff B, D'Antonio C, Dobson A, et al. (2001) Forecasting Agriculturally Driven Global Environmental Change. Science 292: 281–284.
21. Amos GL (1952) Silica in Timbers. Melbourne, Australia: Commonwealth Scientific and Industrial Research Organization.
22. Koeppen RC (1978) Some Anatomical Characteristics of Tropical Woods. U.S. Forest Service Department of Agriculture. Madison, WI 15.
23. Fulweiler RW, Nixon SW (2005) Terrestrial vegetation and the seasonal cycleof dissolved silica in a southern New Englandcoastal river. Biogeochemistry 74: 115–130.
24. Litton CM, Raich JW, Ryan MG (2007) Carbon allocation in forest ecosystems. Global Change Biology 13: 2089–2109.
25. Rockstrom J, Steffen W, Noone K, Persson A, Chapin FS, et al. (2009) A safe operating space for humanity. Nature 461: 472–475.
26. Humborg C, Ittekkot V, Cociasu A, Bodungen Bv (1997) Effect of Danube River dam on Black Sea biogeochemistry and ecosystem structure. Nature 386: 385–388.
27. Conley D, Schelske CL, Stoermer EF (1993) Modification of the biogeochemical cycle of silica with eutrophication. Marine Ecoogy Progress Series 101: 179–192.
28. Turner RE, Qureshi N, Rabalais NN, Dortch Q, Justic D, et al. (1998) Fluctuating silicate:nitrate ratios and coastal plankton food webs. Proceedings of the National Academy of Sciences 95: 13048–13051.
29. Danielsson Å, Papush L, Rahm L (2008) Alterations in nutrient limitations – Scenarios of a changing Baltic Sea. Journal of Marine Systems 73: 263–283.
30. Officer CB, Ryther JH (1980) The possible importance of silicon in marine eutrophication. Marine Ecoogy Progress Series 3: 83–91.
31. Admiraal W, Breugem P, Jacobs DMLHA, Steveninck ED (1990) Fixation of dissolved silicate and sedimentation of biogenic silicate in the lower river Rhine during diatom blooms. Biogeochemistry 9: 175–185.
32. Vandevenne F, Struyf E, Clymans W, Meire P (2011) Agricultural silica harvest: have humans created a new loop in the global silica cycle? Frontiers in Ecology and the Environment 10: 243–248.
33. Carey JC, Fulweiler RW (2011) Human activities directly alter watershed dissolved silica fluxes. Biogeochemistry DOI:10.1007/s10533-011-9671-2.
34. Sferratore A, Garnier J, Billen G, Conley DJ, Pinault S (2006) Diffuse and point sources of silica in the Seine River watershed. Environmental Science & Technology 40: 6630–6635.
35. Struyf E, Smis A, Van Damme S, Garnier J, Govers G, et al. (2010) Historical land use change has lowered terrestrial silica mobilization. Nat Commun 1: 129.
36. Clymans W, Struyf E, Govers G, Vandevenne F, DJ C (2011) Anthropogenic impact on biogenic Si pools in temperate soils. Biogeoscience Discussion 8: 4391–4419.

Growing Season Carbon Dioxide Exchange in Flooded Non-Mulching and Non-Flooded Mulching Cotton

Zhi-guo Li[1,2], Run-hua Zhang[3], Xiu-jun Wang[1], Fang Chen[2], Chang-yan Tian[1]*

1 State Key Laboratory of Desert and Oasis Ecology, Xinjiang Institute of Ecology and Geography, Chinese Academy of Sciences, Urumqi, China, **2** Wuhan Botanical Garden, Chinese Academy of Sciences Key Laboratory of Aquatic Botany and Watershed Ecology, Chinese Academy of Sciences, Wuhan, China, **3** Wuhan Vegetable Research Institute, Wuhan, China

Abstract

There is much interest in the role that agricultural practices might play in sequestering carbon to help offset rising atmospheric CO_2 concentrations. However, limited information exists regarding the potential for increased carbon sequestration of different management strategies. The objective of this study was to quantify and contrast carbon dioxide exchange in traditional non-mulching with flooding irrigation (TF) and plastic film mulching with drip irrigation (PM) cotton (*Gossypium hirsutum* L.) fields in northwest China. Net primary productivity (*NPP*), soil heterotrophic respiration (R_h) and net ecosystem productivity (*NEP*) were measured during the growing seasons in 2009 and 2010. As compared with TF, PM significantly increased the aboveground and belowground biomass and the *NPP* (340 g C m^{-2} season^{-1}) of cotton, and decreased the R_h (89 g C m^{-2} season^{-1}) ($p<0.05$). In a growing season, PM had a higher carbon sequestration in terms of *NEP* of ~ 429 g C m^{-2} season^{-1} than the TF. These results demonstrate that conversion of this type of land use to mulching practices is an effective way to increase carbon sequestration in the short term in cotton systems of arid areas.

Editor: Ben Bond-Lamberty, DOE Pacific Northwest National Laboratory, United States of America

Funding: This study was financially supported by the project of National Natural Science Foundation of China (40971048) and the program for 100 Distinguished Young Scientists of the Chinese Academy of Sciences (0972021001). The funders had no role in study design, data collection and analysis, decision to publish, or preparation of the manuscript.

Competing Interests: The authors have declared that no competing interests exist.

* E-mail: tianchy@ms.xjb.ac.cn

Introduction

Management and policies to increase the carbon (C) sink of agricultural soils have gained more and more attention, and it can be accepted as one of the greatest potential methods to sequester C in terrestrial ecosystems [1,2]. Different agronomic practices have been evaluated and recommended to reduce CO_2 emissions and increase C in soils such as the elimination of tillage, continuous cropping, cover cropping, using legumes in rotation and manure and crop residue application [3–5]. For example, Sperow *et al.* [6] suggested that a change from conventional tillage to no-tillage on all currently annually cropped areas (129 Mha) in the United States could result in 47 Tg C sequestered annually. Vleeshouwers and Verhagen [7] identified animal manure incorporation and the use of straw (or crop residues) could sequester as much as 0.19 and 0.15 Mg C ha^{-1} yr^{-1}, respectively. Although several studies have advanced the notion that changes in soil C stocks are associated with changes in crop management or in land use [1], there is still a lack of information regarding the potential C sequestration resulting from the conversion from traditional non-mulching cultivation to mulching cultivation [8].

Recently, newer cultivation technologies incorporating mulching with plastic film together with drip irrigation (PM) have been shown to increase soil temperatures and conserve soil moisture [9,10], resulting in increased crop production. PM cultivation is now widely applied in China [11,12] with land areas of this type of cultivation reaching 1,000,000 ha in northwestern China in 2009 [13]. It has replaced traditional cultivation (TF) involving flood irrigation with no mulching, which was the main cropping system used in dry land agriculture in northwest China before 1980 [12]. These large land use changes alter the soil microenvironment and have significant effects on the C balance of agro-ecosystems [1]. Understanding the extent to which the conversion of TF to PM contributes to the C cycle is important for developing C management strategies in dry land areas of China.

At present, there are few studies where a complete ecosystem approach (net ecosystem productivity [*NEP*], net primary productivity [*NPP*] and soil heterotrophic respiration [R_h]) is used to compare the C fluxes in traditional and modern cultivations of agro-ecosystems. In this study, a full C balance approach was used to investigate short-term effects (2 years) of cotton (*Gossypium hirsutum* L.) conversion from TF to PM on *NEP*. This study aims to quantify and contrast the amount of C sequestered by the plant biomass and the C lost through soil respiration in PM and TF cotton fields, and to estimate the net gains/losses of C due to the conversion of PM to TF in dry land cotton fields.

Materials and Methods

2.1. Site Description

The study site was located at the Fukang Station of Desert Ecology (87° 56′ E, 44° 17′ N) of the Chinese Academy of Sciences at an elevation of 461 m [14]. The soil type is a clay loam with a bulk density of 1.56 g cm^{-3}. Values of pH, electrical conductivity and total organic carbon at 0–10 cm are 8.42, 2.08 ms cm^{-1} and 5.64 mg C g^{-1}, respectively. In 2009 and 2010, daily mean air pressures ranged from 922.9 to 996.3 hPa. The daily mean air temperature had a maximum of 32°C, a minimum of −21°C, and a mean of 9.28°C. Daily mean air relative humidity showed large fluctuations, varying from 27.2 to 100%. Total annual precipitation was 164.2 mm in 2009 and 166.2 mm in 2010, slightly higher than the long-term average of 160 mm for the last 30 years.

2.2. Field Experiment

The field site has been reclaimed for 15 years from desert, and has been continuously planted cotton under plastic film mulching cultivation. In this field, we designed two treatments during the cotton growing seasons of 2009 and 2010: (1) the traditional cultivation system (TF) of flood irrigation with no mulching; and (2) a modern cultivation system using plastic film mulching with drip irrigation (PM). Each treatment was replicated three times in a randomized complete block design using plots 15×8 m. The cotton seeds (C.V. Xinluzao No. 6) were sown on April 21, 2009, and April 28, 2010. For the PM treatment, a high density and airtight transparent polythene film (0.01–0.02 mm thick, 1.25 m wide) was placed over the soil surface before sowing. The seeder made small holes (0.02×0.02 m) at 0.1 m intervals within a row in the plastic film and seeds were placed into the holes, and then each hole was covered with soil. Four rows were sown on each strip of plastic film. For the TF treatment, the plants were sown as for the PM treatment. For both treatments the planting density was 266 667 plants ha^{-1}. Irrigation models were used: drip irrigation under mulching for the PM treatment and flooded irrigation for the TF treatment, and the timing of irrigation and its volume followed the local commercial practice. Nitrogen fertilizer (110 kg N ha^{-1}) and phosphorus fertilizer (72 kg P_2O_5 ha^{-1}) were applied to both treatments, and weeds were controlled by hand.

2.3. Microclimate

During the growing season of 2009 and 2010, measurements of air temperature at 30 cm and soil temperature at 10 cm in planted sites and non planted sites with and without chambers, were made for each sample by using a SN2202 digital thermo detector (Sinan Instruments Plant of Beijing Normal University, Beijing). At the same time, soil water content at 10 cm in each plot was measured with the oven drying method (105°C for 48 h).

2.4. Growth Analysis

From May 10 to October 20 in 2009 and 2010, five 50×50 cm quadrats were randomly selected from both the PM and the TF every 25 days after planting. The above ground biomass in each quadrat was clipped to ground level and put in large brown paper bags. The below ground biomass was determined by the excavation of 50×50 cm pits at the same time to a depth of 70 cm in 20 cm increments. The live below ground biomass in each quadrat was collected and put in large brown paper bags. Biomass samples were washed in a series of five de-ionized water baths and weighed after oven drying for 5 days at 75°C. Grain yield, for each plot, was determined on three 4-m^2 subplots. Organic C was measured by oxidation with K_2CrO_7 according to the method of Liu *et al.* [15].

2.5. Net Primary Productivity

In this study, the cotton *NPP* was calculated by the measurement of above and below ground biomass during the different cotton periods and the C content using the following equation:

$$NPP_i = (Bb_i - Bb_{i-1}) \times Cb + (Ba_i - Ba_{i-1}) \times Ca \qquad (1)$$

where i is the measurement time of the plant biomass, *Bb* is the below ground biomass (g m^{-2}), *Cb* (g C g^{-1}) is the carbon content of the below ground biomass, *Ba* is the above ground biomass (g m^{-2}), *Ca* (g C g^{-1}) is the carbon content of the above ground biomass.

2.6. Soil Heterotrophic Respiration (R_h) Measurements

Soil heterotrophic respiration was measured at midday using a closed opaque chamber (0.5×0.5×0.5 m), which has been a method used in previous studies [16,17]. The chamber was equipped with a small circulation fan and a gas channel, which was a PVC tube with a three-way stopcock. A stainless steel frame with a water groove (0.5×0.5×0.05 m) was inserted into the soils, and the chamber was put into a groove that was filled with water before each gas sampling to ensure an airtight seal.

Three steel frames were placed between cotton rows in each replicate. Nylon nets (200 cm long×50 cm height, mesh size 0.25 mm) were buried around the frame to prevent roots entering the soil below the frames. When conducting gas sampling, the gas was collected mostly between 9 and 11 am every other week. The fluxes measured during this time in the morning are regarded to be basically representative for the daily average fluxes [18,19]. About 70 mL of gas was aspirated from the chamber at 30-s intervals for 150 s after capping. Gas samples were collected in polyethylene coated aluminum gas bags and taken into the laboratory soon after sampling. CO_2 concentrations were analyzed using a modified gas chromatograph (Agilent 4890D) equipped with an electron capture detector and a hydrogen flame ionization detector. The flux of soil heterotrophic respiration was calculated as follows:

$$F = \rho \times V/A \times \Delta c/\Delta t \times 273/T \times \alpha \qquad (2)$$

where F (g C m^{-2} h^{-1}) refers to flux of soil heterotrophic respiration, ρ is the density of CO_2 (1.98×10^3 g m^{-3}) under standard conditions, V (m^3) and A (m^2) are the volume and bottom area of the chamber, Δc (m^3 m^{-3}) is the change in CO_2 concentration in the chamber during the period Δt (h), T is the absolute temperature and α is the conversion factor for CO_2 to C (12/44).

2.7. Growing Season Net Ecosystem Productivity (NEP) Measurements

NEP is a comprehensive measure of net carbon accumulation by ecosystems. Growing season *NEP* can be calculated as the difference between the season *NPP* and season R_h:

$$\text{season}NEP = \text{season}NPP\text{season}R_h$$
$$= \sum_i^n NPP_i - \sum_j^n \frac{X_j + X_{j+1}}{2}(t_{j+1} - t_j) \qquad (3)$$

where season R_h represent cumulative soil CO_2 emissions throughout the cotton growing season; i and j are the measurement times of the plant biomass and soil respiration; X_j (g C m^{-2} d^{-1}) is the first week CO_2 measurement and X_{j+1} (g C m^{-2} d^{-1}) is the following week CO_2 measurement at times t_j and t_{j+1}, respectively; and n is the final week of CO_2 measurement during the experiment periods.

2.8 Statistical Analysis

Data comparisons between PM and TF sites were performed by one-way ANOVA using SPSS 11.0 software. Averages from all treatments were compared using Fisher's least significant difference at a probability level of 0.05.

Results

3.1. Soil Temperature and Soil Water Content

The mean daily soil temperature showed a marked seasonal trend during the growing season for all years and all sites, and both PM and TF had higher mean annual values in 2010 than in 2009 (Fig. 1). Soil temperature in the PM was significantly higher than in TF during the early growing season (April to July) ($p<0.05$); however, at the mid-late growth stage (after August), the daily average temperature in PM was obviously lower than that in TF, particularly in 2010. Soil water contents for PM and TF were also higher in 2010 than in 2009 during the growing period. The PM had a significant higher soil water content during the growing period than the TC treatment ($p<0.05$), but the difference was larger in 2009 than 2010.

3.2. NPP

Daily *NPP* values relative to the whole growing period (April to November) were computed for both land uses. The time series graphs of *NPP* from PM and TF (Fig. 2) showed large fluctuations ranging from 0.17 to 15.12 g C m^{-2} d^{-1} during the whole growing period. *NPP* increased to the highest levels in July and August but remained at a low level during the early (May–July) and late (October) growing season. Overall, the average daily rates of *NPP* in the PM were significantly higher than in the TF, especially during the main growth periods where PM (15.12 g C m^{-2} d^{-1}) was approximately 1.5 times higher than TF (9.4 g C m^{-2} d^{-1}) ($p<0.05$). Total *NPP* for the PM during the whole growing season was 1030 g C m^{-2} $season^{-1}$ which was significantly higher than the 689 g C m^{-2} $season^{-1}$ for the TF (Table 1) ($p<0.001$). In the 2nd year of measurement (2010), total *NPP* on both the PM and PF sites was less than the 1st year (2009) probably because an insect pest infestation broke out during the main cotton growing period (July). However, PM did show a higher total *NPP* than TF, although the difference was smaller.

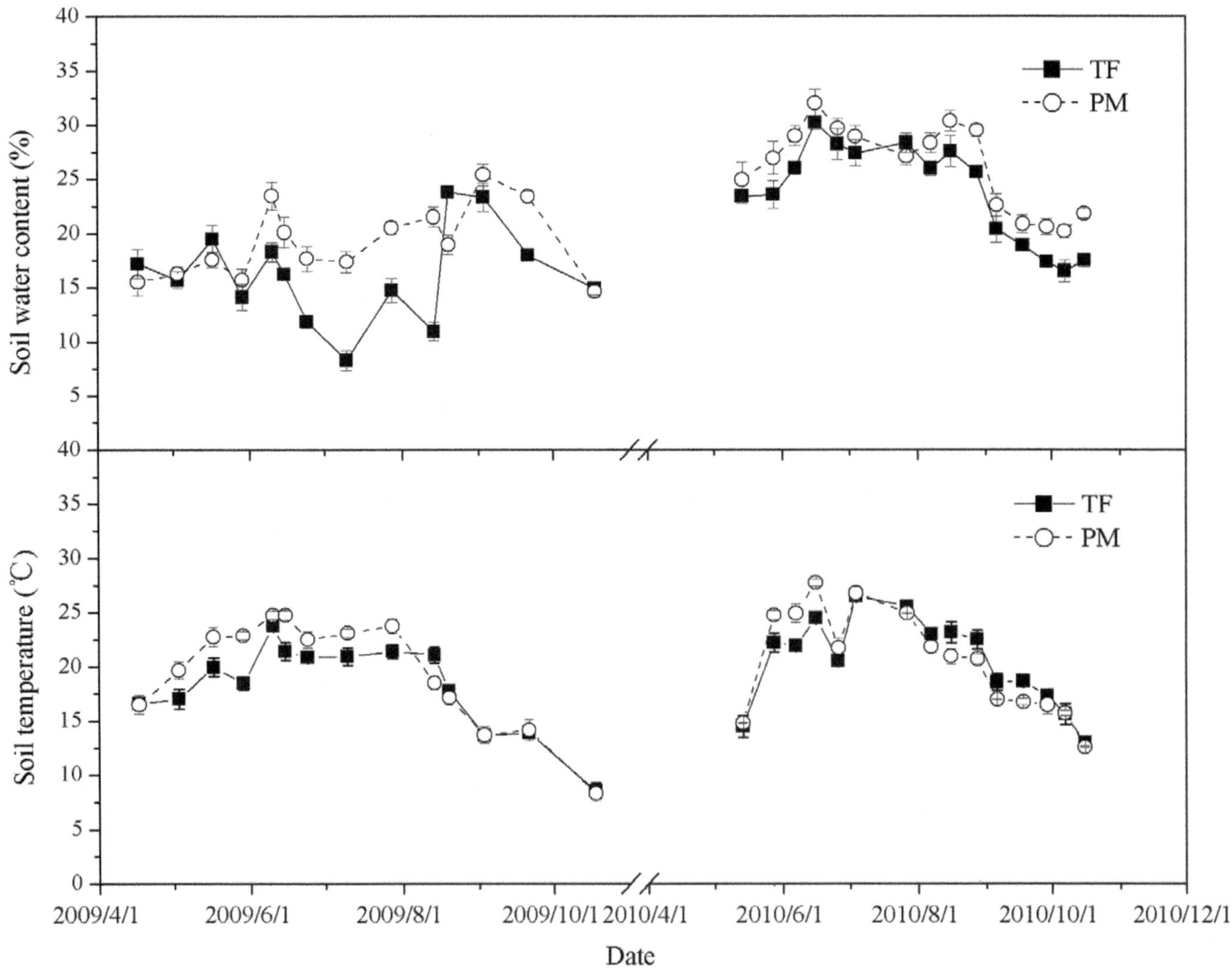

Figure 1. Soil temperature and soil water content measured at 10 cm depth in the traditional cultivation system of flood irrigation with no mulching (TF) and the modern cultivation system using plastic film mulching with drip irrigation (PM) in the cotton growing season of 2009 and 2010.

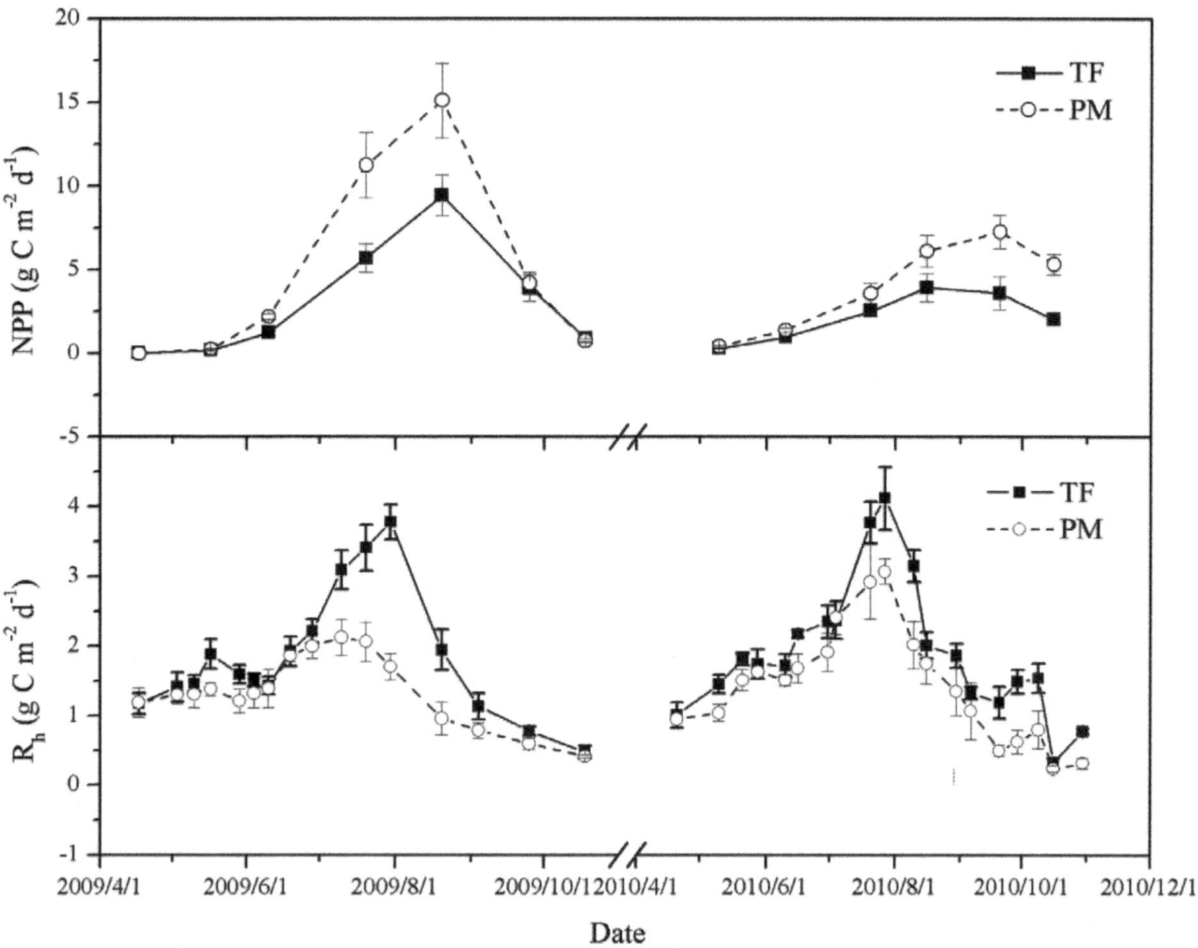

Figure 2. Daily net primary productivity (*NPP*) and heterotrophic respiration (R_h) measured using a closed chamber at the PM site and the TF site in the cotton growing season of 2009 and 2010 (mean±SE).

3.3. Soil Respiration

Soil respiration also showed strong seasonal variations with occasionally high fluxes of 2.12–4.12 g C m^{-2} d^{-1} between July and August (Fig. 2). The average daily rate of R_h was 1.86 g C m^{-2} d^{-1} in the PM, significantly higher than the values of 1.39 g C m^{-2} d^{-1} in the TF ($p<0.05$). During the two growing seasons, the total soil CO_2 flux in the PM was lower than that in the TF (485 and 663 g C m^{-2}, respectively). Thus, the conversion from TF to PM caused a decrease in soil C emissions of approximately 89 g C m^{-2} in one growing season.

3.4. Growing Season NEP

Between years, the values of *NEP* from PM and TF varied largely owing to the large fluctuation of *NPP* for both sites. Comparing *NEP* from the two sites, PM was 479 and 379 g C m^{-2} $season^{-1}$ higher than TF in 2009 and 2010 growing season, respectively. Over the two growing periods, PM had a net C uptake of 1234 g C m^{-2} and TF had a net C uptake of 376 g C m^{-2} from the atmosphere (Table 2).

Table 1. Crop management details for the two cultivation systems in the cotton growing season of 2009 and 2010 (±standard deviation where computed, n = 3).

Treatment[a]	Year	Tillage date	Planting date	Applied N (kg N ha^{-1})	Harvest date	Cotton yield (Mg ha^{-1})
PM	2009	Dec.20, 2008	Apr. 21, 2009	110	Oct. 18	4.41±0.37
	2010	Dec. 15, 2009	Apr. 29, 2009	110	Oct. 20	3.89±0.58
TF	2009	Dec.20, 2008	Apr. 21, 2009	110	Oct. 18	3.34±0.64
	2010	Dec. 15, 2009	Apr. 29, 2009	110	Oct. 20	2.85±0.23

[a]PM = plastic film mulching with drip irrigation; TF = non-mulching with flooding irrigation.

Table 2. Carbon balance components for the PM and TF fields in the cotton growing season of 2009 and 2010 (±SD, n = 3).

Date	Site	NPP^a g C m^{-2}	$R_h{}^b$ g C m^{-2}	NEP^c g C m^{-2}
2009.4–2009.11	PM	1030±155 a	214±33 b	816±49 a
	TF	648±96 b	311±29 a	337±21 b
2010.4–2010.11	PM	689±97 a	271±19 b	418±15 a
	TF	391±73 b	352±32 a	39±19 b
Total	PM	1719±121 a	485±31 b	1234±72 a
	TF	1039±78 b	663±49 a	376±49 b

[a] *NPP* = net primary productivity;
[b] R_h = heterotrophic respiration;
[c] *NEP* = net ecosystem productivity; In a row, values with the different letters are significant different at 5% level based on Fisher's least significant difference tests.

Discussion

In cropland, improved practice can reduce net CO_2 emissions from soils and stabilize and/or increase C sequestration in the soil, and is being accepted as the best potential method to assist in meeting the demands of an international C credit system [1–4]. In this study, the consequences of modern and traditional cultivation management options for C fluxes were investigated using a plastic film mulching management with drip irrigation, versus a non-mulching management with flood-irrigation. It was assumed that all differences between the mulching and non-mulching cropland plots resulted from the contrast in management.

The *NPP* (g C dry matter produced pre unit leaf area and time), referred to as the net CO_2 uptake by canopy, is a primary determinant of the C flux into the terrestrial biosphere [20]. PM significantly increased cotton *NPP* compared with TF, although the difference between PM and TF was smaller in the 2nd year owing to an insect pest infestation. This PM effect was also found in the earlier results for cotton in the Yellow River Delta China [21], and other crops such as wheat in New Delhi, India [22], and tomato in Mancha, Spain [23]. On the PM site, suitable soil conditions could be established with rational fertilizer use [24], high soil temperature and soil moisture [10], and better weed control [25], which could increase the photosynthetic capacity of crop leaves and lead to an increase in plant C content. Li *et al.* [11] and Chakraboyty *et al.* [22] attributed the higher *NPP* in the PM to the improvement of water use efficiency compared with TF. The increase of water use efficiency in PM is reflected in two aspects: prolonging the soil water retention time in soil by reducing soil water evaporation, and exploiting deep soil water. These changes reduced the water stress, triggered deep rooting growth and improved plant photosynthesis, which resulted in the obvious increase in *NPP* at the PM sites. Liu *et al.* [26] and Li *et al.* [11] also reported that the higher soil temperature of PM, particular at early seedling growth, reduced cell damage from early season chilling and produced larger and more vigorous seedlings compared with TF, and these seedlings then sequestered more CO_2 from the atmosphere during the growth period.

Besides the obvious increase in *NPP* (C input), PM also changed soil CO_2 fluxes (C output). Conversion from PM to TF greatly altered soil environment factors such as temperature, plant and microorganism activity, and soil organic carbon content. These factors regulate soil CO_2 production through biochemical action, autotrophic and heterotrophic respiration [27,28], and other soil properties such as soil water content and soil porosity, which affects the transfer of CO_2 in the soil profile [29]. In this study, the soil surface CO_2 flux was significantly lower in PM than in TF during the whole growth period. The result was opposite to what was expected, because previous studies reported that higher soil temperature and soil water content due to mulching increased soil respiration. The lower CO_2 in PM may be related to the soil properties. The soils in the present study sites were rich in calcium ions and had a high pH of 8.42 [30], together with a higher soil water content, which possibly led to a favorable soil environment for the formation of carbonate or organo-mineral complexes [31]. Moreover, mulching decreases wind disturbance, which can induce pressure pumping and vertical advection and increase rates of CO_2 exchange between soil and the atmosphere over and above rates permitted by molecular diffusion alone [32].

NEP, an index of ecosystem function related to CO_2 uptake depends on two contrasting processes: *NPP* and R_h. Due to lower R_h than *NPP* in arid cropland, the mean *NEP* values of 188–617 g C m^{-2} season^{-1} in these cotton systems were both positive and much greater than those observed at forest sites (Borden forest, Ontario: 65–140 g C m^{-2} [33]; European forest: 175 g C m^{-2} [34]; Harvard forest, MA: 174 g C m^{-2} [35]), and grassland sites ([northern temperate grassland in Alberta, Canada: −18 to 20 g C m^{-2} [36]; Mediterranean annual grassland in California: −30 to 130 g C m^{-2} [37]). As, we reported the *NEP* values with regard to single cotton growing season in this study, not surprisingly, the *NEP* values were higher than those reported on yearly basis in other studies. The growing season *NEP* values on the irrigated and rainfed maize sites at the University of Nebraska Agricultural Research and Development Center range from 600–800 g C m^{-2}, which approach those of cotton in our study [38]. In addition, in this region, there is plenty of sunshine during growing season, and hence we used the high density cultivation mode. This mode could produce a higher level of *NPP*. These reasons eventually led to higher *NEP* than in other studies.

However, as a consequence of an increase in *NPP* and a relative reduction in R_h rates, PM showed a higher *NEP* (C sequestration potential) and sequestered an average of 429 g C m^{-2} season^{-1} more than TF. Similarly, Saroa *et al.* [39], Jia *et al.* [40] and Liu *et al.* [41] reported that net C content in soil significantly increased in soils following the initial conversion to PM. However, these results are in contrast with Zhang *et al.* [42], who measured significant net losses of SOC in PM soil due to accelerated decomposition of organic matter under the prevailing conditions of higher soil temperature and moisture. The different C content results after conversion to PM were probably due to differences in soil properties and crop species. In the present study, crop soils are rich in calcium ions, have a high pH (8.42) and have a lower SOC content [30], which provided conditions for the formation of carbonate or organo-mineral complexes [31]. Also, cotton has a larger biomass and deeper root systems that was more suitable for sequestering larger amounts of C into the soil profile [43,44].

Despite these analyses, mulching cultivation still has a relatively finite C sequestration potential. According to results reported by Li *et al.* [45] and Chen *et al.* [46], mulching increased the amount of CO_2 released from soil because the higher soil temperature accelerated the decomposition of soil organic matter. To improve the performance of C sequestration by the PM ecosystem, we suggest that the mulching cultivation system should be integrated with other highly advanced technological methods such as using no-till practices [47], planting biofuel crops with a greater biomass [48], adopting practices that encourage deep rooting of crops that allocate more C and nutrients deep into the soil profile [49] or adding black biochar into the soil to fix more CO_2 from soil respiration.

Conclusion

This study shows that conversion from mulching cultivation to non mulching cultivation in the short term increased the share of cotton *NPP* (C input) and decreased the soil surface CO_2 flux (C output) by ~30% for the whole growing season. As a consequence, there was a net increase of C in terms of *NEP* of ~429 g C m^{-2} $season^{-1}$ after converting to PM cultivation. Nevertheless, to make the mulching cultivation sequestrate more C from the atmosphere into the soil, it was suggested that other C fixing practices are used in conjunction with the mulching cultivation. Furthermore, in order to assess the overall greenhouse gases balance after the conversion from PM to TF, experimental measurements on N_2O and CH_4 fluxes are needed.

Author Contributions

Conceived and designed the experiments: ZGL XJW FC CYT. Performed the experiments: ZGL RHZ. Analyzed the data: ZGL XJW FC. Contributed reagents/materials/analysis tools: ZGL RHZ CYT. Wrote the paper: ZGL.

References

1. Johnson JMF, Franzluebbers AJ, Weyers SL, Reicosky DC (2007) Agricultural opportunities to mitigate greenhouse gas emissions. Environmental Pollution 150: 107–124.
2. Desjardins RL, Smith W, Grant B, Campbell C, Riznek R (2005) Management strategies to sequester carbon in agricultural soils and to mitigate greenhouse gas emissions. Climatic Change 70: 283–297.
3. Yan HM, Cao MK, Liu JY, Tao B (2007) Potential and sustainability for carbon sequestration with improved soil management in agricultural soils of China. Agriculture Ecosystems & Environment 121: 325–335.
4. Paustian K, Six J, Elliott ET, Hunt HW (2000) Management options for reducing CO_2 emissions from agricultural soils. Biogeochemistry 48: 147–163.
5. Li CF, Zhou DN, Kou ZK, Zhang ZS, Wang JP, et al. (2012) Effects of tillage and nitrogen fertilizers on CH_4 and CO_2 emissions and soil organic carbon in paddy fields of central China. Plos One 7: 1–9.
6. Sperow M, Eve M, Paustian K (2003) Potential soil C sequestration on US agricultural soils. Climatic Change 57: 319–339.
7. Vleeshouwers LM, Verhagen A (2002) Carbon emission and sequestration by agricultural land use: a model study for Europe. Global Change Biology 8: 519–530.
8. Okuda H, Noda K, Sawamoto T, Tsuruta H, Hirabayashi T, et al. (2007) Emission of N_2O and CO_2 and uptake of CH_4 in soil from a satsuma mandarin orchard under mulching cultivation in central Japan. Journal of the Japanese Society for Horticultural Science 76: 279–287.
9. Ibarra-Jimenez L, Zermeno-Gonzalez A, Lozano-Del Rio J, Cedeno-Rubalcava B, Ortega-Ortiz H (2008) Changes in soil temperature, yield and photosynthetic response of potato (Solanum tuberosum L.) under coloured plastic mulch. Agrochimica 52: 263–272.
10. Zhang SL, Lovdahl L, Grip H, Tong YN, Yang XY, et al. (2009) Effects of mulching and catch cropping on soil temperature, soil moisture and wheat yield on the Loess Plateau of China. Soil & Tillage Research 102: 78–86.
11. Li FM, Wang J, Xu JZ, Xu HL (2004) Productivity and soil response to plastic film mulching durations for spring wheat on entisols in the semiarid Loess Plateau of China. Soil & Tillage Research 78: 9–20.
12. Zhang ZC, Zhang SF, Yang JC, Zhang JH (2008) Yield, grain quality and water use efficiency of rice under non-flooded mulching cultivation. Field Crops Research 108: 71–81.
13. He WP, Yan CR, Zhao CX, Chang RR, Liu Q, et al. (2009) Study on the pollution by plastic film mulch film and its countermeasure in China. Journal Agro-Environment Science 28: 533–538.
14. Wang YG, Xiao DN, Li Y, Li XY (2008) Soil salinity evolution and its relationship with dynamics of groundwater in the oasis of inland river basins: case study from the Fubei region of Xinjiang Province, China. Environmental Monitoring and Assessment 140: 291–302.
15. Liu WH, Zhao JZ, Ouyang ZY, Soderlund L, Liu GH (2005) Impacts of sewage irrigation on heavy metal distribution and contamination in Beijing, China. Environment International 31: 805–812.
16. Lohila A, Aurela M, Regina K, Laurila T (2003) Soil and total ecosystem respiration in agricultural fields: effect of soil and crop type. Plant and Soil 251: 303–317.
17. Bekku YS, Sakata T, Nakano T, Koizumi H (2009) Midday depression in root respiration of Quercus crispula and Chamaecyparis obtusa: its implication for estimating carbon cycling in forest ecosystems. Ecological Research 24: 865–871.
18. Mosier AR, Delgado JA (1997) Methane and nitrous oxide fluxes in grasslands in western Puerto Rico. Chemosphere 35: 2059–2082.
19. Kessavalou A, Mosier AR, Doran JW, Drijber RA, Lyon DJ, et al. (1998) Fluxes of carbon dioxide, nitrous oxide, and methane in grass sod and winter wheat-fallow tillage management. Journal of Environmental Quality 27: 1094–1104.
20. Ju WM, Gao P, Zhou YL, Chen JM, Chen S, et al. (2010) Prediction of summer grain crop yield with a process-based ecosystem model and remote sensing data for the northern area of the Jiangsu Province, China. International Journal of Remote Sensing 31: 1573–1587.
21. Dong HZ, Li WJ, Tang W, Zhang DM (2009) Early plastic mulching increases stand establishment and lint yield of cotton in saline fields. Field Crops Research 111: 269–275.
22. Chakraboyty D, Nagarajan S, Aggarwal P, Gupta VK, Tomar RK, et al. (2008) Effect of mulching on soil and plant water status, and the growth and yield of wheat (Triticum aestivum L.) in a semi-arid environment. Agricultural Water Management 95: 1323–1334.
23. Moreno MM, Moreno A (2008) Effect of different biodegradable and polyethylene mulches on soil properties and production in a tomato crop. Scientia Horticulturae 116: 256–263.
24. Gao YJ, Li Y, Zhang JC, Liu WG, Dang ZP, et al. (2009) Effects of mulch, N fertilizer, and plant density on wheat yield, wheat nitrogen uptake, and residual soil nitrate in a dryland area of China. Nutrient Cycling in Agroecosystems 85: 109–121.
25. Ahmed ZI, Ansar M, Iqbal M, Minhas NM (2007) Effect of planting geometry and mulching on moisture conservation, weed control and wheat growth under rainfed conditions. Pakistan Journal of Botany 39: 1189–1195.
26. Liu XJ, Wang JC, Lu SH, Zhang FS, Zeng XZ, et al. (2003) Effects of non-flooded mulching cultivation on crop yield, nutrient uptake and nutrient balance in rice-wheat cropping systems. Field Crops Research 83: 297–311.
27. Billings SA, Ziegler SE (2008) Altered patterns of soil carbon substrate usage and heterotrophic respiration in a pine forest with elevated CO_2 and N fertilization. Global Change Biology 14: 1025–1036.
28. Saito M, Kato T, Tang Y (2009) Temperature controls ecosystem CO_2 exchange of an alpine meadow on the northeastern Tibetan Plateau. Global Change Biology 15: 221–228.
29. Reth S, Gockede M, Falge E (2005) CO_2 efflux from agricultural soils in Eastern Germany - comparison of a closed chamber system with eddy covariance measurements. Theoretical and Applied Climatology 80: 105–120.
30. Xie JX, Li Y, Zhai CX (2009) CO_2 absorption by alkaline soils and its implication to the global carbon cycle. Environmental Geology 56: 953–961.
31. Mi N, Wang SQ, Liu JY, Yu GR, Zhang WJ, et al. (2008) Soil inorganic carbon storage pattern in China. Global Change Biology 14: 2380–2387.
32. Reicosky DC, Gesch RW, Wagner SW, Gilbert RA, Wente CD, et al. (2008) Tillage and wind effects on soil CO_2 concentrations in muck soils. Soil & Tillage Research 99: 221–231.
33. Teklemariam T, Staebler RM, Barr AG (2009) Eight years of carbon dioxide exchange above a mixed forest at Borden, Ontario. Agricultural and Forest Meteorology 149: 2040–2053.
34. Bellassen V, Viovy N, Luyssaert S, Le Maire G, Schelhaas MJ, et al. (2011) Reconstruction and attribution of the carbon sink of European forests between 1950 and 2000. Global Change Biology 17: 3274–3292.
35. Hollinger DY, Aber J, Dail B (2004) Spatial and temporal variability in forest-atmosphere CO2 exchange (vol 10, pg 1689, 2004). Global Change Biology 10: 1961–1961.
36. Flanagan LB, Wever LA, Carlson PJ (2002) Seasonal and interannual variation in carbon dioxide exchange and carbon balance in a northern temperate grassland. Global Change Biology 8: 599–615.
37. Xu LK, Baldocchi DD (2004) Seasonal variation in carbon dioxide exchange over a Mediterranean annual grassland in California. Agricultural and Forest Meteorology 123: 79–96.
38. Verma SB, Dobermann A, Cassman KG, Walters DT, Knops JM, et al. (2005) Annual carbon dioxide exchange in irrigated and rainfed maize-based agroecosystems. Agricultural and Forest Meteorology 131: 77–96.
39. Saroa GS, Lal R (2003) Soil restorative effects of mulching on aggregation and carbon sequestration in a miamian soil in central Ohio. Land Degradation & Development 14: 481–493.
40. Jia Y, Li FM, Wang XL (2006) Soil quality responses to alfalfa watered with a field micro-catchment technique in the Loess Plateau of China. Field Crops Research 95: 64–74.
41. Liu CA, Jin SL, Zhou LM, Jia Y, Li FM, et al. (2009) Effects of plastic film mulch and tillage on maize productivity and soil parameters. European Journal of Agronomy 31: 241–249.
42. Zhang GS, Chan KY, Li GD, Huang GB (2008) Effect of straw and plastic film management under contrasting tillage practices on the physical properties of an erodible loess soil. Soil & Tillage Research 98: 113–119.
43. McMichael BL, Oosterhuis DM, Zak JC, Beyrouty CA (2010) Growth and development of root systems. Physiology of Cotton: 57–71.

44. Tum M, Gunther KP (2011) Validating modelled NPP using statistical yield data. Biomass & Bioenergy 35: 4665–4674.
45. Li SQ, Li FM, Song QH, Wang J (2001) Effects of plastic film mulching periods on the soil nitrogen availability in semiarid areas. Acta Ecologica Sinica 21: 1519–1526.
46. Chen XS, Guo S, Wang JK, Zhang J (1998) Effect of mulching cultivation with plastic film on soil microbial population and biological activit. Chinese Journal of Applined Ecology 9: 435.
47. Sombrero A, de Benito A (2010) Carbon accumulation in soil. Ten-year study of conservation tillage and crop rotation in a semi-arid area of Castile-Leon, Spain. Soil & Tillage Research 107: 64–70.
48. Heaton E, Voigt T, Long SP (2004) A quantitative review comparing the yields of two candidate C-4 perennial biomass crops in relation to nitrogen, temperature and water. Biomass & Bioenergy 27: 21–30.
49. Bernacchi CJ, Hollinger SE, Meyers T (2005) The conversion of the corn/soybean ecosystem to no-till agriculture may result in a carbon sink. Global Change Biology 11: 1867–1872.

20

Soil Properties, Nutrient Dynamics, and Soil Enzyme Activities Associated with Garlic Stalk Decomposition under Various Conditions

Xu Han, Zhihui Cheng*, Huanwen Meng

State Key Laboratory of Crop Stress Biology in Arid Areas and College of Horticulture, Northwest A&F University, Yangling, Shaanxi, China

Abstract

The garlic stalk is a byproduct of garlic production and normally abandoned or burned, both of which cause environmental pollution. It is therefore appropriate to determine the conditions of efficient decomposition, and equally appropriate to determine the impact of this decomposition on soil properties. In this study, the soil properties, enzyme activities and nutrient dynamics associated with the decomposition of garlic stalk at different temperatures, concentrations and durations were investigated. Stalk decomposition significantly increased the values of soil pH and electrical conductivity. In addition, total nitrogen and organic carbon concentration were significantly increased by decomposing stalks at 40°C, with a 5:100 ratio and for 10 or 60 days. The highest activities of sucrase, urease and alkaline phosphatase in soil were detected when stalk decomposition was performed at the lowest temperature (10°C), highest concentration (5:100), and shortest duration (10 or 20 days). The evidence presented here suggests that garlic stalk decomposition improves the quality of soil by altering the value of soil pH and electrical conductivity and by changing nutrient dynamics and soil enzyme activity, compared to the soil decomposition without garlic stalks.

Editor: Jack Anthony Gilbert, Argonne National Laboratory, United States of America

Funding: This research was supported by a State Commonwealth (Agriculture) Scientific Research project (No.200903018) and a State Natural Science Foundation project (No.31171949). The funders had no role in study design, data collection and analysis, decision to publish, or preparation of the manuscript.

Competing Interests: The authors have declared that no competing interests exist.

* E-mail: chengzh@nwsuaf.edu.cn

Introduction

The three primary functions of most ecosystems are the production, accumulation, and decomposition of organic matter [1]. The nutrient cycle has an important role to play in the maintenance and organization of ecosystems; it includes both inputs and outputs [1]. Saura-Masa et al. [2] indicated that litter fall was an output of nutrients from the aerial parts of the plants and an input of nutrients to the soil. Decomposition and nutrient cycling have been found to affect soil functions [3,4].

Plant residues are abandoned or burned as waste under traditional farming systems. Burning and disposing of plant residues and continuous tillage have both contributed to excessive soil erosion [5,6]. Increased environmental awareness and the resultant need for efficient agricultural practices have made the use of crop residues increasingly attractive. Because the management of crop residues is a vital component of sustainable agricultural systems, it has attracted much interest in recent years as a method for increasing soil organic matter and the nutrient supply capacity of the soil and of reducing the damage caused by the burning of residue [7].

Cover crops are crops planted to prevent soil erosion and provide green manure. They are considered highly suitable for use in an integrated approach to agriculture, because they can supplement agro-ecosystems, improve soil properties, promote nutrient cycling, and facilitate pest management [8]. Cover crop decomposition is an essential part of ecosystem functioning [9]. Certain factors can regulate the decomposition rate. These include plant residue composition, soil nutrients composition, and the physical environment [10,11]. Early research has indicated that the effects of plant residues are complicated and rely on a delicate balance between promotive and inhibitory effects [12]. In addition to determining the recognised role of crop residues as a source of nutrient [13], studies performed in natural ecosystems, laboratories, and different farming systems have indicated that the addition of plant residues and other types of organic matter to the soil may decrease plant growth and influence the activities of enzymes and the microbial populations in the soil [13–16].

The ecological strategies of plant species are determined by functional traits that respond to ground conditions and influence the ecosystem [17–19]. The study of functional traits therefore plays an important role in research into the physiology and ecology of in plant residue decomposition. Moreover, the use of various vegetative residues to amend soil can improve the soil biological environment and plant growth conditions [20,21]. The use of auxiliary crops during the fallow period in degraded and unfertile fields can reduce nutrient loss, prevent soil erosion, improve the biological environment, and enhance crop productivity [22–24]. Previous studies have reported that crop residues affect soil nitrogen dynamics and ensure the availability of inorganic nitrogen in intensive production systems [22,25]. Some early studies have also shown that the decomposed products of crop residues can promote plant growth and others can inhibit it.

Garlic (*Allium sativum* L.) is used as both a vegetable and in medicinal applications. It is not only rich in nutrients but also possesses antibacterial properties and can be used to prevent and control disease. In vegetable production, it is generally considered a good previous crop. Garlic stalks are a byproduct of garlic production, and they can be a good biological resource [26]. In practice, garlic stalks are usually abandoned or burned as waste. As a result, not only is the rate of utilization of garlic stalks very low, but environmental pollution is also produced. Using this resource efficiently and reasonably would reduce environmental pollution, change ecosystems, and foster sustainable agriculture.

Previous studies on garlic stalks have mainly focused on the allelopathy of garlic straw extracts, such as ultrasonic and aqueous extracts [27,28]. Our recent studies indicated the importance of studying crop litter decomposition under natural and near-natural conditions using a quantitative size–density method, instead of confining litter in mesh bags because these mesh bags can create a microenvironment different from the one created by placing litter in direct contact with soil [29]. The overall objective of this study was to investigate the effects of garlic stalk decomposition in soil under different conditions on soil properties, nutrient dynamics, and soil enzyme activities. In this context, different durations of decomposition, different decomposition temperatures and different ratios of garlic stalk to soil were used to adjust the decomposition conditions.

Materials and Methods

Site Description

The field experiment was conducted in a plastic tunnel at the Horticulture Experimental Station (34° 16' N, 108° 4' E) of the College of Horticulture, Northwest A&F University, Yangling, Shaanxi Province, China. Garlic (cv. G064) stalks were collected from a local garlic production base field in June 2010. The harvested garlic stalks were dried under natural field conditions, ground into homogenised powder using a grinding machine, and then stored in the dark at room temperature until use. The basic characteristics of original experimental soil and garlic stalks were assayed and are shown in Table 1.

Experimental Design

Experiment 1 Effects of different durations of garlic stalk decomposition on soil properties, dynamic changes in nutrients, and enzyme activity. Garlic stalk powder (GSP) was mixed with soil at a ratio of 3 g dry weight per 100 g soil, which was then used to fill pots (15 cm × 15 cm × 15 cm). GSP mixed with soil was allowed to decompose naturally for 0, 10, 20, 30, 40, 50 or 60 days starting on February 15, 2011. Enough water was added to each pot to maintain the soil water content at or above 40% throughout the decomposition periods. Plastic sheets were placed under each pot in order to prevent moisture loss. Experiments were performed in triplicate and a total of 105 pots were placed in a plastic tunnel and arranged in a randomised block design. After the specified decomposition period, the soil samples from all the pots in each treatment group were homogenized, dried in a dark room, and sieved through a 1 mm or 0.149 mm sieve to determine soil properties, and soil enzyme activity, and nutrient dynamics.

Experiment 2 Effects of different garlic stalk decomposition temperatures on soil properties, dynamic changes in nutrients, and enzyme activity. GSP was mixed with soil at a ratio of 3 g dry weight per 100 g soil, which was then used to fill pots (15 cm × 15 cm × 15 cm). GSP mixed with soil was allowed to decompose at 10°C, 25°C and 40°C in the growth chamber (RXZ, Ningbo Jiangnan Instrument Factory, Ningbo, China) for 30 days starting on February 15, 2011. Enough water was added to each pot to keep the soil water content at least 40% throughout the decomposition period. Plastic sheets were placed under each pot to prevent moisture loss. The experiments were performed in triplicate and a total of 45 pots were arranged in a randomized block design. After the specified decomposition treatments, the soil samples from all the pots in each treatment were homogenized, dried in a dark room, and sieved through a 1 mm or 0.149 mm sieve to determine soil properties, and soil enzyme activity, and nutrient dynamics.

Experiment 3 Effects of different concentrations of decomposed garlic stalk on soil properties, dynamic changes in nutrients, and enzyme activity. GSP was mixed with soil at three concentrations (1, 3, or 5 g dry weight per 100 g soil) and filled in the pots (15 cm × 15 cm × 15 cm). Blank control pots were filled with the same amount of soil by weight but no garlic stalks. The experiment was performed in triplicate; a total of 60 pots were placed in a plastic tunnel and arranged in a randomised block design. Decomposition was carried out naturally for 30 days starting on February 15, 2011. Water was added to each pot to keep soil water content at least 40% throughout the decomposition periods. Plastic sheets were placed under each pot to prevent moisture loss. After the specified decomposition treatments, the soil samples from all the pots in each treatment were homogenized, dried in a dark room and sieved through a 1 mm or 0.149 mm sieve to determine soil properties, and soil enzyme activity, and nutrient dynamics.

Determination of Soil Properties

Soil samples were extracted with distilled water (1:1 water to soil suspension) and analyzed for soil pH value by a pH meter (PHS-3C, LIDA, Shanghai, China) [30]. Soil electrical conductivity (EC) value (5:1 distilled water to soil suspension) was determined using a microprocessor conductivity meter (DDS-12DW, Xiaoshan, China) [30].

Determination of Nutrient Dynamics

Total nitrogen (N) content was analyzed by boiling away concentrated sulfuric acid boiled using an Automatic Kieldahl Apparatus (KDY-9830, Beijing, China) [30]. Total organic carbon (C) content was assayed using the potassium dichromate method

Table 1. Basic characteristics of original soil and garlic stalk.

Samples	pH value	EC value ($\mu s \cdot cm^{-1}$)	Total nitrogen ($g \cdot kg^{-1}$)	Total organic carbon ($g \cdot kg^{-1}$)	C/N	Organic matter ($g \cdot kg^{-1}$)	Total phosphorus ($g \cdot kg^{-1}$)	Total potassium ($g \cdot kg^{-1}$)	Cellulose content (%)
Original soil	7.75	267.0	0.99	10.55	10.61	18.19	0.89	7.96	0
Garlic stalk	7.44	653.0	0.24	353.51	146.87	609.46	16.46	0.27	0.68

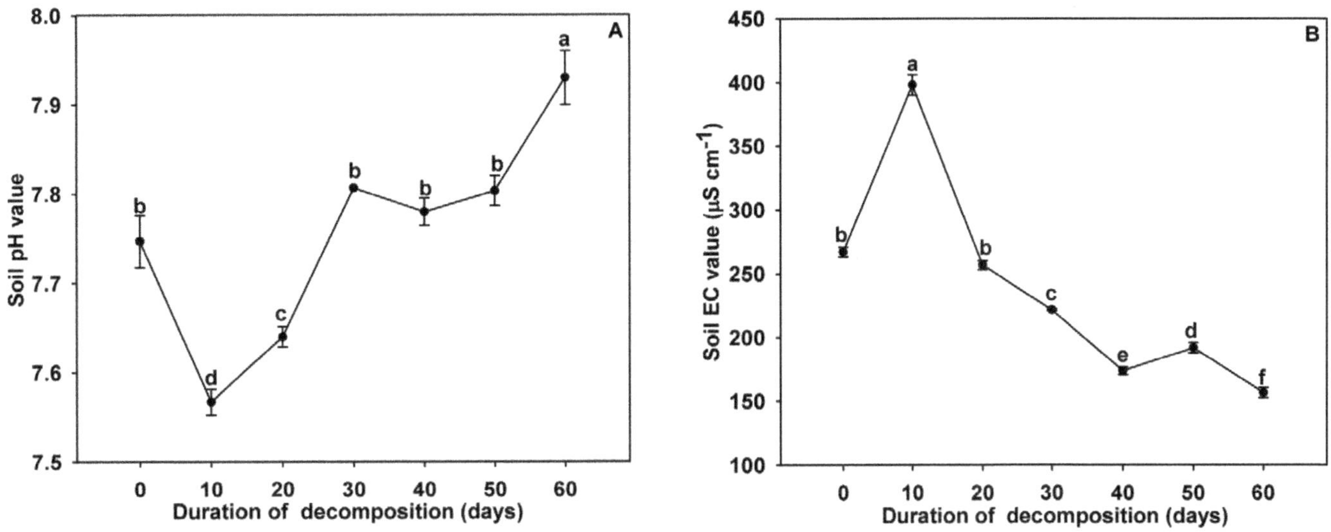

Figure 1. Effects of the duration of garlic stalk decomposition on soil pH (A) and EC (B). Error bars represent as the standard error of the mean. Different letters above the error bars indicate significant differences at the 0.05 level (ANOVA and Duncan's multiple range test), n = 3.

[30].For determination the contents of total phosphorus (P) and potassium (K), the soil was wet-digested with a mixture HNO_3–HCl–$HClO_4$ (1:3:1). Total P in the wet digestion was measured colorimetrically using the molybdenum antimony resistance method and a spectrophotometer at 880 nm, and total K was measured using an atomic absorption spectrophotometer (Z-2000, Hitachi, Japan) [30].

Determination of Soil Enzyme Activities

The activity levels of sucrase, urease and alkaline phosphatase in soil were assayed on the basis of the release and quantitative determination of the products of glucose, NH_3-N, and P_2O_5. Soil samples were incubated with an 8% sucrose solution, a 10% urea solution, and a 0.5% disodium phenyl phosphate solution in a suitable buffer solution for 24 hours at 37°C and spectrophotometric measurements were performed at 508 nm, 578 nm and 660 nm [31].

Statistical Analysis

The data were assessed using one-way analysis of variance (ANOVA) with the SPSS 17.0 software package (SPSS Inc., Chicago, U.S.). Mean separations were performed using Duncan's multiple range tests. Differences at $P = 0.05$ were considered significant.

Results

Soil Properties, Dynamic Changes in Nutrients, and Enzyme Activities Associated with Different Durations of Garlic Stalk Decomposition

Treatments of longer durations (60 days) were associated with significantly higher soil pH values. However, significantly lower soil pH values were associated with 10 days of garlic stalk decomposition (Fig 1A). Soil EC values varied significantly with the duration of garlic stalk decomposition (Fig 1B). The highest soil EC value was associated with garlic stalk decomposition lasting 10 days, and the lowest value was associated with garlic stalk decomposition lasting 60 days.

The test results of soil nutrient dynamic changes associated with different decomposition durations are summarized in Table 2. The shorter durations (10 and 20 days) were associated with significantly higher levels of total organic carbon and

Table 2. Effects of different garlic stalk decomposition durations on dynamic changes in soil nutrients.

Decomposing duration (days)	Total nitrogen (g · kg^{-1})	Total organic carbon (g · kg^{-1})	C/N	Organic matter (g · kg^{-1})	Total phosphorus (g · kg^{-1})	Total potassium (g · kg^{-1})
0	0.99 e	10.55 bc	10.61 abc	18.19 bc	0.89 a	7.96 b
10	1.15 b	14.06 a	12.24 ab	24.23 a	0.90 a	13.74 a
20	1.12 bc	15.20 a	13.50 a	26.20 a	0.88 ab	13.98 a
30	1.01 e	12.33 ab	12.16 ab	21.25 ab	0.77 bc	13.93 a
40	1.10 c	10.33 bc	9.38 bc	17.80 bc	0.76 bc	14.15 a
50	1.06 d	8.43 c	7.93 c	14.53 c	0.70 c	14.03 a
60	1.21 a	10.61 bc	8.76 c	18.30 bc	0.85 ab	13.97 a

The data are presented as soil nutrient dynamics. Different letters in the same column indicate significant differences at the 0.05 level (ANOVA and Duncan's multiple range test), n = 3.

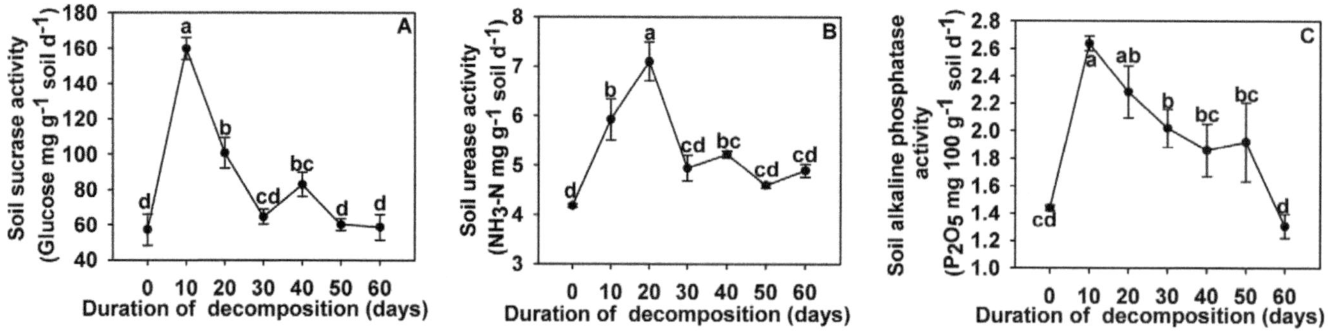

Figure 2. Effects of duration of garlic stalk decomposition on soil enzyme activity. Effects of different durations of garlic stalk decomposition on soil sucrase activity (A), soil urease activity (B) and soil alkaline phosphatase activity (C). Error bars represent the standard error of the mean. Different letters above the error bars indicate significant differences at the 0.05 level (ANOVA and Duncan's multiple range test), n = 3.

organic matter. The highest total nitrogen content was observed in soil in which garlic stalks had decomposed for 60 days. However, the highest C/N ratio was associated with 20 days of garlic stalk decomposition. Total phosphorus content was significantly higher in groups subjected to 0 and 10 days of garlic stalk decomposition. Total potassium content was significantly higher after 10 to 60 days of decomposition than after 0 days of decomposition.

Enzyme activities in soil in which garlic stalks had decomposed for different periods of time are given in Fig. 2. Garlic stalk decomposition 10 days was associated with significantly higher levels of sucrase and alkaline phosphatase activity (Fig 2A and 2C). The highest urease activity was recorded in the soil in which garlic stalks had been allowed to decompose for 20 days (Fig 2B).

Soil Properties, Dynamic Changes in Nutrients, and Enzyme Activities, Associated with Different Garlic Stalk Decomposition Temperatures

The highest soil pH values were associated with garlic stalk decomposition at 25°C, but no significant difference was observed between the decomposition at 10°C and at 40°C (Fig 3A). Garlic stalk decomposition at 25°C was associated with significant lower soil EC values than decomposition at 10°C or 40°C. No significant difference was observed between decomposition at 10°C and at 40°C (Fig 3B).

The effects of different garlic stalk decomposition temperatures on the dynamic changes in soil nutrients were assessed (Table 3). Total nitrogen content increased as decomposition temperature increased, and the highest value was observed at 40°C of decomposition. Garlic stalk decomposition at 25°C and 40°C was found to significantly promote total organic carbon and organic matter content over values observed at 10°C. However, the C/N ratio and total phosphorus content remained relatively constant across all three treatments groups. Total potassium content was significantly inhibited at 40°C but no significant difference was observed between decomposition at 10°C and at 25°C.

Garlic stalk decomposition at 10°C was associated with significantly higher sucrase activity than decomposition 25°C or 40°C, but there were no significant difference between decomposition at 25°C and at 40°C (Fig 4A). Garlic stalk decomposition at 10°C significantly promoted urease activity. However, urease activity was significantly inhibited by decomposition at 25°C (Fig 4B). The lowest alkaline phosphatase

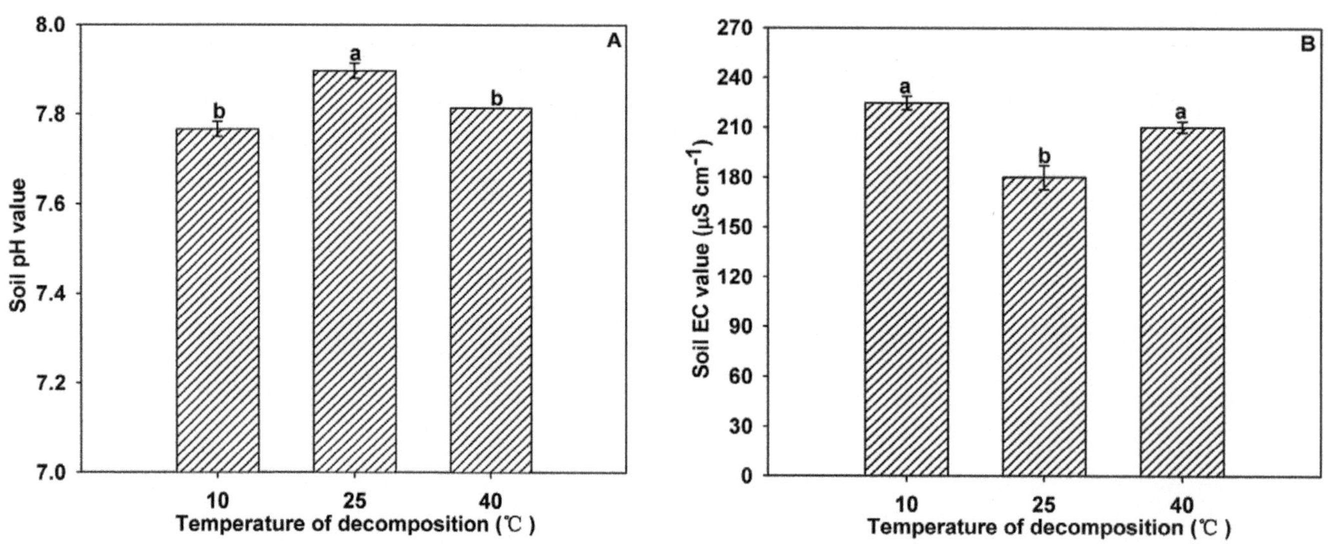

Figure 3. Effects of different garlic stalk decomposition temperatures on soil pH (A) and EC (B). Error bars represent the standard error of the mean. Different letters above the histograms indicate significant differences at the 0.05 level (ANOVA and Duncan's multiple range test), n = 3.

Table 3. Effects of different garlic stalk decomposition temperatures on dynamic changes in soil nutrients.

Decomposing temperature (°C)	Total nitrogen (g • kg^{-1})	Total organic carbon (g • kg^{-1})	C/N	Organic matter (g • kg^{-1})	Total phosphorus (g • kg^{-1})	Total potassium (g • kg^{-1})
10	0.86 c	8.85 b	10.59 a	15.26 b	0.58 a	13.35 a
25	0.90 b	10.00 a	11.17 a	17.24 a	0.59 a	13.93 a
40	0.97 a	10.11 a	10.42 a	17.43 a	0.44 a	9.49 b

The data are presented as soil nutrient dynamics. Different letters in the same column indicate significant differences at the 0.05 level (ANOVA and Duncan's multiple range test), n = 3.

activity was associated with decomposition at 40°C, and the higher values were associated with decomposition at 10°C and 25°C (Fig 4C).

Soil Properties, Dynamic Changes of Nutrients, and Enzyme Activities Associated with Different Concentrations of Decomposed Garlic Stalk

Soil pH (Fig 5A) and EC values (Fig 5B) varied significantly with the concentration of decomposed garlic stalk. The lowest non-zero concentration (1:100) of decomposed garlic stalk was associated with the highest soil pH value and lowest soil EC value. These reverse was true of the highest concentration (5:100).

Effects of different concentrations of decomposed garlic stalk on dynamic changes in soil nutrients are presented in Table 4. The highest concentrations of garlic stalk to soil (5:100) significantly increased the concentration of total nitrogen, total organic carbon and organic matter relative to the control (0:100). However, the C/N ratio did not differ across the treatment groups. The total phosphorus content decreased as the ratio of garlic stalk to soil increased. Total soil potassium content of all treatments was significantly higher than control.

The effects of decomposed garlic stalks on the activity levels of sucrase (Fig 6A), urease (Fig 6B) and alkaline phosphatase (Fig 6C) varied with the ratio of garlic stalk to soil. The highest activities of all three soil enzymes were observed at the highest ratio (5:100).

Discussion

The decomposition of plant residues is one of the important processes in the nutrient cycle of almost any ecosystem [32]. Decomposition dynamics are impacted by several main factors, such as the duration of decomposition, environmental conditions (especially temperature), the quantity of the decomposers, and the quality of plant residues as the substrate for the decomposers [33,34]. For these reasons, three different sets of decomposition conditions, such as duration, temperature and concentrations of garlic stalks were chosen for this investigation of the effects of decomposed garlic stalks on soil properties, nutrient dynamics, and soil enzyme activities.

Soil pH and EC value are the primary factors indicating the chemical properties of the soil. The burning of plant residues and continuous cultivation of the same plant could lead to soil acidification or salinization. Soil pH can also affect nutrient availability, such as soil ion exchange capacity, nutrient adsorption and solubility, soil microbial activities, and the root uptake process [35–37]. Several studies have reported that high soil pH is usually associated with low soil nutrient availability [38–42]. In the current study, the highest soil pH value and total nitrogen content were observed when garlic stalks decomposed over 60 days. Decomposition times of 10 days, temperatures of 10°C and 40°C, and garlic stalk to soil ratios of 5:100 were associated with significantly higher soil EC values.

Plant residue decomposition is a vital pathway in the carbon cycle and nutrient regeneration [43,44]. Plant residue is a major source of carbon for microbes in the soil and a major route of input of soil organic carbon [45,46]. In this way, plant residue can regulate essential biological functions in soil ecosystems. Our results showed that garlic stalk decomposition lasting 10 days or 20 days significantly increased soil total organic carbon and organic matter. These results are similar to those of one previous study that showed that many elements in plant

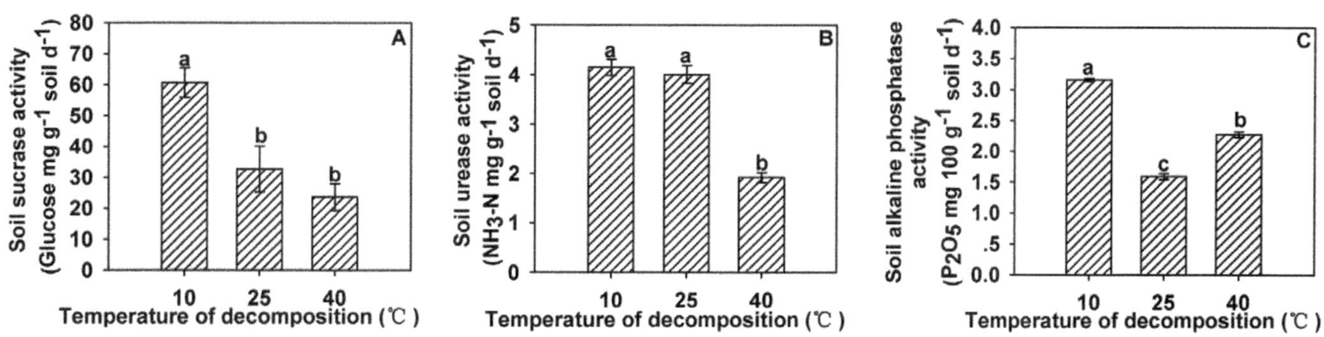

Figure 4. Effects of different garlic stalk decomposition temperatures on soil enzyme activities. Effect of different decomposition temperature of garlic stalk on soil sucrase activity (A), soil urease activity (B) and soil alkaline phosphatase activity (C). Error bars represent the standard error of the mean. Different letters above the histograms indicate significant differences at the 0.05 level (ANOVA and Duncan's multiple range test), n = 3.

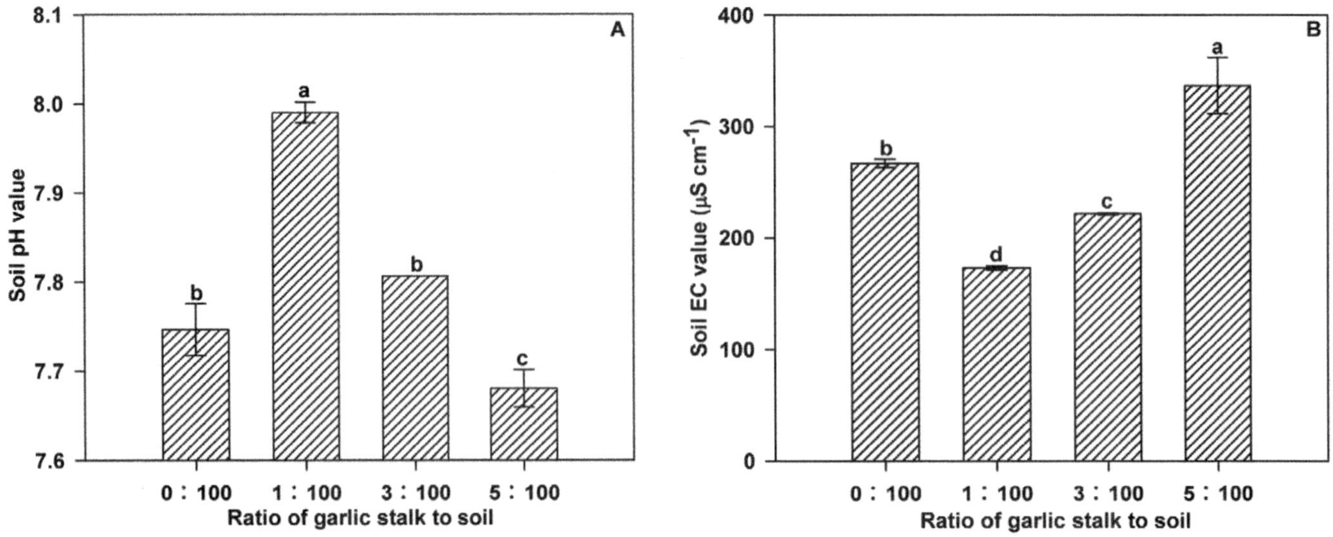

Figure 5. Effects of different concentrations of decomposed garlic stalk on soil pH (A) and EC (B). Error bars represent the standard error of the mean. Different letters above the error bars indicate significant differences at the 0.05 level (ANOVA and Duncan's multiple range test), n = 3.

materials were released faster after the plant materials were placed in the soil and decomposed immediately [47]. The highest values of soil total organic carbon and organic matter were observed at decomposition temperatures of 25°C and 40°C. This was consistent with the results of other studies, which reported that higher temperatures were associated with faster nutrient cycling and that air temperature could impact the degree of decomposition [48,49]. In our study, the highest levels of total nitrogen were observed in soil where garlic stalks had decomposed at 40°C, for 60 days and concentrations of 5:100. Several studies have reported that crops grown on legume residues experienced increased nitrogen uptake [50,51]. In the present study, a 5:100 stalk-to-soil ratio also significantly increased the level of total soil organic carbon and organic matter.

The activity of soil enzymes plays a key role in many ecosystem and soil management processes, and it may influence soil quality and function [52,53]. It is usually higher in plant residues vicinity than in bulk soil. In this study, the peak sucrase and alkaline phosphatase activity was observed in soil where garlic stalks had decomposed for 10 days and peak urease activity was observed where garlic stalks had decomposed for 20 days. Soil urease activity was found to be highly responsive to nutritional conditions [54]. Dilly *et al.* [55] reported that the highest activity of urease was observed at later stages of decomposition. However, our results were not consistent with this. The results of our study showed that garlic stalk decomposition at 10°C significantly increased soil sucrase and urease activities and stalk decomposition at 10°C and 25°C significantly promoted soil alkaline phosphatase activity. These results are consistent with the previous studies stating that bogs, fens, and other wetlands exhibited less soil phosphatase activity because of the cool temperatures [56]. The highest activities of these three soil enzymes detected in this study were observed at garlic-stalk-to-soil ratios of 5:100.

Conclusions

The results of this study showed that different garlic stalk decomposition conditions can have a significant impact on soil properties, such as pH and EC value. Soil nutrient dynamics were also found to vary with garlic stalk decomposing conditions, such as the duration of decomposition, temperature, and the ratio of garlic stalk to soil. The highest soil enzyme activity appeared at shorter durations (10 days or 20 days), lower temperatures (10°C), and higher concentrations of stalk (5:100).

Table 4. Effects of different concentrations of decomposed garlic stalk on dynamic changes in soil nutrients.

Ratio of garlic stalk to soil	Total nitrogen ($g \cdot kg^{-1}$)	Total organic carbon ($g \cdot kg^{-1}$)	C/N	Organic matter ($g \cdot kg^{-1}$)	Total phosphorus ($g \cdot kg^{-1}$)	Total potassium ($g \cdot kg^{-1}$)
0 : 100	0.99 b	10.55 b	10.61 a	18.19 b	0.89 a	7.96 b
1 : 100	0.94 c	10.45 b	11.10 a	18.02 b	0.82 ab	13.18 a
3 : 100	1.01 b	12.37 b	12.20 a	21.32 b	0.77 b	13.93 a
5 : 100	1.21 a	15.09 a	12.44 a	26.01 a	0.66 c	13.11 a

The data are presented as the soil nutrient dynamics. Different letters in the same column indicate significant differences at the 0.05 level (ANOVA and Duncan's multiple range test), n = 3.

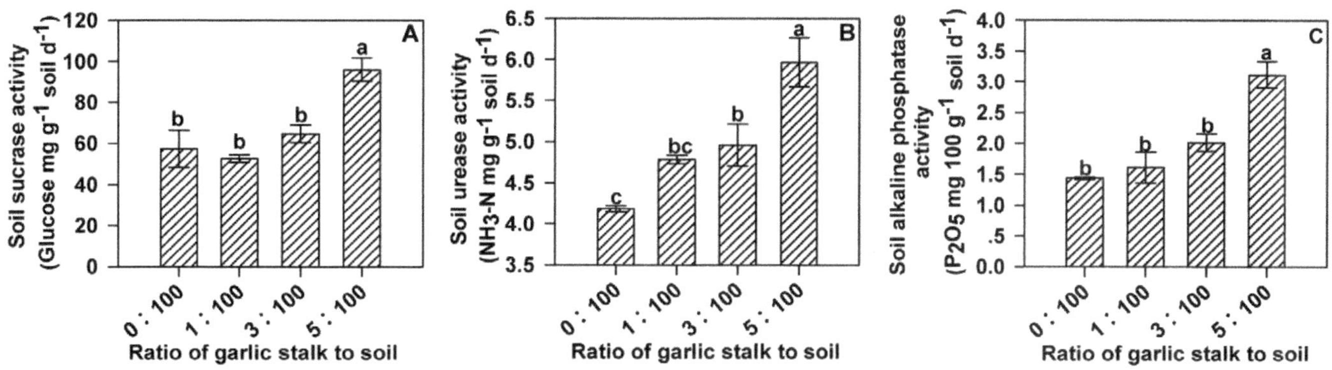

Figure 6. Effects of different concentrations of decomposed garlic stalk on soil enzyme activities. Effects of different concentrations of decomposed garlic stalk on soil sucrase activity (A), soil urease activity (B) and soil alkaline phosphatase activity (C). Error bars represent the standard error of the mean. Different letters above the error bars indicate significant differences at the 0.05 level (ANOVA and Duncan's multiple range test), n = 3.

Author Contributions

Conceived and designed the experiments: XH ZC. Performed the experiments: XH. Analyzed the data: XH HM. Wrote the paper: XH.

References

1. Jacobson TKB, Bustamante MMC, Kozovits AR (2011) Diversity of shrub tree layer, leaf litter decomposition and N release in a Brazilian Cerrado under N, P and N plus P additions. Environ Pollut 159: 2236–2242.
2. Saura-Masa S, Estiarte M, Peñuelasb J, Lloret F (2012) Effects of climate change on leaf litter decomposition across post-fire plant regenerative groups. Environ Exp Bot 77: 274–282.
3. Muthukaruppan G, Janardhanan S, Vijayalakshmi G (2004) Sublethal toxicity of the herbicide butachlor on the earthworm Perionyx sansibaricus and its histological changes. J Soils Sediments 5: 82–86.
4. Römbke J, Förster B, Jänsch S, Scheffczyk A, Garcia M (2005) Terrestrische ökotoxikologische Testmethoden für die Tropen d Teil 2: Halbfreiland-und Freilandtests sowie Risikobeurteilung. UWSF 17: 85–93.
5. Laker MC (2004) Advances in soil erosion, soil conservation, land suitability evaluation and land use planning research in South Africa 1978–2003. S Afr J Plant Soil 21(5): 345–368.
6. Mills AJ, Fey MV (2004) Declining soil quality in South Africa: effects of land use on soil organic matter and surface crusting. S Afr J Plant Soil 21: 388–398.
7. Singh Y, Gupta RK, Singh J, Singh G, Singh G, et al. (2010) Placement effects on rice residue decomposition and nutrient dynamics on two soil types during wheat cropping in rice–wheat system in northwestern India. Nutr Cycl Agroecosys 88: 471–480.
8. Sarrantonio M, Gallandt E (2003) The role of cover crops in North American cropping systems. J Crop Production 8: 53–74.
9. Giulia M, Marco AD, Meola A, Sessa L, Santo AVD (2011) Nutrient dynamics in litter mixtures of four Mediterranean maquis species decomposing in situ. Soil Biol Biochem 43: 520–530.
10. Kang H, Xin Z, Berg B, Burgess PJ, Liu Q, et al. (2010) Global pattern of leaf litter nitrogen and phosphorus in woody plants. Ann For Sci 67: 811p1–811p8.
11. Prescott CE (2010) Litter decomposition: what controls it and how can we alter it to sequester more carbon in forest soils? Biogeochemistry 101: 133–149.
12. Holmgren M, Scheffer M, Huston MA (1997) The interplay of facilitation and competition in plant communities. Ecol 78: 1966–1975.
13. Vitousek PM, Sanford RLJ (1986) Nutrient cycling in moist tropical forest. Annu Rev Ecol Syst 17: 137–167.
14. Meier CL, Keyserling K, Bowman WD (2009) Fine root inputs to soil reduce growth of a neighbouring plant via distinct mechanisms dependent on root carbon chemistry. J Ecol 97: 941–949.
15. Hodge H, Stewart J, Robinson D, Griffiths BS, Fitter AH (1998) Root proliferation, soil fauna and plant nitrogen capture from nutrient-rich patches in soil. New Phytol 139: 479–494.
16. Putnam AR (1994) Phytotoxicity of plant residues, In: Unger PW. (Ed.), Managing agricultural residues. Lewis Publishers, Boca Raton, FL, USA. 285–314.
17. Lavorel S, Garnier E (2002) Predicting changes in community composition and ecosystem functioning from plant traits: revisiting the Holy Grail. Funct Ecol 16: 545–556.
18. Garnier E, Lavorel S, Ansquer P, Castro H, Cruz P, et al. (2007) Assessing the effects of land-use change on plant traits, communities and ecosystem functioning in grasslands: a standardized methodology and lessons from an application to 11 European sites. Ann Bot 99: 967–985.
19. Suding KN, Lavorel S, Chapin FS, Cornelissen JHC, Diaz S, et al. (2008) Scaling environmental change through the community-level: a trait-based response-and-effect framework for plants. Glob Chang Biol 14: 1125–1140.
20. McSorley R, Gallaher RN (1994) Effect of tillage and crop residue management on nematode densities on corn. J Nematol 26: 669–674.
21. Ilieva-Makulec K, Olejniczak I, Szanser M (2006) Response of soil micro- and mesofauna to diversity and quality of plant litter. Eur. J. Soil Biol. 42: S244–S249.
22. Guo RY, Li XL, Christie P, Chen Q, Jiang RF, et al. (2008) Influence of root zone nitrogen management and a summer catch crop on cucumber yield and soil mineral nitrogen dynamics in intensive production systems. Plant Soil 313: 55–70.
23. DuPont ST, Ferris H, Horn MV (2009) Effects of cover crop quality and quantity on nematode-based soil food webs and nutrient cycling. Appl Soil Ecol 41: 157–167.
24. Tian Y, Liu J, Zhang X, Gao L (2010) Effects of summer catch crop, residue management, soil temperature and water on the succeeding cucumber rhizosphere nitrogen mineralization in intensive production systems. Nutr Cycl Agroecosys 88(3): 429–446.
25. Williams SM, Weil RR (2004) Crop cover root channels may alleviate soil compaction effects on soybean crop. Soil Sci Soc Am J 68: 1403–1409.
26. Cheng ZH, Tong F, Jin R (2008) Primary study on the inhibitory effects on plant pathogen and the inhibitory ingredient of garlic straw aqueous extracts. Acta Bot.Boreal.-Occident.Sin. 28 (2): 0324–0330.
27. Wang CH, Cheng ZH, Niu Q, Liang JN, Xue SH (2009) Allelopathy of ultrasonic extract of garlic plant on different receiver vegetable crops. J Northwest A &F Univ Natur Sci Edit 37(7): 103–109.
28. Wei L, Cheng ZH, Zhang L (2008) Allelopathy of straw aqueous extracts of different garlic varieties on tomato (*Lycopersicon esculentum*). J Northwest A &F Univ Natur Sci Edit 36 (10): 139–145.
29. Bhupinderpal-Singh, Rengel Z, Bowden JW (2009) A quantitative size-density separation method to recover and characterise decomposing crop residues added to soil. Biol. Fertil. Soils 45: 423–434.
30. Bao SD (2000) Soil agricultural chemistry analysis. China: China Agriculture Press.
31. Guan SY, Shen GQ (1984) Enzyme activities in main soil in China. Tu Rang Xue Bao 21(4): 368–381.
32. Strakova P, Penttila T, Laine J, Laiho R (2012) Disentangling direct and indirect effects of water table drawdown on above-and belowground plant litter decomposition: consequences for accumulation of organic matter in boreal peatlands. Glob Chang Biol 18: 322–335.
33. Belyea LR (1996) Separating the effects of litter quality and microenvironment on decomposition rates in a patterned peatland. Oikos 77: 529–539.
34. Laiho R (2006) Decomposition in peatlands: Reconciling seemingly contrasting results on the impacts of lowered water levels. Soil Biol Biochem 38: 2011–2024.
35. Marschner H (1995) Mineral nutrition of higher plants, 2nd edn. New York: Academic Press.
36. Larcher W (2003) Physiological plant ecology: ecophysiology and stress physiology of functional groups, 4th edn. Springer, Berlin.

37. Cheng L, Zhu J, Chen G, Zheng X, Oh NH, et al. (2010) Atmospheric CO_2 enrichment facilitates cation release from soil. Ecol Lett 13: 284–291.
38. Gough L, Shaver GR, Carroll J, Royer DL, Laundre JA (2000) Vascular plant species richness in Alaskan arctic tundra: the importance of soil pH. J Ecol 88: 54–66.
39. Shaver GR, Giblin AE, Nadelhoffer KJ, Thieler KK, Downs MR, et al. (2006) Carbon turnover in Alaskan tundra soils: effects of organic matter quality, temperature, moisture and fertilizer. J Ecol 94: 740–753.
40. Zak DR, Kling GW (2006) Microbial community composition and function across an arctic tundra landscape. Ecol 87: 1659–1670.
41. Eskelinen A (2008) Herbivore and neighbour effects on tundra plants depend on species identity, nutrient availability and local environmental conditions. J Ecol 96: 155–165.
42. Eskelinen A, Stark S, Männistö M (2009) Links between plant community composition, soil organic matter quality and microbial communities in contrasting tundra habitats. Oecologia 161: 113–123.
43. Jordan TE, Whigham DF, Correll DL (1989) The role of litter in nutrient cycling in a brackish tidal marsh. Ecol 70: 1906–1915.
44. Aerts R, Chapin FS (2000) The mineral nutrition of wild plants revisited: A re-evaluation of processes and patterns. Adv. Ecol. Res. 30: 1–67.
45. Icoz I, Stotzky G (2008) Fate and effects of insect-resistant Bt crops in soil ecosystems. Soil Biol Biochem 40: 559–586.
46. Zibilske LM, Materon L (2005) Biochemical properties of decomposing cotton and corn stem and root residues. Soil Sci Soc Am J 69: 375–386.
47. Ibewiro B, Sanginga N, Vanlauwe B, Merckx R (2000) Nitrogen contributions from decomposing cover crop residues to maize in a tropical derived savanna. Nutr Cycl Agroecosys 57: 131–140.
48. MacDonald NW, Zak DR, Pregitzer KS (1995) Temperature effects on kinetics of microbial respiration and net nitrogen and sulphur mineralization. Soil Sci Soc Am J 59: 233–240.
49. Ruffo ML, Bollero GA (2003) Modeling rye and hairy vetch residue decomposition as a function of degree days and decomposition days. Agron J 95: 900–907.
50. Kuo S, Jellum EJ (2002) Influence of winter cover crop residue management on soil nitrogen availability and corn. Agron J 94: 501–508.
51. Miguez FE, Bollero GA (2005) A review of corn yield response under winter cover cropping systems using metaanalytic methods. Crop Sci 45: 2318–2329.
52. Yao HY, Jiao XD, Wu FZ (2008) Effects of continuous cucumber cropping and alternative rotations under protected cultivation on soil microbial community diversity. Plant Soil 284: 195–203.
53. Lagomarsin A, Benedetti A, Marinari S, Pompili L, Moscatelli MC, et al. (2011) Soil organic C variability and microbial functions in a Mediterranean agro-forest ecosystem. Biolo. Fertil. Soils 47: 283–291.
54. Dilly O, Nannipieri P (2001) Response of ATP content, respiration rate and enzyme activities in an arable and a forest soil to nutrient addition. Biol. Fertil. Soils 34: 64–72.
55. Dilly O, Munch JC, Pfeiffer EM (2007) Enzyme activities and litter decomposition in agricultural soils in northern, central, and southern Germany. J Plant Nutr Soil Sci 70: 197–204.
56. Kang H, Freeman C (2009) Soil Enzyme Analysis for Leaf Litter Decomposition in Global Wetlands. Commun Soil Sci Plant Anal 40: 3323–3334.

21

Pyrosequencing Reveals the Influence of Organic and Conventional Farming Systems on Bacterial Communities

Ru Li[1], Ehsan Khafipour[2,3]*, Denis O. Krause[2,3], Martin H. Entz[1], Teresa R. de Kievit[4], W. G. Dilantha Fernando[1]*

1 Department of Plant Science, University of Manitoba, Winnipeg, Manitoba Canada, **2** Department of Animal Science, University of Manitoba, Winnipeg, Manitoba, Canada, **3** Department of Medical Microbiology and Infectious Diseases, Winnipeg, Manitoba, Canada, **4** Department of Microbiology, University of Manitoba, Winnipeg, Manitoba, Canada

Abstract

It has been debated how different farming systems influence the composition of soil bacterial communities, which are crucial for maintaining soil health. In this research, we applied high-throughput pyrosequencing of V1 to V3 regions of bacterial 16S rRNA genes to gain further insight into how organic and conventional farming systems and crop rotation influence bulk soil bacterial communities. A 2×2 factorial experiment consisted of two agriculture management systems (organic versus conventional) and two crop rotations (flax-oat-fababean-wheat versus flax-alfalfa-alfalfa-wheat) was conducted at the Glenlea Long-Term Crop Rotation and Management Station, which is Canada's oldest organic-conventional management study field. Results revealed that there is a significant difference in the composition of bacterial genera between organic and conventional management systems but crop rotation was not a discriminator factor. Organic farming was associated with higher relative abundance of *Proteobacteria*, while *Actinobacteria* and *Chloroflexi* were more abundant in conventional farming. The dominant genera including *Blastococcus, Microlunatus, Pseudonocardia, Solirubrobacter, Brevundimonas, Pseudomonas*, and *Stenotrophomonas* exhibited significant variation between the organic and conventional farming systems. The relative abundance of bacterial communities at the phylum and class level was correlated to soil pH rather than other edaphic properties. In addition, it was found that *Proteobacteria* and *Actinobacteria* were more sensitive to pH variation.

Editor: Niyaz Ahmed, University of Hyderabad, India

Funding: The research was funded by Natural Sciences and Engineering Research Council of Canada discovery grants to WGDF and TRdK. The funders had no role in the study design, data collection and analysis, or preparation of the manuscript.

Competing Interests: The authors have declared that no competing interests exist.

* E-mail: Ehsan.Khafipour@ad.umanitoba.ca (EK); Dilantha.Fernando@ad.umanitoba.ca (WGDF)

Introduction

It has long been recognized that maintaining biodiversity of soil microbes is crucial to soil health, which has been defined as soil with the capacity of resilience to stress, sustaining high biological diversity, productivity and high level of internal nutrient cycling, maintaining environmental quality and promoting plant health [1,2]. Bacterial communities are responsible for multifaceted biological functions in soils [3,4,5], and exert an important role in maintaining plant health [6,7,8,9,10]. In turn, soil has a direct impact on the structure and function of soil bacterial communities through perturbations caused by natural or human activities [11,12,13]. It was reported that agricultural soil, perturbed by human activities, has different bacterial diversity, compared to non-disturbed forest and grassland soil [14,15]. However, there is still lack of the detailed information about the bacterial diversity affected by agriculture perturbation.

Over the past decades, conventional agricultural management practices have involved the use of artificial chemical fertilizers and pesticides to increase crop yields. This has led to severe environmental problems such as soil degradation, emission and leaching of fertilizer and pesticide, and the emergence of pesticide resistant species [16,17], resulting in an unsustainable practice [18]. The aim in sustainable management systems is to maintain the biological function of the soil and to promote plant health. Organic farming contributes to these factors using techniques such as crop rotation, green manure, and biological pest control instead of chemical fertilizers and pesticides [19]. Consequently, organic farming systems may have a strong potential for restoring soil health and increase agro-ecosystem resilience to stress [18].

Few studies have evaluated the impact of fertilizer, crop rotation, and crop varieties on microbial community structure when conventional and organic farming systems are compared [20,21,22,23,24]. These studies have found that fertilizers, and crop varieties and rotation could shape the size and structure of soil microbial communities. However, these studies were based on field experiments where the above-mentioned factors varied at the same time between conventional and organic soil management. Therefore, the main discriminator between conventional and organic farming could not be defined. Moreover, the results of these studies were not consistent,

Table 1. Physico-chemical characteristics of Glenlea soil (0–15 cm) on the different treatments (Welsh, 2009; Bell 2012).

Rotation	Management	Total C (g/kg)	Carbonate C (g/kg)	Organic matter (%)	Total N (g/kg)	Olsen P (mg/kg)	pH	C: N Ratio
Grain-Only	Organic	3.0	2.8	6.7	2.6	23.4	7.0	11.3
Grain-Only	Conventional	3.2	2.6	7.2	2.7	21.7	6.6	11.5
Forage-Grain	Organic	3.1	0.4	7.9	2.7	3.6	7.0	11.3
Forage-Grain	Conventional	3.0	1.7	7.9	2.7	14.2	6.6	10.9
SEM		0.06	0.19	0.50	0.008	1.63	0.32	0.19
P-value								
Rotation		NS	<0.0001	0.005	NS	<0.0001	NS	NS
Management		NS	0.0039	NS	NS	0.0035	0.023	NS
Management × Rotation		0.01	0.0001	NS	NS	0.0001	NS	NS

NS = not significant ($P>0.05$).

which could be due to different analytical methodologies that varied in resolution [14,24,25,26].

Previous studies in our group had investigated the effects of different agriculture management practices (organic versus converntional) and crop rotation systems [(flax-oat-fababean-wheat (Grain-Only rotation) versus flax-alfalfa-alfalfa-wheat (Forage-Grain rotation)] on crop yield, soil edaphic traits, such as nitrate, phosphorus, pH, and organic matter. Results showed that rotation rather than farming managements affect most soil nutrient traits including nitrate-N, Oslen P, and organic matter. Whereas, farming system affect soil pH, with lower pH in conventional farming system than that in organic farming system. However, the effects of farming management and rotation on bacterial communitiy structure of the soil was not evaluated [27,46].

We hypothesized that microbial composition of soil differs between organic and conventional farming sytstems or in different crop rotations. As such, some microorganisms might be present in organic farms while absent or less frequent in conventional farms and vice versa. Similarly, there might be some bacteria that are unique to a specific crop rotation system. Using Roche 454 pyrosequencing methodology, the objective of this study was to identify bacterial populations that are associated with specific farming practice that could potentially influence soil and plant health. This manuscript provides a detailed framework of the soil bacterial composition at the genus level and its possible connection to farming practices.

Materials and Methods

Soil Sampling, Sampling Site and Experimental Design

Soil samples were collected at Glenlea Long-term Crop Rotation and Management Station (GLCRMS) at southern Manitoba, which is Canada's longest running organic-conventional management comparison station commenced in 1992. A detailed description of the location and site management was described previously [27]. In brief, the study site is located 20 km south of Winnipeg, Manitoba, Canada (N 49,39,0/W 97,7,0). The soil type is Rego Black Chernozem and the soil texture is clay (9% sand, 26% silt, and 66% clay) with an organic matter content of 7.7%. The experiment was a randomized complete block design in a split-plot arrangement with three replicates. Two crop rotations, that is, flax-oat-fababean-wheat (Grain-Only rotation) and wheat-alfalfa-alfalfa-flax (Forage-Grain rotation) were used as main plots, and certificated organic and conventional methods served as subplots. The 2×2 combinations of treatments included: Grain-Only Organic (GO), Grain-Only Conventional (GC), Forage-Grain Organic (FO) and Forage-Grain Conventional (FC). All rotation crops appeared in the rotation each year. Both organic and conventional experiments were managed using conventional tillage and plots were tilled with a disc and a field cultivator prior to sowing. Pesticides and chemical fertilizers were applied on the conventional plots but not on the organic plots. Eighteen kg P_2O_5 was banded when wheat was seeded in conventional plots, and 65 kg Nitrogen/ha was

Table 2. Pearson correlation coefficients between soil edaphic factors[1].

Variables	pH	Olsen P	Total N	Total C	Carbonate C	Organic matter	C: N ratio
pH	1.000	−0.380	−0.414	−0.429	−0.527	−0.424	−0.205
Olsen P		1.000	**0.996**	**0.993**	**0.615**	**0.994**	0.256
Total N			1.000	**0.998**	**0.606**	**0.998**	0.258
Total C				1.000	**0.622**	**1.000**	0.297
Carbonate C					1.000	**0.611**	0.135
Organic matter						1.000	0.297
C: N ratio							1.000

[1]Significant correlations between edaphic factors are indicated in bold type when $P<0.05$.

Table 3. Summary statistics of pyrosequencing 16S rRNA sequences of soil samples.

Rotation	Management	Number of trimmed sequences	Mean (SEM) results for indicated variable[1]						
			OTU[2] (95% distance)	Coverage (%)	Richness[3]		Diversity[4]		
					Chao1	ACE	Shannon	Simpson	Effective number of species
Grain-Only	Organic	30,482	1,917	78.8[a,b]	3,660.0	4,936.4[a,b]	7.2	0.0012	562.8
	Conventional	20,473	1,628	84.5[a]	2,889.7	3,044.2[b]	6.8	0.0022	504.4
Forage-Grain	Organic	23,923	2,118	73.2[b]	4,308.7	6,147.9[a]	7.6	0.0005	681.3
	Conventional	31,552	1,860	80.3[a,b]	3,494.4	4,533.6[a,b]	7.2	0.0012	592.2
SEM			236.5	2.7	512.1	677.1	0.22	0.01	122.1
P-value									
Rotation			0.35	0.11	0.20	0.08	0.20	0.38	0.43
Management			0.30	0.04	0.13	0.03	0.20	0.41	0.57
Rotation × Management (*P*-value)			0.95	0.83	0.96	0.83	0.97	0.86	0.91

[a, b, c]Means with different letters are significantly different for management at $P<0.05$.
[1]Mean are from statistical models based on 5 to 6 replicate samples.
[2]OTU = operational taxonomic units.
[3]Based on Chao1 and abundance based coverage estimation (ACE) richness indices.
[4]Based on Shannon and Simpson diversity estimators.

Table 4. Phylogenetic composition of bacterial phyla from pyrosequenced 16S rRNA sequences.

Phylum	Rotation (Grain-Only)		Rotation (Forage-Grain)		SEM	*P*-value		
	Management					Rotation	Management	Rotation × Management
	Organic	Conventional	Organic	Conventional				
	Abundant phyla[1]							
Proteobacteria	44.5^{a}	32.2^{a,b}	34.1^{a,b}	27.3^{b}	3.13	0.10	0.05	0.70
Actinobacteria	32.5^{a,b}	39.2^{a,b}	28.4^{b}	43.1^{a}	2.21	0.97	0.0002	0.08
Acidobacteria	8.5	10.3	13.8	12.1	2.35	0.15	0.88	0.44
Gemmatinomadetes	3.5^{B}	3.6^{B}	8.6^{A}	2.9^{B}	1.16	0.15	0.06	0.04
Chloroflexi	3.4^{b}	6.1^{a}	5.2^{a,b}	6.8^{a}	0.98	0.18	0.04	0.42
Bacteroidetes	3.3	2.7	2.7	2.2	0.98	0.54	0.52	0.99
Planctomycetes	1.5	1.7	2.1	1.8	0.35	0.30	0.81	0.49
	Low-abundance phyla[2]							
Firmicutes	0.6	0.9	0.2	1.1	0.34	0.47	0.11	0.32
Fibrobacteres	0.1^{B}	0.2^{B}	0.3^{A}	0.1^{B}	0.04	0.05	0.07	0.01
Nitrospirae	0.2^{b}	0.2^{a,b}	0.3^{a,b}	0.5^{a}	0.07	0.01	0.21	0.35
Verrucomicrobia	0.2^{C}	0.7^{A}	0.8^{A}	0.5^{B}	0.14	0.09	0.42	0.006
OP10	0.2^{B}	0.4^{A,B}	0.5^{A}	0.2^{B}	0.08	0.44	0.36	0.04
TM7	0.1	*	0.1	*	0.13	0.98	0.62	0.86
WS3	*	0.2	0.2	0.1	0.15	0.59	0.85	0.56
Unclassified	1.1	1.3	2.1	1.3	0.52	0.45	0.66	0.40

^{a,b,c}Means for main effects (rotation or management) are significantly different at $P<0.05$.
$^{A,\ B,\ C}$Means for the interaction between rotation and system are significantly different at $P<0.05$.
[1]Percentage of sequences larger than 1.
[2]Percentage of sequences smaller than 1.
*Percentage of sequences below 0.1.

broadcasted in conventional flax plots. One L/ha Buctril M and 0.235 L/ha Horizon were sprayed on conventional wheat plots and 0.2 L/ha Select and 1 L/ha Buctril were sprayed on conventional flax plots. These plots were managed with no external input of manure.

Bulk soil samples were randomly collected from the top level (0–15 cm) throughout wheat and flax plots in June and August 2008. Part of each soil sample was kept at −20°C prior to DNA extraction after sieving (2 mm) to remove roots and stones, while the rest was kept at 4°C for chemical analyses. Samples were analyzed using an elemental analyzer at a commercial soil analysis laboratory (AGVISE, Northwood, ND) for total soil carbon and total soil nitrogen (Vario MAX Carbon-Nitrogen analyzer, Elemetar, Germany). Soil Carbonate carbon was analyzed with a modified pressure technique [28]. Organic matter, soil pH, and Olsen phosphorus (P_{olsen}): sodium bicarbonate-extractable phosphorus was measured as described by Welsh et al. [27].

DNA Extraction

To remove PCR inhibitors, such as humic acids, covalent cations and other easily dissolved organic compounds, from soil samples a pre-lysis washing procedure was introduced before DNA extraction [29]. Soil samples of 0.25 g were mixed with 1.25 ml sodium phosphate (0.1 M, pH 7.5), then incubated in a shaker for 1 hr at room temperature, followed by centrifuging for 10 min at 16000×*g*. Supernatant was discarded. DNA was extracted from pre-washed samples using the PowerSoil DNA Isolate kit, which included a bead-beating step, according to the manufacturer's specifications (Mobio Laboratories, Solana Beach, CA). The DNA purity and quantity were tested by using spectrophotometer (Du 800 Spectrophotometer, BECKMAN COULTER). The average ratio of 260:280 was 1.7. The average DNA yield was 10 ng/μL. The variable regions of V1–V2 of the 16S rRNA genes were successfully amplified using forward primer 27F (AGAGTTTGATCMTGGCTCAG) and reverse primer 342R (CTGCTGCSYCCCGTAG), indicating that the quality of extracted DNA was sufficient for further PCR application [30]. In order to test the long-term effect of farming practices on the bacterial communities in the soil and reduce the temporal effects of different sampling times, DNA samples of the same treatment collected at different sampling times were pooled before pyrosequencing.

Pyrosequencing

A total of 23 pooled DNA samples were pyrosequenced using the bacterial tag-encoded GS FLX-Titanium amplicon as described by Dowd et al. [31] and Khafipour et al. [32]. In brief, a mixture of Hot Start, HotStar high fidelity Taq polymerases, and Titanium reagents were used to perform a one-step PCR (35 cycles) with primers 28F (GAGTTTGATCMTGGCTCAG) and 519R (GTNTTACNGCGGCKGCTG), which covered the variable regions V1–V3 of the bacterial 16S rRNA genes [31]. The pyrosequencing procedures were carried out at the Research and Testing Laboratory (Lubbock, TX; http://www.Researchandtesting.com).

Table 5. Bacterial taxa showing significant variation under different farming systems generated using pyrosequenced 16S rRNA sequences.

Taxa (family and genus within each phylum or class)	Rotation (Grain-Only)		Rotation (Forage-Grain)		SEM	*P*-value		
	Management					Rotation	Management	Rotation × Management
	Organic	Conventional	Organic	Conventional				
Actinobacteria	32.5^{a,b}	39.2^{a,b}	28.4^{b}	43.1^{a}	2.21	0.97	0.0002	0.08
Geodermatophilaceae; *Blastococcus*	0.8^{b}	1.7^{a}	1.2^{a,b}	1.7^{a}	0.19	0.27	0.002	0.23
Intrasporangiaceae; *Lapillicoccus*	0.3^{b}	0.4^{a,b}	0.2^{b}	0.6^{a}	0.09	0.36	0.01	0.11
Propionibacteriaceae; *Microlunatus*	1.1^{a,b}	2.0^{a}	0.8^{b}	2.1^{a}	0.31	0.78	0.004	0.66
Pseudonocardiaceae; *Pseudonocardia*	0.9^{b}	1.7^{a,b}	1.3^{b}	2.6^{a}	0.24	0.02	0.001	0.31
Solirubrobacteriaceae; *Solirubrobacter*	0.6^{b}	1.7^{a}	0.8^{b}	1.6^{a}	0.29	0.88	0.003	0.48
Rubrobacteriaceae; *Rubrobacter*	0.5^{b}	1.0^{a}	0.3^{b}	1.4^{a}	0.24	0.57	0.005	0.30
Unclassified bacteria	11.9^{a,b}	15.6^{a,b}	9.9^{b}	17.4^{a}	1.81	0.97	0.01	0.33
Proteobacteria	44.5^{a}	32.2^{a,b}	34.1^{a,b}	27.3^{b}	3.13	0.10	0.05	0.70
Alphaproteobacteria	20.5	15.9	18.2	15.0	2.39	0.55	0.15	0.85
Caulobacteraceae; *Brevundimonas*	1.7^{a}	0.1^{b}	0.2^{b}	0.01^{b}	0.11	0.15	0.03	0.78
Xanthobacteraceae; uncultured	0.5^{b}	0.8^{a,b}	0.7^{a,b}	1.1^{a}	0.15	0.10	0.05	0.64
Rhodospirillaceae; *Skermanella*	0.8^{b}	1.5^{a,b}	1.3^{ab}	2.6^{a}	0.30	0.02	0.005	0.36
Gammaproteobacteria	11.4	7.1	7.7	3.6	2.72	0.13	0.08	0.67
Pseudomonadaceae; *Pseudomonas*	4.3	3.9	1.7	0.5	2.02	0.10	0.52	0.44
Xanthomonadaceae; *Stenotrophomonas*	0.7^{a}	0.0^{b}	0.3^{a}	0.0^{b}	0.25	0.67	0.04	0.71
Betaproteobacteria	10.3^{a}	6.3^{b}	8.7^{a,b}	5.1^{b}	2.00	0.42	0.04	0.93
Burkholderiaceae; *Burkholderia*	0.6	0.0	0.1	0.0	0.19	0.29	0.14	0.99
Deltaproteobacteria	1.0^{b}	2.7^{a,b}	3.4^{a}	3.5^{a}	0.05	0.009	0.10	0.13
Chloroflexi	3.4^{b}	6.1^{a}	5.2^{a,b}	6.8^{a}	0.98	0.18	0.04	0.42
Chloroflexaceae; *Roseiflexus*	0.6	1.5	1.1	2.3	0.28	0.02	0.002	0.52

^{a,b,c}Means for main effects (rotation or management) are significantly different at $P<0.05$.
$^{A, B, C}$Means for the interaction between rotation and system are significantly different at $P<0.05$.
[1]Percentage of sequences larger than 1.
[2]Percentage of sequences smaller than 1.
*Percentage of sequences below 0.1.

Bioinformatics of Pyrosequencing Data

Sequence editing, categorical transformation/ classification. Pyrosequencing data were edited, categorically transformed and classified as described by Khafipour et al. [32]. Briefly, all low quality sequences, tags, non-bacterial ribosomal sequences, and chimeras were removed from the database. In total, 123, 316 sequences were generated in this step. Then, the mothur software package [33] was utilized to perform the second round of sequence quality control and assignments of operational taxonomic units (OTU). All sequences shorter than 200 bp, or sequences having one or more ambiguous base, or containing a homopolymer length equal or greater than 8bp were removed from the dataset. The minimum, median and maximum lengths of sequences were 200, 471 and 647 bp, respectively. The unique sequences were then identified and aligned against a database of high quality 16S rRNA bacterial sequences derived from Silva (version 106) [34]. Through screening, filtering, and pre-clustering processes, columns containing a gap were removed in all sequences to reduce noise from pyrosequencing data. The remaining 987 columns (with an actual sequence length varying from 203 to 342 bp) and 39,283 sequences were used to build a distance matrix with a distance threshold of 0.1. Using the furthest neighbor algorithm with a cutoff of 95% similarity, these sequences were clustered to OTU. Representative sequences from each OTU were taxonomically classified with a confidence level of 60% using RDP Bayesian approach [35].

Alfa diversity analysis. An OTU-based approach was performed to calculate the richness, diversity and coverage at OTU cutoff of 0.05, which characterizes the biodiversity of the bacterial population in the soil samples at the genus level. Richness indices, Chao1 and abundance based coverage estimation (ACE), were calculated to estimate the number of species or OTU that were present in the sampling assemblage. The diversity within each individual sample, which is made up of richness and species abundance, was estimated using Simpson and non-parametric Shannon diversity indices. Good's non-parametric coverage estimator was used to estimate the percentage of the total species that were sequenced in each sample. Rarefaction curves for treatment groups were created in mothur [33], based on a re-sampling without replacement approach.

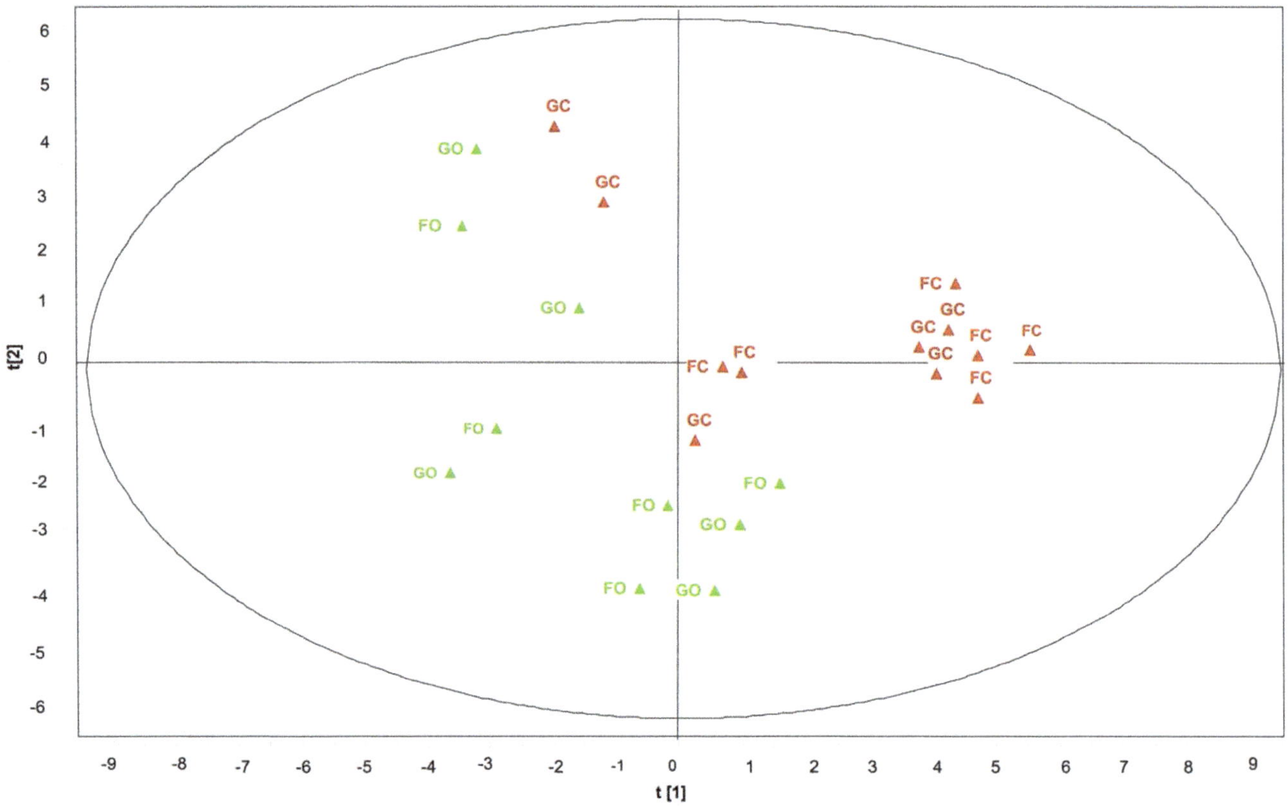

Figure 1. Partial least square discriminant score plot of soil bacteria under organic and conventional treatments. GO: Grain-Only organic; GC: Grain-Only conventional; FO: Forage-Grain organic; FC: Forage-Grain conventional. Model indicated a significant difference in the composition of putative bacterial genera between organic and conventional managements ($R^2X = 0.427$, $R^2Y = 0.882$, $Q^2 = 0.159$). Only genera with VIP>0.35 is included in the model.

Statistical Hypothesis Testing

The UNIVARIATE procedure of SAS [36] was used to test the normality of residuals for Alfa diversity indices. Non-normally distributed data were power transformed using Box-Cox power transformation macro (http://www.datavis.ca/sasmac/boxcox.html) in SAS based on the following models: BoxCox $(y) = (y^\lambda - 1)/\lambda$, if $\lambda \neq 0$ OR BoxCox $(y) = \log(y)$, if $\lambda = 0$. The range of λ for each parameter was adjusted by trial and error and the best fitting value of λ was identified using maximum likelihood methods. Normalized data were then used to assess the effect of treatment using MIXED procedure of SAS [36]. The effect of replicates was treated as random in the model.

Percentage data was used to evaluate statistical differences among treatments at the phylum and genus levels. To do so, the count data for each taxon was first transformed to the percentage of that taxon in an individual sample. Then UNIVARIATE procedure of SAS was used to test the normality of residuals for percentage data at each taxonomic level. For non-normally distributed data, Poisson and negative binomial distributions were fitted in the GLIMMIX procedure of SAS [36] to assess the effect of treatment. A log link function was specified for Poisson and negative binomial distributions. The goodness of fit for different distributions was compared using Pearson chi-square/DF (closer to 1 is better). Taxa were categorized as abundant and low-abundant in order to characterize them within each treatment. All taxa above 1% of the population were considered abundant and those below 1% were classified as less-abundant [32]. The differences between treatments were considered significant at $P<0.05$ while trends were observed at $P<0.1$.

Partial Least Square Discriminant Analysis and Redundancy Analysis

Partial least square discriminant analysis (PLS-DA; SIMCA-P+12.0.1, Umetrics, Umea, Sweden) [37] was performed on genus data to identify the effects of crop rotation and management on the bacterial community. The PLS-DA is a particular case of partial least square regression analysis in which Y is a set of binary (0 versus 1) variables describing the categories of a categorical variable on X. In this case, X variables were bacterial genera and binary Y was observations of organic (1) versus conventional (0), or Grain-Only (1) versus Forage-Grain (0) treatments. For this analysis, data were scaled using Unit Variance in SIMCA-P+ [37]. Cross-validation was then performed to determine the number of significant PLS components and a permutation testing was conducted to validate the model. To avoid over parameterization of the model, variable influence on projection value (VIP) was estimated for each genus and genera with VIP<0.35 were removed from the final model [38,39]. R^2X and R^2Y estimates were then used to evaluate the goodness of fit and Q^2 estimate was used to evaluate the predictive value of the model. Scatter- and score-plots were generated only for treatments that were significantly differentiated by the model. The PLS regression coefficients were used to identify genera that were most characteristic of each treatment group. The positive or negative correlations were considered significant when there was no overlap

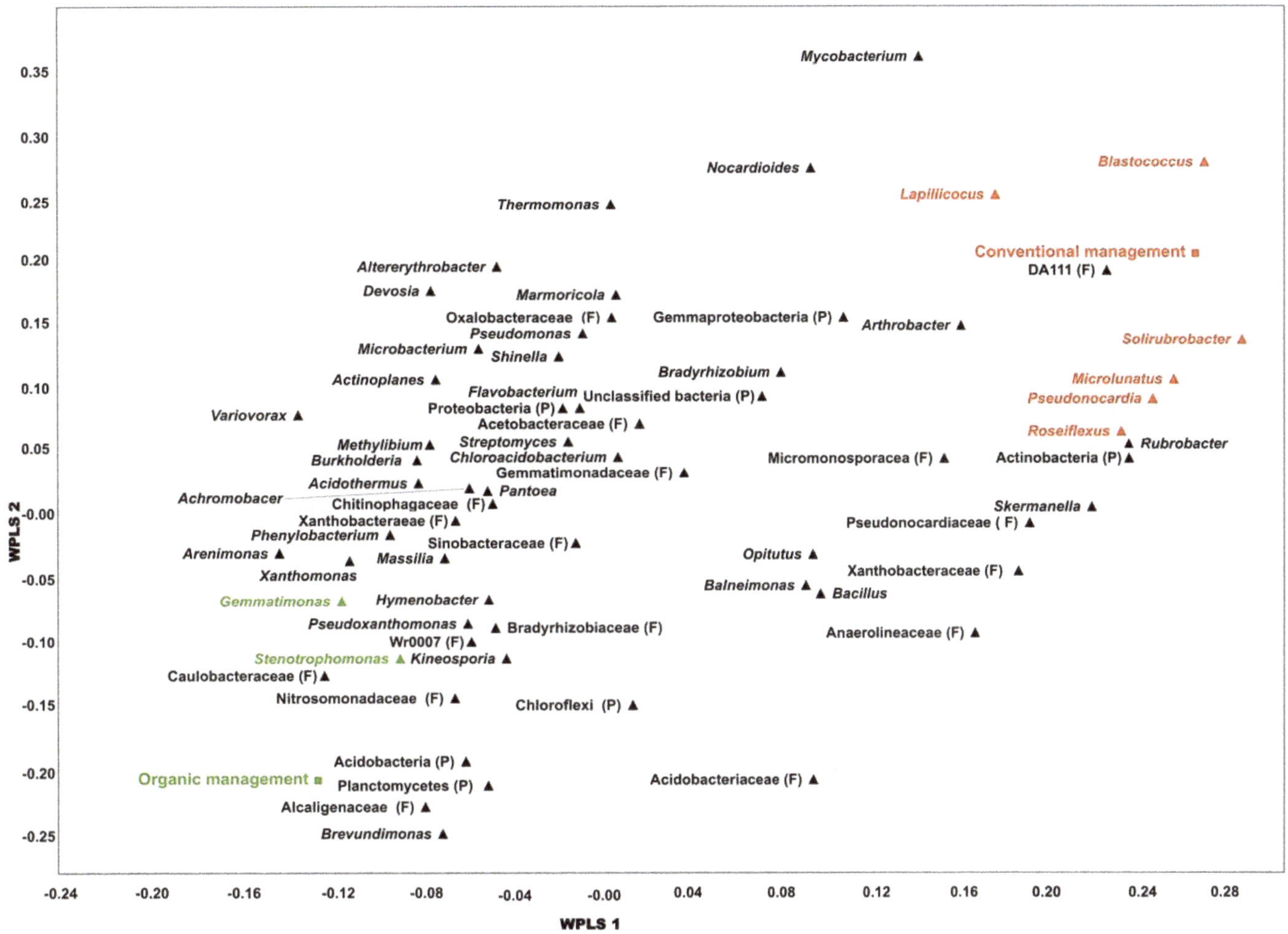

Figure 2. Partial least square discriminant analysis (PLS-DA) loading plot based on the relative abundance of the putative bacterial genera in soil microbiome and their association with organic or conventional treatments. Bacterial genera closer to organic or conventional are highly correlated to either treatment. PLS1 ($R^2X = 0.27$, $R^2Y = 0.525$, $Q^2 = 0.186$) and PLS2 ($R^2X = 0.127$, $R^2Y = 0.218$, $Q^2 = -0.081$). Some sequences could only be affiliated to phylum (P) or family (F) levels.

between the genus 95% confidence interval and the horizontal axis in the PLS regression coefficients graph.

Redundancy analysis (RDA) was carried out using canonical community ordination (CANOCO; Plant Research International BV, Wageningen, The Netherlands) to examine the relationship between abundant phyla and environment variables. Spearman's rank correlations were used to correlate abundant phyla and soil properties using SAS [36].

Results

Cropping Systems and Edaphic Soil Properties

The total soil C, N, and C: N ratio did not vary significantly under different cropping systems, while pH, organic matter, carbonate C and Olsen P were affected to varying degree by cropping systems. Total soil C was 3.0 g/kg, 3.2 g/kg 3.1 g/kg and 3.0 g/kg under GO, GC, FO, and FC farming systems, respectively. Total soil N was 2.7 g/kg in all of four treatments. C: N ratio was 11.3 under GO and FO farming systems, while it was 11.5 and 10.9 under GC and FC farming systems, respectively. In contrast, pH was 7.0 under organic management systems, higher than that of conventional systems with 6.7 ($P = 0.023$). Organic matter was 7.9 under Forage-Grain rotation systems, significantly higher than that of Grain-only rotation systems with 6.7 and 7.2 ($P = 0.005$). Both rotation and management affected carbonate C and Olsen P (Table 1). Carbonate C and Olsen P were much lower under Forage-grain rotation, compared to Grain-only system. Significant correlations existed between total N, C, carbonate C, organic matter and Olsen P, as well as between organic matter, carbonate C, total C and total N (Table 2).

Bacterial α-diversity

Bacterial diversity and richness in individual samples under different treatments were calculated (Table 3). Statistical differences in richness and diversity were only observed for coverage and ACE at the management level. Percentage of coverage for conventional treatment was higher than that of organic treatment ($P = 0.04$). The GC had the highest percentage of coverage (84.5%), followed by FC (80.3%), GO (78.8%), and FO (73.2%). The ACE richness was highest for FO (6,147.9), and lowest for GC (3,044.2). The rarefaction curve (Figure S1) generated with mothur demonstrated that observed numbers of OTU of FO and GO groups were higher than that of FC and GC groups, with FO having the highest number of observed OTU.

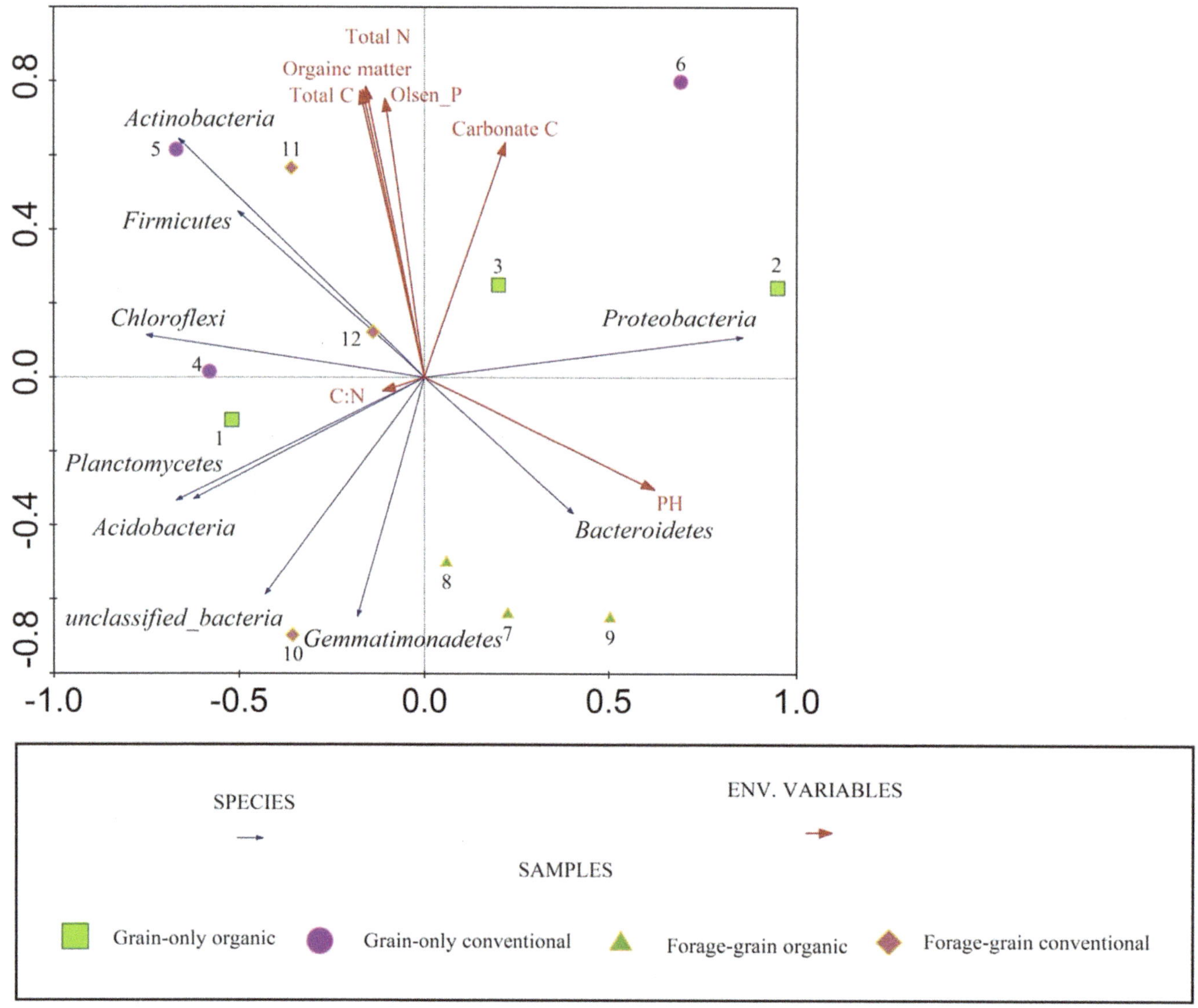

Figure 3. Redundancy analysis ordination plots of abundant phyla for individual sample.

Bacterial Community Composition

A total of 14 bacterial phyla were found in all the samples, of which seven were abundant (>1%) (Table 4). Ninety six percent of soil bacterial sequences belonged to these abundant phyla including: *Proteobacteria*, *Actinobacteria*, *Acidobacteria*, *Gemmatinomadetes*, *Chloroflexi*, *Bacteroidetes* and *Planctomycetes*. *Firmicutes*, *Fibrobacteres*, *Nitrospirae*, *Verrucomicrobia*, *P10*, *TM7* and *WS3* were in low abundance. The phylum distribution fluctuated under different farming disturbances. *Proteobacteria* accounted for 44.5% of total bacterial communities under the GO system, while it was only 27.3% under FC. In contrast, *Actinobacteria* made up 43.1% of total bacterial communities under the FC system, but were present in lower percentage (32.5%) under GO. Phylum *Chloroflexi* was also significantly influenced by management, with the highest percentage (6.8%) found in FC compared to the lowest (3.5%) in GO. A significant interaction between rotation and system was observed for *Gemmatinomadetes*, *Fibrobacteres*, *Verrucomicrobia*, and *P10*. Percentage of *Nitrospirae* was higher under Forage-Grain farming system. Other phyla did not show significant differences under different treatments. Unassigned bacterial sequences at the phylum level were approximately 1% of the total. In total, eight out of 14 phyla showed significant differences under different farming systems.

The relative abundance of different genera showing significant difference under different treatments was listed in Table 5. In phylum *Actinobacteria*, several putative genera including *Blastococcus*, *Lapillicoccus*, *Microlunatus*, *Pseudonocardia*, *Solirubrobacter*, and *Rubrobacter* showed significant differences among the treatments. The relative abundance of these genera was highest in the FC farming system, followed by GC, FO, and GO. The percentage of different class and genera belonging to the phylum *Proteobacteria* was higher under organic farming system compared to the conventional farming conditions with the exception of *Skermanella*, which was 2.6% and 1.5% under FC and GC farming systems, respectively. Classes *Alphaproteobacteria*, *Betaproteobacteria* and *Gammaproteobacteria* were higher in the Grain-Only organic farming system, although the difference was not statistically significant. Class *Deltaproteobacteria* showed the opposite pattern, being highest in Forage-Grain conventional systems, and lowest in Grain-Only organic system. *Pseudomonas* was the predominant genus in *Gammaproteobacteria* with

Table 6. Spearman's rank correlations between abundant phyla with soil properties[1].

Abundant phyla	Correlation						
	pH	Olsen P	Total N	Total C	Carbonate C	Organic matter	C: N ratio
Proteobacteria	**0.61**	0.07	−0.04	−0.03	−0.06	−0.03	0.05
Alphaproteobacteria	0.46	−0.47	−0.09	−0.12	−0.17	−0.11	−0.17
Betaproteobacteria	**0.62**	−0.18	−0.24	−0.38	−0.36	−0.20	−0.36
Gammaproteobacteria	0.25	−0.004	−0.05	−0.15	−0.13	0.04	−0.13
Deltaproteobacteria	−0.47	0.18	−0.21	−0.15	−0.13	−0.04	−0.13
Actinobacteria	**−0.65**	0.44	0.51	0.50	0.41	0.50	−0.17
Bacteroidetes	0.33	−0.37	−0.37	−0.36	−0.53	−0.36	0.38
Chloroflexi	−0.53	0.33	0.40	0.39	0.40	0.39	0.09
Firmicutes	−0.21	0.47	0.53	0.52	0.33	0.52	0.35
Gemmatimonadetes	−0.08	−0.46	−0.45	−0.45	−0.30	−0.45	−0.03
Planctomycetes	−0.18	−0.28	−0.21	−0.24	−0.31	−0.24	−0.16
Acidobacteria	−0.06	−0.23	−0.16	−0.20	−0.35	−0.20	−0.15
Nitrospirae	−0.004	−0.24	−0.19	−0.27	−0.40	−0.27	−0.23
Unclassified	−0.39	−0.47	−0.41	−0.41	−0.19	−0.41	−0.14

[1]Significant correlations between edaphic factors are indicated in bold type when $P<0.05$.

4.3% in GO, 3.9% in GC, 1.7% in FO, and 0.5% in FC. Within Phylum *Chloroflexi*, genus *Roseiflexus* was significantly influenced by interaction of rotation and management. Other genera in the phyla of *Actinobacteria* and *Proteobacteria* did not show statistical variation among the treatments (Table S1 and Table S2). There was no significant fluctuation under different farming systems in other genera within *Acidobacteria*, *Bacteroidetes*, *Firmicutes*, and *Planctomycetes* (Table S3).

The PLS-DA analysis showed that there is a significant difference in the composition of bacterial genera between organic and conventional managements ($R^2X = 0.427$, $R^2Y = 0.882$, $Q^2 = 0.159$) (Figure 1). However, crop rotation (Forage-Grain versus Grain-Only) was not a discriminator factor. Genera that were the most characteristic of each management system were identified using a scatter-plot (Figure 2). Among the putative bacterial genera included in the model, *Blastococcus* spp. *Microlunatus* spp. *Pseudonocardia* spp. and *Solirubrobacter* spp. were significantly correlated with conventional treatment, whereas *Gemmatimonas* spp. and *Stenotrophomonas* spp. were highly associated with organic management (Figure 2 and Figure S2).

Effect of Soil Edaphic Properties on Abundant Phyla

Canonical correspondence analysis tested the effect of soil edaphic properties on samples and bacterial populations by using an unconstrained analysis (RDA) (Figure 3). pH explained 24% of the variance ($P = 0.06$), CaCO3 C explained 19% ($P = 0.02$), and the C: N ratio accounted for less than 5% of the variance ($P = 0.52$). Other soil edaphic variables were highly correlated with each other and were not able to explain variance separately. We also used Spearman's rank order correlation to evaluate relationships between abundant phyla and soil edaphic properties (Table 6). It was found that the relative abundance of *Proteobacteria* phylum and *Betaproteobacteria* class was positively correlated with soil pH, while the abundance of *Actinobacteria* was negatively correlated with soil pH. Other phyla did not show significant correlations with soil edaphic properties.

Discussion

In this study, crop rotation and management strategies did not alter total C, total N and C:N ratio but significantly affected organic matter, soil pH, carbonate C and Olsen P (Table 1). This indicates that farming systems gradually but not dramatically change soil edaphic properties [40]. We observed that organic management led to a neutral soil pH compared to conventional practices (7.0 versus 6.6; Table 1). This might be due to the application of synthetic fertilizer that could acidify the soil in the conventional systems [41]. Other reports indicated that soil pH could be influenced by other soil traits such as C:N ratio [42], vegetation, or soil type [43]. However, in our study, the field trials were run under identical condition, and the soil pH was not significantly correlated with other soil edaphic characteristics (Table 2). Therefore, the farming system was the sole factor to change the soil pH. As we expected, soil organic matter was higher under Forage-Grain rotations (Table 1) [42,44]. However, organic farming system did not increase organic matter in the soil surface, compared to conventional farming system. In contrast, other studies have shown that organic matter was higher in the top 0.3 m of soil under organic management [21,45]. This discrepancy could be due to different crops contributing to different amount of biomass and no additional manure added to our trials [46].

We used high-resolution power of 454-pyrosequencing to obtain insight into the effects of farming management styles (organic, conventional) and crop rotations (Grain-Only, Forage-Grain) on the diversity, richness and composition of soil bacterial communities. In total, pyrosequencing identified 14 phyla and 178 putative genera of bacteria in different soil samples. We found that organic and conventional farming management had major influence on soil bacterial communities while the effects of crop rotation were of smaller magnitude. We were also able to identify putative genera that were correlated with either organic or conventional farming management.

It has been reported that organic farming systems enhance microbial diversity in soil compared to the conventional systems

[21]. Although not statistically significant, we found a similar trend in this study (Table 3). Previous research indicated that soil pH might be the primary factor influencing richness and diversity of bacterial communities [47] with the highest richness and diversity found to be near the neutral pH. Lauber et al. [48] proposed that bacterial diversity had a strong negative relationship with soil pH when it was lower than 6.5. In this study, soil pH ranged from 6.6 to 7.0, and was significantly higher for organic compared to conventional system. This indirectly indicates that organic management might favor higher bacterial diversity. And it is important to notice that standard diversity parameters only based on OTU without taxonomic identity of the different groups is not sensitive enough to detect the influence of agriculture management on the soil bacterial community, because changes in some taxonomic groups might be compensated by changes in others [49]. Lending support to this hypothesis, we detected a significant shift in soil bacterial communities due to farming systems when sequences were taxonomically ranked (Figure 1).

When sequences were affiliated to taxonomic level, bacterial populations fluctuated under different farming systems. An interesting observation in this study was the greater percentage of phylum *Proteobacteria*, including classes *Alphaproteobacteria*, *Betaproteobacteria* and *Gammaproteobacteria* in organic farming management (39.3%) compared to the conventional system (29.7%). To interpret these findings in an ecological context and to explain why some bacterial phyla are more abundant in soil than others, some researchers have used the concept of copiotrophic versus oligotrophic bacteria [50,51]. Copiotrophic bacteria (fast growing) flourish in soils with large amounts of available nutrients, while oligotrophic groups (slow growing) predominate in soil having low nutrient availability. It has been proposed that oligotrophic bacteria are more associated with organically than conventionally farmed soils due to low availability of organic carbon and nitrogen [2,40]. Among *Proteobacteria*, *Betaproteobacteria* are considered as copiotrophic [40,50], and thus, their population is expected to be lower in organic farming. There is no indication if other classes within *Proteobacteria* can be classified into copiotrophic-oligotrophic scheme [50]. In our study we found higher *Betaproteobacteria* in organic farmed soil. As organic farming system did not contribute to higher amount of top bulk soil total C, total N and organic matter compared to conventional system, higher relative abundance of these bacteria could be due to other factors. It was found that *Proteobacteria* and *Betaproteobacteria* was highly correlated with pH in this study ($P<0.05$, Table 6), we assumed that neutral pH could increase the abundance of these bacteria in soil.

We believe that because of enormous phylogenetic and physiological diversity within each bacterial phyla, it is unlikely that an entire phylum demonstrate same ecological characteristics. An example would be *Burkholderia*, a genus in *Betaproteobacteria* that exhibits oligotrophic traits due to their catabolically versatility that enables them to degrade recalcitrant compounds and survive in environments with limited nutrient availability [52]. Thus, the hypothesis that oligotrophic bacteria are more associated with organic farmed soil could simplify the ecological categories of bacterial communities in soil.

In this study, we found a higher population of *Brevundimonas* spp., *Burkholderia* spp., *Pseudomonas* spp., and *Stenotrophomonas* spp. in organic farming systems (Table 5). These genera are ubiquitously in the soil and several of their species have important ecological roles in nutrient cycling and suppression of plant diseases [52,53,54]. For instance, members of *Stenotophomonas, Pseudomonas and Burkholderia* genera can fix nitrogen [52,55]. Higher relative abundance of these genera might help maintaining total N level in organic farming soil without fertilizer supplementation. In addition, many plant growth-promoting bacteria (PGPB) belong to *Burkholeria*, *Stenotrophomonas* and *Pseudomonas* genera, which were more abundant in organic farming system (Table 5). Interestingly, these genera were abundant in soils planted with alfalfa, wheat, oilseed rape and various weeds [53,54]. Because organic farming systems support more weeds than the conventional farming systems, it might promote these PGPB populations [56]. However, it is important to notice that not all species in these genera are PGPB and there are species, which are pathogenic to humans, animals and plants (i.e. *P. aerugionsa*, *P. syringae*, and *S. maltophilia* K279a*)*. The 16S rRNA marker genes have limitation for identification of bacteria up to the species level, and thus other methodologies with high resolution including metagenomic shotgun sequencing [57] must be applied in order to differentiate PGPB from pathogenic species in the soil bacterial community.

The percentage of *Actinobacteria* and *Chloroflexi* were lower in organic (30.4% and 4.3%, respectively) compared to the conventional system (41.1% and 6.4%, respectively). Our results show that the conventional farming system increases the actinobacterial proportion in the community with no change in their composition, compared to the organic farming system. The PLS-DA loading scatter and coefficient plots (Figure 2 and Figure S2) indicated that conventional farming system supported higher population of several genera within *Actinobacteria*, including *Blastococcus* spp., *Microlunatus* spp., *Pseudonocardia* spp. and *Solirubrobacter* spp. *Actinobacteria* are able to degrade a variety of organic compounds including some herbicides and pesticides [58]. *Pseudonocardia* spp. has been reported to degrade environmental contaminants, particularly aromatic hydrocarbons or compounds that contain aromatic rings [59]. As such, herbicides and pesticides sprayed containing aromatic rings may have favored bacteria, such as *Pseudonocardia* spp., with specific metabolic capabilities that can degrade them. Some *Microlunatus* spp. has high levels of phosphorus accumulating function and phosphate uptake/release activities [60]. Therefore, in a conventional farming system where pesticides and inorganic fertilizers are commonly used to increase the crop yield, high availability of substrate for *Microlunatus* spp. and other actinobacterial species could boost their population.

Actinobacteria also play a major role in organic matter turnover and carbon cycling. They can decompose some recalcitrant carbon sources including cellulose and chitin [15,61]. Organically farmed soils have been reported to be rich in recalcitrant carbon sources [62], and the diversity of *Actinobacteria* would be expected to be higher in those soils than in conventionally farmed soils. However, in our organic farming fields the recalcitrant carbon sources were not higher in the organically surface soil than the conventional one [46]. Therefore, recalcitrant carbon sources could not drive the increasing diversity of *Actinobacteria* in our study.

A number of studies have shown that soil edaphic factors shaped microbial communities [43,63,64,65]. In our study, we found that proportions of abundant phyla were highly affected by soil pH. Our observations were consistent with other studies that demonstrated pH was one of the main drivers of change in soil bacterial communities from continental scale [48] to small landscape [64,65]. At the phylum level, *Proteobacteria* were positively correlated with soil pH, while *Actinobacteria* were negatively correlated and *Acidobacteria* had a very weak correlation with soil pH. Our results are in contrast to some studies that showed *Actinobacteria* significantly increased with higher pH values, and *Acidobacteria* was dependent on soil pH [65,66]. The soil pH value varied significantly from 3 to 8 in other studies, while the soil pH in our experiments only varied from 6.6 to 7 which could be the reason for lack of change in *Acidobacteria* populations. In our study,

Betaproteobacteria and *Alphaproteobacteria* populations increased with higher soil pH, while *Deltaproteobacteria* declined. This result was concomitant with the study by Nacke et al. [65]. Our studies demonstrated that *Proteobacteria* and *Actinobacteria* were more sensitive to pH variation than other bacterial phyla.

Finally, it is important to acknowledge that the choice of target variable regions of 16S rRNA may have affected the outcome of species richness and diversity analyses because the sequence divergence is not distributed evenly along the 16S rRNA gene [67,68]. We deep sequenced the V1–V3 regions of the bacterial 16S rRNA, which covered V2–V3 region, most suitable for distinguishing most bacterial species ranging from the phylum level to the genus level [68,69]. Therefore, even if some bacterial communities might have been missed or overestimated, the overall shifting in the phylogenetic composition of bacterial communities under different treatments have been assessed.

Conclusion

We demonstrated that different farming practices significantly changed the relative abundances of *Proteobacteria* and *Actinobacteria*. Farming management practices (organic versus conventional) rather than crop rotation (Grain-Only versus Forage-Grain) appeared to have a strong impact on shifting the abundance of soil bacterial communities, which could translate to changes in soil quality and productivity. Some bacterial groups, such as *Gemmatinomadetes*, *Fibrobacteres*, *Verrucomicrobia* and *OP10* were influenced by the interaction of crop rotation and management. Most soil properties including C: N ratio, total N, total C, Olsen P, and organic matter, did not play a major role in shaping bacterial communities. However, pH had the strongest effect on the bacterial community structure. Organic farming systems led to a neutral pH, which might be beneficial to *Proteobacteria*. On the other hand, conventional farming systems supported a higher percentage of *Actinobacteria*. Therefore, neither organic farming nor conventional farming can address all the aspects of beneficial soil bacterial communities, which is crucial to soil quality and productivity. Further research is required to investigate the shifts in diversity of beneficial bacterial and fungal pathogens under different farming systems in the long run.

Supporting Information

Figure S1 Rarefaction curves for pooled samples within each treatment at OTU cutoff of 0.05 distance.

Figure S2 Coefficient plot of the bacterial profiles of organic and conventional treatments. Partial least squares discriminant analysis (PLS-DA) coefficient plot based on the relative abundant of the bacterial genera in the microbiome profile of organic and conventional treatments. Genera with significantly positive (>0) or negative (<0) PLS regression coefficients (i.e. no overlap between the 95% confidence interval indicated and the horizontal axis) contribute significantly to the prediction of the organic (green bars) or conventional samples (red bars).

Table S1 Phylogenetic composition of putative bacterial genera in *Actinobacteria* phylum determined using 16S rRNA pyrosequencing

Table S2 Phylogenetic composition of putative bacterial genera in *Proteobacteria* phylum determined using 16S rRNA pyrosequencing

Table S3 Phylogenetic composition of putative bacterial genera in *Acidobacteria*, *Bacteroidetes*, *Chloroflexi*, *Firmicutes*, *Gemmatimonadetes*, and *Planctomycetes* phyla determined using 16S rRNA pyrosequencing

Acknowledgments

Authors would like to thank Bell LW and Welsh C for testing soil edaphic traits.

Author Contributions

Conceived and designed the experiments: RL WGDF MHE TRdK. Performed the experiments: RL. Analyzed the data: RL EK. Contributed reagents/materials/analysis tools: WGDF EK DOK. Wrote the paper: RL EK WGDF.

References

1. Doran JW, Parkin TB (1994) Defining and assessing soil quality; Doran JW, Coleman DC, Bezdicek DF, Stewart BA, editors. Madison, Wisconsin USA: Soil Science Society of America, America Society of Agronomy. pp 3–21.
2. van Bruggen AHC, Semenov AM (2000) In search of biological indicators for soil health and disease suppression. Appl Soil Ecol 15: 13–24.
3. Sprent JI (1979) The biology of nitrogen-fixing organisms. London; New York: McGraw-Hill. 196 p.
4. Dorn E, Hellwig M, Reineke W, Knackmuss HJ (1974) Isolation and characterization of a 3-chlorobenzoate degrading pseudomonad. Arch Microbiol 99: 61–70.
5. Torstensson L (1980) Role of microorganisms in decomposition. In: Hance RJ, editor. Interaction between hercibides and the soil London: Academic Press. 159–177.
6. Ongena M, Duby F, Rossignol F, Fauconnier M-L, Jacques D, et al. (2004) Stimulation of the lipoxygenase pathway is associated with systemic resistance induced in bean by a nonpathogenic *Pseudomonas* strain. Mol Plant Microbe Interact 17: 1009–1018.
7. Zehnder GW, Murphy JF, Sikora EJ, Kloepper JW (2001) Application of rhizobacteria for induced resistance. Eur J Plant Pathol 107: 39–50.
8. Weller DM, Raaijmakers JM, Gardener BBM, Thomashow LS (2002) Microbial populations responsible for specific soil suppressiveness to plant pathogens. Annu Rev Phytopathol 40: 309–348.
9. Suman A, Gaur A, Shrivastava AK, Yadav RL (2005) Improving sugarcane growth and nutrient uptake by inoculating *Gluconacetobacter diazotrophicus*. Plant Growth Regul 47: 155–162.
10. Basak BB, Biswas DR (2010) Co-inoculation of potassium solubilizing and nitrogen fixing bacteria on solubilization of waste mica and their effect on growth promotion and nutrient acquisition by a forage crop. Biol Fertil Soils 46: 641–648.
11. Upchurch R, Chiu CY, Everett K, Dyszynski G, Coleman DC, et al. (2008) Differences in the composition and diversity of bacterial communities from agricultural and forest soils. Soil Bio Biochem 40: 1294–1305.
12. Gattinger A, Ruser R, Schloter M, Munch JC (2002) Microbial community structure varies in different soil zones of a potato field. J Plant Nutr Soil Sci 165: 421–428.
13. Smalla K, Wieland G, Buchner A, Zock A, Parzy J, et al. (2001) Bulk and rhizosphere soil bacterial communities studied by denaturing gradient gel electrophoresis: plant-dependent enrichment and seasonal shifts revealed. Appl Environ Microbiol 67: 4742–4751.
14. Roesch LF, Fulthorpe RR, Riva A, Casella G, Hadwin AK, et al. (2007) Pyrosequencing enumerates and contrasts soil microbial diversity. ISME J 1: 283–290.
15. Acostamartinez V, Dowd S, Sun Y, Allen V (2008) Tag-encoded pyrosequencing analysis of bacterial diversity in a single soil type as affected by management and land use. Soil Biol Biochem 40: 2762–2770.
16. Shafiani S, Malik A (2003) Tolerance of pesticides and antibiotic resistance in bacteria isolated from wastewater- irrigated soil. World J Microb Biot 19.
17. Wasi S, Jeelani G, Ahmad M (2008) Biochemical characterization of a multiple heavy metal, pesticides and phenol resistant Pseudomonas fluorescens strain. Chemosphere 71: 1348–1355.
18. Azadi H, Schoonbeek S, Mahmoudi H, Derudder B, De Maeyer P, et al. (2011) Organic agriculture and sustainable food production system: main potentials. Agric Ecosyst & Environ 144: 92–94.
19. Zhengfei G (2005) Damage control inputs: a comparison of conventional and organic farming systems. Eur Rev Agric Eco 32: 167–189.

20. Gunapala N, Scow KM (1998) Dynamics of soil microbial biomass and activity in conventional and organic farming systems. Soil Biol Biochem 30: 805–816.
21. Mader P, Fliessbach A, Dubois D, Gunst L, Fried P, et al. (2002) Soil fertitity and biodiversity in organic farming. Science 296: 1694–1697.
22. Hartmann M, Fliessbach A, Oberholzer HR, Widmer F (2006) Ranking the magnitude of crop and farming system effects on soil microbial biomass and genetic structure of bacterial communities. FEMS Microbiol Ecol 57: 378–388.
23. Esperschutz J, Gattinger A, Mader P, Schloter M, FlieBbach A (2007) Response of soil microbial biomass and community structures to conventional and organic farming systems under identical crop rotations. FEMS Microbiol Ecol 61: 26–37.
24. Sugiyama A, Vivanco JM, Jayanty SS, Manter DK (2010) Pyrosequencing assessment of soil microbial communities in organic and conventional potato farms. Plant Dis 94: 1329–1335.
25. Wu T, Chellemi DO, Graham JH, Martin KJ, Rosskopf EN (2007) Comparison of soil bacterial communities under diverse agricultural land management and crop production practices. Microbial Ecol 55: 293–310.
26. Joergensen RG, Mäder P, Fließbach A (2010) Long-term effects of organic farming on fungal and bacterial residues in relation to microbial energy metabolism. Biol Fert Soils 46: 303–307.
27. Welsh C, Tenuta M, Flaten DN, Thiessen-Martens JR, Entz MH (2009) High yielding organic crop management decreases plant-available but not recalcitrant soil phosphorus. Agron J 101: 1027.
28. Williams DE (1948) A rapid manometer method for the determination of carbonate in soils. Soil Sci Soc of Am J 13: 127–129.
29. He J, Xu Z, Hughes J (2005) Pre-lysis washing improves DNA extraction from a forest soil. Soil Biol Biochem 37: 2337–2341.
30. Khafipour E, Li S, Plaizier JC, Krause DO (2009) Rumen microbiome composition determined using two nutritional models of subacute ruminal acidosis. Appl Environ Microbiol 75: 7115–7124.
31. Dowd SE, Callaway TR, Wolcott RD, Sun Y, McKeehan T, et al. (2008) Evaluation of the bacterial diversity in the feces of cattle using 16S rDNA bacterial tag-encoded FLX amplicon pyrosequencing (bTEFAP). BMC Microbiol 8: 125.
32. Khafipour E, Berard N, Little A, Tkachuk V, Dowd AS (2012) Microbiome changes during sub-acute ruminal acidosis using high-throughput pyrosequencing and statistical evaluation of sequence data. Appl Environ Microbiol (Accepted with revision).
33. Schloss PD, Westcott SL, Ryabin T, Hall JR, Hartmann M, et al. (2009) Introducing mothur: open-source, platform-independent, community-supported software for describing and comparing microbial communities. Appl Environ Microbiol 75: 7537–7541.
34. Pruesse E, Quast C, Knittel K, Fuchs BM, Ludwig W, et al. (2007) SILVA: a comprehensive online resource for quality checked and aligned ribosomal RNA sequence data compatible with ARB. Nucleic Acids Res 35: 7188–7196.
35. Wang Q, Garrity GM, Tiedje JM, Cole JR (2007) Naive bayesian classifier for rapid assignment of rRNA sequences into the new bacterial taxonomy. Appl Environ Microbiol 73: 5261–5267.
36. SAS (2008) SAS/STAT user guide, release 9.2 Cary, NC.: SAS Institute Inc.
37. SIMCA-P+_12 (2008) SIMCA-P+_12 user guide. Malmo, Sweden: MKS Umetrics AB.
38. Verhulst NO, Qiu YT, Beijleveld H, Maliepaard C, Knights D, et al. (2011) Composition of human skin microbiota affects attractiveness to malaria mosquitoes. PLoS One 6: e28991.
39. Pérez-Enciso M, Tenehaus M (2003) Prediction of clinical outcome with microarray data: a Partial least squares discriminant analysis (PLS-DA) approach. Hum Genet 112: 581–592.
40. Vandiepeningen A, Devos O, Korthals G, Vanbruggen A (2006) Effects of organic versus conventional management on chemical and biological parameters in agricultural soils. Appl Soil Ecol 31: 120–135.
41. Barak P, Jobe A, Krueger A, Peterson A, Laird A (1997) Effects of long-term soil acidification due to nitrogen fertilizer inputs in Wisconsin. Plant Soil 197: 61–69.
42. Kuramae EE, Yergeau E, Wong LC, Pijl AS, Veen JA, et al. (2012) Soil characteristics more strongly influence soil bacterial communities than land-use type. FEMS Microbiol Ecol 79: 12–24.
43. Ziadi N, Sen Tran T (2008) Lime requirement; Carter MR, Gregorich EG, editors. Boca Raton, FL: Taylor & Francis Group. pp 129–134.
44. Su Y (2007) Soil carbon and nitrogen sequestration following the conversion of cropland to alfalfa forage land in northwest China. Soil Till Res 92: 181–189.
45. Pimentel D, Hepperly P, Hanson J, Douds D, Seidel R (2005) Environmental, energetic, and economic comparisons of organic and conventional farming systems. BioScience 55: 573–582.
46. Bell LW, Sparling B, Tenuta M, Entz MH (2012) Soil profile carbon and nutrient stocks under long-term conventional and organic crop and Alfalfa-crop rotations and re-established grassland. Agr Ecosyst Environ 158: 156–163.
47. Ramirez KS, Lauber CL, Knight R, Bradford MA, Fierer N (2010) Consistent effects of nitrogen fertilization on soil bacterial communities in contrasting systems. Ecology 91: 3463–3470.
48. Lauber CL, Hamady M, Knight R, Fierer N (2009) Pyrosequencing-based assessment of soil pH as a predictor of soil bacterial community structure at the continental scale. Appl Environ Microbiol 75: 5111–5120.
49. Hartmann M, Widmer F (2006) Community structure analyses are more sensitive to differences in soil bacterial communities than anonymous diversity indices. Appl Environ Microbiol 72: 7804–7812.
50. Fierer N, Bradford MA, Jackson RB (2007) Toward an ecological classificatin of soil bacterial. Ecology 88: 1354–1364.
51. Meyer O (1994) Functional groups of microorganisms; Schulze E, Mooney H, editors. New York, USA: Springer-verlag. pp 67–96.
52. Suárez-Moreno ZR, Caballero-Mellado J, Coutinho BG, Mendonça-Previato L, James EK, et al. (2011) Common features of environmental and potentially beneficial plant-associated *Burkholderia*. Microb Ecol 63: 249–266.
53. Ryan RP, Monchy S, Cardinale M, Taghavi S, Crossman L, et al. (2009) The versatility and adaptation of bacteria from the genus *Stenotrophomonas*. Nat Rev Microbiol 7: 514–525.
54. Haas D, Défago G (2005) Biological control of soil-borne pathogens by fluorescent pseudomonads. Nat Rev Microbiol 3: 307–319.
55. Park M, Kim C, Yang J, Lee H, Shin W, et al. (2005) Isolation and characterization of diazotrophic growth promoting bacteria from rhizosphere of agricultural crops of Korea. Microbiol Res 160: 127–133.
56. Entz M, Penner K, Vessey J, Zelmer C, Thiessen-Martens J (2004) Mycorrhizal colonization of flax under long-term organic and conventional management. Can J Plant Sci 84.
57. Segata N, Waldron L, Ballarini A, Narasimhan V, Huttenhower C. et al. (2012) Metagenomic microbial community profiling using unique clade-specific marker genes. Nat Methods 9: 811–814.
58. De Schrijver A, De Mot R (1999) Degradation of pesticides by Actinomycetes. Crit Rev Microbiol 25: 85–119.
59. Lee SB (2004) *Pseudonocardia chloroethenivorans* sp. nov., a chloroethene-degrading actinomycete. Int J Syst Evol Micr 54: 131–139.
60. Akar A, Akkaya EU, Yesiladali SK, Çelikyilmaz G, Çokgor EU, et al. (2005) Accumulation of polyhydroxyalkanoates by *Microlunatus phosphovorus* under various growth conditions. J Ind Microbiol Biot 33: 215–220.
61. Jenkins S, Waite I, Blackburn A, Husband R, Rushton S, et al. (2010) Actinobacterial community dynamics in long term managed grasslands; Brisbane, Australia.
62. Fließbach A, Oberholzer HR, Gunst L, Mäder P (2007) Soil organic matter and biological soil quality indicators after 21 years of organic and conventional farming. Agr, Ecosyst Environ 118: 273–284.
63. Lauber CL, Strickland MS, Bradford MA, Fierer N (2008) The influence of soil properties on the structure of bacterial and fungal communities across land-use types. Soil Biol Biochem 40: 2407–2415.
64. Singh BK, Dawson LA, Macdonald CA, Buckland SM (2009) Impact of biotic and abiotic interaction on soil microbial communities and functions: A field study. Appl Soil Ecol 41: 239–248.
65. Nacke H, Thürmer A, Wollherr A, Will C, Daniel R, et al. (2011) Pyrosequencing-based assessment of bacterial community structure along different management types in German forest and grassland soils. PLoS One 6: 1–12.
66. Jones RT, Robeson MS, Lauber CL, Hamady M, Knight R, et al. (2009) A comprehensive survey of soil acidobacterial diversity using pyrosequencing and clone library analyses. ISME J 3: 442–453.
67. Liu Z, DeSantis TZ, Andersen GL, Knight R (2008) Accurate taxonomy assignments from 16S rRNA sequences produced by highly parallel pyrosequencers. Nucleic Acids Res 36: e120–e120.
68. Kim M, Morrison M, Yu Z (2011) Evaluation of different partial 16S rRNA gene sequence regions for phylogenetic analysis of microbiomes. J Microbiol Meth 84: 81–87.
69. Chakravorty S, Helb D, Burday M, Connell N, Alland D (2007) A detailed analysis of 16S ribosomal RNA gene segments for the diagnosis of pathogenic bacteria. J Microbiol Meth 69: 330–339.

22

Annual Glyphosate Treatments Alter Growth of Unaffected Bentgrass (*Agrostis*) Weeds and Plant Community Composition

Collin W. Ahrens, Carol A. Auer*

Agricultural Biotechnology Lab, Department of Plant Science, University of Connecticut, Storrs, Connecticut, United States of America

Abstract

Herbicide resistance is becoming more common in weed ecotypes and crop species including turfgrasses, but current gaps in knowledge limit predictive ecological risk assessments and risk management plans. This project examined the effect of annual glyphosate applications on the vegetative growth and reproductive potential of two weedy bentgrasses, creeping bentgrass (CB) and redtop (RT), where the glyphosate resistance (GR) trait was mimicked by covering the bentgrass plants during glyphosate application. Five field plots were studied in habitats commonly inhabited by weedy bentgrasses including an agricultural hayfield, natural meadow, and wasteland. Results showed that annual glyphosate treatment improved bentgrass survivorship, vegetative growth, and reproductive potential compared with bentgrass in unsprayed subplots. In the second year of growth, RT plants had an 86-fold increase in flower number in glyphosate-treated subplots versus controls, while CB plants had a 20-fold increase. At the end of the three year study, plant community composition had changed in glyphosate-treated subplots in hayfield and meadow plots compared to controls. Soils in subplots receiving glyphosate had higher nitrate concentrations than controls. This is the first study to mimic the GR trait in bentgrass plants with the goal of quantifying bentgrass response to glyphosate selection pressure and understanding the impacts on surrounding plant communities.

Editor: Luis Herrera-Estrella, Centro de Investigación y de Estudios Avanzados del IPN, Mexico

Funding: This project was supported by the Biotechnology Risk Assessment Program Competitive Grant no. 2008-39211-19566 from the USDA National Institute of Food and Agriculture (http://www.csrees.usda.gov/fo/biotechnologyriskassessment.cfm) to CAA. Funds were also provided by the Storrs Agricultural Experiment Station and the University of Connecticut. The funders had no role in study design, data collection and analysis, decision to publish, or preparation of the manuscript.

Competing Interests: The authors have declared that no competing interests exist.

* E-mail: carol.auer@uconn.edu

Introduction

Herbicide resistance has increased in weed species concurrent with the expanding acreage used for herbicide resistant crops in the U.S., but the long-term environmental impact of these traits on plant communities is largely unknown [1]. Crops and weeds have become resistant to the herbicide glyphosate [*N*(phosphonomethyl)glycine] through two mechanisms: 1) genetic engineering with the 5-enolpyruvylshikimate-3-phosphate synthase (EPSPS) transgene or glyphosate oxidoreductase (GOX) transgene, or 2) repeated glyphosate selection pressure coupled with physiological and genetic changes [2,3,4]. Genetic engineering has been used in the U.S. to commercialize glyphosate resistance (GR) in important agronomic crops (e.g. maize, cotton, soybeans, alfalfa, canola, and sugarbeet) and the trait has been found in feral GR crops (e.g. canola, volunteer corn) [5,4]. Genetic engineering has been used to add the GR trait to two perennial turfgrasses, creeping bentgrass (*Agrostis stolonifera*) and Kentucky bluegrass (*Poa pratensis* L.). While GR creeping bentgrass remains under regulatory review in the U.S., the USDA recently declined to regulate GR Kentucky bluegrass because it was generated without using any DNA sequences from an organism deemed a plant pest [6,7].

Unfortunately, the presence of the GR trait in diverse weed species is negatively affecting the utility of GR crops [8,9]. Selection pressure has created GR and glyphosate tolerant ecotypes in diverse weed species such as Canadian horseweed (*Conyza canadensis*), carelessweed (*Amaranthus palmeri*), ryegrass (*Lolium* species), Indian goosegrass (*Eleusine indica*), annual ragweed (*Ambrosia artemisiifolia*), Johnsongrass (*Sorghum halapense*), Mexican fireplant (*Euphorbia heterophylla*) and narrowleaf plantain (*Plantago lanceolata*) [10,1,4]. The mechanism for GR in weeds appears to vary with evidence for EPSPS gene amplification, EPSPS gene mutation, vacuolar sequestration and other mechanisms [11,12,13]. Regardless of the mechanism, glyphosate selection pressure will cause the heritable GR trait to spread in weed populations via pollen-mediated gene flow, seed dispersal, and vegetative propagules. Without selection pressure, the GR trait will be governed by genetic drift and gene flow. However, a study of field mustard (*Brassica rapa*) showed that the GR trait can persist without selection pressure [14].

Plant populations and community assemblages are affected by natural forces such as plant competition, facilitation, and stochastic processes [15]. In addition, many plant communities are also intentionally or unintentionally impacted by management regimes including herbicides and herbicide drift. Glyphosate usage has increased dramatically in the U.S., and recent experiments have shown that glyphosate can be detected in some air and surface water samples [16,17]. Deposition of sub-lethal glyphosate concentrations (drift) can negatively affect plant reproductive fitness [5]. However, there is no clear consensus about the effects of glyphosate on plant community assemblages or biodiversity. While some studies have reported that glyphosate applications had little impact on species richness or diversity in plant communities [18], other studies have reported significant changes [19,20]. A U.K. study by Heard et al. [21] showed that herbicides reduced weed abundance in GR beet (*Beta*) and canola (*Brassica*) fields, but weed species diversity and species richness within the fields was affected.

Risk assessments for the ecological impact of herbicide resistance trait require knowledge about host plant biology, crop-to-wild gene flow, and the ecology of the site for environmental release [22,23,24]. GR creeping bentgrass (*Agrostis stolonifera* L., CB) has generated concern because it is a cold-hardy perennial, a weed in many natural and managed habitats, part of a complex of closely-related species that hybridize through pollen-mediated gene flow, and able to spread vegetatively [25,26,27]. *Agrostis* species (bentgrasses) are common in early to mid-successional grasslands [28], implying that they readily establish in habitats with a variety of plant competion levels [29]. CB is a non-native species in the U.S. and a common weed outside of cultivation [30,31,32,33]. Our previous research showed that that CB is common and abundant as a weed in various managed and disturbed habitats (e.g. roadsides, wastelands) in two ecoregions in the Northeastern U.S. [31]. Furthermore, CB frequently co-occurs with a closely-related, non-native *Agrostis* called redtop (*Agrostis gigantea* Roth., RT). This co-occurrence suggests a potential for pollen-mediated gene flow [31]. In another study, CB, four other bentgrass species, and interspecific CB hybrids were able to establish and compete in residential and roadside habitats [34]. CB has also been shown to persist in sand dune, salt marsh, meadow and polder habitats [35]. In the Western U.S., pollen-mediated transgene flow between CB and RT populations moved the GR trait long distances [36], and removal of the escaped GR CB plants has been problematic along irrigation ditches because glyphosate is the only herbicide approved for these waterways [37]. Thus, the complex of native and non-native *Agrostis* species already present in the U.S. support intraspecific and interspecific gene flow, although the temporal and spatial scales have not been defined [38,37,27,39].

It has been reported that GR CB plants were not significantly different to their non-transformed counterparts with regard to infloresence development, reproductive morphology, pollen longevity, seed-set capacity or growth and establishment [40,41]. However, these studies did not examine CB growth under glyphosate selection pressure which would reduce competition between CB and other plants in natural, cultural or agricultural landscapes. It is important to understand GR CB behavior with glyphosate selection pressure in order to predict its behavior as a weed, create predictive models of gene flow and dispersal, and quantify impacts on plant community assemblages. Therefore, the primary goals of this project were to: 1) examine the impact of annual glyphosate applications on established plant communities in agricultural and natural areas, and 2) determine if one glyphosate application per year could alter the survivorship, growth, and reproductive capacity of introduced CB and RT plants. In this project, the GR trait in the bentgrass plants had to be mimicked due to a variety of factors including the potential for transgene escape during flowering and seed production. Our first hypothesis was that plant community composition would be altered in subplots receiving glyphosate compared to control (unsprayed) subplots. Second, we predicted that CB and RT plants with mimicked GR in glyphosate-treated subplots would have increased growth and reproductive potential compared to control subplots. To our knowledge, this is the first study to examine the effects of glyphosate on weedy bentgrass plants in the context of natural and agricultural plant communities.

Materials and Methods

Field Plots and Glyphosate Treatment

Five field plots (12 m×15 m) were established in Mansfield, Connecticut in the Lower New England Ecoregion, subecoregion

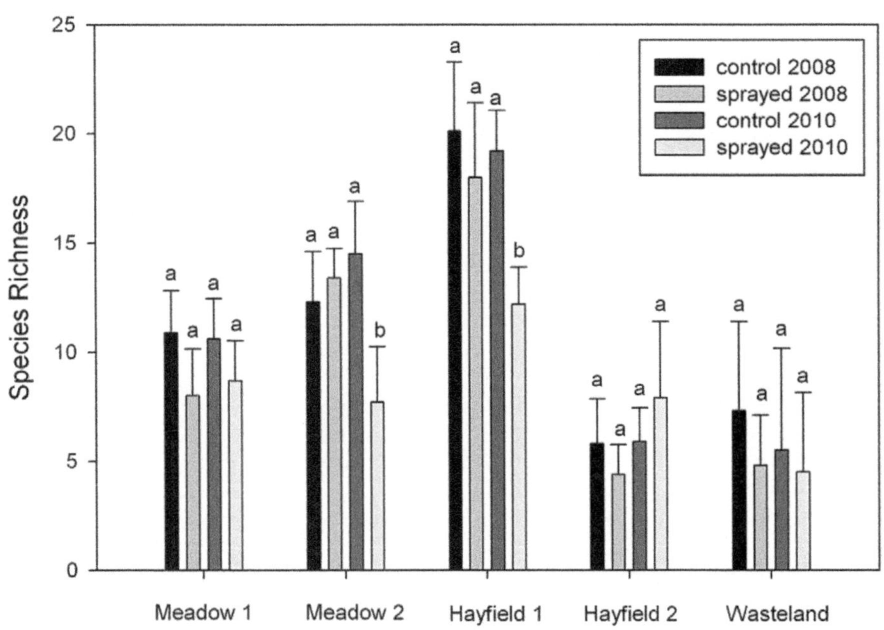

Figure 1. Species richness in subplots before the study (2008) and at the end of the experiment (2010). Error bars indicate SD and letters indicate differences within plot locations according to Fisher's LSD test ($\alpha<0.05$).

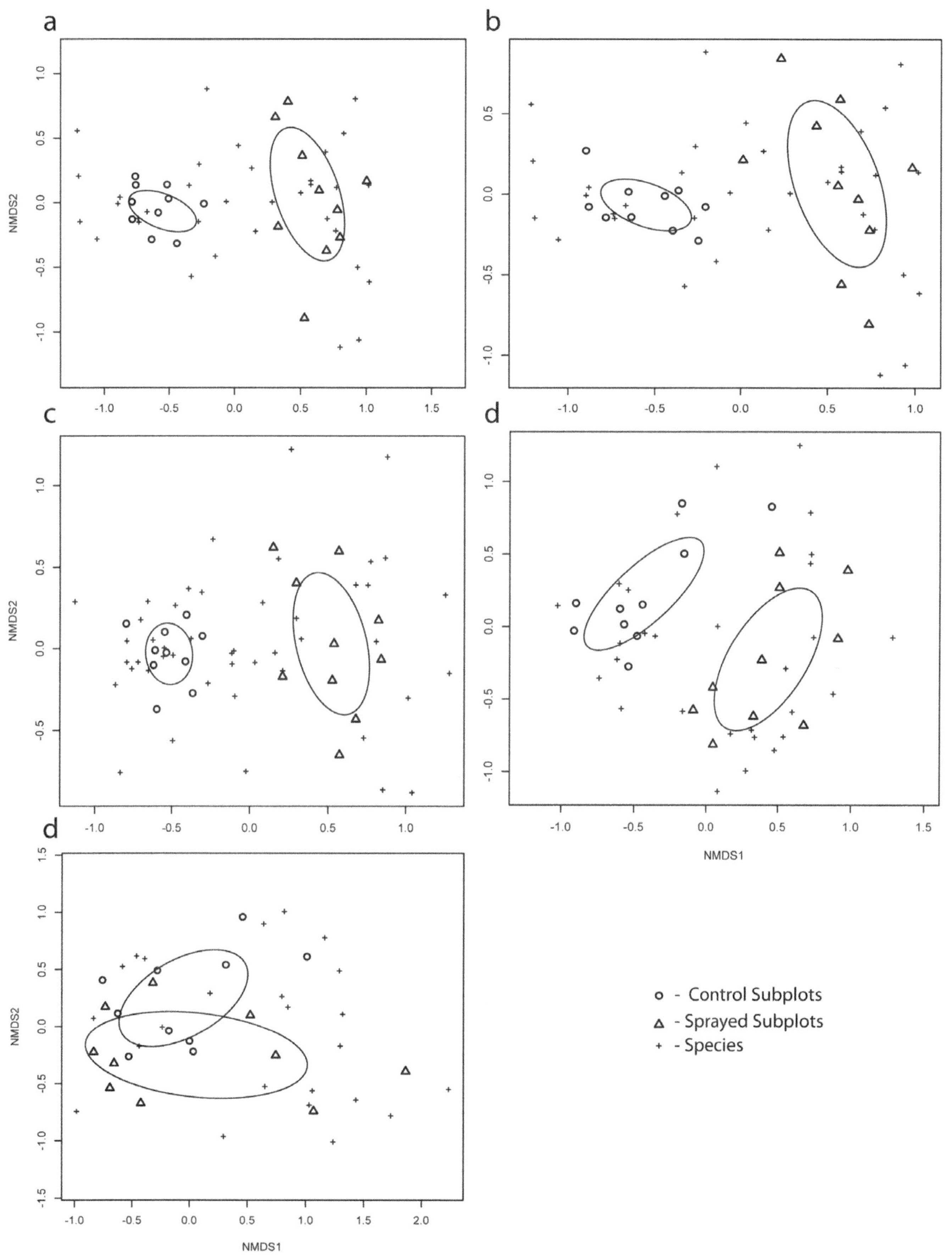
a
b
c
d
d
NMDS1
NMDS2
o - Control Subplots
Δ - Sprayed Subplots
+ - Species

Figure 2. Non-multidimensional scaling of each plot using Bray-Curtis Dissimilarity. a) meadow 1, b) meadow 2, c) hayfield 1, d) hayfield 2, and e) wasteland. The ellipses enclose the 99% confidence intervals.

Southern New England Coastal Hills and Plains. The plot names and locations were: Meadow 1, 41 49′ 30.30″ N, 72 14′ 12.91″ W; Meadow 2, 41 49′ 30.71″ N, 72 14′ 18.43″ W; Hayfield 1, 41 46′ 58.30″ N, 72 13′ 07.59″ W; Hayfield 2, 41 46′ 55.80″ N, 72 13′ 10.56″ W; and Wasteland, 41 49′ 01.25″ N, 72 13′ 10.56″ W. Meadow 1 and 2 were established in a natural meadow along the Fenton River. Hayfield 1 and 2 were placed on the upper and lower edges of an agricultural hayfield. The Wasteland plot was established in an area bordered by a road, cow pasture and wetlands, and dominated by the invasive grass canary reedgrass (*Phalaris arundinaceae* L.). Plots were not fertilized, irrigated or mowed during the 3 years of the study, and there was no known history of herbicide use at these sites. The overall intent was to mimic the localized application of glyphosate to control invasive plants or serious weeds in conservation areas, hayfields, or roadside wastelands. No permits were required for the project. Glyphosate was applied according to label specifications using appropriate personal protective equipment. All field plots were located on property owned by the University of Connecticut and did not contain any species of special concern.

A 2×2 factorial experiment was set up in each plot using a randomized complete block design with two management regimes (annual glyphosate spray or non-sprayed control) and two weedy, non-native bentgrass species (CB or RT). Each treatment group in the factorial design was replicated five times creating 20 randomized subplots with a size of 3 m^2. Soil samples (combination of two soil cores 2.2 cm×10.2 cm depth) were taken in each subplot prior to the study and 3 years later at the same time of year (Oct.-Nov.). Samples were analyzed for pH and nutrients using a modified Morgan test and a soil test for nitrates (University of Connecticut Soils Testing Lab, Storrs, CT). The soil texture for all plots was either sandy loam or loamy sand. Weather data were obtained from the National Oceanic and Atmospheric Administration (NOAA, (www.noaa.gov)). Yearly rainfall totals from November 1 to October 31 were: 2008, 375 cm; 2009, 310 cm; 2010, 304 cm. Mid-summer rainfall was: June, 2008, 10.8 cm; July, 2008, 11.5 cm; June, 2009, 14.5 cm; July, 2009, 19.5 cm; June, 2010, 11.1 cm; July, 2010, 8.1 cm. Mid-summer mean maximum temperatures (C°) were: June, 2008, 24.2; July, 2008, 26.4; June, 2009, 21.1; July, 2009, 24.1; June, 2010, 24.8; July, 2010, 27.8.

In 2007, conventional (non-genetically engineered) CB 'Penn A-4' and RT 'unspecified' (Des Moines Forage and Turf Seed Corporation, Ankeny, IA) plants were grown at the university research farm with biweekly mowing (1.27 cm height) and two fertilization treatments at a rate of 0.45 kg per 0.0093 ha for N, P and K (19-19-19, Lesco fertilizer, Troy, MI). Genetically engineered GR CB were not used in the study because: 1) we did not have permission to use commercially-developed, patented, transgenic GR CB seeds, 2) the U.S. government would be unlikely to provide permits for experimental field trials since the bentgrass plants needed to flower, release pollen, and produce seed in order to measure reproductive potential, and 3) to our knowledge, RT has not been engineered to contain the GR trait.

Table 1. Soil nutrients and pH in five field plots prior to the experiment.

Plot Name	Ca	Mg	P	K	NO_3	pH
	kg/ha				mg/kg	
Meadow 1	3226[a]	433[a]	1[b]	164[b]	3.49	5.78
Meadow 2	2839[a]	519[a]	1[b]	147[b]	4.34	5.64
Hayfield 1	1101[b]	147[b]	0[b]	160[b]	1.26	5.82
Hayfield 2	1572[b]	283[a]	2[b]	247[a]	4.11	6.07
Wasteland	1466[b]	322[a]	44[a]	276[a]	3.92	5.76

Letters indicate significant differences within a column according to Fisher's LSD test (α<0.05). No differences were observed for nitrate-N or pH.

In May, 2008, small plugs (3 cm×3 cm) of acclimated CB or RT were transplanted into the center of 100 subplots (20 subplots×5 plots). Measurements of bentgrass growth were taken monthly (May-October) for three years including: surface area covered, maximum leaf height, number of tillers (stolons), number of panicles, and number of flowers. All vascular plant species in each subplot were identified and monitored during the three year study. At the end of the third year (November, 2010), bentgrass survival and above-ground plant biomass were recorded. Subplots with CB or RT plants were randomly assigned to two treatment groups: annual glyphosate treatment or control (no herbicide). Subplots receiving glyphosate (1.69 kg per hectare, Roundup Pro ®, Monsanto Company, St. Louis, MO) were sprayed once per year on June 24–26th using a hand-held wand with a single boom Teejet XR8004VS nozzle (Wheaton, IL) and a backpack sprayer (SP system backpack, Santa Monica, CA). *Agrostis* plants within the sprayed plots were temporarily covered with white plastic bags (0.5 mm thickness) for 10–15 minutes to mimic the herbicide resistance trait. Control subplots received no weed management during the experiment.

At the end of the experiment, dependent variables were examined using analyses of variance (SAS ver. 9). Soil nutrient levels were analyzed as possible covariates. Data from all subplots were combined to analyze treatment, year (treated as a random variable) and bentgrass species interaction effects. Survivorship was analyzed using a probit model to determine the significance of glyphosate application. Probit analysis was chosen because this type of regression analysis was developed for biological systems involving quantal responses (e.g. death) to differing dosages of a toxin or a stress (e.g. glyphosate application) [42,43].

To assess changes in plant community composition, subplots were graphed on a two-axis graph using non-multidimensional scaling. The Bray-Curtis Dissimilarity between each subplot was calculated and the layout was based on the lowest amount of stress between actual dissimilarity and graphed dissimilarity. Confidence intervals (99%) were calculated and drawn for both control and sprayed subplot groups within each plot. If the confidence intervals did not overlap, then the communities in the subplot treatments (control vs. sprayed) were considered to be different [44].

Results

Effects of Glyphosate on Plant Community Composition and Soil Nitrogen

The five plot sites were typical examples of a natural meadow, hayfield or wasteland plant community that frequently contain non-native bentgrasses in our ecoregion. The high plant community diversity in Hayfield 1 could be attributed to cultivation of agronomic species such as smooth brome (*Bromus inermis* Leyss.)

Table 2. Mean bentgrass biomass (g dry weight) and survivorship (percent) for introduced creeping bentgrass (CB) and redtop (RT) plants at the end of the field experiment (2010).

Species	Treatment	Biomass	Meadow 1	Meadow 2	Hayfield 1	Hayfield 2	Wasteland	Total
CB	sprayed	5.85[a] (7.85)	100	60	80	40	0	56
	control	0.0[b] (0)	0	0	0	0	0	0
RT	sprayed	25.30[a] (22.13)	100	20	60	60	40	56
	control	0.1[b] (0)	0	0	20	0	0	4

Standard deviation is given in parentheses for plant biomass and letters indicate differences within bentgrass species according to Fisher's LSD test ($\alpha<0.05$).

and timothy (*Phleum pratense* L.), plus the naturalization of native and non-native species such as RT, CB, velvet bentgrass (*A. canina* L.), colonial bentgrass (*A. capillaris* L.), rough bentgrass (*A. scabra* Willd.), rough goldenrod (*Solidago rugosa* Mill.), white panical aster (*Symphyotrichum lanceolatum* (Willd.) G.L. Nesom), poverty oatgrass (*Danthonia spicata* (L.) P. Beauv. Ex Roem. & Schult.), Kentucky bluegrass, fowl bluegrass (*Poa palustris* L.) and annual bluegrass (*Poa annua* L.). The Hayfield 2 plot had a lower species richness value because it contained mostly cultivated species including smooth brome and oat (*Avena sativa* L.). Meadow 1 was dominated by coastal plain joe pye weed (*Eupatoriadelphus dubius* (Wild. Ex Poir.) King & H. Rob.), rough goldenrod, giant goldenrod (*Solidago gigantea* Aiton), flat top goldenrod (*Euthamia graminifolia* (L.) Nutt.) and arrowleaf tearthumb (*Polygonum sagittatum* L.). Meadow 2 contained plants such as black raspberry (*Rubus occidentalis* L.), American red raspberry (*Rubus idaeus* L.), flat top goldenrod and rough goldenrod. The low species richness in the Wasteland plot was due to the predominance of reed canarygrass, although chicory (*Cichorium intybus*), Canada thistle (*Cirsium arvense*), bull thistle (*Cirsium vulgare*), Kentucky bluegrass and prickly lettuce (*Lactuca serriola*) were also identified.

After three years in the study, the glyphosate-treated subplots in Meadow 2 and Hayfield 1 had lower plant species richness (14.5 to 7.7 species per subplot, and 19.2 to 12.2 species, respectively) and altered plant community assemblages compared to control subplots (Fig. 1). These two plots also had the highest number of species prior to the study ($P<0.0001$). This result contrasts with the meta-analysis performed by Sullivan and Sullivan [18] which found that glyphosate did not change species richness. However, while glyphosate treatment in Meadow 1, Hayfield 2, and Wasteland subplots did not alter the metric of plant species richness, changes in plant community assemblages were observed. Plant community composition shifted in four plot sites due to glyphosate treatment (Fig. 2). This change could largely be attributed to the increased occurrence of early and late annuals such as fall panicum (*Panicum dichotomiflorum* Michx.), witchgrass (*Panicum capillare* L.), large crabgrass (*Digitaria sanguinalis* (L.) Scop.), Virginia pepperweed (*Lepidium virginicum* L.) and yellow rocket (*Barbarea vulgaris* W.T. Aiton) in sprayed subplots. The Wasteland plot was an exception to this trend (Fig. 2) due to the localized dominance of the invasive reed canarygrass, which quickly re-established itself after glyphosate treatment.

Soil samples taken before the experiment showed differences in soil fertility between plots (Table 1). The high concentration of P and K in the Wasteland plot was most likely due to

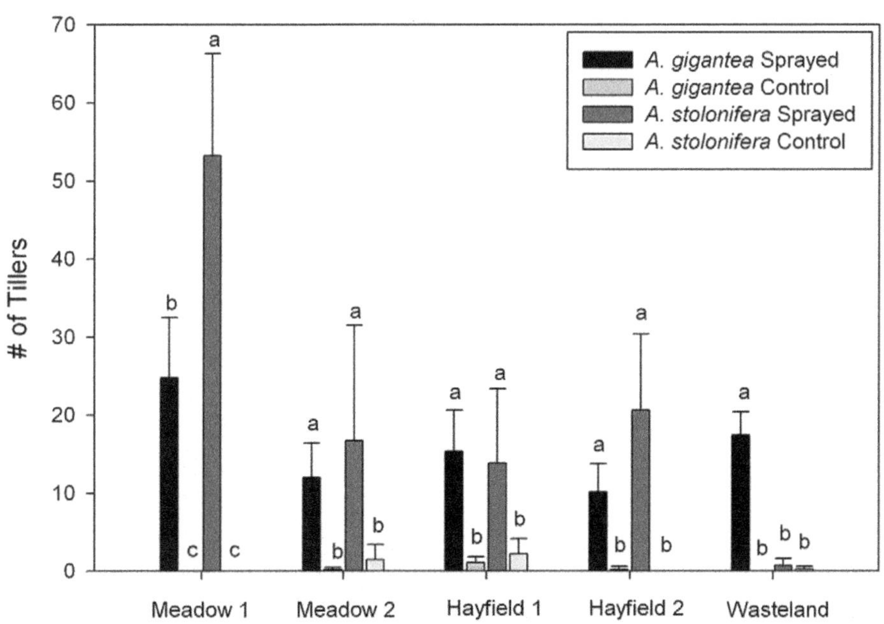

Figure 3. Number of tillers on creeping bentgrass (CB) and redtop (RT) plants in the last year of the field study (May, 2010). Error bars indicate SD and letters indicate differences within plot location at according to Fisher's LSD test ($\alpha<0.05$).

Table 3. Mean number of flowers produced by *A. gigantea* (RT) and *A. stolonifera* (CB) plants in glyphosate-treated subplots or unsprayed controls.

Species	Treatment	2008	2009	2010
RT	Control	0	213^b (92)	0^b (0)
	Glyphosate	0	18,377^a (16,032)	7959^a (8551)
CB	Control	0	104^b (46)	0^b (0)
	Glyphosate	0	2,042^a (1310)	1346^a (789)

Standard deviation is given in parentheses and letters indicate differences within bentgrass species and year according to Fisher's LSD test (α<0.05).

previous land use for crop production, manure storage and/or manure application (personnal communication, Tom Morris). Meadow plots 1 and 2 had higher calcium concentrations than other plots, and Hayfield 1 had a lower Mg level (Table 1). However, analysis of covariance indicated that the differences in soil Mg, Ca, P, and K did not affect bentgrass growth or survivorship. At the end of the third year, soils in subplots receiving glyphosate had higher nitrate concentrations as follows (Nitrate-N shown as mg/Kg in sprayed/control subplots): Meadow 1, 11.8/3.5 (P<0.0001); Meadow 2, 9.2/3.9 (P=0.0024); Hayfield 1, 6.1/1.3 (P=0.0021); Hayfield 2, 6.8/4.2 (P=0.167); Wasteland, 13.6/3.8 (P<0.0001). The nitrate levels in sprayed and control subplots in Hayfield 2 did not change. Other soil nutrients (P, K, Ca, Mg) were not affected by glyphosate application.

Effect of Glyphosate on Bentgrass Survivorship, Vegetative Growth and Reproductive Potential

Introduced CB and RT plants had higher survivorship in subplots receiving glyphosate (56%) than control subplots (4%) (Table 2). This was confirmed by probit model analysis (P<0.001). RT and CB plants from sprayed subplots had higher final biomass than bentgrass in control subplots, and plot location did not affect final biomass (Table 2). The number of CB or RT tillers increased in sprayed subplots compared to controls (Fig. 3). The only exception was CB plants in the Wasteland plot where competition with invasive reed canarygrass negatively affected growth in all subplots (Fig. 3). CB plants had significantly more tillers than RT plants in sprayed subplots in Meadow 1; a similar but non-significant trend was observed in Meadow 2 and Hayfield 2.

Reproductive potential of CB and RT plants was measured in each year of the study. There was no significant plot effect (P=0.524), so data were combined for analysis. As expected, CB and RT plants did not produce panicles or flowers in the first year (2008). In the second year (2009), RT plants in glyphosate subplots had an 86-fold increase in flower number compared to those in control subplots (Table 3). Thirteen RT plants produced a total of 238,896 flowers compared with 640 flowers on 3 RT control plants. Results in the third year (2010) were similar, although total flower numbers were probably reduced due to dry weather; 2009 had cooler temperatures and more rain during May and June which favored the cool-season grasses. RT seed collected in 2010 had 80% germination in the laboratory. Throughout the study, CB plants had a somewhat lower reproductive potential (flower number) than RT plants, but CB seeds had a relatively high germination rate of 74%.

Discussion and Conclusion

Effects of Glyphosate on Plant Community Composition and Soil Nitrate-N

Annual glyphosate application over a three-year period altered plant community composition in agricultural hayfields and natural meadows supporting our original hypothesis. This result is relevant to long-term risk assessment for HR crops that can escape cultivation because changes in plant community assemblages can produce negative impacts on other trophic levels [45]. Our results contrast with some previous reports that herbicides have little effect on plant species richness [20,46,47,48,18]. This discrepancy may be due, at least in part, to plant community types, differences in glyphosate application, and the metric (species richness) used to characterize changes in plant community composition over time. In our study, species richness declined in glyphosate-treated subplots in two plot locations (Meadow 2 and Hayfield 1). However, species richness data alone failed to capture the glyphosate-induced plant community changes in two other plots (Meadow 1 and Hayfield 2). This was because the number of new plant species appearing in the glyphosate-treated subplots was approximately equal to the number lost due to herbicide treatment. For example, the loss of some perennial plant species was often concurrent with an increase in weeds such as witchgrass (an annual warm-season C4 grass) or yellow rocket (a cool-season biennial), which were not present in subplots without glyphosate treatment. Furthermore, native perennial plants that declined in sprayed subplots included goldenrods such as tall goldenrod (*Solidago altissima* L.), rough goldenrod, giant goldenrod, and narrowleaf goldenrod (*Euthamia graminifolia* (L.) Nutt.), a trend that has been observed by other researchers [20]. Thus, measuring community composition over time is important for describing the response of complex agricultural or natural plant communities to annual glyphosate applications.

The only plot in our study that showed no effect from glyphosate on community composition was the Wasteland plot (Fig. 2). This plot was different because it had high soil P and K from past agricultural use and was dominated by one invasive plant species (reed canarygrass). This invasive regrew quickly after glyphosate application and competed with the introduced bentgrass plants, leading to very low survival rates. Nearly all of the introduced bentgrass plants died except for two RT plants in glyphosate-treated subplots. Based on our observations, it would be difficult for weedy bentgrass species to compete in this particular type of plant community regardless of the GR trait or glyphosate treatment.

The observation of increased nitrate-N in subplots sprayed with glyphosate was unexpected and difficult to explain, in part because the study did not include studies on microbial activity, soil nitrogen fluxes, and other minerals. A study in the Hubbard Brook ecosystem observed an increase in nitrate-N in a watershed where all vegetation was cut and suppressed for two years by herbicide application [49]. A study in young loblolly pine forests showed increased soil nitrogen in response to glyphosate, and the authors postulated the involvement of microbial mineralization [50,51]. Increased soil nitrate-N was observed when turfgrass was removed with glyphosate, but nitrate-N levels returned to normal six weeks after reseeding [52]. Previous research has also suggested that the removal of living plants coupled with the deposition of dead plant material in the sprayed subplots could affect the nitrogen competition dynamics between microbes and plants [53]. In our plots, increased microbial mineralization of dead plant tissues in sprayed subplots may have lead to the increased nitrate-N observed in October-November. Alternatively, nitrate levels in our plots may have been affected by changes in the C:N balance because of shifts in the plant community and the amount or type of plant detritus in the sprayed subplots. In this study, high nitrate-N levels were dependant on glyphosate application and may have contributed to the growth of the introduced bentgrass plants because nitrogen is often the limiting element for plant growth [8]. However, the effect of nitrate-N on *Agrostis* growth and survivorship is difficult to separate from other glyphosate effects such as the removal of plant competitors. Further studies are needed to understand glyphosate effects on GR plant growth, competition and soil nutrient dynamics.

Effect of Glyphosate on Bentgrass Growth and Reproductive Capacity

Our results supported the hypothesis that CB and RT plants with mimicked GR trait would have higher levels of survivorship, vegetative growth and reproductive potential with selection pressure. Transgenic GR bentgrasses could not used in this study for a variety of reasons including: 1) it would be necessary to obtain a Memorandum of Understanding (MOU) and approval from the company holding the intellectual property rights and patents for the plant product, 2) the experimental design required quantifying flower and seed number and this would have allowed transgene escape because bentgrasses were already present in the study sites, and 3) the GR trait has not been inserted into RT. Unintended transgene flow has been well documented in bentgrasses in Oregon [36]. Given our special attention to mimicking the GR trait, we feel confident that the results are applicable to transgenic bentgrasses in the habitats and plant communities studied.

CB and RT plants had more tillers, above-ground biomass, and flowers in sprayed subplots. The most likely explanation for this is that they experienced less competition from neighboring vegetation for resources such as light, soil nutrients, or soil moisture. This contrasts with the inference from some previous bentgrass studies [40,41], but agrees with a study on 'mock feral' GR canola that showed increased crop fitness when exposed to glyphosate drift selection pressure [5] and a study that found glyphosate resistant *Brassica* species persisted as the dominant species after two annual applications of glyphosate drift [54].

CB and RT produced a high number of flowers and viable seeds in the second and third seasons of growth. RT plants in glyphosate-sprayed subplots had an 86-fold higher number of flowers compared to control subplots in 2009 (Table 3). CB plants in glyphosate-treated subplots also produced more flowers than controls. In addition, the germination rates for seed collected from CB and RT plants were above 70%. If the number of viable seeds is proportional to the number of flowers observed in 2009 and 2010, then seeds represent a significant pathway for dispersal of weedy bentgrasses carrying the GR trait. In addition, the large number of flowers suggests a strong potential for pollen-mediated transgene flow through either intraspecific or interspecific hybridization. Forced crosses and landscape-scale experiments have already demonstrated pollen-mediated gene flow between bentgrass species in the U.S. [38,36,39,55]. In our study area, seven bentgrass species can be found in natural and cultural landscapes, and all of these species are able to hybridize [30].

The removal of vegetation by glyphosate is a disturbance that creates gaps in plant communities. Research has shown that, without gaps, seed recruitment in herbaceous plant communities (e.g. tallgrass prairie and pastures) is rare and episodic [56], but disturbed sites have a higher rate of seedling establishment [57]. A positive correlation between gap diameter, grass population growth rate, and panicle production has also been demonstrated [58]. These observations are consistent with our study where subplots with glyphosate disturbance had higher CB and RT survival and growth (Table 2). In addition, the gaps created by glyphosate were postively correlated with increased bentgrass panicle and tiller production. Thus, preventing or reducing gaps in plant communities could be one strategy for preventing the establishment of GR bentgrasses in natural and agricultural habitats in the Northeastern U.S.

Acknowledgments

We thank Thomas Morris (Univ. of Connecticut) for help in analyzing soil fertility data, and Karl Guillard (Univ. of Connecticut) for helpful comments on the manuscript.

Author Contributions

Conceived and designed the experiments: CWA CAA. Performed the experiments: CWA. Analyzed the data: CWA. Contributed reagents/materials/analysis tools: CWA CAA. Wrote the paper: CWA CAA.

References

1. Heap I (2012) The International Survey of Herbicide Resistant Weeds. www.weedscience.com. Accessed: May 02, 2012.
2. Cerdeira AL, Duke SO (2006) The current status and environmental impacts of glyphosate-resistant crops: a review. J Environ Qual 35: 1633–1658.
3. Duke SO, Powles SB (2008) Glyphosate: a once-in-a-century herbicide. Pest Manag Sci 64: 319–325.
4. Powles S (2008) Evolved glyphosate-resistant weeds around the world: lessons to be learnt. Pest Manag Sci 64: 360–365.
5. Londo JP, Bautista NS, Sagers CL, Lee EH, Watrud LS (2010) Glyphosate drift promotes changes in fitness and transgene gene flow in canola (*Brassica napus*) and hybrids. Ann Bot-London 106: 957–965.
6. APHIS (2011) Animal and Plant Health Inspection Service. http://www.aphis.usda.gov/brs/aphisdocs/scotts_kbg_resp.pdf Accessed: May 2, 2012.
7. Ledford H (2011) Transgenic grass skirts regulators: Technological advances remove basis for government oversight of genetically modified crops. Nature 475: 274–275.
8. Service RF (2007) Agbiotech: A growing threat down on the farm. Science 316: 1114–1117.
9. Shaner DL, Lindenmeyer RB, Ostlie MH (2011) What have the mechanisms of resistance to glyphosate taught us? Pest Manag Sci. 68: 3–9.
10. Baucom RS, Mauricio R (2008) Constraints on the evolution of tolerance to herbicide in the common morning glory: resistance and tolerance are mutually exclusive. Evolution 62: 2842–2854.
11. Baerson SR, Rodriguez DJ, Tran M, Feng YM, Biest NA, et al. (2002) Glyphosate-Resistant Goosegrass. Identification of a mutation in the target enzyme 5-Enolpyruvylshikimate-3-Phosphate Synthase. Plant Physiol 129: 1265–1275.
12. Gaines TA, Shaner DL, Ward SM, Leach JE, Preston C, et al. (2011) Mechanism of resistance of evolved glyphosate-resistant palmer amaranth (*Amaranthus palmeri*). J Agr Food Chem 59: 5886–5889.
13. Ge X, d'Avignon DA, Ackerman JJH, Sammons RD (2010) Rapid vacuolar sequestration: the horseweed glyphosate resistance mechanism. Pest Manag Sci 66: 345–348.
14. Warwick SI, Légère A, Simard M-J, James T (2008) Do escaped transgenes persist in nature? The case of an herbicide resistance transgene in a weedy *Brassica rapa* population. Mol Ecol 17: 1387–1395.
15. Lortie CJ, Brooker RW, Choler P, Kikvidze Z, Michalet R, et al. (2004) Rethinking plant community theory. Oikos 107: 433–438.
16. Chang F, Simicik MF, Capel P (2011) Occurrence and fate of the herbicide glyphosate and its degradate aminomethylphosphonic acid in the atmosphere. Environ Toxicol Chem 30: 548–555.
17. Coupe RH, Kalkhoff SJ, Capel PD, Gregoire C (2011) Fate and transport of glyphosate and aminomethylphosphonic acid in surface waters of agricultural basins. Pest Manag Sci 68: 16–30.
18. Sullivan TP, Sullivan DS (2003) Vegetation management and ecosystem disturbance: impact of glyphosate herbicide on plant and animal diversity in terrestrial systems. Environ Rev 11: 37–59.
19. Boyd RS, Freeman JD, Miller JH, Edwards MB (1995) Forest herbicide influences on floristic diversity seven years after broadcast pine release treatments in central Georgia, USA. New Forest 10: 17–37.

20. Freedman B, Morash R, MacKinnon D (1993) Short-term changes in vegetation after the silvicultural spraying of glyphosate herbicide onto regenerating clearcuts in Nova Scotia, Canada. Can J Forest Res 23: 2300–2311.
21. Heard MS, Hawes C, Champion GT, Clark SJ, Firbank LG, et al. (2003) Weeds in fields with contrasting conventional and genetically modified herbicide-tolerant crops. I. Effects on abundance and diversity. Philos T R Soc Biol 358: 1819–1832.
22. Marvier M, Van Acker R (2005) Can crop transgenes be kept on a leash? Front Ecol Environ 3: 99–106.
23. NASS (2011) National Agricultural Statistics Service. http://www.ers.usda.gov/data/biotechcrops/Accessed August 2011.
24. Warwick S, Beckie HJ, Hall LM (2009) Gene flow, invasiveness, and ecological impact of genetically modified crops. Ann NY Acad Sci 1168: 72–99.
25. Behrendt S, HanfM (1979) *Grass Weeds in World Agriculture*. BASF Aktiengesellschaft, Ludwigshafen am Rhein, Germany, 73–80.
26. IPA (2011) Invasive Plant Atlas. http://www.invasiveplantatlas.org/grass.html Accessed: August 2011.
27. MacBryde B (2006) White Paper: Perspectives on creeping bentgrass, *Agrostis stolonifera* L. USDA/APHIS/BRS, Riverdale, Maryland, USA. http://www.aphis.usda.gov/peer_review/downloads/perspectiveCBG-wp.pdf. Accessed: November 3, 2009.
28. Wedin DA, Tilman D (1990) Species effects on nitrogen cycling: a test with perennial grasses. Oecologia 84: 433–441.
29. Gurevitch J, Scheiner SM, Fox GA (2006) The Ecology of Plants. 2nd ed. Sunderland, MA: Sinauer Associates, Inc.
30. Ahrens C, Chung J, Meyer T, Auer C (2011) Bentgrass distribution surveys and habitat suitability maps support ecological risk assessment in cultural landscapes. Weed Sci 59: 145–154.
31. Ahrens C, Ecker G, Auer C (2011) The intersection of ecological risk assessment and plant communities: an analysis of *Agrostis* and *Panicum* species in northeastern U.S. Plant Ecol 212: 1629–1642.
32. Beard JB (2002) Turf management for golf courses. 2nd ed. Chelsea, MI, USA: Ann Arbor Press,. 137–138, 716–726.
33. Magee DW, Ahles HE (2007) Flora of the northeast: a manual of the vascular flora of New England and adjacent New York. Amherst, MA: University of MA press.
34. Hart SE, Belanger FC, McCullough PE, Rotter D (2009) Competitiveness of *Agrostis* interspecific hybrids in turfgrass swards. Crop Sci 49: 2275–2284.
35. Kik C, Van Andel J, W. Joenjes W (1990) Life-history variation in ecologically contrasting populations of *Agrostis stolonifera*. Ecology 78: 962–973.
36. Watrud LS, Lee EH, Fairbrother A, Burdick C, Reichman JR, et al. (2004) Evidence for landscape-level, pollen-mediated gene flow from genetically modified creeping bentgrass with CP4 EPSPS as a marker. P Natl Acad Sci USA 101: 14533–14538.
37. Charles D (2011) Scientist in the middle of the GM-Organic Wars. Science 332: 168.
38. Belanger FC, Meagher TR, Day PR, Plumley K, Meyer WA (2003) Interspecific hybridization between *Agrostis stolonifera* and related *Agrostis* species under field conditions. Crop Sci 43: 240–246.
39. Wipff JK, Fricker CR (2000) Determining gene flow of transgenic creeping bentgrass and gene transfer to other bentgrass species. Diversity 16: 36–39.
40. Fei S, Nelson E (2004) Greenhouse evaluation of fitness-related reproductive traits in roundup®-tolerant transgenic creeping bentgrass (*Agrostis Stolonifera* L.). In Vitro Cell Dev B 40: 266–273.
41. Gardner DS, Nelson EK, Waldecker MA, Tarter WR (2006) Establishment and lateral growth of glyphosate-resistant creeping bentgrass in bare soil. HortTechnology 16: 590–594.
42. Bliss CL (1934) Methods of probits. Science 79: 38–39.
43. Kuykendall KL, Mahoney CJ, Conroy GC (1992) Probit and survival analysis of tooth emergence ages in mixed-longitudinal sample of chimpanzees (*Pan troglodytes*). Am J Phys Anthropol 89: 379–399.
44. Oksanen J (2009) Ordination and analysis of dissimilarities: tutorial with R and Vegan. www.bio.utk.edu/fesin/MSA2009/R/veganbird.pdf Accessed: April 2012.
45. Flanders AA, Kuvlesky WP, Ruthven DC, Zaiglin RE, Bingham RL, et al. (2006) Effects of invasive exotic grasses on south Texas rangeland breeding birds. Auk 123: 171–182.
46. Krueger-Mangold J, Sheley R, Roos B (2002) Maintaining plant community diversity in a waterfowl production area by controlling canada thistle (*Cirsium arvense*) using glyphosate. Weed Technol 16: 4 57–463.
47. Puricelli E, Tuesca D (2005) Weed density and diversity under glyphosate-resistant crop sequences. Crop Prot 24: 533–542.
48. Rice PM, Toney JC, Bedunah DJ, Carlson CE (1997) Plant community diversity and growth form responses to herbicide applications for control of *Centaurea maculosa*. J Appl Ecol 34: 1397–1412.
49. Likens GE, Bormann FH, Johnson NM, Fisher DW, Pierce RS (1970) Effects of forest cutting and herbicide treatment on nutrient budgets in the Hubbard Brook Watershed-Ecosystem. Ecol Monogr 40: 23–47.
50. Vitousek PM, Matson PA (1985) Disturbance, nitrogen availability, and nitrogen losses in an intensively managed loblolly pine plantation. Ecology 66: 1360–1376.
51. Vitousek PM, Andarlese SW, Matson PA, Morris L, Sanford RL (1992) Effects of harvest intensity, site preperation, and herbicide use on soil nitrogen transformations in a young loblolly pine plantation. Forest Ecol Manag 49: 277–292.
52. Bushoven JT, Jiang Z, Ford HJ, Sawyer CD, Hull RJ, et al. (2000) Stabilization of soil nitrate by reseeding with perenial ryegrass following sudden turf death. J Environ Qual 29: 1657–1661.
53. Schimel JP, Bennet J (2004) Nitrogen mineralization: challenges of a changing paradigm. Ecology 85: 591–602.
54. Watrud LS, King GA, Londo JP, Colasanti R, Smith BM, et al. (2011) Changes in constructed *Brassica* communities treated with glyphosate drift. Ecol Appl 21: 525–538.
55. Zapiola ML, Campbell CK, Butler MD, Mallory-Smith CA (2008) Escape and establishment of transgenic glyphosate-resistant creeping bentgrass *Agrostis stolonifera* in Oregon, U.S.A: a 4-year study. J Appl Ecol 45: 486–494.
56. Benson EJ, Hartnett DC (2006) The role of seed and vegetative reproduction in plant recuritment and demography in tallgrass prairie. Plant Ecol 187: 163–178.
57. Reader RJ, Buck J (1991) Control of seedling density on disturbed ground: role of seedling establishment for some midsuccessional, old-field species. Can J Bot 69: 773–777.
58. Tozer KN, Chapman DF, Quigley PE, Dowling PM, Cousens RD, et al. 2008. Controlling invasive annual grasses in grazed pastures: population dynamics and critical gap sizes. J Appl Ecol 45: 1152–1159.

The *In Vitro* Mass-Produced Model Mycorrhizal Fungus, *Rhizophagus irregularis*, Significantly Increases Yields of the Globally Important Food Security Crop Cassava

Isabel Ceballos[1], Michael Ruiz[1], Cristhian Fernández[2], Ricardo Peña[2], Alia Rodríguez[1]☯¶, Ian R. Sanders[3]*☯¶

1 Soil Microbiology, Universidad Nacional de Colombia, Bogotá, Colombia, **2** Utopía, Universidad de La Salle, Yopal, Colombia, **3** Department of Ecology and Evolution, University of Lausanne, Lausanne, Switzerland

Abstract

The arbuscular mycorrhizal symbiosis is formed between arbuscular mycorrhizal fungi (AMF) and plant roots. The fungi provide the plant with inorganic phosphate (P). The symbiosis can result in increased plant growth. Although most global food crops naturally form this symbiosis, very few studies have shown that their practical application can lead to large-scale increases in food production. Application of AMF to crops in the tropics is potentially effective for improving yields. However, a main problem of using AMF on a large-scale is producing cheap inoculum in a clean sterile carrier and sufficiently concentrated to cheaply transport. Recently, mass-produced *in vitro* inoculum of the model mycorrhizal fungus *Rhizophagus irregularis* became available, potentially making its use viable in tropical agriculture. One of the most globally important food plants in the tropics is cassava. We evaluated the effect of *in vitro* mass-produced *R. irregularis* inoculum on the yield of cassava crops at two locations in Colombia. A significant effect of *R. irregularis* inoculation on yield occurred at both sites. At one site, yield increases were observed irrespective of P fertilization. At the other site, inoculation with AMF and 50% of the normally applied P gave the highest yield. Despite that AMF inoculation resulted in greater food production, economic analyses revealed that AMF inoculation did not give greater return on investment than with conventional cultivation. However, the amount of AMF inoculum used was double the recommended dose and was calculated with European, not Colombian, inoculum prices. *R. irregularis* can also be manipulated genetically *in vitro*, leading to improved plant growth. We conclude that application of *in vitro R. irregularis* is currently a way of increasing cassava yields, that there is a strong potential for it to be economically profitable and that there is enormous potential to improve this efficiency further in the future.

Editor: Matthias Rillig, Freie Universität Berlin, Germany

Funding: The authors thank the Swiss National Science Foundation (http://www.snf.ch) for funding this study with a Joint Research Project awarded to AR and IRS (project number: IZ70Z0-131311/1). The funders had no role in study design, data collection and analysis, decision to publish, or preparation of the manuscript.

Competing Interests: The authors have declared that no competing interests exist.

* E-mail: ian.sanders@unil.ch

☯ These authors contributed equally to this work.

¶ These authors are joint senior authors on this work.

Introduction

The global human population is expected to reach over 9 billion by 2050, according to United Nations estimates [1]. Feeding 9 billion people represents a major challenge that requires global yield increases of approximately 70% to 100% [2]. However, most of these yield increases will be necessary in globally important food crops in developing countries in tropical and sub-tropical regions, where population growth is high. Developing both existing and new technologies for more efficient production of globally important food crops is key to achieving these goals [2,3]. One of the most potentially useful technologies is to apply the fungi involved in the arbuscular mycorrhizal symbiosis. The symbiosis occurs between plant roots and arbuscular mycorrhizal fungi (AMF). It is potentially useful because most plants, including all the major globally important food crops, naturally form this symbiosis and the symbiosis can increase plant biomass because the fungi help plants obtain phosphate from the soil [4]. Phosphate is an essential nutrient for plant growth. While stocks of phosphate fertilizer are rapidly being depleted there is a simultaneous increase in demand for phosphate to help feed the growing population [5,6]. Thus, the mycorrhizal symbiosis is potentially applicable at a global scale for increasing food production and more efficiently using phosphate reserves.

The potential benefits of inoculating crops with AMF have been known for decades. Yet there are very few published examples clearly demonstrating that large-scale inoculation of globally important crops in a real agricultural situation results in significant increases in food production that are either economically viable or practically achievable. There are three main reasons for this. Firstly, AMF are present in almost all soils. Crops naturally become colonized by native AMF. An overwhelming majority of published studies demonstrating the benefits of AMF inoculation on plant growth and phosphate acquisition have been conducted in sterile soil and compare the growth of mycorrhizal plants with non-mycorrhizal plants. However, effective AMF inoculation in agriculture requires inoculating plants with AMF that results in yield increases over those attained in the uninoculated crop that that is naturally colonized by the native AMF community. Secondly, most field trials with AMF have been conducted in temperate cropping systems. In many temperate agricultural regions, soil nutrient availability is relatively high compared with tropical soils. In most temperate agricultural soils inorganic phosphate fertilization is an effective way to increase yields. In contrast, most tropical regions are comprised of highly acidic Oxisols and Ultisols that typically have low total and available phosphate contents and a very high phosphate retention capacity [7]. In such soils farmers have to add large amounts of phosphate fertilizer to achieve significant yield increases. It is in these soils where the application of AMF could potentially increase food production and reduce application of phosphate fertilizer. Thirdly, AMF inoculum can only be grown with plants. Culturing AMF has traditionally been labour-intensive, requiring large-scale production of plants, from which the AMF inoculum can be harvested. Often this AMF inoculum is in the form of soil containing AMF propagules. It is difficult to ensure consistent inoculum quality, impossible to ensure that the soil is free from other potentially harmful microorganisms and the weight and volume of substrate for inoculum can make transport costs prohibitively high. However, these problems have recently been partially resolved by the development of a biotechnological mass production system for AMF. The fungi are grown *in vitro* on *Agrobacterium rhizogenes*-transformed carrot roots in a sterile artificial medium. Efficient production systems and the ability to concentrate very large numbers of AMF propagules into a small volume of sterile medium make the product easy to transport, free of unwanted microorganisms and potentially economically viable for large-scale application to important food crops.

One obvious target crop for the application of biotechnologically produced AMF is cassava (*Manihot esculenta* Crantz). Cassava is a traditional crop in latin America that has become globally important, annually feeding almost a billion people in 105 countries and representing almost a third of their daily caloric contribution [8]. The Food and Agriculture Organization (FAO) of the UN promotes cassava cropping in developing countries. It is considered vital for food security and as an alternative energy source [9].

Cassava is highly dependent on AMF for its growth and nutrition. Large growth increases due to mycorrhizal inoculation have been observed in cassava. Plants inoculated with AMF were 10–20 times larger than non-mycorrhizal plants growing in sterile soil [10] [11]. While cassava plants are never non-mycorrhizal in the field, these studies demonstrate that cassava is dependent on AMF for its growth. In a series of pioneering experiments by Sieverding and co-workers (summarized in [12] impressive effects of AMF inoculation were observed in the field in Colombia, where yields of cassava roots could be increased by up to approximately 5 tons ha^{-1}.

Numerous studies show that different AMF species have highly variable effects on plant growth [13,14] [15,16]. Indeed, yields of cassava in the field were highly variable following inoculation with different AMF species, ranging from no effect, compared to uninoculated plants colonized by the native AMF community, up to an approximate 20% yield increase [12]. This shows the need to consider the AMF identity for inoculation of cassava. Thus, it is not a foregone conclusion that a given biotechnologically-produced AMF species will significantly enhance cassava yields.

The most efficient large-scale biotechnological production of an AMF that can be highly concentrated in a sterile carrier has been achieved for one AMF species *Rhizophagus irregularis*, making it a strong candidate for use in cassava cropping. An additional strong incentive to use this species to increase cassava production is that *R. irregularis* has become the model AMF species for researchers as it can easily be grown in an *in vitro* culture system in the laboratory that is very similar to that used in large-scale commercial *in vitro* AMF cultivation. Consequently, more information is known about the genome and transcriptome of *R. irregularis* than any other AMF species [17,18]. The availability of only one *in vitro*-cultivated AMF species may appear very limited. However, populations of this fungus exhibit high natural genetic variation [19,20]. Recent studies showed that crossing genetically different *R. irregularis* isolates *in vitro* gave rise to genetically novel lines [21]. Allowing segregation of genetic material in subsequent generations by so-called single spore culturing also gave rise to a large number number of genetically novel *R. irregularis* lines [22]. Both crossed and segregated *R. irregularis* lines were shown to have differential effects on the growth of rice, in the greenhouse, with up to a five-fold increase in rice growth with some lines [23,24]. The ability to easily generate a large number of genetically novel *R. irregularis* lines *in vitro* that can have differential effects on plant growth, and then rapidly produce them in commercial quantities, would make this fungus an ideal candidate for a future molecular-assisted AMF breeding and improvement program, if *in vitro*-produced *R. irregularis* has the capacity to actually increase cassava yield in the field over uninoculated cassava. Therefore, it is highly pertinent to know whether *in vitro*-produced inoculum of this particular model AMF species can significantly increase cassava yields in commercial cassava cropping.

We, therefore, tested whether *in vitro*-produced *R. irregularis* increases cassava yield over uninoculated cassava in large-scale commercial cassava cropping in Colombia. We performed the experiment at two different locations with different soils, with no phosphate fertilization and with two different levels of phosphate fertilization. The highest application of phosphate represented the amount normally used locally by farmers in commercial cassava cropping in Colombia. A full economic analysis of the results allowed us to test not only whether *in vitro* produced *R. irregularis* can be used to significantly increase food production but also at what point such technology could become economically viable to produce food more cheaply. Thus, we were able to address the potential of using this technology in the short term to help reduce both hunger and poverty; the two main goals of the FAO of the United Nations. Additionally we were able to assess whether this fungus is a potentially good candidate for a molecular assisted AMF improvement program in an attempt to increase cassava yields over that which is currently possible and reduce food production costs even more in the future.

Materials and Methods

Field Sites

Two field experiments were established to test the effects of *in vitro*-produced AMF inoculum on the growth of cassava. The first was at the Utopía campus of La Salle University (at Yopal, Casanare) in the Los Llanos region of Colombia (72° 17′ 48″ W, 5° 19′ 31″ N). The second was on a commercial farm located in Santana, Boyacá in a mountainous region of Colombia (73° 29′ 59″ W, 6° 03′ 27″ N). Physical and chemical soil properties at both field sites are shown in Table S1.

The climate in Yopal is tropical with average temperatures of 18°C (night) to 28°C (day), with an average air humidity of 75% and total annual precipitation of 2335 mm with 172 rain days.

The field in Yopal had not been cultivated for 14 years prior to this experiment.

In Santana the temperature ranges between 15°C (night) and 23°C. The average air humidity is 78% and the total annual precipitation is 1901 mm with 221 rainy days. The plot in Santana had been regularly cultivated with sugar cane before cassava crops were established five years ago.

Plant and Fungal Material

Cassava (*Manihot esculenta* Crantz) varieties used in Yopal and Santana were MCOL2737 and COL2215, respectively. These are the most frequently used varieties by local farmers for cassava as a food crop.

The AMF used for these experiments was *Rhizophagus irregularis* produced in an artificial *in-vitro* AMF culture system with *Agrobacterium rhizogenes*-transformed carrot roots. The commercial product used is known as Glomygel® Hortalizas (http://www.mycovitro.com). Most isolates were previously ascribed to the species *Glomus intraradices* but were then found phylogenetically to be a separate species which was subsequently named *G. irregulare* [25]. Ribosomal DNA sequences obtained for this fungus allowed us to verify that the fungus used in Glomygel® fitted to the *G. irregulare* group described by Stockinger et al. (2009). *Glomus irregulare* has, however, recently been renamed as *Rhizophagus irregularis* [26].

Design and Establishment of the Field Experiment

The experiment in Yopal was a two-factor design arranged into blocks using a split-plot layout, with mycorrhizal treatment as the main plot and phosphorus treatment as the sub-plot. There were four blocks, each containing three replicate minor plots of 6 treatment combinations. The blocks were arranged perpendicular to the slope of the field. There were two mycorrhizal treatments, inoculated with AMF (+AMF) or non-inoculated (−AMF). There were 3 phosphate treatments; no phosphate fertilization (0 P), 50% phosphate fertilization (50% P) and 100% phosphate fertilization (100% P). Thus, there were 6 treatment combinations. The treatment 100% P represented the normal amount to P fertilizer applied to cassava crops in the region. There were 12 replicates of each of the six treatments with 25 plants within each minor plot. Two rows of plants were planted around each minor plot to reduce edge effects. The experimental design was the same in Santana except that there were five replicates of each of the six treatments, with 20 plants inside each minor plot.

Cassava was planted as stem cuttings (stakes). These were taken from the middle third of parental plants and had two or three vegetative buds and a thickness of between 1.5 and 2.0 cm. Ends of the stakes were cut diagonally to improve plant emergency. In Yopal (May 2011), 20 cm deep holes were made at an angle of 25°–30° and cassava stakes were completely buried with vegetative buds facing upward. In Santana (April 2011), cassava stakes were completely buried horizontally, 5 cm below ground. Total plant density was 7600 ha^{-1}. For both experiments, there was no artificial irrigation and conventional crop management for the region was applied depending on pests, diseases and weed incidence.

Fertilizers were applied in Yopal and Santana at 45 and 80 days after planting, respectively. The amount of fertilizer applied was determined by the initial soil nutrient content, nutritional requirements of the cassava variety and fertilizer efficiency. Plants in the 100% P treatments in Yopal received 104 Kg ha^{-1} urea, 201 Kg ha^{-1} di-ammonium phosphate, 54 Kg ha^{-1} potassium chloride (KCl), 42 Kg ha^{-1} of Kieserite (a fertilizer comprising 3% soluble potassium, 24% magnesium and 19% sulphur) and 40 Kg ha^{-1} of Vicor (a granular fertilizer comprising 15% nitrogen, 5% calcium, 3% magnesium, 2% sulphur, 0.02% boron, 0.02% copper, 0.02% manganese, 0.005% molybdenum, 2.5% zinc). In Santana, plants in the 100% P treatment received 380 Kg ha^{-1} urea, 330 Kg ha^{-1} di-ammonium phosphate, 240 Kg ha^{-1} potassium chloride (KCl) and 100 Kg ha^{-1} of Kieserite (a fertilizer comprising 3% soluble potassium, 24% magnesium and 19% sulphur).

Inoculation of cassava plants in the inoculated treatment (+AMF) was carried out at planting. Plants were inoculated with approximately 12500 AMF propagules per plant, which corresponded to double recommended dose. The product was diluted in water (1:200) in a 150 l container. Cassava stakes were completely submerged in the diluted inoculum before planting. Stakes were then planted and the excess inoculum in was distributed in an equal amount in the hole where each plant in the +AMF treatment was placed. Non-inoculated plants received the same amount of water.

Plant and Fungal Growth Measurements

In the experiment in Yopal, plant growth variables were measured every 45 days. One plant was collected from each minor plot of each treatment. The number of roots per plant, root diameter and root fresh weight per plant were measured. Total leaf area per cassava plant (m^2/plant) was determined by measuring the leaf area of all leaves on a plant using a Li-Cor leaf area meter Li-Cor, Nebraska, USA). Total plant dry mass and the separate dry mass of leaves, stems, petioles and tuberous roots was measured for each plant. Roots were cut into pieces to facilitate drying. Plant material was dried at 70°C until constant weight (approximately 49 hours).

In the experiment in Yopal, total AMF colonization in roots was measured in fine roots with a thickness of ≤2 mm. Fungal structures were visualised with Shaeffer black ink acording to [27]. The percentage of root colonization was determined by the grid line-intersect method [28], on roots of one plant per minor plot. Values of colonization in the roots of the three plants from minor plots per block were pooled for further statistical analysis.

At the final harvest in Yopal, all the same measurements were made as at each previous time point. In addition, the fresh weight (yield) of cassava roots was measured. Final yield was calculated as cassava root fresh weight per hectare. In Santana, the only measurements made were at the final harvest, 14 months after planting, and these were root fresh weight (yield) and the colonization of the roots by AMF.

Statistical Analyses

All data were analysed using the JMP® statistical discovery software (Statistical Analysis Systems Institute, version 10). To test for significant differences in the cassava growth variables or colonization of the roots by AMF, analysis of variance (ANOVA) was performed using a split-split plot model in JMP®. Prior to ANOVA, data were first checked for equal variances and normal distribution. When significant differences were observed in ANOVA, a Tukey honest significant difference (HSD) test was performed to determine which treatments were different from each other at $P \leq 0.05$.

Economic Analysis

The economic analysis included a full assessment of all production costs in the different treatments, profits, cash flow and profitability (return on investment - ROI%). The cost-benefit and ROI %, of the crop under the six treatments (+AMF, 0 P; +AMF, 50% P; +AMF, 100% P; −AMF, 0 P; −AMF, 50% P;

−AMF, 100% P), were calculated at both locations. Agricultural interest rates were used for calculations involving monetary value, corresponding to an effective rate of 11% annually. This represents the most conservative interest rate for such analyses.

An economic simulation was performed using different inoculum prices to calculate the economic feasibility of using the product and also to find the maximum inoculum price at which the farmer would create more profit than by using the traditional crop management. Additionally, the same simulation was made with varying phosphate fertilizer prices, to calculate the economic feasibility of using this product when considering potential future increases in the price of P fertilizer. The inoculum price used in these calculations is the European price of the product as it is not currently available on the Colombian market.

Ethics Statement

According to Colombian law and with advice from the Colombian Environmental Licensing Authority, it was found that for this project, permission was not required for use of Glomygel® in scientific research, because the inoculum was used in agricultural research activities without involving fauna or flora specimens. CITIES permission signed by Colombia for export and import of inoculum was not required because the species in the study did not belong to any of the lists of endangered species. Both field experiments were conducted on private land and permission was given by the owner of the land at each location. Furthermore, no additional special permission was required by Colombian law as the organisms in this study also occur in Colombia. None of the species used in this study are endangered or protected.

Results

There was a significant effect of inoculation with *in vitro*-produced AMF on cassava yields at both sites. Additionally, the yields were higher than the locally expected yields at both sites.

Cassava Growth and Yield in Yopal

In Yopal, growth of the plants was measured at 45 day intervals throughout the experiment. As expected, the growth of cassava increased over time for each of the variables and for overall plant dry mass. However, there was no significant overall AMF effect on the whole dataset over all time periods (data not shown). However, at the final harvest, both root fresh weight and root dry weight were significantly affected by AMF inoculation. Inoculation with AMF significantly increased root dry biomass at the final harvest in Yopal with an 18.3% increase in root dry mass and an increase of over 2 tons more cassava roots per hectare (ANOVA F ratio = $F_{(1,3)}$ 22.84, $P \leq 0.017$; Fig. 1a). At the final harvest in Yopal, increasing P fertilization levels also significantly altered root dry weight, irrespective of AMF inoculation (data not shown). However, there was no AMF inoculation x P treatment interaction meaning that AMF inoculation significantly influenced cassava dry weight at all levels of P fertilization.

Cassava root fresh weight (yield) in Yopal was also significantly affected by AMF inoculation with a highly significant effect (ANOVA F ratio = $F_{(1,3)}$ 16104.27, $P \leq 0.001$). Overall, inoculated plants were 20.4% heavier than roots of non-inoculated plants. However, this was not the same at each of the 3 P fertilization levels as indicated by a significant AMF inoculation x P fertilization interaction (ANOVA F ratio = $F_{(2,60)}$ 3.16, $P \leq 0.049$; Fig. 1b). The combination of inoculation with AMF and 100% P resulted in the significantly highest yield. The significantly lowest yields were in the treatment combinations without AMF inoculation and with 0 P and 100% P. Surprisingly, the treatment combination with AMF inoculation and 0 P resulted in yields as high, or higher, than non-inoculated plants with either 50% P or 100% P (Fig. 1b).

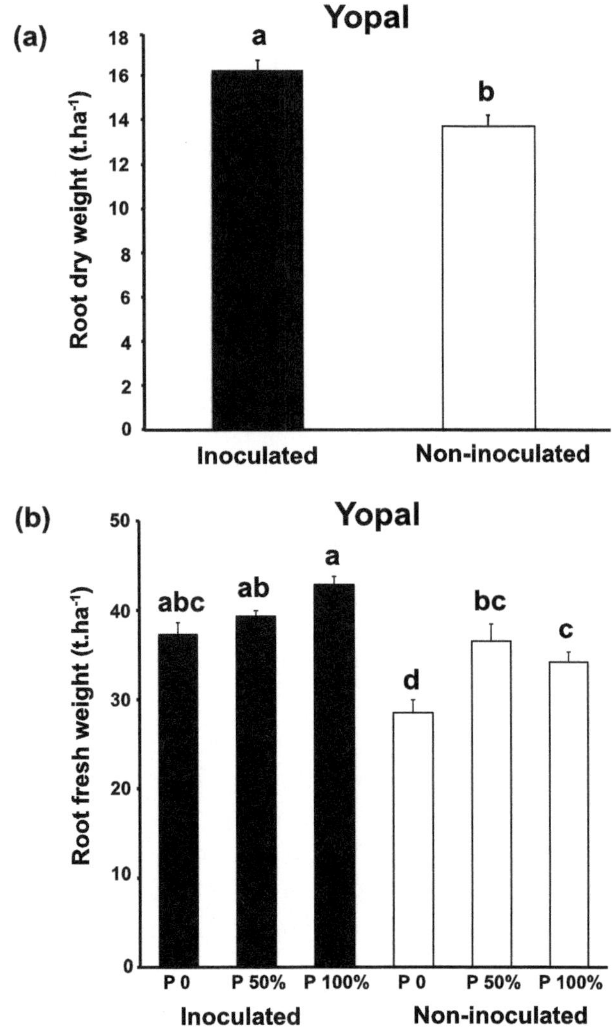

Figure 1. Effects of inoculation with AMF on cassava root growth. (**a**) Effects of inoculation with AMF on cassava root dry weight ($t.ha^{-1}$) in Yopal. (**b**) Effects of inoculation with AMF and P fertilization on cassava root fresh weight or yield ($t.ha^{-1}$) in Yopal. Black shaded bars represent the weight of inoculated cassava and white bars represent the weight of non-inoculated cassava. Error bars represent +1 S.E. Different letters above bars represent significant differences at $P \leq 0.05$.

Only root weight was affected by inoculation with AMF. None of the aboveground cassava growth variables were significantly affected by inoculation with AMF at the final harvest (data not shown).

Cassava Yield in Santana

In Santana, cassava yield (root fresh weight) was only measured at the final harvest. Inoculation with AMF had a significant effect on cassava yield but this was not the same at the different P fertilization levels, as indicated by an AMF inoculation x P fertilization interaction (ANOVA F ratio = $F_{(2,16)}$ 4.12, $P<0.036$; Fig. 2). The highest yields were obtained with cassava inoculated with AMF and at 50% P fertilization. The highest yield in non-inoculated plants was obtained at 100% P. Meanwhile, the lowest

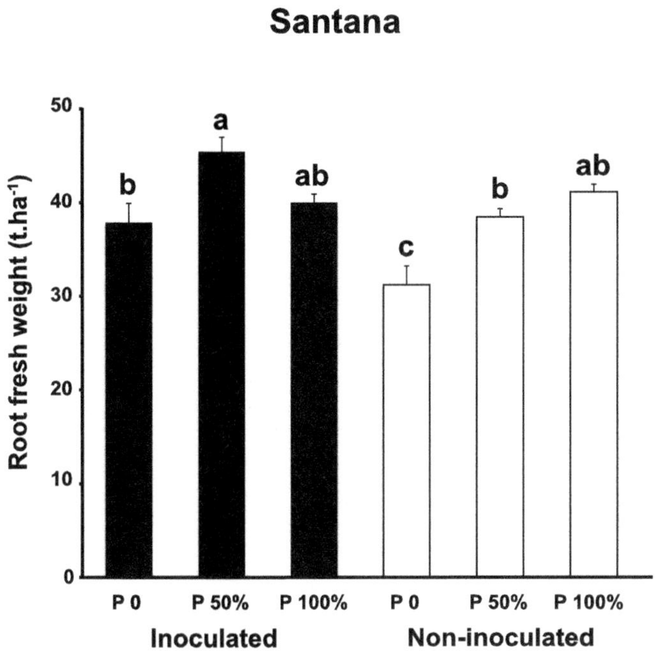

Figure 2. Effects of inoculation with AMF and P fertilization on cassava root fresh weight or yield (t.ha^{-1}) in Santana. Black shaded bars represent the weight of AMF-inoculated cassava and white bars represent the weight of non-inoculated cassava. Error bars represent +1 S.E. Different letters above bars represent significant differences at $P \leq 0.05$.

yields were obtained with non-inoculated plants at 0 P fertilization. Plants that were inoculated with AMF but received no P fertilizer achieved a yield that was not significantly different that of non-inoculated plants that received either 50% or 100% P fertilization.

AMF Colonization in Yopal and Santana

Cassava naturally becomes colonized by mycorrhizal fungi. Therefore, in both AMF inoculated and non-inoculated treatments, we expected plants to be colonized by AMF. In Yopal, AMF colonization of cassava roots increased over the duration of the experiment, reaching a peak of colonization by the end of the experiment (Fig. 3). There was, however, a decline in AMF colonization in all the treatments at 225 days after planting that coincided with a particularly dry period. There was no significant difference in colonization in cassava plants that were inoculated with AMF or non-inoculated for most of the experiment. At the final harvest AMF colonization differed in inoculated and non-inoculated plants but this was not the same effect at each level of P fertilization (ANOVA for the AMF inoculation x P fertilization interaction was F ratio = $F_{(2,12)}$ 3.9, $P<0.049$). At the final harvest, AMF colonization was higher in non-inoculated plants at 50% and 100% P fertilization than non-inoculated plants. The opposite effect on colonization occurred at 0 P fertilization (data not shown).

In Santana, AMF inoculation and P fertilization also had a combined effect on the colonization of cassava roots by AMF at harvest (ANOVA for the AMF inoculation x P fertilization interaction was F ratio = $F_{(2,16)}$ 18.16, $P \leq 0.001$; Fig. 4). The highest AMF colonization was observed in inoculated plants with no added P fertilizer. The lowest colonization values were obtained with non-inoculated and plants that received no phosphate fertilizer (Fig. 4).

The methodology used to measure colonization in this experiment does not allow us to differentiate between colonization by the local the AMF community and the inoculated AMF.

Economic Analyses

All production costs were used to calculate the return on investment (ROI) for production of cassava in one year in each of the six different treatments. The return on investment, represented as a percentage, is shown in Table 1. With the amount of inoculum used (which was double the recommended dose), and using a European inoculum price for the analysis, the highest ROI was observed in uninoculated treatments. The highest ROI in Yopal was in the treatment with no AMF inoculation and 50% P fertilizer for both cassava root fresh weight and dry weight (Table 1). In Santana, the highest ROI was achieved in the treatment with no AMF inoculation and 100% P fertilizer (Table 1). This was the case even though AMF inoculation significantly increased cassava yield at both sites. Interestingly, at both sites inoculation with AMF, in the absence of any P fertilizer, resulted in a higher ROI than when cassava was not inoculated (Table 1). The traditional practice is for the farmer to apply 100% P fertilizer but no AMF inoculum even though this is not necessarily the most profitable practice. Thus, in Yopal, inoculation with AMF and 100% P fertilization gave a higher return with root fresh weight than the traditional practice.

Because we used an unrealistic inoculum price and double the recommended dose, we made a simulation to calculate and project the following:

1. At what price the inoculum should be sold to make inoculation profitable for the farmers at the dose used in this experiment;
2. Project how profitable AMF inoculation will be in the event of future increasing P fertilizer prices;
3. Project how profitable AMF inoculation could be for farmers if the recommended dose would give similar yields.

To do this we used a baseline ROI of the most profitable treatment which was no inoculation with 100% P fertilizer in Santana and no inoculation with 50% P fertilizer in Yopal (Fig. 5). We simulated how the ROI would be affected at both sites when inoculating with AMF of varying price. We also calculated how profitable inoculation with the different priced inoculum would be in the event of rising P fertilizer prices (Fig. 5).

In Yopal, and with the sale of dry cassava roots, inoculation with AMF would become profitable if the same amount of inoculum required for one hectare of cassava were sold below 1 million Colombian pesos (COP) (Fig. 5a). Small price increases P fertilizer would not greatly affect the profitability of inoculating with AMF. This is also true for the sale of fresh cassava roots (Fig. 5b). In Santana, a small inoculum price reduction, to below COP 1200000, makes the use of AMF inoculum economically viable and any small prices increases in P fertilizer would have a large effect on the profitability of inoculating plants with AMF (Fig. 5c).

Because we used double the amount of AMF inoculum than that recommended by the inoculum producer, we also calculated the return on investment if the yield remained the same with only half the dose of inoculum that was used in this experiment (shown by the vertical dashed line in Fig. 5 representing half the inoculum cost). In all cases, this would make inoculation with AMF highly profitable, even at European inoculum prices and at current P fertilizer prices.

It is important to note that all the economic analyses were made on yields from one inoculation with AMF and for one crop of

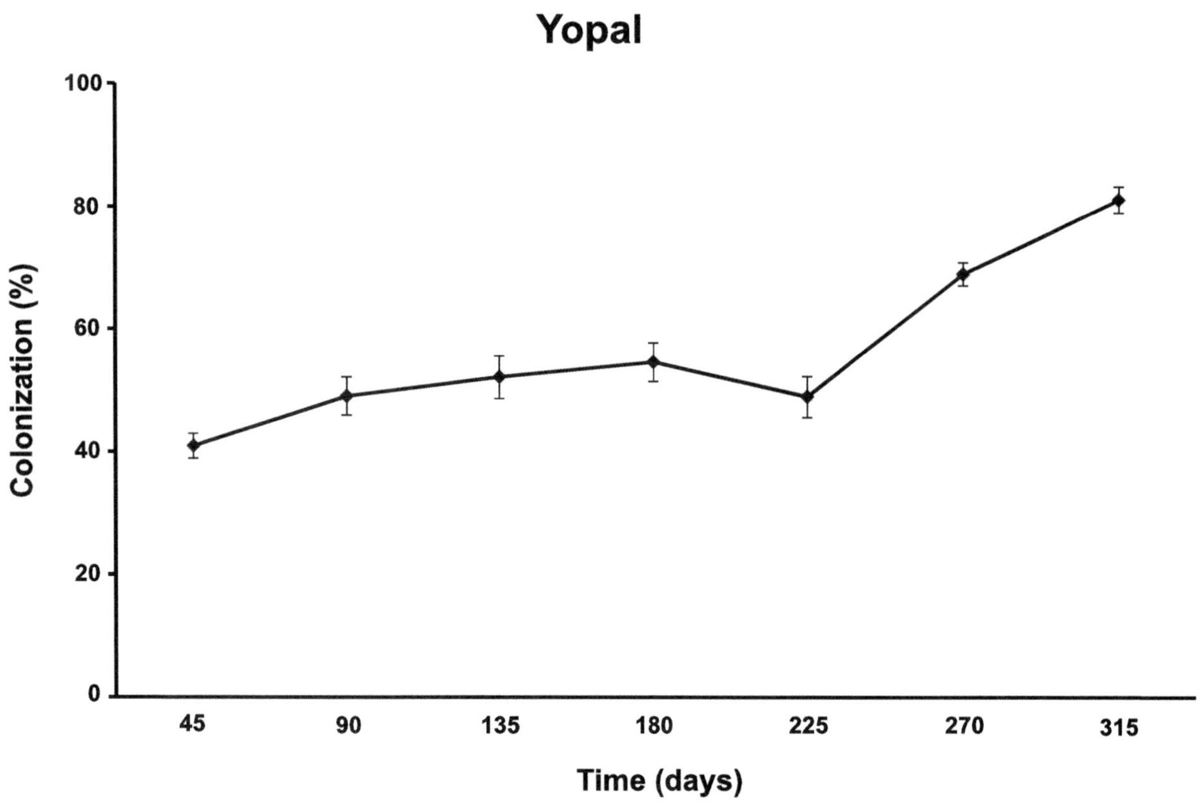

Figure 3. Colonization (represented as a % of root length) by AMF in the roots of cassava for the duration of the experiment in Yopal. The axis Time represents the number of days after planting. Error bars represent ±1 S.E.

cassava. Further yield increases in the next cassava crop due to inoculation in the previous year are not accounted for in these economic analyses.

Discussion

The results of this study show that inoculating the globally important food security crop cassava with AMF significantly improves cassava yields. The differences between this study and previous studies are that: 1. The fungus has been produced biotechnologically in an *in vitro* system that affords many advantages. 2. The fungal species was one that can very easily be grown and genetically manipulated for improvement *in vitro* and has become the model AMF for molecular and genomic research. 3. A detailed economic analysis allows us to set realistic targets for how to make the application of *in vitro* produced AMF economically viable for cassava production. The goal of the FAO is to increase food security and reduce hunger and poverty. Below we address how our results could help to realistically address these three goals in the future and potentially in a large number of developing countries.

Applying *in vitro*-produced AMF Leads to Higher Food Production

Mycorrhizal effects on food production and food security. Applying *in vitro*-produced AMF to cassava was effective in increasing cassava yields in both regions in Colombia, even though the soils, climate and cassava variety were different. The effects were not exactly the same in the two regions.

In Yopal, the most effective treatment was to add the same amount of P fertilizer as a farmer would normally use (P 100% treatment) and inoculate with AMF. However, an important result is that inoculating with AMF gave significantly greater dry mass yield than uninoculated cassava at all P fertilization levels (Figure 1a). Furthermore, a higher cassava yield was achieved with AMF, without P fertilization, in comparison with plants with the full P dose. This is an important result as it indicates that the farmer can achieve significant increases in food production by using *in vitro* produced *R. irregularis* even if P fertilizer becomes more scarce or fluctuates greatly in price. Thus, the use of AMF in a cropping system, like cassava, represents an increment in the efficiency of phosphate fertilizer use, since yields achieved in inoculated plants exceeded yields obtained with maximum P fertilizer level. Thus, in these soils applying AMF can significantly increase food production and may increase food security by ensuring stable yields in the event of rising or fluctuating P fertilizer prices/availability.

In Santana, the significantly highest cassava yields were obtained by applying *in vitro* produced AMF inoculum in combination with 50% of the P fertilizer that is normally used by farmers in the region. This is a particularly interesting result as it shows that in these soils, it is possible to reduce P fertilizer applications and simultaneously increase food production. This is also particularly important as it means that more food can be produced more securely as the farmer would rely less on P fertilizer price fluctuations and availability.

There are some notable features of the data, particularly in Yopal, where measurements of plant growth and AMF colonization were made at 45-day intervals throughout the growing season. Inoculation with AMF in Yopal only had a significant positive effect on cassava root production; the part that is harvested for food. Inoculation did not have any effect on above-ground cassava

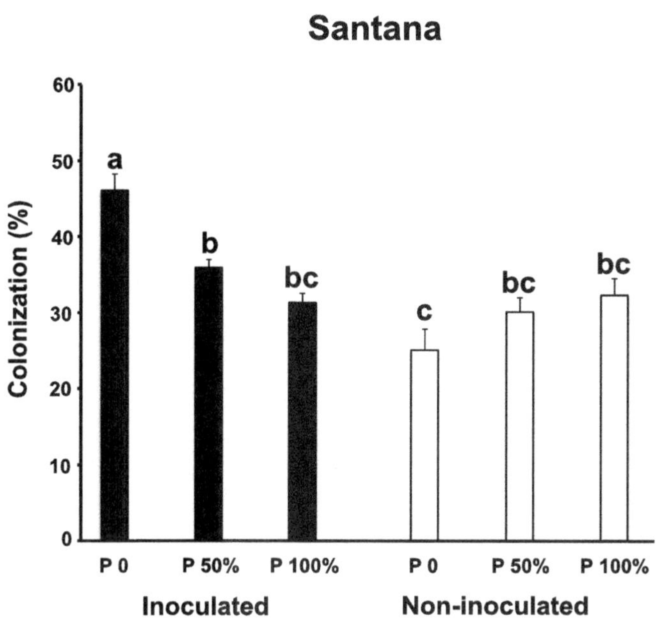

Figure 4. Colonization (represented as a % of root length) by AMF in the roots of inoculated and non-inoculated cassava and with different levels of P fertilization at the final harvest in Santana. Error bars represent +1 S.E. Different letters above bars represent significant differences at $P \leq 0.05$.

Table 1. Comparison of return on investment (ROI %) among the six treatments in Yopal and Santana.

Treatment		ROI (%)		
		Root Fresh Weight (yield)		Root Dry Weight
% P	AMF inoculation	Santana	Yopal	Yopal
0	Without	210.4	82.8	2.5
50	Without	265.5	126.8	15.1
100	Without	276.6	113.2	14.7
0	With	220.9	95.9	1.9
50	With	270.7	103.1	10.5
100	With	218.0	117.0	11.0

growth. Additionally, the positive effect of AMF inoculation was only observed at the end of the growing season when the roots were filling out with starch. Both of these results are consistent with previous findings from some field trials in Colombia on AMF and cassava [12]. In Yopal, inoculating cassava with AMF did not have any effect on overall AMF colonization levels in the roots of cassava during most of the experiment. In Santana, AMF colonization levels were overall higher when cassava was inoculated with AMF. However, how AMF colonization responded to P fertilization in the inoculated and uninoculated cassava in Santana followed different patterns. There are several reasons that could explain these differences. *Rhizophagus irregularis* may be more efficient at colonizing one cassava variety than the other, or the colonization levels could represent a difference in ability of *R. irregularis* to co-exist with the native AMF community at the two sites. From our data, we cannot distinguish between these different scenarios.

It is not possible to directly compare the effects of AMF inoculation at the two sites. The soils and cassava varieties differed at the two sites. Additionally, farmers typically use very different amounts of fertilizer in the two regions. To make the experiments agriculturally realistic the 100% P treatment represented the amount of P fertilizer typically applied by farmers in the region. Consequently, 100% P in Yopal is not an equivalent amount of P to 100% P in Santana.

Mycorrhizal effects in a realistic agricultural management system. Increased plant growth following AMF inoculation has been observed in many plant species and in numerous scientific publications [4]. However, most of these demonstrations have not been performed in an agriculturally realistic situation with unsterilized soil that already contains AMF. Furthermore, there are very few examples of where AMF inoculation has produced significant yield increases in a realistic agricultural management system over that which can be achieved using conventional fertilizer. In this respect, cassava is one of the only crops that has been shown to benefit from AMF inoculation and that feeds a very significant number of the global population (approx. 1 billion people in 105 countries). However, many previous applications of AMF were not very applicable on a wide-scale to many different farmers as they required inoculating plants with a large amount of inoculum contained in soil with a farmer managed on-site inoculum production system. For example, in the pioneering experiments documented by Sieverding [12], 500 g of soil containing AMF had to be added to the roots of each cassava plant meaning that on-site fungus production and local specialist knowledge was necessary. We argue that such systems are susceptible to poor quality inoculum production and are unlikely to be sustained as they require a greatly different management practice to the farmers traditional management system. Also, the amount of inoculum required for a farming unit (i.e. per hectare) would result in an expensive practice in view of the amount of material needed to be transported/stored and applied in real farming conditions. In contrast, adding inoculum of an assured quality that is easy to transport and which is very easy to apply at planting is more likely to be adopted by farmers. The *in vitro*-produced fungus used in these experiments was shipped in a concentration where 0.25 ml volume of non-diluted liquid product was needed per plant. Furthermore, the cost of labour for applying AMF at planting was negligible.

The Importance of using the AMF Species *R. irregularis*

One major importance of this study is that we used the fungus *R. irregularis*. This species has become the model AMF studied by molecular biologists because, unlike many other AMF species, it can be efficiently produced *in vitro*. Also, the genome of this fungus has now been sequenced (F. Martin, personal communication). Thus, laboratory-based studies on this species, have rendered much of the knowledge we have about AM fungi. More importantly, studies have shown that natural genetic variation is very high in this fungus. Such naturally occurring genetic variation can be used to give rise to genetically novel strains of the fungus by crossing the fungi *in vitro* [29]. Previously, researchers had assumed that this was not possible. Inoculation of rice with genetically novel varieties of *R. irregularis* resulted in up to five-fold differences in rice growth in the greenhouse [23]. Even though non-bred strains of the fungi actually reduced rice growth several of the genetically novel lines greatly increased rice biomass. While we show that this fungus can be used now to increase cassava yields, the results of genetic studies on this species highlight the enormous potential for using such a breeding program to greatly improve the effect of this

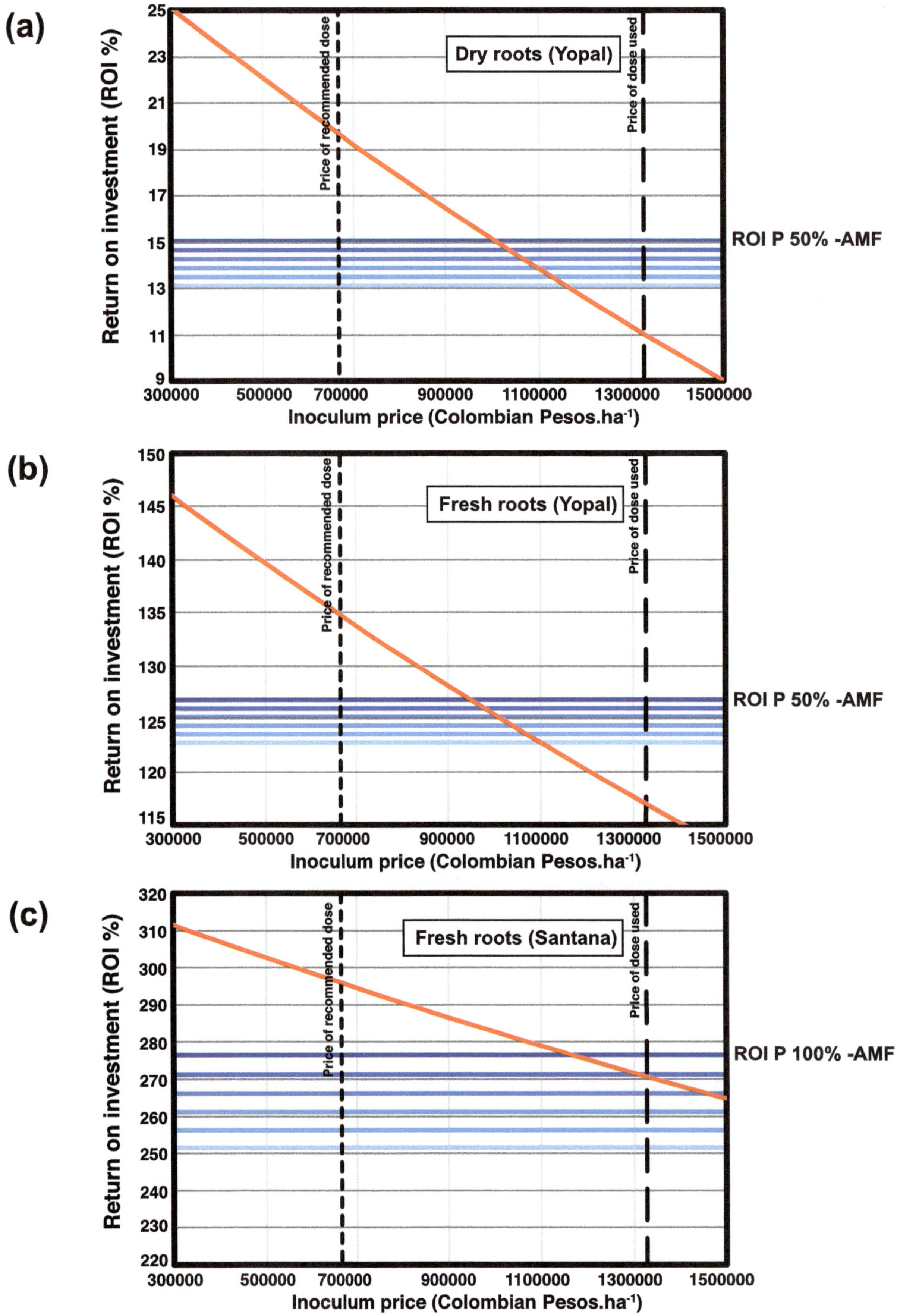
(a)
Return on investment (ROI %)
25
23
21
19
17
15
13
11
9
Price of recommended dose
Price of dose used
Dry roots (Yopal)
ROI P 50% -AMF
300000 500000 700000 900000 1100000 1300000 1500000
Inoculum price (Colombian Pesos.ha-1)
(b)
Return on investment (ROI %)
150
145
140
135
130
125
120
115
Price of recommended dose
Price of dose used
Fresh roots (Yopal)
ROI P 50% -AMF
300000 500000 700000 900000 1100000 1300000 1500000
Inoculum price (Colombian Pesos.ha-1)
(c)
Return on investment (ROI %)
320
310
300
290
280
270
260
250
240
230
220
Price of recommended dose
Price of dose used
Fresh roots (Santana)
ROI P 100% -AMF
300000 500000 700000 900000 1100000 1300000 1500000
Inoculum price (Colombian Pesos.ha-1)

Figure 5. Simulation of return on investment (ROI %) with varying prices of *in vitro*- produced AMF inoculum and increasing P fertilizer prices. (a) Simulation based on sale of dry cassava root produced in Yopal. (b) Simulation based on sale of fresh cassava root produced in Yopal. (c) Simulation based on sale of fresh cassava root produced in Santana. The diagonal red line represents the ROI for the most profitable treatment with AMF inoculation, at that site, when simulating varying inoculum price. The top horizontal blue line represents the ROI for the most profitable non-inoculated treatment at current P fertilizer prices, even though this may not necessarily be what farmers normally practice. Blue horizontal lines below represent a simulation of ROI for this treatment with a scenario of increasing P fertilizer prices (+20%, +40%, +60%, +80% and +100% price increase). Vertical dashed lines represent the price for the amount of inoculum used in these experiments (right) and the price for half of the amount of inoculum which represents the recommended dose (left).

fungus on cassava yields in the future and be able to expand these results to other parts of the world.

Environmental Impact of Adding a Non-local AM Fungus

One point which has received almost no attention in the commercial application of AMF is the potential dangers of adding any AMF inoculum into a region from which it was not isolated [30]. In this study, we did not address the potential impact of such a practice. However, we consider two important tests are made before such inocula are used commercially. First, an assessment of whether the addition of the AMF inoculum alters the structure and diversity of the native AMF community. Second, a population genetics-based assessment of whether adding the commercial inoculum introduces a significant number of new alleles into the local *R. irregularis* population. To our knowledge, this type of environmental impact assessment for AMF application has never been undertaken.

Economics of using *in vitro* Produced AMF in Cassava Cropping

The economic analyses were performed for cassava production in two regions of Colombia. Although the experiments documented here were not set up to test the economic viability of applying this inoculum to cassava, the economic analyses in this study allow us to formulate clear targets for achieving this in the future. While it was not economically viable to use *in vitro* produced AMF with the amounts of inoculum we used, the economic analyses indicate that such inoculum application could easily be economically viable in the near future, and effectively alleviate strong dependency on P fertilizers, in the cassava cropping system. We highlight realistic predictions from our analysis that indicate how to make such applications economically viable:

Inoculum amount. In these experiments we applied double the amount of inoculum to that recommended by the manufacturer. This means that each plant received 12500 propagules of the fungus. This is a very large amount given that in the greenhouse we inoculate plants with 300 spores of *R. irregularis* to obtain consistent colonization in the roots. Previous experiments in the field with cassava have demonstrated that inoculating cassava with about 1500–2500 propagules gives significant yield increases that cannot be increased more by adding larger numbers of effective propagules [12]. Therefore, it is very likely that the amount of inoculum used in these experiments could have been greatly reduced. Using 2500 propagules per plant would have reduced the inoculum price to a fifth of that which we used. This would make the use of *in vitro* AMF extremely profitable for farmers in both regions of Colombia (see Fig. 5).

Effects over multiple seasons. These experiments were conducted for only one field season. Therefore, the economic analysis only takes into account the effects on profitability from inoculating once and harvesting once. Previous detailed experiments over 2 cropping periods in Colombia have shown that one inoculation in the first year consistently gives rise to significant crop yields, in the first and second cropping [12]. In fact, in those experiments only one inoculation in the first year resulted in cassava yields that were higher in the second year than the first year. Thus, it is possible that our economic analyses greatly underestimate the profitability of inoculating cassava with *in vitro* produced AMF. However, it must be recognized that in this study, we have not measured the effects in a second cropping and therefore, it is also possible that there could be no carry over effect or that it could be negative. This needs to be tested in further experiments.

Inoculum price. The inoculum used was bought at European market price as the product is not distributed in Colombia. Therefore, the economic analyses use European inoculum prices. It is highly likely that inoculum prices on the market in a developing country could be greatly reduced in several ways: 1. The inoculum could be produced under licence locally, thereby, taking advantage of local salary levels that would be much lower than those in Europe; 2. Currently the inoculum price in Europe is based on relatively low sales for a specialised market. If this technology were adopted by farmers, the market for very widely-grown food crops such as cassava in the tropics would be much larger than on specialized crops in Europe, thus allowing a significant price decrease of the product because of much larger effective product sales. 3. Inoculum producers could lower their prices for sales of the product in developing countries.

Fluctuating P fertilizer prices and availability. Phosphate fertilizer prices and availability are expected to change over the next decades [5,6]. The economic analyses also indicate that relatively small changes in P fertilizer prices would make the use of this inoculum economically viable, even at the high price and inoculum quantity used in these experiments and even if there was no inoculation effect in a second year.

Conclusions

We conclude that application of the *in vitro*-produced AMF *R. irregularis* in cassava cropping can help in achieving the FAOs main goals of reducing hunger and poverty by significantly increasing cassava yields in a way is very likely to be easily made economically viable and profitable for farmers in developing countries. We also predict that its application could lead to economically more secure food production because cassava yield can be increased at very different levels of P fertilization making the farmer much less dependent on P fertilizer availability or price. Because the inoculation was observed in different soils, and with different cassava varieties, our study shows the potential to use this technology to increase cassava production in other parts of the world, especially in Africa where many cassava-cropped soils are very similar to those in Yopal. Finally, while we show that *in vitro*-produced *R. irregularis* can immediately be used to significantly increase cassava yields, the fact that the effects of this fungus can be increased by *in vitro* crossing makes it a very strong candidate for an improvement program to further increase cassava yields in the future over that which can be achieved with these fungi at present.

Supporting Information

Table S1 Mean physical and chemical soil properties at the two study sites Yopal and Santana, Colombia.

Acknowledgments

We thank the students at the Utopía campus for their help in the planting, maintenance and measurement of plants, and in particular Leonela Yised Nañes Cortés, Maria Idaly Ibañez, Victor Fernando Camacho, Camilo Alberto Gutiérrez Rodríguez, Leidy Andrea Barrera Franco, José Irenarco Tapias Fernández, Diego Alexander Arias Zorro, Erika Tatiana Naranjo Barragán, Alenis Guillermina Chaparro Pulido.

Author Contributions

Conceived and designed the experiments: IRS AR IC MR CF RP. Performed the experiments: IC MR. Analyzed the data: IC MR. Contributed reagents/materials/analysis tools: IRS AR MR RP. Wrote the paper: IRS AR IC.

References

1. FAO (2013) FAOSTAT. Rome, Italy: FAO publications. Available: http://faostat3.fao.org/home/index.html. Accessed 2013 June 17.
2. Godfray HCJ, Beddington JR, Crute IR, Haddad L, Lawrence D, et al. (2010) Food security: The challenge of feeding 9 billion people. Science 327: 812–818.
3. Foresight: The future of farming (2011) Final project report. London, UK: The Government Office for Science. 211 p.
4. Smith SE, Read DJ (2008) The mycorrhizal symbiosis. San Diego, USA: Academic Press.
5. Gilbert N (2009) The disappearing nutrient. Nature 461: 716–718.
6. Gross G (2010) Fears over phosphorus supplies. Curr Biol 20: 386–387.
7. Friesen DK, Rao IM, Thomas RJ, Oberson A, Sanz JI (1997) Phosphorus acquisition and cycling in crop and pasture systems in low fertility tropical soils. Plant Soil 196: 289–294.
8. FAO (2010) Available: http://www.fao.org/ag/AGP/AGPC/gcds/index_en.html.Accessed 2013 June 17.
9. FAO (2005) A review of cassava in Africa with country case studies on Nigeria, Ghana, the United Republic of Tanzania, Uganda and Benin: Proceedings of the validation forum on the global cassava development strategy. Rome, Italy: Food and Agriculture Organization of the United Nations.
10. Howeler RH, Sieverding E (1983) Potentials and limitations of mycorrhizal inoculation illustrated by experiments with field-grown cassava. Plant Soil 75: 245–261.
11. Sieverding E, Howeler RH (1985) Influence of species of VA mycorrhizal fungi on cassava yield response to phosphorus fertilization. Plant Soil 88: 213–221.
12. Sieverding E (1991) Vesicular-arbuscular mycorrhiza management in tropical agrosystems. Eschborn, Germany: Deutche Gesellschaft für Technische Zusammenarbeit (GTZ). 371 p.
13. van der Heijden MGA, Klironomos JN, Ursic M, Moutoglis P, Streitwolf-Engel R, et al. (1998) Mycorrhizal fungal diversity determines plant biodiversity, ecosystem variability and productivity. Nature 396: 69–72.
14. van der Heijden MGA, Boller T, Wiemken A, Sanders IR (1998) Different arbuscular mycorrhizal fungal species are potential determinants of plant community structure. Ecology 79: 2082–2091.
15. Bever JD, Schultz PA, Pringle A, Morton JB (2001) Arbuscular mycorrhizal fungi: More diverse than meets the eye, and the ecological tale of why. Bioscience 51: 923–931.
16. Vogelsang KM, Reynolds HL, Bever JD (2006) Mycorrhizal fungal identity and richness determine the diversity and productivity of a tallgrass prairie system. New Phytol 172: 554–562.
17. Martin F, Gianinazzi-Pearson, Hijri M, Lammers P, Requena N, et al. (2008) The long hard road to a completed *Glomus intraradices* genome. New Phytol 180: 747–750.
18. Tisserant E, Kohler A, Dozolme-Seddas P, Balestrini R, Benabdellah K, et al. (2012) The transcriptome of the arbuscular mycorrhizal fungus *Glomus intraradices* (DAOM 197198) reveals functional tradeoffs in an obligate symbiont. New Phytol 193: 755–769.
19. Croll D, Wille L, Gamper HA, Mathimaran N, Lammers PJ, et al. (2008) Genetic diversity and host plant preferences revealed by simple sequence repeat and mitochondrial markers in a population of the arbuscular mycorrhizal fungus *Glomus intraradices*. New Phytol 178: 672–687.
20. Börstler B, Raab PA, Thiéry O, Morton JB, Redecker D (2008) Genetic diversity of the arbuscular mycorrhizal fungus *Glomus intraradices* as determined by mitochondrial large subunit rRNA gene sequences is considerably higher than previously expected. New Phytol: 452–465.
21. Croll D, Giovannetti M, Koch AM, Sbrana C, Ehinger M, et al. (2009) Nonself vegetative fusion and genetic exchange in the arbuscular mycorrhizal fungus *Glomus intraradices*. New Phytol 181: 924–937.
22. Ehinger M, Croll D, Koch MA, Sanders IR (2012) Significant genetic and phenotypic changes arising from clonal growth of a single spore of an arbuscular mycorrhizal fungus over multiple generations. New Phytol 196: 853–861.
23. Angelard C, Colard A, Niculita-Hirzel H, Croll D, Sanders IR (2010) Segregation in a mycorrhizal fungus alters rice growth and symbiosis-specific gene transcription. Current Biology 20: 1216–1221.
24. Colard A, Angelard C, Sanders IR (2011) Genetic exchange in an arbuscular mycorrhizal fungus results in increased rice growth and altered mycorrhiza-specific gene transcription. Appl Environ Microbiol 77: 6510–6515.
25. Stockinger H, Walker C, Schussler A (2009) '*Glomus intraradices* DAOM197198', a model fungus in arbuscular mycorrhizal research, is not *Glomus intraradices*. New Phytol 183: 1176–1187.
26. Krüger M, Krüger C, Walker C, Stockinger H, Schüssler A (2012) Phylogenetic reference data for systematics and phylotaxonomy of arbuscular mycorrhizal fungi from phylum to species level. New Phytol 193: 970–984.
27. Vierheilig H, Coughlan AP, Wyss U, Piche Y (1998) Ink and vinegar, a simple staining technique for arbuscular- mycorrhizal fungi. Appl Environ Microbiol 64: 5004–5007.
28. Giovannetti M, Mosse B (1980) Evaluation of techniques for measuring vesicular arbuscular mycorrhizal infection in roots. New Phytol 84: 489–500.
29. Sanders IR, Croll D (2010) Arbuscular mycorrhiza: The challenge to understand the genetics of the fungal partner. Ann Rev Gen 44: 271–292.
30. Schwartz MW, Hoeksema JD, Gehring CA, Klironomos JN, Johnson NC, et al. (2006) The promise and the potential consequences of the global transport of mycorrhizal fungal inoculum. Ecol Lett 9: 501–515.

24

Extensive Management Promotes Plant and Microbial Nitrogen Retention in Temperate Grassland

Franciska T. de Vries[1]*, Jaap Bloem[2], Helen Quirk[1], Carly J. Stevens[1], Roland Bol[3], Richard D. Bardgett[1]

1 Lancaster Environment Centre, Lancaster University, Lancaster, United Kingdom, **2** Alterra, Wageningen University and Research Centre, Wageningen, The Netherlands, **3** Institute of Bio- and Geosciences, IBG-3: Agrosphere, Forschungszentrum Jülich GmbH, Jülich, Germany

Abstract

Leaching losses of nitrogen (N) from soil and atmospheric N deposition have led to widespread changes in plant community and microbial community composition, but our knowledge of the factors that determine ecosystem N retention is limited. A common feature of extensively managed, species-rich grasslands is that they have fungal-dominated microbial communities, which might reduce soil N losses and increase ecosystem N retention, which is pivotal for pollution mitigation and sustainable food production. However, the mechanisms that underpin improved N retention in extensively managed, species-rich grasslands are unclear. We combined a landscape-scale field study and glasshouse experiment to test how grassland management affects plant and soil N retention. Specifically, we hypothesised that extensively managed, species-rich grasslands of high conservation value would have lower N loss and greater N retention than intensively managed, species-poor grasslands, and that this would be due to a greater immobilisation of N by a more fungal-dominated microbial community. In the field study, we found that extensively managed, species-rich grasslands had lower N leaching losses. Soil inorganic N availability decreased with increasing abundance of fungi relative to bacteria, although the best predictor of soil N leaching was the C/N ratio of aboveground plant biomass. In the associated glasshouse experiment we found that retention of added ^{15}N was greater in extensively than in intensively managed grasslands, which was attributed to a combination of greater root uptake and microbial immobilisation of ^{15}N in the former, and that microbial immobilisation increased with increasing biomass and abundance of fungi. These findings show that grassland management affects mechanisms of N retention in soil through changes in root and microbial uptake of N. Moreover, they support the notion that microbial communities might be the key to improved N retention through tightening linkages between plants and microbes and reducing N availability.

Editor: Han Y.H. Chen, Lakehead University, Canada

Funding: This project was part of the EU 7th Framework funded SOILSERVICE project, led by Katarina Hedlund. The contribution of Wageningen University and Research Centre was supported by the research program KB IV "Innovative scientific research for sustainable green and blue environment" funded by the Netherlands Ministry of Economic Affairs, Agriculture and Innovation. CJS is funded by a Leverhulme Early Career fellowship. The funders had no role in study design, data collection and analysis, decision to publish, or preparation of the manuscript.

Competing Interests: The authors have declared that no competing interests exist.

* E-mail: f.devries@lancaster.ac.uk

Introduction

Humans have doubled the input of nitrogen (N) to the Earth's land surface. The excessive use of fertiliser N has caused severe environmental problems as a result of increased gaseous N emissions from agricultural soils. This increased gaseous N loss due to denitrification contributes to climate change, as N_2O is an approximately 300 times stronger greenhouse gas than CO_2 [1]. Moreover, it also results in increased atmospheric N deposition and excessive N leaching from soils, which cause eutrophication of ground and surface waters, and have led to widespread changes in plant community composition and loss of plant species diversity [2–4]. In addition, although less extensively studied, N enrichment through atmospheric deposition or agricultural management can affect the structure and function of soil microbial communities. For example, chronic N addition has been shown to reduce soil microbial biomass and alter microbial community composition across ecosystems and biomes [5–7], and typically reduce the biomass of decomposer [8,9], arbuscular mycorrhizal [10] and ectomycorrhizal fungi [11], and the abundance of fungi relative to bacteria [6]. Because soil microbes play a major role in regulating processes of N cycling [12,13], such changes in microbial communities will have consequences for the capacity of soils to retain N, and might thus feed back to the N cycle, potentially further increasing N loss from soil. However, our knowledge of the factors that determine soil N retention, and hence the mitigation of soil N loss, is limited, despite the importance of such information for sustainable food production [13].

A long standing notion in soil microbial ecology is that ecosystems with a soil microbial community dominated by fungi have more efficient N cycling than bacterial-dominated systems [14,15]. This concept is based on the general pattern that fungi dominate soils of undisturbed, late-successional systems of low N availability [16], and the knowledge that fungi are more efficient in their resource use than are bacteria [17], thereby slowing down rates of N cycling. Also, because of their filamentous growth form, fungi can access spatially separated C and N [18], and soils with microbial communities dominated by fungi have been shown to immobilise more added N than soils with bacterial-dominated microbial communities [19,20]. However, results from controlled experiments assessing differences in C use efficiency and N immobilisation raise questions about the idea that fungi are more efficient in their C and N use than are bacteria [17,21,22], and it is possible that increased immobilisation of C and N in fungal-dominated soil microbial communities might be more a consequence of the soil conditions that they occur under in the field, rather than a physiological difference between fungi and bacteria. Land use intensification, and especially the application of fertiliser N and tillage, generally leads to a shift from fungal to bacterial dominated soil systems, although this shift is sometimes restricted to top soil [23–26]. In contrast, land use extensification, for instance through the cessation of fertiliser use, reductions in grazing pressure and adoption of no-tillage farming, can cause a shift from bacterial to fungal dominated systems, albeit in the long term [9,25–27]. These increases in the abundance of fungi relative to bacteria due to land use extensification have been linked to more efficient N cycling and lower soil N losses [9,20,28], but direct support for this is lacking and the mechanisms behind the relationship between fungal abundance and soil N retention remain unclear.

Species-rich, extensively managed hay meadows are highly valued ecosystems, and the restoration of grassland biodiversity is an important aim of European Agri-Environmental schemes. A common feature of extensively managed, species-rich grasslands is that they have fungal-based food webs, which is in contrast to more intensively managed grasslands that have bacterial-based food webs [25,29]. Grassland restoration practices such as seed addition, reduced grazing and cutting, and cessation of fertiliser application have been shown to promote the abundance of fungi relative to bacteria [30,31]. In addition, pot experiments based on mesotrophic grasslands have shown that high plant diversity promotes soil fungal biomass with associated increases in soil N retention and reduced N loss [32,33]. The mechanisms for these plant community composition effects on N cycling are unclear, but they likely act through linkages between plants and soil microbes [13]; plant species differentially impact on belowground microbial communities through altering the quality and quantity of organic matter entering soil via root and leaf litter and root exudates [15]. For example, it has been shown that fast-growing plant species characteristic of N-rich grasslands, which produce N rich litter and exudates, select for more bacterial-dominated microbial communities, whereas slow growing species of N-poor conditions, which produce less decomposable litter, select for fungal-dominated microbial communities, both on the level of individual plant species [34], but also on a landscape-scale [35].

Here, we investigated how changes in soil microbial communities resulting from differences in grassland management intensity affect the capacity of plants and soil to retain N. We hypothesised that extensively managed, species-rich grasslands of high conservation value would have lower N loss and greater N retention than intensively managed, species-poor grasslands, and that this would be due to a greater immobilisation of N by the biomass of a more fungal-dominated microbial community in the former. This was tested using mesotrophic grasslands of contrasting management intensity in northern England, and involved a combination of field and laboratory experiments. The field study tested whether extensively managed grasslands have lower soil N leaching losses at the landscape scale, whereas the glasshouse experiment was done to identify, through the addition of ^{15}N-labelled inorganic N, the mechanisms for improved N retention in extensively managed grasslands.

Materials and Methods

Field Sites and Sampling

We sampled mesotrophic grasslands from eleven sites, each with an intensively managed (fertilised and grazed), species poor grassland, and adjacent traditionally managed (unfertilised, extensively grazed and cut), species-rich haymeadow of high conservation value, on the same soils (sandy silt loam) and of similar topography. The 22 mesotrophic grasslands we used were located in northern England in the region of the Yorkshire Dales (mean annual temperature 7.3°C, mean annual precipitation 1382 mm). For details of all sites used, see Table 1. Extensively managed grasslands received no inorganic fertiliser, whereas intensively managed grasslands received >100 kg N ha^{-1} y^{-1}. In general, plant communities of species-rich grasslands were *Anthoxanthum odoratum-Geranium sylvaticum* grassland (MG3 or subcategories), and plant communities of intensively managed grasslands were *Lolium perenne-Cynosurus cristatus* grassland (MG6, MG7, and subcategories), according to the UK National Vegetation Classification of Rodwell [36]. In each of the 22 fields, three 1 m^2 plots were randomly chosen within a central 25×25 m plot to avoid any edge effects. From these 1 m^2 plots, a composite bulk soil sample was taken for assessment of moisture content, microbial biomass and community, carbon (C) and N availability, and total C and N pools. From all fields sampled, one intact soil column (18 cm height, 12 cm diameter) was taken from the centre of each plot for a field-based leaching measurement.

Permission for sampling locations was obtained from all land owners. These field studies did not involve endangered or protected species.

Glasshouse Experiment and Sampling

Four intact soil columns (one for each treatment–control and ^{15}N addition, destructive sampling after 48 hours and two months) were taken per plot from a subset of eight grasslands (four intensively and four extensively managed, Table 1) for the glasshouse experiment (with three plots per grassland this resulted in 96 columns). Columns were arranged in a randomized block design and kept in the glasshouse for one month prior to the start of the experiment, during which time they were kept at 60% water holding capacity to standardise initial soil moisture content. Two weeks before the start of the experiment, a vegetation survey was done for all columns. One week before ^{15}N addition, aboveground vegetation was cut to 4 cm, and the bottom of the columns was sealed to prevent unwanted leaching. Twenty-five ml of $^{15}NH_4{}^{15}NO_3$ solution (99.5% enriched, 24.5 mg ^{15}N $column^{-1}$, equals 30 kg ^{15}N ha^{-1}), and demineralised water for control treatments, was injected in the top 5 cm of each column at five, evenly spaced, locations (5 ml each). Forty-eight hours and 60 days after ^{15}N vs. demineralised water injection, columns were leached in the same way as field columns, weighed, and dismantled, after which all aboveground vegetation was clipped. Columns were then divided in two: one half was used to determine root biomass, whereas the other half was used for soil and microbial analyses.

Leachate, Soil, Vegetation, and Microbial Analyses

Columns used for the field-based leaching measurement were kept cool and immediately leached on return to the laboratory, by slowly adding 330 ml of demineralised water (equal to a 40 mm rainfall event or a heavy summer thunderstorm (MetOffice, 2012)). Columns from the glasshouse experiment were leached in the same way, 48 hours and 60 days after ^{15}N addition. Leachate volumes were recorded. Vegetation was clipped and columns were divided in two for root and soil analyses. Leachates, field soil, and

Table 1. Characteristics of the field sites used for the sampling and the glasshouse experiment; data from De Vries, et al [35].

Site	Management	Soil type	Vegetation[1]	Latitude	Longitude	Altitude (m.a.s.)	pH[2]	^{15}N exp.
Askrigg Bottoms	Extensive	Sandy silt loam	MG3b	54.308398	−2.074833	307	6.0	Yes
	Intensive	Sandy silt loam	MG7c	54.308404	−2.071801	292	6.0	Yes
Waldendale	Extensive	Homose sandy silt loam	MG3b	54.206064	−2.144537	398	6.3	Yes
	Intensive	Sandy silt loam	U4b	54.244598	−1.987352	335	5.3	Yes
Yockenthwaite	Extensive	Sandy silt loam	MG3b	54.237597	−1.991726	336	5.1	Yes
	Intensive	Sandy silt loam	MG7b	54.204159	−2.1443	398	5.1	Yes
Muker	Extensive	Sandy silt loam		54.379048	−2.138109	490	5.2	Yes
	Intensive	Sandy silt loam	MG3a	54.377735	−2.138936	490	5.5	Yes
Thornton Rust	Extensive	Sandy silt loam	MG4	54.237597	−1.991726	307	5.62	
	Intensive	Sandy silt loam	MG7a	54.204159	−2.1443	307	5.7	
Church and Middlethorpe Ings	Extensive	Sandy silt loam	MG4	53.902989	−1.098635	10	6.94	
	Intensive	Sandy silt loam	OV29	53.905474	−1.093162	10	5.99	
Wheldrake and Storwood Ings	Extensive	Sandy silt loam	MG4	53.890228	−0.926907	33	6.4	
	Intensive	Clay loam	MG6a	53.880685	−0.922468	31	7.44	
Melbourne and Thornton Ings	Extensive	Sandy silt loam	MG4	53.891644	−0.847229	41	7.63	
	Intensive	Sandy loam	MG6a	53.891616	−0.847169	41	7.28	
East Cottingwith Ings	Extensive	Clay loam	MG4	53.857938	−0.944019	25	5.43	
	Intensive	Clay loam	MG7c	53.856805	−0.942937	25	6.63	
Thorganby and East Cottingwith Ings	Extensive	Clay loam	MG4	53.862402	−0.940607	28	5.16	
	Intensive	Clay loam	OV29	53.867402	−0.94702	28	5.26	
Selside	Extensive	Homose sandy silt loam	MG5b	54.168173	−2.340677	354	6.71	
	Intensive	Clay loam	MG6a	54.17053	−2.3397	334	5.1	

[1]According to Rodwell [36].
[2]Measured on a field level in 2005.

column soil were analysed for concentrations of inorganic N and dissolved organic carbon (DOC) and total N, as described by Gordon et al. [28]. Vegetation samples were dried at 60°C, weighed, ground, and analysed for total C and N content using an Elementar Vario EL elemental analyzer (Hanau, Germany). For the glasshouse experiment, 200 ml of leachate was freeze dried for ^{15}N analysis. Microbial biomass C and N in field and column soil was determined by fumigation extraction as described by Brookes et al. [37]. For the glasshouse experiment, all N in microbial extracts was converted into ammonium by Kjeldahl digestion, which were then used to diffuse N onto an acid trap. Leachate (after freeze-drying), shoot, root, total soil, and microbial biomass ^{15}N (determined by diffusing microbial derived N onto an acid trap) were analysed using a Carlo Erba NA2000 analyser (CE Instruments, Wigan, UK) and a SerCon 20–20 isotope ratio mass spectrometer (SerCon Ltd, Crewe, UK) at Rothamsted Research, North Wyke. A dried and ground grass herbage sample labelled with ^{15}N (2.79 atom % ^{15}N) or natural abundance wheat flour (0.368 atom % ^{15}N), both calibrated against IAEA-N-1 by Iso-Analytical, Crewe, UK, were used as the references for enriched or natural abundance samples respectively. ^{15}N excess atom percent values for enriched samples were calculated using mean ^{15}N atom percent values of unlabelled samples [38]. These values were then used to calculate ^{15}N concentrations in samples, and total amounts of ^{15}N in pools were calculated using total pool sizes in columns [19], which were then scaled to kilograms per hectare using the surface area of columns.

In field samples, the biomass and structure of the soil microbial community was assessed by PLFA analysis. The fatty acids i-15:0, a-15:0, 15:0, i-16:0, 17:0, cyclo-17:0, 18:1ω7 and cyclo-19:0 were chosen to represent bacterial PLFA, 10-methyl 18:0 and 10-methyl 16:0 represent actinomycetes, 18:1ω9 represents eukaryotes, and PLFA 18:2ω6 was used as an indicator of fungal biomass [39]. The ratio of fungal to bacterial PLFA was used as an indicator of changes in the relative abundance of these two microbial groups [40]. Shifts in microbial community composition were assessed using PCA of relative abundances of all PLFAs, and Simpson's evenness was calculated of PLFA profiles [41]. In addition, fungal and bacterial biomass were determined by epifluorescence microscopy [20]. Briefly, microscopic slides for counting fungi were stained with Differential Fluorescent Stain solution, and slides for counting bacteria were stained with the fluorescent protein dye 5-(4,6-dichlorotriazin-2-yl) aminofluorescein. Hyphal length was measured using an epifluorescence microscope at 400× magnification. Total hyphal length was calculated using the grid intersection method [42]. Fungal biomass was calculated assuming a mean hyphal diameter (width) of 2.5 mm and a specific C content of 1.3 $\times 10^{13}$ g C mm^3 [43,44]. Bacterial numbers, cell volumes and number of dividing cells were measured automatically with a confocal laser-scanning microscope (Leica TCS SP2) combined with image analysis software (Leica Qwin Pro), as described by [45]. Bacterial biomass (C) was estimated from the biovolume using a specific C content of 3.1 $\times 10^{13}$ g C mm^3 [46].

Statistical Analysis

Data were log-transformed to meet assumptions of normality. Field data were analysed using linear mixed effects models with a field level random effect to account for the nesting in fields, and true significant values were obtained by a likelihood ratio test (LRT) [47]. Apart from testing for effects of management on N and C leaching, we selected the models best explaining field N and C leaching. Parameters added to the models for nitrate availability and N leaching included management, soil properties (total C, total N, C/N ratio), plant and root properties (biomass, total N, total C, C/N ratio), and microbial properties (microbial biomass C and N, fungal PLFA, bacterial PLFA, F/B PLFA ratio, PC axis 1 (PC1) scores of all PLFAs, PC axis 2 (PC2) scores of all PLFAs, PLFA evenness). Parameters retained in the models were selected based on Akaike's Information Criterion (AIC) and estimated significance values, after which true significance of included parameters was obtained by an LRT [48]. Glasshouse data were analysed for treatment effects using ANOVA with an error term to account for the nesting in block and field. In addition, glasshouse data were analysed using linear mixed effects models to find the parameters that best explained ^{15}N leaching. Parameters added to this model were management, soil properties (total C, total N, C/N ratio), plant and root properties (biomass, total N, total C, C/N ratio), and microbial properties (microbial biomass C and N, fungal PLFA, bacterial PLFA, F/B PLFA ratio, PC1 scores, PC2 scores, PLFA evenness). Similar to the field data, parameters included were selected based on AIC and estimated significance values, after which true significance of included parameters was obtained by an LRT. R-squared values were obtained by regressing saved model predictions against observed values. For all statistical analyses, *P*-values smaller than 0.05 were considered significant. All analyses were done in R version 2.12.1 (R Development Core Team 2009).

Results

Field Sampling

First, we examined differences in soil and microbial properties, and C and N leaching, between the different management intensities. Intensively and extensively managed grasslands differed in some soil, microbial, and vegetation properties, but were similar for others. Inorganic N leaching was greater in intensively than in extensively managed grasslands (L-ratio = 8.51, $P = 0.0035$), whereas total N leaching, DOC leaching, and soil inorganic N did not differ between the two grassland types (Table 2). Total soil N content, and soil, root, and shoot C/N ratios did not differ between grasslands, although total soil C content and microbial biomass C and N tended to be greater in extensively than in intensively managed grasslands, albeit non significantly.

Bacterial biomass measured as PLFA tended to be greater in extensively managed grasslands (L-ratio = 2.77, $P = 0.096$), whereas fungal PLFA was significantly (L-ratio = 17.7, $P<0.0001$) greater, and more than twice as high, in extensively than in intensively managed grasslands. As a consequence, the F/B PLFA ratio was greater in extensively than in intensively managed grasslands (L-ratio = 24.9, $P<0.0001$, Table 2). Surprisingly, microbial community composition, as assessed by PCA of all PLFAs, was not affected by management intensity, and neither was evenness of PLFA profiles (Fig. 1, Table 2). Bacterial biomass, as measured by microscopy, tended to be greater in extensively managed fields (L-ratio = 2.64, $P = 0.105$). However, fungal biomass measured by microscopy did not differ between extensive and intensive management, and as a result the F/B biomass ratio did not differ (Table 2).

Table 2. Differences in C and N leaching, and soil, vegetation, and microbial properties between intensively and extensively managed grasslands.

	Intensive	Extensive	L-ratio	*P*
Inorganic N leached (kg ha^{-1})	0.38 (0.07)	0.08 (0.02)	8.51	0.0035
Total N leached (kg ha^{-1})	1.29 (0.12)	1.00 (0.14)	1.93	0.165
DON leached (kg ha^{-1})	0.91 (0.08)	0.93 (0.14)	0.91	0.340
DOC leached (kg ha^{-1})	2.12 (0.19)	1.82 (0.16)	0.78	0.376
Soil inorganic N (mg kg^{-1})	16.9 (1.5)	16.5 (3.3)	1.58	0.209
Total soil C (mg g^{-1})	72.8 (2.5)	84.7 (5.7)	2.87	0.090
Total soil N (mg g^{-1})	6.74 (0.26)	7.59 (0.56)	1.72	0.190
Soil C/N ratio	10.8 (0.1)	11.4 (0.2)	1.72	0.189
Root C/N ratio	22.5 (1.1)	25.3 (1.3)	1.79	0.181
Shoot C/N ratio	19.7 (1.4)	22.5 (1.2)	1.22	0.269
Root biomass (kg m^{-2})	1.44 (0.15)	2.21 (0.17)	4.68	0.031
Microbial biomass C (μg g^{-1})	803 (96)	1175 (122)	2.54	0.111
Microbial biomass N (μg g^{-1})	277 (25)	387 (34)	2.82	0.093
Bacterial PLFA (nmol g^{-1})	72.2 (4.4)	92.8 (6.3)	2.77	0.096
Fungal PLFA (nmol g^{-1})	2.02 (0.22)	5.28 (0.40)	17.65	<0.0001
F/B PLFA ratio	0.027 (0.002)	0.061 (0.004)	24.93	<0.0001
PLFA evenness	0.813 (0.002)	0.815 (0.002)	0.20	0.658
Bacterial biomass (μg C g^{-1})	89.7 (5.7)	111.7 (7.5)	2.64	0.105
Fungal biomass (μg C g^{-1})	61.7 (6.9)	78.0 (7.6)	0.03	0.860
F/B biomass ratio	0.78 (0.11)	0.77 (0.08)	0.03	0.860

Values denote means (1SE), n = 66.

Second, we used model selection (see Methods) to explore which parameters best explained C and N leaching in the field. This resulted in two satisfactory models for explaining leaching losses of N and C from soils. Inorganic N leaching was strongly explained by a model containing management intensity, (log-transformed) shoot C/N ratio, and the interaction term between the two: inorganic N leaching decreased with greater shoot C/N ratio, but only in extensively managed grasslands (Table 3, Fig. 2A). Although the models selected for DON leaching and total N leaching were significant, their explanatory power was very low ($R^2 = 0.04$ and $R^2 = 0.09$ respectively, Table 2). Leaching of DON was explained by a combination of management intensity, soil C/N ratio and shoot C/N ratio, and total N leached was explained by only one factor, namely soil C/N ratio. The model for DOC leaching was similar to the model explaining inorganic N leaching, with the exception that a term for microbial community was included, namely microbial biomass C (Table 3). Although inorganic N leaching was not explained by microbial community characteristics, soil NO_3^- concentration was strongly explained by a combination of soil C/N ratio (Parameter Value (PV) = −2.16, $P = 0.0025$) (log-transformed), F/B PLFA ratio (PV = −0.54, $P = 0.01$), and shoot C/N ratio (PV = −0.65, $P = 0.0004$) ($R^2 = 0.48$, Fig. 2B). A second model also including soil C/N ratio (PV = −2.84, P = 0.0004), shoot C/N ratio (PV = −0.73,

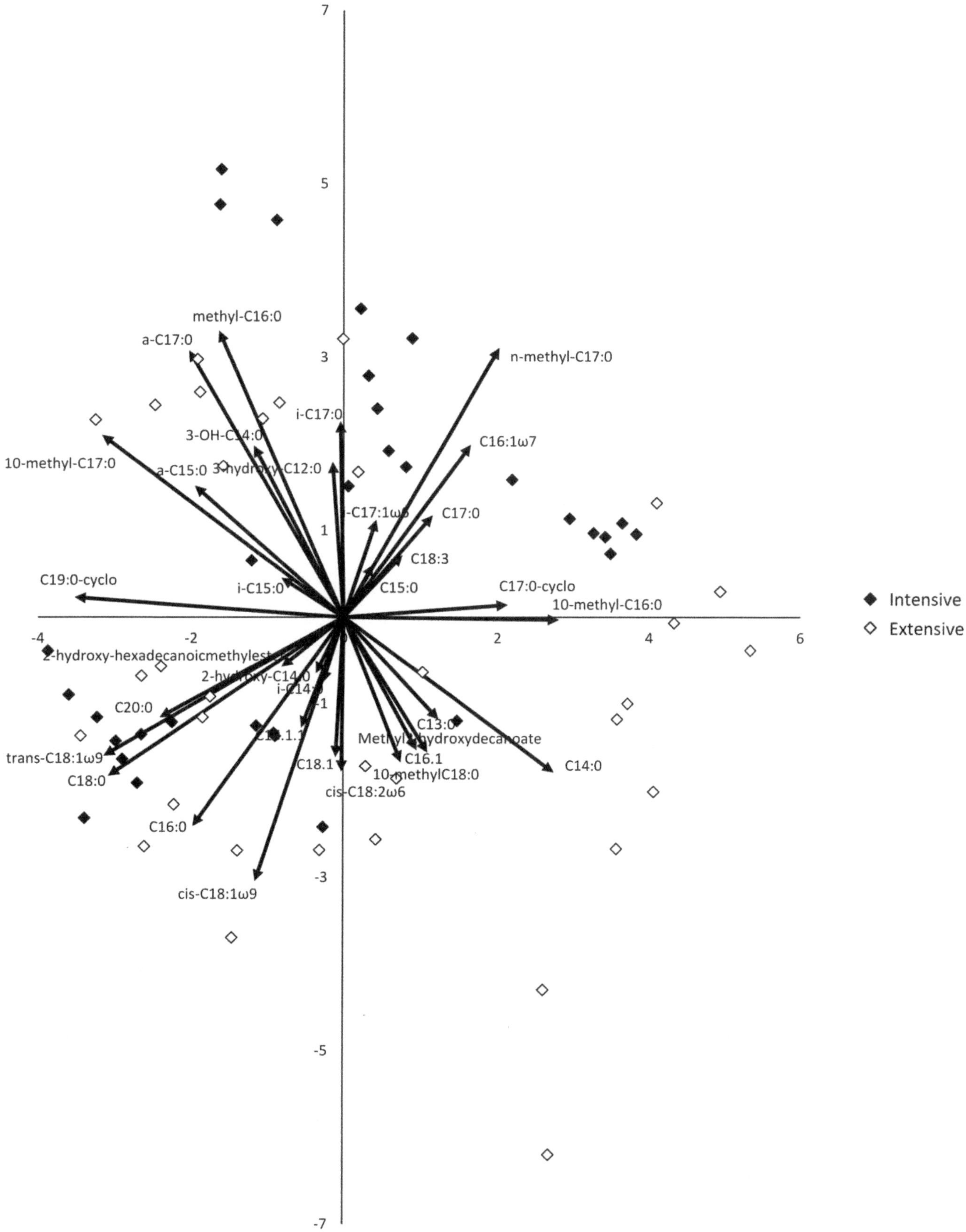

Figure 1. Principal components analysis (PCA) of the relative abundance of all PLFAs. PCA axis 1 explains 19.5% and PCA axis 2 explains 16% of variation in microbial community composition. Microbial community composition was not affected by grassland management.

$P = 0.0002$), but including PCA axis 1 scores for all PLFAs ($PV = 0.05$, $P = 0.008$), along which the relative abundance of actinomycetes increased (Fig. 1), explained a similar amount of variation ($R^2 = 0.47$). In addition, inorganic N leaching was strongly explained by soil NO_3^- concentration ($PV = 1.06$, $P = 0.0029$, $R^2 = 0.31$).

Glasshouse Experiment

As for the field sampling, we first tested whether N leaching losses, and uptake of added ^{15}N into different pools, differed between management intensities. Leaching of inorganic N in the glasshouse experiment was lower ($F_{1,86} = 17.7$, $P < 0.0001$) from columns from extensively than from intensively managed grassland (Fig. 3). Total N leached from columns during the glasshouse experiment was strongly related to total N leaching in the field ($P < 0.001$, $R^2 = 0.50$). There was a weak trend ($F_{1,42} = 2.16$, $P = 0.149$) towards lower ^{15}N loss from columns from extensively managed grasslands (Fig. 4A), but the total amount of added ^{15}N leached did not differ between management types. However, both 48 hours and two months after addition of ^{15}N, significantly ($F_{1,40} = 7.5$, $P = 0.003$) more added ^{15}N was immobilised by the microbial biomass in extensive than intensive management (Fig. 4B). Roots took up the largest amount of ^{15}N, and significantly ($F_{1,42} = 6.9$, $P = 0.01$) more so in columns from extensively than intensively managed grasslands; this pool only decreased slightly over time (Fig. 4C). In contrast, shoot uptake did not differ between the two management intensities and increased towards the end of the experiment (Fig. 4D). Taken together, the amount of added ^{15}N retained in microbial, soil, and aboveground and belowground vegetation pools was greatest in extensively managed grasslands ($F_{1,40} = 5.7$, $P = 0.02$, Fig. 4E). In both systems, total retention of ^{15}N did not decrease towards the end of the experiment.

Second, we selected the models that best explained leaching and retention of ^{15}N in the soil columns of the glasshouse experiment. Leaching of ^{15}N was found to decline with increasing (log-transformed) abundance of fungi relative to bacteria across all samples at both sampling dates (Sampling date $P = 0.0006$, Fungal/Bacterial (F/B) ratio $P = 0.0003$, Sampling date × F/B ratio $P < 0.0001$, $R^2 = 0.72$, Fig. 5A). A model including sampling date and PC2 scores explained less variation, but showed a similar pattern (Sampling date $P < 0.0001$, PC2 $P = 0.011$, Sampling date × PC2 $P = 0.06$, $R^2 = 0.58$); leaching of ^{15}N increased with increasing PC2 scores, along which the relative abundance of fungal PLFA 18:2ω6 decreased (Fig. 2). Immobilisation of added ^{15}N into microbial biomass increased with increasing (log-transformed) fungal biomass (Sampling date $P = 0.03$, Fungal PLFA $P < 0.0001$, Sampling date × Fungal PLFA $P = 0.0001$, $R^2 = 0.58$, Fig. 5B). In addition, the retention of added ^{15}N increased with greater (log-transformed) fungal biomass (Fig. 5C). Similar to the field sampling, leaching of ^{15}N was explained by the (log-transformed) C/N ratio of aboveground biomass, but this model explained a smaller part of the variation in ^{15}N leached than the model that included F/B ratio ($R^2 = 0.67$, Fig. 5D).

Total recovery of added ^{15}N in soil, vegetation and leachates was greater in columns from extensive than from intensive management (76±5% vs. 64±5%, respectively), but was not affected by sampling date. Although grassland communities in the two management types were different, the number of plant species, and the abundance of legumes, grasses, and herbs, did not differ between the columns taken from the two grassland types in the glasshouse experiment (data not shown).

Discussion

We hypothesised that N leaching would be lower from extensively managed, species-rich grasslands than from intensively managed, species-poor grasslands, and that this would be because of a greater immobilisation of available N into microbial biomass in more fungal-dominated soils of the former. In the field experiment, extensively managed grasslands showed less inorganic N leaching than intensively managed grasslands, and the amount of inorganic N leached was best explained by a combination of grassland management and the C/N ratio of aboveground vegetation. In the glasshouse experiment, we found that extensively managed grasslands had greater retention of added ^{15}N than

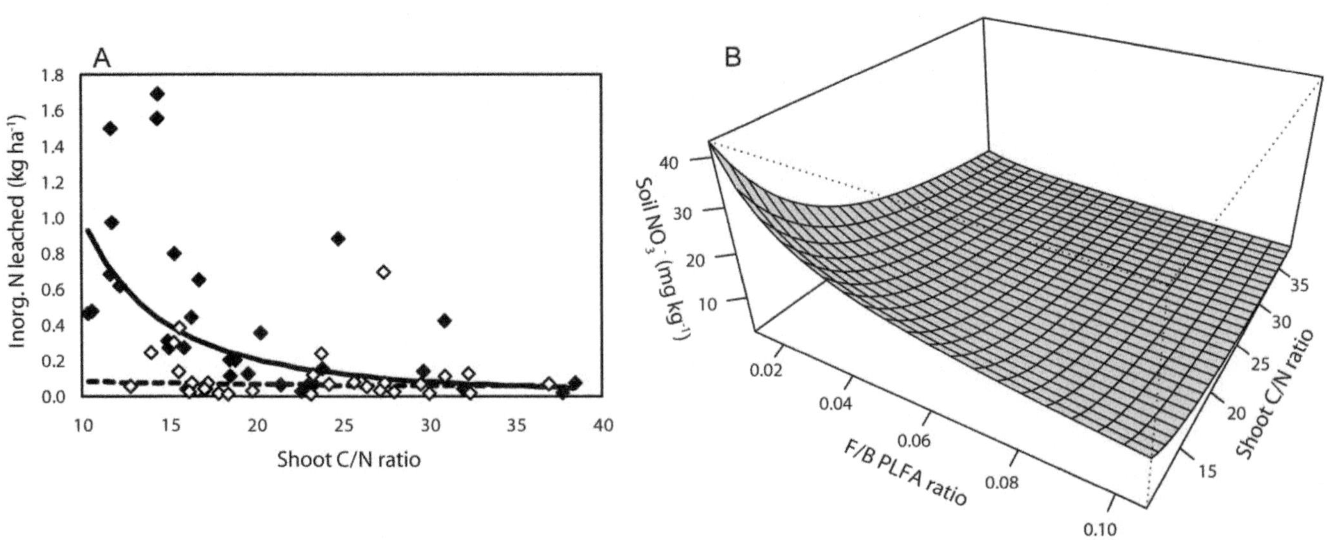

Figure 2. Soil inorganic N availability and inorganic N leaching from soil in the field sampling. A, Inorganic N leached in the field as explained by shoot C/N ratio in intensive (filled symbols) and extensive (open symbols) grasslands. B, Modelled relationship between soil nitrate availability, shoot C/N ratio and F/B ratio in the field. Soil C/N ratio was kept constant in the model. Soil C/N ratio $P = 0.0025$, Shoot C/N ratio $P = 0.0004$, F/B ratio $P = 0.01$. Variables were log-transformed, but axes represent true values.

Table 3. Selected models for inorganic N leached, DON leached, total N leached, and DOC leached in the field sampling.

	Inorganic N leached (kg ha^{-1})		DON leached (kg ha^{-1})		Total N leached (kg ha^{-1})		DOC leached (kg ha^{-1})	
	Parameter Value	***P***	**Parameter Value**	***P***	**Parameter Value**	***P***	**Parameter Value**	***P***
Intercept	+5.27	0.0059	−3.24	0.68	+7.55	0.037	+0.77	0.43
Management	−7.07*E	0.019	+20.2*E	0.029			−3.3*E	0.012
Soil properties			+1.42*soil C/N	0.64	−3.19*soil C/N	0.03		
			−11.6*E*soil C/N	0.0016				
Vegetation properties	−2.29*shoot C/N	0.0006	−0.13*shoot C/N	0.82			−0.73*shoot C/N	0.0087
	+1.99*E*shoot C/N	0.044	+2.4*E*shoot C/N	0.004			+1.0*E*shoot C/N	0.017
Microbial community							+0.31*microbial C	0.032
R-squared	0.46		0.04		0.09		0.28	

E = extensive management.

intensive grasslands, because of a combination of greater root uptake and microbial immobilisation of ^{15}N.

In the field sampling, and in accordance with our hypothesis, the fungal to bacterial PLFA ratio, and PCA axis 1 scores along which the relative abundance of actinomycetes increased, explained a significant portion of variation in soil NO_3^- concentration, which is highly prone to leaching. However, inorganic N leaching was best explained by the C/N ratio of aboveground plant biomass, but only in intensively managed grasslands, which were more variable in their N leaching (Table 1, Fig. 1A). The C/N ratio of aboveground plant biomass most likely reflects differences in management within the improved grasslands, rather than differences in plant community composition, which was consistent within grassland type. There is growing evidence that plant traits, such as leaf N content, exert a strong control on belowground processes through altering the quality and quantity of organic matter entering soil [49,50]. For instance, slow-growing plants with low leaf N content that are adapted to low-fertility conditions have been shown to decrease rates of nitrification [51] and N mineralisation [34], and to select for a more fungal-dominated microbial community [34,35], which should decrease rates of N cycling even more [15]. Our results, therefore, point to an indirect link between plant traits, in this case leaf C/N ratio, and processes that govern N availability and leaching from soil at the field-scale. Although leaching of DOC did not differ between the two grassland types, it decreased with increasing C/N ratio of aboveground biomass similar to inorganic N leaching, which further points to the importance of plant traits for processes of C cycling. Moreover, DOC leaching increased with greater microbial biomass C, which is consistent with the notion that the microbial biomass stimulates decomposition, and thus the

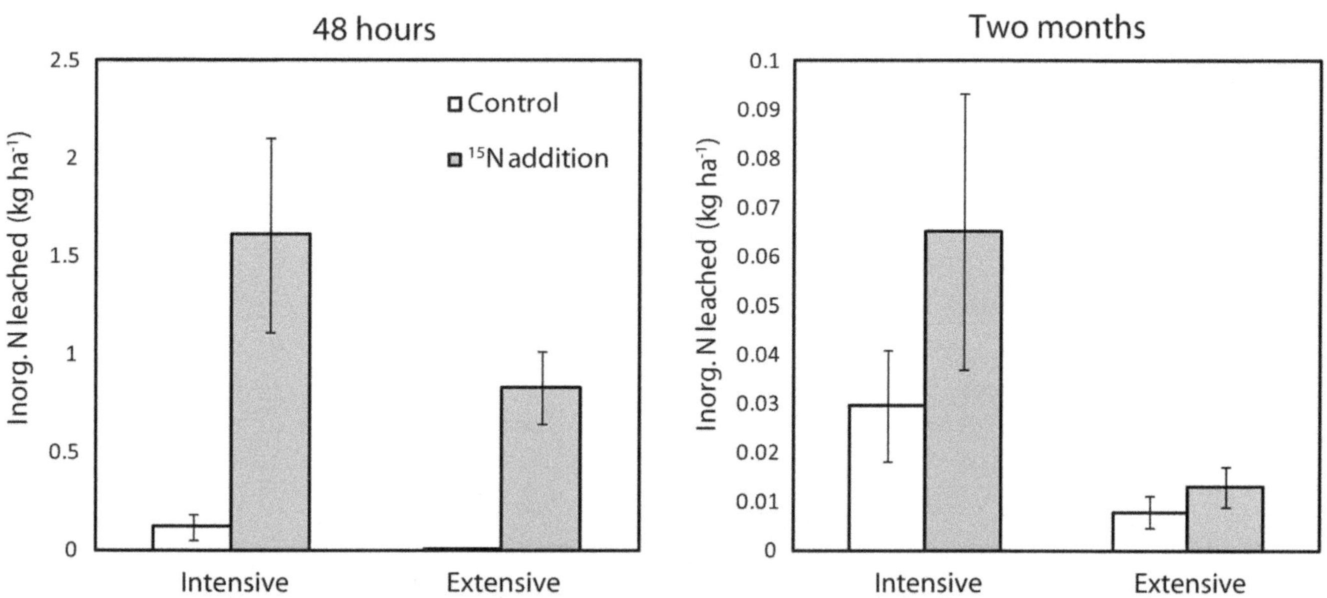

Figure 3. Total amounts of inorganic N leached in intensive vs. extensive soils in the glasshouse experiment, as affected by ^{15}N addition and sampling date. Management $F_{1,86} = 17.7$, $P<0.0001$, N addition $F_{1,86} = 75.7$, $P<0.0001$, Sampling date $F_{1,86} = 21.9$, $P<0.0001$, N addition × Sampling date $F_{1,86} = 45.4$, $P<0.0001$. Bars represent means (n = 12) ±1SE.

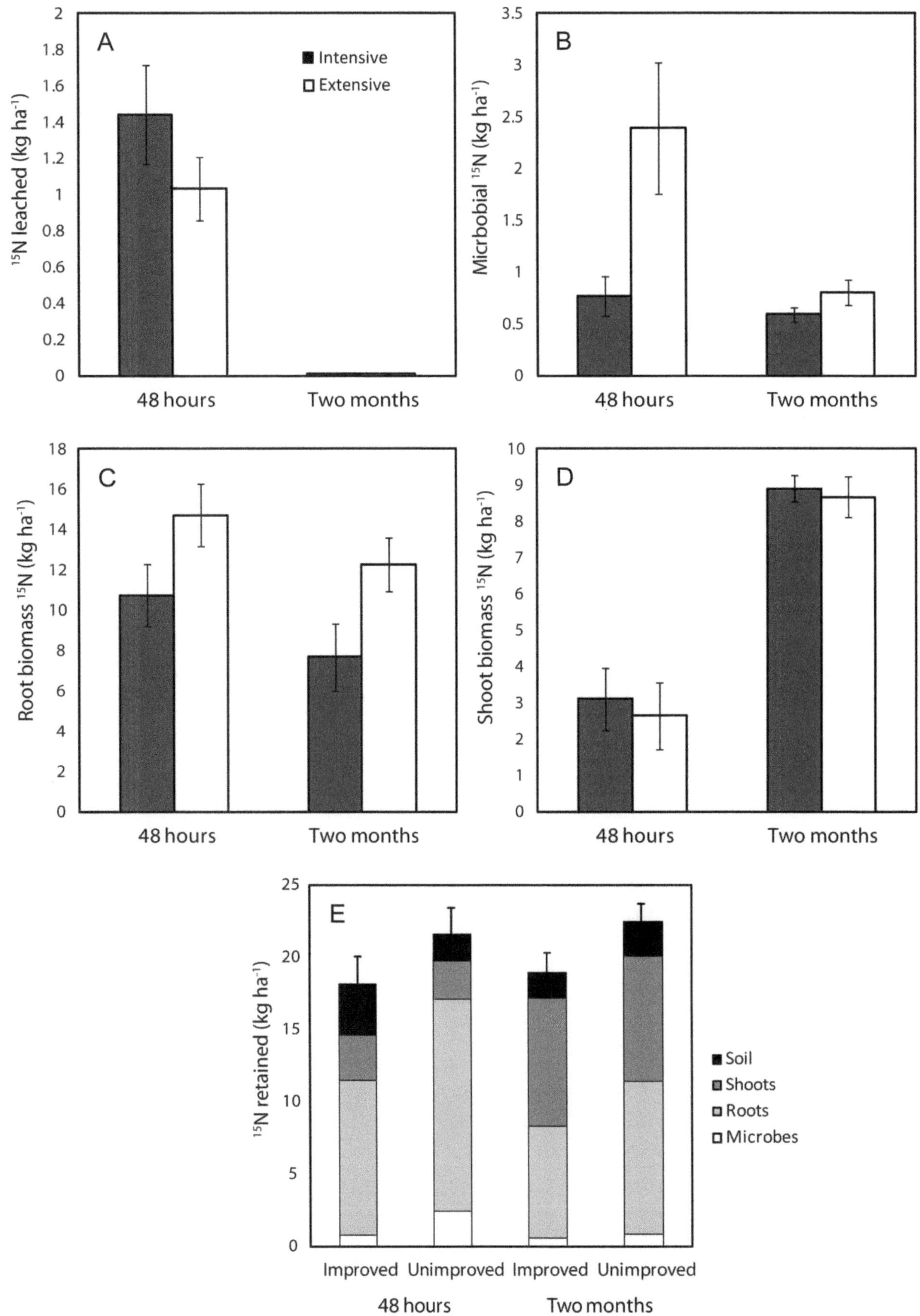
A
Intensive
Extensive
15N leached (kg ha-1)
48 hours
Two months
B
Micrbobial 15N (kg ha-1)
C
Root biomass 15N (kg ha-1)
D
Shoot biomass 15N (kg ha-1)
E
15N retained (kg ha-1)
Soil
Shoots
Roots
Microbes
Improved
Unimproved
48 hours
Two months

Figure 4. ^{15}N pools in intensive (black bars) vs. extensive grasslands, 48 hours and two months after ^{15}N addition. A, ^{15}N leached (Management $F_{1,42}=2.15$, $P=0.15$, Sampling date $F_{1,42}=58.1$, $P<0.0001$, Management × Sampling date $F_{1,42}=1.61$, $P=0.21$); B, ^{15}N uptake in microbial biomass (Management $F_{1,40}=7.5$, $P=0.003$, Sampling date $F_{1,40}=9.7$, $P=0.009$, Management × Sampling date $F_{1,40}=5.2$, $P=0.03$); C, ^{15}N in roots (Management $F_{1,42}=6.9$, $P=0.01$, Sampling date $F_{1,42}=3.1$, $P=0.08$, Management × Sampling date $F_{1,42}=0.03$, $P=0.85$); D, ^{15}N in aboveground plant biomass (Management $F_{1,42}=0.06$, $P=0.80$, Sampling date $F_{1,42}=59.6$, $P<0.0001$, Management × Sampling date $F_{1,42}=0.03$, $P=0.87$). E, amount of ^{15}N retained in the different pools, after 48 hours and two months (Management $F_{1,40}=5.7$, $P=0.02$, Sampling date $F_{1,40}=0.2$, $P=0.69$, Management × Sampling date $F_{1,40}=0.005$, $P=0.94$). Bars represent means (n = 12) ±1SE.

availability of DOC in soil, or that a C-rich soil sustains a greater microbial biomass [52,53].

Surprisingly, although fungal PLFA and the F/B PLFA ratio were significantly greater in extensively managed than intensive grasslands, as has been shown previously [25,29], fungal biomass and the F/B ratio, measured by microscopy, were not (Table 1). In contrast, bacterial PLFA and bacterial biomass, as measured by microscopy, both tended to be greater in extensive grasslands, and by the same magnitude (25%). An explanation for the discrepancy between fungal PLFA and microscopic counts of fungal hyphae could be that a large part of hyphae visible through a microscope might be inactive or dead [54], although the assumption that PLFAs degrade more rapidly than cell walls, and thus represent active biomass more accurately than microscopy, has been challenged [55]. Although all of our samples were treated and stored in a similar way, storage might have resulted in differences in decay of fungal hyphae and PLFAs, although this is hard to judge given that very little is known about the impacts of pre-

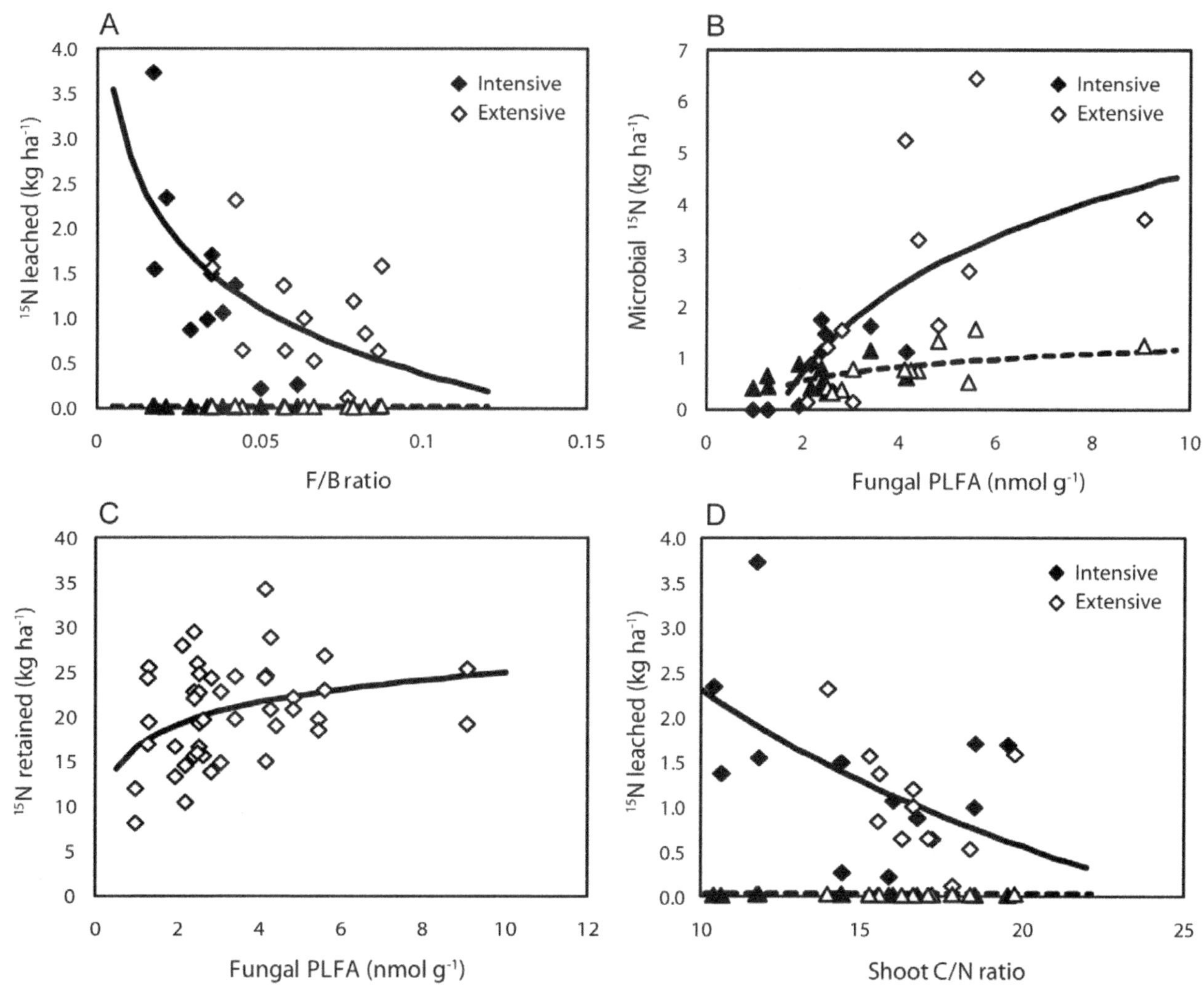

Figure 5. ^{15}N leaching and microbial ^{15}N immobilisation in the glasshouse experiment. A, ^{15}N leaching in the glasshouse experiment as explained by F/B ratio. Sampling date P = 0.0006, F/B ratio P = 0.0003, Sampling date × F/B ratio P<0.0001, $R^2=0.72$. B, Microbial ^{15}N uptake as explained by fungal PLFA. Sampling date P = 0.03, Fungal PLFA P<0.0001, Sampling date × Fungal PLFA P = 0.0001, $R^2=0.58$. C, ^{15}N retention in the glasshouse experiment across both sampling dates as explained by fungal PLFA (P = 0.03, $R^2=0.12$). D, ^{15}N leaching in the glasshouse experiment as explained by shoot C/N ratio. Sampling date P = 0.0006, Shoot C/N ratio P = 0.0001, Sampling date × Shoot C/N ratio P<0.0029, $R^2=0.67$. Analyses were done on log-transformed data, but axes represent true values. Filled symbols represent improved grasslands, open symbols unimproved grasslands; diamonds represent 48-hour-sampling (except for 4C, where sampling dates are pooled), triangles two-month-sampling. Solid lines are the predicted relationship for 48-hour-sampling, dashed lines are predicted relationships for two-month-sampling.

treatment of soil samples on these methods [55,56]. Furthermore, a general problem with PLFAs is that species composition within groups cannot be detected (for example within decomposer fungi, which are all represented by PLFA 18:2ω6), while different species within a group might differ in their PLFA content [56], and fungal communities are likely to be impacted by grasslands management [57]. Another possibility is that fungi in extensive grassland had thicker hyphae, and thus greater membrane surface and PLFA; however, then also greater microbial biomass C would have been found. Furthermore, the PLFA 18:2ω6 only includes decomposer fungi, while the microscopic measure also includes mycorrhiza. Although not measured here, arbuscular mycorrhizal fungi can be measured by quantifying the PLFA 16:1ω5, although this PLFA also occurs in Gram-negative bacteria [58]. Thus, a combination of different methods is needed for a complete picture.

In the glasshouse experiment, and in support of our hypothesis, significantly more added ^{15}N was immobilised into microbial biomass in extensively managed than intensive grassland soil, and ^{15}N immobilisation into microbial biomass increased with increasing fungal biomass (measured as PLFA). In addition, leaching of ^{15}N declined with increasing abundance of fungi relative to bacteria (F/B ratio, measured as PLFA). Although it has been suggested that fungi would immobilise available N more efficiently than bacteria, it is not possible to distinguish between ^{15}N immobilised by bacteria and fungi. Therefore, although in the current study we cannot elucidate the exact mechanism, our results suggest that a greater fungal abundance is linked to increased soil N retention, a key ecosystem service in grassland. Greater microbial immobilisation of added N in extensively compared to intensively managed grasslands [19], and in fungal-dominated microbial communities compared to bacterial-dominated microbial communities [20], has been shown previously; but here we provide the first evidence that this greater microbial immobilisation of N is linked to smaller N leaching losses across a range of grassland sites.

The amount of ^{15}N immobilised by microbes in extensively managed grassland soil was twice as high as the amount leached after 48 hours, which shows that microbes can be a significant short-term N sink in grassland (Bardgett et al. 2003). Roots took up the largest amount of added ^{15}N, however, and significantly more so in columns from extensively managed than intensive grasslands; this pool only decreased slightly over time. Root biomass did not differ between the two grassland types in the glasshouse experiment (whereas it did in the field sampling). Therefore, arbuscular mycorrhizal fungi, which were not measured in this study, may have contributed to greater root N uptake, and also to greater N immobilisation in microbial biomass, in the extensively managed grassland. Indeed, it is known that arbuscular mycorrhizal fungi are adversely affected by intensive grassland management, including liming and fertilisation [59], and they have been shown to reduce N leaching, albeit under highly artificial conditions, and have been suggested to significantly contribute to ecosystem N retention [13]. However, as far as we are aware, there is no experimental evidence that AMF reduce N leaching under field conditions, so more work is needed to quantify their role in N uptake and recycling in grasslands [13,29,60,61].

The retention of ^{15}N was significantly greater in extensively managed grasslands than in intensive grasslands. Importantly, in both systems, the total retention of ^{15}N did not decrease towards the end of the experiment, which suggests that the immediate N uptake in the different pools determines longer-term ecosystem N retention in mesotrophic grasslands. This is in sharp contrast with earlier results from a forest ecosystem [62], where the added N retained in soil pools after 16 weeks was only a quarter of the amount retained immediately after addition, although here aboveground N uptake was not measured. Similarly, in a study comparing N retention in urban lawns and forest, the amount of added N retained in the system after 70 days was significantly lower than the retention after one day [63]. In our experiment, ^{15}N in aboveground plant biomass showed a three-fold increase during the two months of our experiment, indicating a transfer of retained ^{15}N from belowground to aboveground pools. Differences in soil N retention and recovery of added ^{15}N can also be a consequence of differences in gaseous N losses, which can make up a substantial amount of total N lost from soil. Our results of greater recovery of added ^{15}N in extensively managed grassland soil are in line with previous findings of smaller recovery of ^{15}N and greater N loss through denitrification in soils with bacterial-dominated microbial communities [20].

In conclusion, the results from our field sampling show that extensively managed, species-rich grasslands of high conservation value have lower leaching of inorganic N than agriculturally improved, species poor grasslands. Our linked glasshouse experiment showed that both roots and microbes form a stronger sink for added N in extensively managed grasslands, and that the strength of the microbial sink is related to a greater abundance of decomposer fungi relative to bacteria. This greater root and microbial uptake of N contributes to smaller N leaching losses and greater soil N retention in extensively managed grasslands. Our results advance understanding of the mechanisms of N retention in terrestrial ecosystems and how the capacity to retain N is affected by grassland management. Moreover, they support the notion that microbial communities might be the key to improved N retention through tightening linkages between plants and microbes and reducing N availability [13]. However, more detailed experiments are needed to elucidate the role of arbuscular mycorrhizal and decomposer fungi, and specific bacterial groups, in controlling N cycling processes. Pressures on land for production of food, feed and biofuel are increasing, and this has led to an urgent need to make managed systems more sustainable. Here we show that extensification of grassland management has the potential to optimize the delivery of ecosystem services like N retention, which is of central importance to sustainable food production [64,65] and pollution mitigation [2,3].

Acknowledgments

We thank Marina Louzada, Victor van Velzen, Nicola Thompson, An Vos, and Louise Walker for help in the field and in the laboratory. Liz Dixon of Rothamsted Research North Wyke performed, and gave invaluable advice on, ^{15}N analyses. We are very grateful to the landowners and farmers for allowing us to sample their fields, and to the DIGFOR project members for help in selecting the sites. Dave Johnson of Aberdeen University provided us with helpful comments on the manuscript.

Author Contributions

Conceived and designed the experiments: FTdV RDB HQ. Performed the experiments: FTdV HQ CJS RB. Analyzed the data: FTdV. Contributed reagents/materials/analysis tools: HQ RB JB CJS. Wrote the paper: FTdV RDB.

References

1. IPCC (2007) Climate Change 2007: The Physical Science Basis; Solomon S, Qin D, Manning M, Chen Z, Marquis M et al., editors. Cambridge, United Kingdom and New York, NY, USA: Cambridge University Press.
2. Galloway JN, Townsend AR, Erisman JW, Bekunda M, Cai ZC, et al. (2008) Transformation of the nitrogen cycle: Recent trends, questions, and potential solutions. Science 320: 889–892.
3. Schlesinger WH (2009) On the fate of anthropogenic nitrogen. P Natl Acad Sci USA 106: 203–208.
4. Stevens CJ, Dise NB, Mountford JO, Gowing DJ (2004) Impact of nitrogen deposition on the species richness of grasslands. Science 303: 1876–1879.
5. Lu M, Yang YH, Luo YQ, Fang CM, Zhou XH, et al. (2011) Responses of ecosystem nitrogen cycle to nitrogen addition: a meta-analysis. New Phytol 189: 1040–1050.
6. Treseder KK (2008) Nitrogen additions and microbial biomass: a meta-analysis of ecosystem studies. Ecol Lett 11: 1111–1120.
7. Ramirez KS, Craine JM, Fierer N (2012) Consistent effects of nitrogen amendments on soil microbial communities and processes across biomes. Glob Change Biol: n/a-n/a.
8. Frey SD, Knorr M, Parrent JL, Simpson RT (2004) Chronic nitrogen enrichment affects the structure and function of the soil microbial community in temperate hardwood and pine forests. For Ecol Manage 196: 159–171.
9. De Vries FT, Hoffland E, van Eekeren N, Brussaard L, Bloem J (2006) Fungal/bacterial ratios in grasslands with contrasting nitrogen management. Soil Biol Biochem 38: 2092–2103.
10. Bradley K, Drijber RA, Knops J (2006) Increased N availability in grassland soils modifies their microbial communities and decreases the abundance of arbuscular mycorrhizal fungi. Soil Biol Biochem 38: 1583–1595.
11. Nilsson LO, Baath E, Falkengren-Grerup U, Wallander H (2007) Growth of ectomycorrhizal mycelia and composition of soil microbial communities in oak forest soils along a nitrogen deposition gradient. Oecologia 153: 375–384.
12. Canfield DE, Glazer AN, Falkowski PG (2010) The Evolution and Future of Earth's Nitrogen Cycle. Science 330: 192–196.
13. De Vries FT, Bardgett RD (2012) Plant-microbial linkages and ecosystem N retention: lessons for sustainable agriculture. Front Ecol Environ 10: 425–432.
14. Van der Heijden MGA, Bardgett RD, van Straalen NM (2008) The unseen majority: soil microbes as drivers of plant diversity and productivity in terrestrial ecosystems. Ecol Lett 11: 296–310.
15. Wardle DA, Bardgett RD, Klironomos JN, Setälä H, van der Putten WH, et al. (2004) Ecological linkages between aboveground and belowground biota. Science 304: 1629–1633.
16. Coleman DC, Reid CPP, Cole CV (1983) Biological strategies of nutrient cycling in soil systems. Adv Ecol Res 13: 1–55.
17. Six J, Frey SD, Thiet RK, Batten KM (2006) Bacterial and fungal contributions to carbon sequestration in agroecosystems. Soil Sci Soc Am J 70: 555–569.
18. Frey SD, Six J, Elliott ET (2003) Reciprocal transfer of carbon and nitrogen by decomposer fungi at the soil-litter interface. Soil Biol Biochem 35: 1001–1004.
19. Bardgett RD, Streeter TC, Bol R (2003) Soil microbes compete effectively with plants for organic-nitrogen inputs to temperate grasslands. Ecology 84: 1277–1287.
20. De Vries FT, Van Groenigen JW, Hoffland E, Bloem J (2011) Nitrogen losses from two grassland soils with different fungal biomass. Soil Biol Biochem 43: 997–1005.
21. Strickland MS, Rousk J (2010) Considering fungal:bacterial dominance in soils - Methods, controls, and ecosystem implications. Soil Biol Biochem 42: 1385–1395.
22. Thiet RK, Frey SD, Six J (2006) Do growth yield efficiencies differ between soil microbial communities differing in fungal: bacterial ratios? Reality check and methodological issues. Soil Biol Biochem 38: 837–844.
23. Frey SD, Elliott ET, Paustian K (1999) Bacterial and fungal abundance and biomass in conventional and no-tillage agroecosystems along two climatic gradients. Soil Biol Biochem 31: 573–585.
24. Helgason BL, Walley FL, Germida JJ (2009) Fungal and Bacterial Abundance in Long-Term No-Till and Intensive-Till Soils of the Northern Great Plains. Soil Sci Soc Am J 73: 120–127.
25. Bardgett RD, McAlister E (1999) The measurement of soil fungal:bacterial biomass ratios as an indicator of ecosystem self-regulation in temperate meadow grasslands. Biol Fertil Soils 29: 282–290.
26. De Vries FT, Bloem J, van Eekeren N, Brusaard L, Hoffland E (2007) Fungal biomass in pastures increases with age and reduced N input. Soil Biol Biochem 39: 1620–1630.
27. Postma-Blaauw MB, de Goede RGM, Bloem J, Faber JH, Brussaard L (2010) Soil biota community structure and abundance under agricultural intensification and extensification. Ecology 91: 460–473.
28. Gordon H, Haygarth PM, Bardgett RD (2008) Drying and rewetting effects on soil microbial community composition and nutrient leaching. Soil Biol Biochem 40: 302–311.
29. Donnison LM, Griffith GS, Hedger J, Hobbs PJ, Bardgett RD (2000) Management influences on soil microbial communities and their function in botanically diverse haymeadows of northern England and Wales. Soil Biol Biochem 32: 253–263.
30. Smith RS, Shiel RS, Bardgett RD, Millward D, Corkhill P, et al. (2008) Long-term change in vegetation and soil microbial communities during the phased restoration of traditional meadow grassland. J Appl Ecol 45: 670–679.
31. Smith RS, Shiel RS, Bardgett RD, Millward D, Corkhill P, et al. (2003) Soil microbial community, fertility, vegetation and diversity as targets in the restoration management of a meadow grassland. J Appl Ecol 40: 51–64.
32. De Deyn GB, Quirk H, Yi Z, Oakley S, Ostle NJ, et al. (2009) Vegetation composition promotes carbon and nitrogen storage in model grassland communities of contrasting soil fertility. J Ecol 97: 864–875.
33. De Deyn GB, Quirk H, Bardgett RD (2011) Plant species richness, identity and productivity differentially influence key groups of microbes in grassland soils of contrasting fertility. Biology Letters 7: 75–78.
34. Orwin KH, Buckland SM, Johnson D, Turner BL, Smart S, et al. (2010) Linkages of plant traits to soil properties and the functioning of temperate grassland. J Ecol 98: 1074–1083.
35. De Vries FT, Manning P, Tallowin J, Mortimer SR, Pilgrim E, et al. (2012) Abiotic drivers and plant traits explain landscape-scale patterns in soil microbial communities. Ecol Lett 15: 1230–1239.
36. Rodwell JS (1992) British Plant Communities, Volume 3. Grassland and montane communities. Cambridge: Cambridge University Press.
37. Brookes PC, Landman A, Pruden G, Jenkinson DS (1985) Chloroform fumigation and the release of soil-nitrogen - a rapid direct extraction method to measure microbial biomass nitrogen in soil. Soil Biol Biochem 17: 837–842.
38. Harrison KA, Bol R, Bardgett RD (2007) Preferences for different nitrogen forms by coexisting plant species and soil microbes. Ecology 88: 989–999.
39. Harrison KA, Bardgett RD (2010) Influence of plant species and soil conditions on plant-soil feedback in mixed grassland communities. J Ecol 98: 384–395.
40. Bardgett RD, Hobbs PJ, Frostegård Å (1996) Changes in soil fungal:bacterial biomass ratios following reductions in the intensity of management of an upland grassland. Biol Fertil Soils 22: 261–264.
41. De Vries FT, Liiri M, Bjørnlund L, Bowker M, Christensen S, et al. (2012) Land use alters the resistance and resilience of soil food webs to drought. Nat Clim Change 2: 276–280.
42. Bloem J, Vos A (2004) Fluorescent staining of microbes for total direct counts. In: Kowalchuk GA, De Bruijn FJ, Head IM, Akkermans ADL, Van Elsas JD, editors. Molecular Microbial Ecology Manual, 2nd edition. Dordrecht, The Netherlands: Kluwer Academic Publishers. 861–874.
43. Bakken LR, Olsen RA (1983) Buoyant densities and dry-matter contents of microorganisms: conversion of a measured biovolume into biomass bacteria and fungi from soil. Appl Env Microbiol 45: 1188–1195.
44. Van Veen JA, Paul EA (1979) Conversion of biovolume measurements of soil organisms, grown under various moisture tensions, to biomass and their nutrient content. Appl Env Microbiol 37: 686–692.
45. Bloem J, Veninga M, Shepherd J (1995) Fully automatic determination of soil bacterium numbers, cell volumes, and frequencies of dividing cells by confocal laser scanning microscopy and image analysis. Appl Env Microbiol 61: 926–936.
46. Fry JC (1990) Direct methods and biomass estimation. In: Grigorova R, Norris JR, editors. Methods in Microbiology. London: Academic Press. 41–85.
47. Crawley MJ (2007) The R book. Chichester: John Wiley & Sons Ltd.
48. Pinheiro JC, Bates DM (2000) Mixed effects models in S and S-PLUS. New York: Springer-Verlag.
49. Diaz S, Hodgson JG, Thompson K, Cabido M, Cornelissen JHC, et al. (2004) The plant traits that drive ecosystems: Evidence from three continents. J Veg Sci 15: 295–304.
50. De Deyn GB, Cornelissen JHC, Bardgett RD (2008) Plant functional traits and soil carbon sequestration in contrasting biomes. Ecol Lett 11: 516–531.
51. Laughlin DC (2011) Nitrification is linked to dominant leaf traits rather than functional diversity. J Ecol 99: 1091–1099.
52. Alvarez CR, Alvarez R, Grigera MS, Lavado RS (1998) Associations between organic matter fractions and the active soil microbial biomass. Soil Biol Biochem 30: 767–773.
53. Hopkins DW, Waite IS, O'Donnell AG (2010) Microbial biomass, organic matter mineralization and nitrogen in soils from long-term experimental grassland plots (Palace Leas meadow hay plots, UK). European Journal of Soil Science 62: 95–104.
54. Bardgett RD, Frankland JC, Whittaker JB (1993) The effects of agricultural management on the soil biota of some upland grasslands. Agr Ecosyst Environ 45: 25–45.
55. De Vries FT, Bååth E, Kuyper TW, Bloem J (2009) High turnover of fungal hyphae in incubation experiments. FEMS Microbiol Ecol 67: 389–396.
56. Joergensen RG, Wichern F (2008) Quantitative assessment of the fungal contribution to microbial tissue in soil. Soil Biology and Biochemistry 40: 2977–2991.
57. Kennedy N, Edwards S, Clipson N (2005) Soil bacterial and fungal community structure across a range of unimproved and semi-improved upland grasslands. Microb Ecol 50: 463–473.
58. Zelles L, Palojarvi A, Kandeler E, Von LM, Winter K, et al. (1997) Changes in soil microbial properties and phospholipid fatty acid fractions after chloroform fumigation. Soil Biol Biochem 29: 1325–1336.

59. Johnson D, Leake JR, Read DJ (2005) Liming and nitrogen fertilization affects phosphatase activities, microbial biomass and mycorrhizal colonisation in upland grassland. Plant Soil 271: 157–164.
60. van der Heijden MGA, Horton TR (2009) Socialism in soil? The importance of mycorrhizal fungal networks for facilitation in natural ecosystems. J Ecol 97: 1139–1150.
61. Van der Heijden MGA (2010) Mycorrhizal fungi reduce nutrient loss from model grassland ecosystems. Ecology 91: 1163–1171.
62. Zogg GP, Zak DR, Pregitzer KS, Burton AJ (2000) Microbial immobilization and the retention of anthropogenic nitrate in a northern hardwood forest. Ecology 81: 1858–1866.
63. Raciti SM, Groffman PM, Fahey TJ (2008) Nitrogen retention in urban lawns and forests. Ecol Appl 18: 1615–1626.
64. Godfray HCJ, Beddington JR, Crute IR, Haddad L, Lawrence D, et al. (2010) Food security: the challenge of feeding 9 billion people. Science 327: 812–818.
65. Power AG (2010) Ecosystem services and agriculture: tradeoffs and synergies. Philos T R Soc B 365: 2959–2971.

PERMISSIONS

All chapters in this book were first published in PLOS ONE, by The Public Library of Science; hereby published with permission under the Creative Commons Attribution License or equivalent. Every chapter published in this book has been scrutinized by our experts. Their significance has been extensively debated. The topics covered herein carry significant findings which will fuel the growth of the discipline. They may even be implemented as practical applications or may be referred to as a beginning point for another development.

The contributors of this book come from diverse backgrounds, making this book a truly international effort. This book will bring forth new frontiers with its revolutionizing research information and detailed analysis of the nascent developments around the world.

We would like to thank all the contributing authors for lending their expertise to make the book truly unique. They have played a crucial role in the development of this book. Without their invaluable contributions this book wouldn't have been possible. They have made vital efforts to compile up to date information on the varied aspects of this subject to make this book a valuable addition to the collection of many professionals and students.

This book was conceptualized with the vision of imparting up-to-date information and advanced data in this field. To ensure the same, a matchless editorial board was set up. Every individual on the board went through rigorous rounds of assessment to prove their worth. After which they invested a large part of their time researching and compiling the most relevant data for our readers.

The editorial board has been involved in producing this book since its inception. They have spent rigorous hours researching and exploring the diverse topics which have resulted in the successful publishing of this book. They have passed on their knowledge of decades through this book. To expedite this challenging task, the publisher supported the team at every step. A small team of assistant editors was also appointed to further simplify the editing procedure and attain best results for the readers.

Apart from the editorial board, the designing team has also invested a significant amount of their time in understanding the subject and creating the most relevant covers. They scrutinized every image to scout for the most suitable representation of the subject and create an appropriate cover for the book.

The publishing team has been an ardent support to the editorial, designing and production team. Their endless efforts to recruit the best for this project, has resulted in the accomplishment of this book. They are a veteran in the field of academics and their pool of knowledge is as vast as their experience in printing. Their expertise and guidance has proved useful at every step. Their uncompromising quality standards have made this book an exceptional effort. Their encouragement from time to time has been an inspiration for everyone.

The publisher and the editorial board hope that this book will prove to be a valuable piece of knowledge for researchers, students, practitioners and scholars across the globe.

LIST OF CONTRIBUTORS

Heiko Nacke and Christiane Will
Department of Genomic and Applied Microbiology, Institute of Microbiology and Genetics, Georg-August University Göttingen, Göttingen, Germany

Andrea Thürmer and Antje Wollherr
2 Göttingen Genomics Laboratory, Institute of Microbiology and Genetics, Georg-August University Göttingen, Göttingen, Germany

Rolf Daniel
Department of Genomic and Applied Microbiology, Institute of Microbiology and Genetics, Georg-August University Göttingen, Göttingen, Germany
Göttingen Genomics Laboratory, Institute of Microbiology and Genetics, Georg-August University Göttingen, Göttingen, Germany

Ladislav Hodac
Experimental Phycology and Culture Collection of Algae at the University of Göttingen, Göttingen, Germany

Nadine Herold and Marion Schrumpf
Max Planck Institute for Biogeochemistry, Jena, Germany

Ingo Schöning
Institute of Ecology, Friedrich-Schiller-University, Jena, Germany

Carolina Levis, Juliana Schietti and Thaise Emilio
Programa de Pós-Graduação em Ecologia, Instituto Nacional de Pesquisas da Amazônia (INPA), Manaus, Amazonas, Brasil

Priscila Figueira de Souza
Programa de Pós-Graduação em Botânica, Instituto Nacional de Pesquisas da Amazônia (INPA), Manaus, Amazonas, Brasil

José Luiz Purri da Veiga Pinto and Flavia R. C. Costa
Coordenação de Biodiversidade, Instituto Nacional de Pesquisas da Amazônia (INPA), Manaus, Amazonas, Brasil

Charles R. Clement
Coordenação de Tecnologia e Inovação, Instituto Nacional de Pesquisas da Amazônia (INPA), Manaus, Amazonas, Brasil

Runhe Shi and Wei Gao
Key Laboratory of Geographical Information Science, Ministry of Education, East China Normal University, Shanghai, China

Yuan Zhang
Key Laboratory of Geographical Information Science, Ministry of Education, East China Normal University, Shanghai, China
Research Center of Remote Sensing & Geoscience, Northeast Institute of Geography & Agroecology, Chinese Academy of Sciences, Changchun, China

Shiliang Su
College of Environmental & Resource Sciences, Zhejiang University, Hangzhou, China

Feng Zhang
Key Laboratory of Arid and Grassland Agro-Ecology, Lanzhou University, Lanzhou, China

Daniel Collins, Shreya Shah and Vishal Shah
Department of Biology, Dowling College, Oakdale, New York, United States of America

Todd Luxton
National Risk Management Research Laboratory, U.S. Environmental Protection Agency, Cincinnati, Ohio, United States of America

Niraj Kumar and Virginia K. Walker
Department of Biology and School of Environmental Studies, Queen's University, Kingston, Ontario, Canada

James Angus Fraser
Departamentos de Antropología e Historia, Universidad de los Andes, Bogotá, Cundinamarca, Colombia
Department of Anthropology, University of Sussex, East Sussex, United Kingdom

Alessandro Alves-Pereira
Post-graduate Program in Genetics and Plant Breeding, Universidade de São Paulo, Piracicaba, Estado de São Paulo, Brazil
Laboratório de Evolução Aplicada, Universidade Federal do Amazonas, Manaus, Amazonas, Brazil

Charles Roland Clement
Laboratório de Evolução Aplicada, Universidade Federal do Amazonas, Manaus, Amazonas, Brazil

Coordenação de Tecnologia e Inovação, Instituto Nacional de Pesquisas da Amazônia, Manaus, Amazonas, Brazil

André Braga Junqueira
Coordenação de Tecnologia e Inovação, Instituto Nacional de Pesquisas da Amazônia, Manaus, Amazonas, Brazil
Centre for Crop Systems Analysis/Technology and Agrarian Development Group Wageningen University, Wageningen, Gelderland, The Netherlands

Peroni
Departamento de Ecologia e Zoologia, Universidade Federal de Santa Catarina, Florianópolis, Santa Catarina, Brazi

Lei Cheng, Cong Tu, H. David Shew and Shuijin Hu
Department of Plant Pathology, North Carolina State University, Raleigh, North Carolina, United States of America

Fitzgerald L. Booker Kent O. Burkey and Edwin L. Fiscus
Plant Science Research Unit, United States Department of Agriculture, Agricultural Research Service, Raleigh, North Carolina, United States of America
Department of Crop Science, North Carolina State University, Raleigh, North Carolina, United States of America

Thomas W. Rufty
Department of Crop Science, North Carolina State University, Raleigh, North Carolina, United States of America

Jared L. Deforest
Department of Environmental and Plant Biology, Ohio University, Athens, Ohio, United States of America

Jing Huang and Yanzheng Gao
Institute of Organic Contaminant Control and Soil Remediation, College of Resource and Environmental Sciences, Nanjing Agricultural University, Nanjing, Jiangsu Province, P.R. China

Yuanpeng Wang
Institute of Organic Contaminant Control and Soil Remediation, College of Resource and Environmental Sciences, Nanjing Agricultural University, Nanjing, Jiangsu Province, P.R. China
Department of Chemical and Biochemical Engineering, College of Chemistry and Chemical Engineering, Xiamen University, Xiamen, Fujian Province, P.R. China

Kun Meng, Huiying Luo, Yaru Wang, Huoqing Huang, Pengjun Shi, Peilong Yang and Bin Yao
Key Laboratory for Feed Biotechnology of the Ministry of Agriculture, Feed Research Institute, Chinese Academy of Agricultural Sciences, Beijing, People's Republic of China

Zhifang Zhang
Biotechnology Research Institute, Chinese Academy of Agricultural Sciences, Beijing, People's Republic of China

Guozeng Wang
Key Laboratory for Feed Biotechnology of the Ministry of Agriculture, Feed Research Institute, Chinese Academy of Agricultural Sciences, Beijing, People's Republic of China
Biotechnology Research Institute, Chinese Academy of Agricultural Sciences, Beijing, People's Republic of China

Shun-an Zheng, Xiangqun Zheng and Chun Chen
Agro-Environmental Protection Institute, Ministry of Agriculture, Tianjin, People's Republic of China
Key Laboratory of Production Environment and Agro-Product Safety, Ministry of Agriculture, Tianjin, People's Republic of China
Tianjin Key Laboratory of Agro-Environment and Agro-Product Safety, Tianjin, People's Republic of China

Cristina A. Viegas
Department of Bioengineering, Instituto Superior Técnico (IST), Technical University of Lisbon (TUL), Lisboa, Portugal
Institute for Biotechnology and Bioengineering (IBB), Centre for Biological and Chemical Engineering, Instituto Superior Técnico (IST), Technical University of Lisbon (TUL), Lisboa, Portugal

Catarina Costa
Institute for Biotechnology and Bioengineering (IBB), Centre for Biological and Chemical Engineering, Instituto Superior Técnico (IST), Technical University of Lisbon (TUL), Lisboa, Portugal

Rui Ribeiro and Matilde Moreira-Santos
Instituto do Mar (IMAR), Department of Life Sciences, University of Coimbra, Coimbra, Portugal

Sandra André and Paula Viana
Agência Portuguesa do Ambiente (APA), Amadora, Portugal

Clémentine Fritsch, Michaël Coeurdassier, Patrick Giraudoux, Francis Raoul, Dominique Rieffel, Annette de Vaufleury and Renaud Scheifler
Department of Chrono-Environment, UMR UFC/CNRS 6249 USC INRA, University of Franche-Comté, Besançon, France

Francis Douay
Université Lille Nord de France, Lille, France
Laboratoire Génie Civil et géoEnvironnement (LGCgE), EA 4515, Lille, France

Yingxin Huang, Daowei Zhou, Hongxiang Zhang and Wei Zheng
Northeast Institute of Geography and Agroecology, Chinese Academy of Sciences, Changchun, China

Xueyong Zhao
Cold and Arid Regions Environmental and Engineering Research Institute, Chinese Academy of Sciences, Lanzhou, China

Mingna Chen, Xiao Li, Qingli Yang, Xiaoyuan Chi, Lijuan Pan, Na Chen, Zhen Yang, Tong Wang, Mian Wang and Shanlin Yu
Shandong Peanut Research Institute, Qingdao, China

Martina Köberl, Henry Müller and Gabriele Berg
Institute for Environmental Biotechnology, Graz University of Technology, Graz, Austria

Elshahat M. Ramadan
Faculty of Agriculture, SEKEM and Heliopolis University, Cairo, Egypt

Xiujun Wang
State Key Laboratory of Desert and Oasis Ecology, Xinjiang Institute of Ecology and Geography, Chinese Academy of Sciences, Urumqi, Xinjiang, China
Earth System Science Interdisciplinary Center, University of Maryland, College Park, Maryland, United States of America

Jiaping Wang and Juan Zhang
State Key Laboratory of Desert and Oasis Ecology, Xinjiang Institute of Ecology and Geography, Chinese Academy of Sciences, Urumqi, Xinjiang, China
Graduate University of Chinese Academy of Sciences, Beijing, China

Rie Shimizu-Inatsugi, Masaki J. Kobayashi and Kentaro K. Shimizu
Institute of Plant Biology, University of Zurich, Zurich, Switzerland

Yayoi Takeuchi
Institute of Plant Biology, University of Zurich, Zurich, Switzerland
Graduate University for Advanced Studies, School of Advanced Sciences, Hayama, Kanagawa, Japan

Michaela M. Salcher and Jakob Pernthaler
Department of Limnology, Institute of Plant Biology, University of Zurich, Kilchberg, Switzerland

Masayuki Ushio
Center for Ecological Research, Kyoto University, Otsu, Shiga, Japan

Bibian Diway
Botanical Research Centre, Sarawak Forestry Corporation, Kuching, Malaysia

Christian von Mering
Institute of Molecular Life Sciences, University of Zurich, Zurich, Switzerland

Nan Liu, Xiaoya Wang, Yingjun Zhang and Yue Xie
Institute of Grassland Science, China Agricultural University, Beijing, People's Republic of China

Shujuan Chang
Institute of Grassland Science, China Agricultural University, Beijing, People's Republic of China
Inner Mongolia Rangeland Survey & Design Institute, Hohhot, Inner
Mongolia, People's Republic of China

Joanna C. Carey
Department of Earth and Environment, Boston University, Boston, Massachusetts, United States of America

Robinson W. Fulweiler
Department of Earth and Environment, Boston University, Boston, Massachusetts, United States of America
Department of Biology, Boston University, Boston,Massachusetts, United States of America

Xiu-jun Wang and Chang-yan Tian
State Key Laboratory of Desert and Oasis Ecology, Xinjiang Institute of Ecology and Geography, Chinese Academy of Sciences, Urumqi, China

Zhi-guo Li
State Key Laboratory of Desert and Oasis Ecology, Xinjiang Institute of Ecology and Geography, Chinese Academy of Sciences, Urumqi, China
Wuhan Botanical Garden, Chinese Academy of Sciences
Key Laboratory of Aquatic Botany and Watershed Ecology, Chinese Academy of Sciences, Wuhan, China

Fang Chen
Wuhan Botanical Garden, Chinese Academy of Sciences
Key Laboratory of Aquatic Botany and Watershed Ecology, Chinese Academy of Sciences, Wuhan, China

Run-hua Zhang
Wuhan Vegetable Research Institute, Wuhan, China

Xu Han, Zhihui Cheng and Huanwen Meng
State Key Laboratory of Crop Stress Biology in Arid Areas and College of Horticulture, Northwest A&F University, Yangling, Shaanxi, China

Ru Li, W. G. Dilantha Fernando and Martin H. Entz
Department of Plant Science, University of Manitoba, Winnipeg, Manitoba Canada

Ehsan Khafipour and Denis O. Krause
Department of Animal Science, University of Manitoba, Winnipeg, Manitoba, Canada
Department of Medical Microbiology and Infectious Diseases, Winnipeg, Manitoba, Canada

Teresa R. de Kievit
Department of Microbiology, University of Manitoba, Winnipeg, Manitoba, Canada

Collin W. Ahrens and Carol A. Auer
Agricultural Biotechnology Lab, Department of Plant Science, University of Connecticut, Storrs, Connecticut, United States of America

Isabel Ceballos, Michael Ruiz and Alia Rodríguez
Soil Microbiology, Universidad Nacional de Colombia, Bogotá, Colombia

Cristhian Fernández and Ricardo Peña
Utopía, Universidad de La Salle, Yopal, Colombia

Ian R. Sanders
Department of Ecology and Evolution, University of Lausanne, Lausanne, Switzerland

Franciska T. de Vries, Helen Quirk, Carly J. Stevens and Richard D. Bardgett
Lancaster Environment Centre, Lancaster University, Lancaster, United Kingdom

Jaap Bloem
Alterra, Wageningen University and Research Centre, Wageningen, The Netherlands

Roland Bol
Institute of Bio- and Geosciences, IBG-3: Agrosphere, Forschungszentrum Jülich GmbH, Jülich, Germany

Index

P

R

S

T

U

V

W

X

Z